NUMERICAL TAXONOMY

A SERIES OF BOOKS IN BIOLOGY

Editors: *Donald Kennedy*
Roderic B. Park

NUMERICAL TAXONOMY

THE PRINCIPLES AND PRACTICE
OF NUMERICAL CLASSIFICATION

Peter H. A. Sneath

MEDICAL RESEARCH COUNCIL MICROBIAL SYSTEMATICS UNIT
UNIVERSITY OF LEICESTER

Robert R. Sokal

STATE UNIVERSITY OF NEW YORK AT STONY BROOK

W. H. FREEMAN AND COMPANY
San Francisco

Library of Congress Cataloging in Publication Data

Sneath, Peter H   A
        Numerical taxonomy.

        Bibliography: p.
        1. Numerical taxonomy. I. Sokal, Robert R., joint
author. II. Title.
QH83.S58   574'.01'2   72-1552
ISBN 0-7167-0697-0

Printed in the United States of America

The cover design of this book is adapted from a contour diagram of the relationships of
dermanyssid mites. Reproduced with permission of Dr. W. Wayne Moss of The Academy of
Natural Sciences of Philadelphia.

1 2 3 4 5 6 7 8 9

To JOAN and JULIE

*whose endless forbearance
has sustained us through
yet another venture.*

Contents

A Tribute

We are indebted to those numerous younger colleagues all over the world who, stimulated by our first book on this subject, expanded and improved our ideas and methods and built new conceptual constructs on the earlier foundations. We also acknowledge the many established systematists and specialists in other sciences who were willing to make the effort to try out our methodology. We have benefited immeasurably from the activities and enthusiasm of these persons and we hope that their efforts and the intellectual excitement that they have engendered are faithfully reflected in the pages that follow.

Preface

*From the time of Linnaeus to our own, a weak point in
biological science has been the absence of any quantitative
meaning in our classificatory terms. What is a Class, and does
Class A differ from Class B as much as Class C differs from Class D?
The question can be put for the other classificatory grades, such as
Order, Family, Genus, and Species. In no case can it be answered fully,
and in most cases it cannot be answered at all. . . . Until some adequate
reply can be given to such questions as these, our classificatory
schemes can never be satisfactory or "natural." They can be little
better than mnemonics—mere skeletons or frames on which we hang
somewhat disconnected fragments of knowledge. Evolutionary
doctrine, which has been at the back of all classificatory systems of
the last century, has provided no real answer to these difficulties.
Geology has given a fragmentary answer here and there. But to sketch
the manner in which the various groups of living things arose is a very
different thing from ascribing any quantitative value to those groups.*
C. Singer, *A History of Biology* (1959), p. 200.

The rapid, almost explosive development of numerical taxonomy and the increasingly wide interest in this field made it clear to us that a new edition of our *Principles of Numerical Taxonomy*, first published in 1963, was necessary. We soon realized that a mere updating of the contents would not do justice to our task. Not only had much new material been published, which required incorporation, but changes in emphasis and in the theoretical framework of the science have taken place that needed to be presented in proper perspective. In view of this, an entirely new book has been written.

The past decade has witnessed a marked change in outlook and methodology in the fields of biological systematics and population biology. Many of the new approaches are quantitative, employing a variety of mathematical disciplines. In taxonomy there has been a considerable development of numerical methods, many of them implemented by computers. These methods have influenced other disciplines as well, which have in turn provided numerous new concepts and techniques for systematics.

Numerical taxonomy—the grouping by numerical methods of taxonomic units based on their character states—has been aided in its present rapid development by the simultaneous development of computer techniques. Numerical taxonomy aims to develop methods that are objective, explicit, and repeatable, both in evaluation of taxonomic relationships and in the erection of taxa. Moreover, numerical methods have opened up a wide field in the exact measurement of evolutionary rates and in phylogenetic analysis. The success of the intuitive approach of the past lay in the ability of the mind to recognize swiftly, though inexactly, overall similarity in morphological detail. Such recognition is not easy with the ever increasing data bases in taxonomy, now often in tabular form, as is true of microbiological, chemical, or physiological characters; use of numerical methods with these characters becomes a necessity.

The purpose of this book is to present an up-to-date theoretical basis for numerical taxonomy, to acquaint readers with its procedures, to illustrate its advantages over conventional taxonomy, and to report on the status of the field so far.

We cannot treat all forms of numerical analysis that have been used in taxonomy, for which numerous texts on the use of statistical and mathematical methods in biology can be consulted easily. We have restricted the scope of this book instead to methods that demonstrate taxonomic relationships and create taxonomic groupings, although we treat some other techniques briefly for completeness. We have, however, attempted to treat our topics as broadly as possible and to make the book of value to zoologists, botanists, microbiologists, and paleontologists, as well as to scientists in related fields.

Readers familiar with *Principles of Numerical Taxonomy* will note that the sequence of chapters and topics has been reorganized to form, we believe, a better integrated whole. Some earlier sections have been dropped or much abbreviated—others have been liberally increased. For example, there seems by now little need for a detailed criticism and polemic against conventional practices in taxonomy. The point has been made and widely accepted, and the present need is for an expansion and elaboration of the theory and methodology of newer views. By contrast, numerical methods of phylogenetic analysis—undeveloped when the earlier book was written, required considerable space for a balanced treatment in the present text. We have tried to consolidate the discussion of the theoretical foundations of systematics and taxonomy into a cohesive whole, no longer separating our critical review of conventional systematics from the views we advocate.

Considerable changes will be found in the sections on the theory and assumptions of numerical taxonomy. The work of the last few years has led us to realize that some of our earlier hopes and expectations of rapid, clearcut solutions to the problems of taxonomy were premature. Thus, for example, we still do not know a method for an optimal taxonomy (or if one exists) and therefore cannot advocate one. Nevertheless, it is clear to all but the most conservative workers in the field that the taxonomy of the future will be greatly aided if not entirely carried out by computers, and though it is too early in the development of numerical taxonomy to provide "cookbook" recipes for all problems, the systematist who ignores numerical taxonomic methods in his own work does so at his own loss.

Rewriting sections on methods has provided the greatest challenge. A complete account of all that has been done or proposed would result in a book twice the size of this volume, obsolescent at the time of publication, and confusing for the novice. We have had to be eclectic in our choice of methods to be presented, reporting here the most frequently used approaches and providing references to others that have been less often used or that we feel are of less general interest. Although, as we shall point out from time to time, the statistical and conceptual foundations of various aspects of numerical taxonomy are insufficient, we present here even those aspects we know lack rigor because we wish to report on the current state of the art as well as to encourage further work in these areas.

Of necessity, the level of mathematical knowledge needed for mastering the subject matter has increased somewhat since *Principles of Numerical Taxonomy* was published. Fortunately, the increasing complexity of the subject has been matched by the increasing mathematical sophistication of young biologists in systematics and other fields of biology in recent years. The time is rapidly approaching, if not at hand already, when a thorough knowledge of biometric analysis and some acquaintance with computer processing will be an ineluctable prerequisite for the aspiring taxonomist. We would expect readers of this book to have some knowledge of statistics and of elementary set and graph theory, as well as of matrix algebra. Numerous books are now on the market that provide such introductions; a selection of these is given in Appendix B.

In *Principles of Numerical Taxonomy* we attempted to review to the best of our knowledge all applications of numerical taxonomy in the field of biology. The number of papers in the field has since become so numerous that we have been hard-put to keep our fingers on the entire literature and have not tried to furnish an exhaustive bibliography in this book. By referring to a number of key review papers we have tried to point the reader to sources where comprehensive reviews of the literature in particular branches of numerical taxonomy can be found. However, in Appendix A we do furnish a list of all recent papers known to us in which numerical taxonomy has been applied to classifications of various organisms, listing these by broad taxonomic groupings. We hope the list will be of value to taxonomists who wish to survey the field in their area of specialization.

In our account of the applications of numerical taxonomy to fields other than biological systematics, we have had to be more restrictive. Applications have been astonishing in number and diversity. Methods proposed here, and similar ones, have been applied in fields ranging from archaeology to political science, from materials classification to linguistics, and from television programming to biogeography, and though we have tried to give some coverage to the applications of numerical taxonomy to these subjects, limitations of space and our own circumscribed competence in these fields require brevity.

Readers who are not primarily interested in biological systematics but would like to use this book as an introduction to clustering and the principles of numerical classification in general, can limit their reading to the following chapters and sections without losing much by way of continuity: Sections 1.1–5, 1.9; 2.1–2, 2.8; 3.1–3, 3.8–9; 4.1–8, 4.10, 4.12–13; 5.1–2, 5.4–10; 6.4; Chapter 8; and Chapter 11.

We have not attempted to illustrate techniques by means of simple examples, as we did in an appendix to *Principles of Numerical Taxonomy*. It would not have been possible to illustrate the many more numerous methods now extant in taxonomy and still keep the present volume reasonable in size and price. Also, complexity of many of the methods makes computer handling of the data virtually a necessity. To be useful, a discussion of the details of methodology would require explicit advice on computer handling (as well as some mention of character coding and initial processing of the data). This is being done in a forthcoming book by F. J. Rohlf and P. M. Neely (W. H. Freeman and Company).

Limiting ourselves to a brief discussion of computational aspects, we have listed some sources in Appendix B for workers who need to go more deeply into this area and provided general advice and information as well. We have also listed there a number of books and reviews on various other aspects of numerical taxonomy.

Preparation of this book started during the fall semester of 1967 when P.H.A.S. was a Visiting Professor at the University of Kansas. Early drafts of several chapters were read before the Biosystematics and the Numerical Taxonomy Luncheon Groups at that institution and benefited from constructive criticism by their members. Collaboration continued during a visit by P.H.A.S. to the State University of New York at Stony Brook made possible by the Medical Research Council of the United Kingdom. The authors were fortunate to be able to complete the final editing during a summer course in numerical taxonomy at the Institute of Advanced Studies in Oeiras, Portugal, sponsored by the Gulbenkian Foundation. We are indebted to Dr. N. Van Uden for putting all necessary facilities at our disposal.

We are glad to acknowledge three colleagues who have read the entire draft and given us the benefit of extensive constructive criticism. Drs. D. H. Colless (Division of Entomology, C.S.I.R.O., Canberra), J. C. Gower (Rothamsted Experimental Station, Harpenden), and F. J. Rohlf (State University of New York at Stony Brook) contributed greatly to improving our exposition. The chapter on phylogeny was read by Dr. J. S. Farris, the section on kinds of characters by

Dr. J. A. W. Kirsch. Dr. J. H. Strauss read and improved the section on psychiatric classification. Dr. T. J. Crovello contributed to numerous discussions throughout the text. These colleagues were of great assistance in editing the material they reviewed. Various students in several courses on numerical taxonomy at the State University of New York at Stony Brook and at the Estudos Avançados de Oeiras made suggestions for improvements in the text. We are grateful for all this assistance from friends, colleagues, and students and beg their indulgence if we have not always followed their advice.

Our efforts were greatly aided by the meticulous and professional secretarial work of Mrs. Ethel Savarese who saw the manuscript through its many stages of preparation. We are very much in her debt. We would also like to acknowledge the assistance of Mrs. Brenda Jones who typed the Appendixes and Bibliography. The students of the numerical taxonomy course at Stony Brook cooperated in getting a draft copy of the book reproduced in record time. We should single out Richard Stone and Mary Mickevich, who proofread most of the typed copy, and Irving Kornfield, who supervised the assembling. Barbara Torraca and Che Nu Paul were most helpful in checking the indexes.

We are indebted to Abelard-Schuman Ltd., Publishers, for permission to cite a passage from C. Singer, *A History of Biology*. Authors of all sources are identified in the text and are cited in the Bibliography.

We would like to acknowledge here the support given by the Medical Research Council of the United Kingdom (to P.H.A.S.) and by the National Science Foundation of the United States (to R.R.S.) for research in numerical taxonomy. The help of these organizations was crucial during the early development of the subject and their continued support is deeply appreciated.

February 1973   *Peter H. A. Sneath*
                                                        Robert R. Sokal

1

The Aims and Principles of Numerical Taxonomy

The contents of this book fall into three main parts. The chapters of the first part furnish an introduction to taxonomic theory in general and to numerical taxonomy in particular. The purpose of this chapter is to outline in summary form the aims and principles of numerical taxonomy. In Chapter 2, we discuss in some detail the conceptual bases of classification, contrasting conventional views with those espoused by a growing number of taxonomists in recent years.

The central part of the book is arranged on a plan that closely reflects the successive steps followed by taxonomists in performing the classificatory process. In Chapter 3 we discuss taxonomic evidence, which comprises the selection of organisms for study, the choice and definition of taxonomic characters, and criteria for homology. The estimation of taxonomic resemblance between organisms follows in Chapter 4. Chapter 5 considers the grouping of organisms into taxa on the basis of these resemblances.

The final part deals with the implications of numerical taxonomy for systematic research. Numerical approaches to the study of phylogeny are detailed in Chapter 6. The next chapter (7) treats the numerical taxonomy of populations, leading to a discussion of phenetic patterns and evolutionary structure. The implications of numerical taxonomy for keys and identification are discussed in Chapter 8, implications for nomenclature in Chapter 9. A critical examination of numerical taxonomy—its advantages as well as its shortcomings—is featured in Chapter 10. The next chapter (11) is a necessarily brief survey of the application of numerical taxonomy to fields other than biological systematics. The main text concludes in Chapter 12 with a brief outlook on the future of systematics in the computer age.

Two appendixes are provided. A list of applications of numerical taxonomy to different groups of organisms is found in Appendix A; advice on data processing in a numerical taxonomic study is found in Appendix B.

We have made a deliberate attempt to maintain a uniform symbolism for characters, operational taxonomic units (OTU's), and taxa (groups of OTU's), throughout the text. This has meant that in numerous places we have had to report the work of other authors through the use of a symbolism different from that employed in the original publications. We feel that the danger of confusion to the reader who wishes to consult the original sources is more than offset by having a consistent symbolism within the book and we hope that the publication of our book will serve in some small way to retard the development of disparate symbolisms in numerical taxonomy. Even with the best will, it has not been entirely possible to keep symbolism totally consistent, partly because of the limitations on the number of suitable letters in the alphabet. We have attempted to flag all such cases for the reader to further lessen the chances for confusion.

1.1   DEFINITIONS OF SOME TERMS IN TAXONOMY

Adequate definition of taxonomic terms itself would almost require a book. Many terms are used in so many different senses that we have had difficulty in using them in a consistent and exact manner, and no doubt there are still many ambiguities that we have overlooked. The meanings of terms and symbols as they are used in this book can be looked up by way of the Index. There are, however, several that are employed so frequently that we have deemed it desirable to present them here at the outset.

Classification, systematics, and taxonomy are often used interchangeably. In recent years, especially in the United States, there has been a trend toward assigning separate meanings to these terms. In this sense they have been well defined by Simpson (1961), and we follow his usage here.

Systematics.   This is defined by Simpson (1961, p. 7) as "*the scientific study of the kinds and diversity of organisms and of any and all relationships among them.*" This definition uses systematics in its widest sense, concerning itself not only with the arrangement of organisms into taxa and with naming them, but also with the causes and origins of these arrangements.

Classification. We have modified Simpson's definition (1961, p. 9), which restricted itself to zoological classification, to more general usage. *Classification is the ordering of organisms into groups (or sets) on the basis of their relationships.* There may be confusion over the term "relationship," which may imply phylogenetic relationship (analyzed in Section 2.3), or which may simply indicate the resemblance or overall similarity as judged by the characters of the organisms without any implication as to their relationship by ancestry. Relationship based on overall similarity may be distinguished from phylogenetic relationship by calling it *phenetic relationship*, employing the convenient term of H. K. Pusey as used by Cain and Harrison (1960b), to indicate that it is judged from the phenotype of the organism and not from its phylogeny. We have restricted the definition of classification to organisms, since this book is primarily intended for the biological taxonomist. However, there are many methods of classification, including numerical taxonomy, which are equally applicable to concepts and entities other than organisms. Classification as defined above is the name of a process; however, it has also been used for the end product of this process. Thus the result of classification is a classification. The term classification has also been employed, mainly in fields outside biology, in the restricted sense of putting entities into distinct classes as opposed to arraying them in a continuous spectrum, cline, or other arrangement (ordination) showing no distinct divisions. We have not restricted the term in this manner. The term classification should also be distinguished from "identification."

Identification. This is the *allocation or assignment of additional unidentified objects to the correct class once a classification has been established.* Thus, a person using a key to the known wild flowers of Yellowstone National Park "identifies" a given specimen as a goldenrod. Some mathematicians and philosophers would call this process classification also, but we follow conventional usage of biologists in distinguishing strictly between the two.

Taxonomy. Simpson used this term to mean "*the theoretical study of classification, including its bases, principles, procedures and rules*" (Simpson, 1961, p. 11). By this definition the bulk of the subject matter of our book is concerned with taxonomy. We would include the theoretical aspects of identification under taxonomy as well. Taxonomy, like classification, has also been used to designate the end products of the taxonomic process. Since this is a generally accepted usage, we will occasionally employ it in this sense. Taxonomy is frequently made synonymous with classification and we have not attempted to distinguish rigorously between the terms "numerical taxonomy" for theory and "numerical classification" for practice. By contrast, Blackwelder (1967a) understands taxonomy to mean ". . . the day-to-day practice of dealing with the kinds of organisms. This includes the handling and identification of specimens, the publication of the data, the study of the literature, and the analysis of the variation shown by the specimens." Most of these activities

are procedural, related to taxonomy but of no theoretical significance. We shall
have little to say regarding them.

We use the term *taxon* (plural *taxa*) as an abbreviation for taxonomic group
of any nature or rank, as suggested by H. J. Lam (in Lanjouw, 1950) and Rickett
(1958).

1.2 DEFINITION OF NUMERICAL TAXONOMY

Before proceeding, it is necessary that we clearly define our use of the term "numerical taxonomy." We mean by it *the grouping by numerical methods of taxonomic
units into taxa on the basis of their character states*. The term includes the drawing of
phylogenetic inferences from the data by statistical or other mathematical methods
to the extent to which this is possible. These methods require the conversion of
information about taxonomic entities into numerical quantities. We have preferred
"numerical taxonomy" to "quantitative taxonomy," since the latter would include
other methods, such as serology or paper chromatography.

In view of the development of numerical methods for estimating cladistic
relationships subsequent to the publication of *Principles of Numerical Taxonomy*,
it has been suggested (by, among others, W. H. Wagner, personal communication)
that the entire field be renamed "numerical systematics." Because of the generally
accepted breadth of the meaning of the term systematics (see previous section) and
the numerous aspects of systematics that are not suited to the approaches discussed
in this book, we prefer to retain the established definition. Jardine and Sibson (1971)
suggest the name "mathematical taxonomy," and Blackith and Reyment (1971)
employ "multivariate morphometrics." A synonym, "taxometrics," has been suggested for numerical taxonomy (by Mayr, 1966, in lieu of "taximetrics," proposed
by Rogers, 1963), and we shall use the adjective "taxometric" occasionally in place
of the more cumbersome "numerical taxonomic." However, the term numerical
taxonomy seems well established and we retain it here.

We do not wish to widen the concept of numerical taxonomy to include every
application of statistical or other numerical methods in systematic research. Our
approach consists of a variety of numerical techniques; but when such techniques
are not applied to problems of classification they are not included in numerical
taxonomy.

The practice of numerical taxonomy embraces a number of fundamental
assumptions and philosophical attitudes toward taxonomic work, which we shall
discuss and defend in detail in the sections that follow. None of the attitudes and
assumptions is new. They, as well as isolated attempts at a numerical treatment of
taxonomic relationships, date back more than 200 years. However, we would
prefer to restrict the appellation numerical taxonomy to the integrated approach
of recent years, which is bringing about a revision of the theory and practice of
taxonomy.

1.3   THE FUNDAMENTAL POSITION OF NUMERICAL TAXONOMY

The fundamental position of numerical taxonomy may be summarized in the following principles (modified from Sneath, 1958), which embody concepts that can be traced to Michel Adanson (*1727–1806*), a French botanist whose views are discussed in some detail in Section 2.2. The views represented by these principles are therefore frequently called neo-Adansonian.

   1. The greater the content of information in the taxa of a classification and the more characters on which it is based, the better a given classification will be.
   2. A priori, every character is of equal weight in creating natural taxa.
   3. Overall similarity between any two entities is a function of their individual similarities in each of the many characters in which they are being compared.
   4. Distinct taxa can be recognized because correlations of characters differ in the groups of organisms under study.
   5. Phylogenetic inferences can be made from the taxonomic structures of a group and from character correlations, given certain assumptions about evolutionary pathways and mechanisms.
   6. Taxonomy is viewed and practiced as an empirical science.
   7. Classifications are based on phenetic similarity.

   Principles 1 through 3 are elaborated in Section 1.4, "The Estimation of Resemblance"; principle 4 is treated in Section 1.5, "The Construction of Taxa"; principle 5 is discussed in Section 1.6, "The Recognition of Phyletic Relationships"; principles 6 and 7 are dealt with in Section 1.7, "Phenetic and Phylogenetic Taxonomy."

   In practice the operations of numerical taxonomy are carried out in the following sequence: organisms and characters are chosen and recorded; the resemblances between organisms are calculated; taxa are based upon these resemblances; and last, generalizations are made about the taxa (such as inferences about their phylogeny, choice of discriminatory characters, etc.).

   It should be clear that generalizations about the taxa cannot be made before one has recognized the taxa; that taxa cannot be recognized before resemblances between organisms are known; and that these resemblances cannot be estimated before organisms and their characters have been examined. Therefore, although some of these steps may be in effect combined in certain computational methods, or the whole procedure may be repeated a second time for some special reason, the order of the steps within the procedure cannot be changed without destroying the rationale of the classificatory process.

1.4   THE ESTIMATION OF RESEMBLANCE

Estimation of resemblance is the most important and fundamental step in numerical taxonomy. It commences with the collection of information about characters in the taxonomic group to be studied. This information may already exist and merely require extraction from the literature, or it may have to be discovered entirely or partly de novo. In most cases both of these procedures will need to be applied. For the method to be reliable, many characters are needed. All kinds of characters are equally desirable: morphological, physiological, ethological, and sometimes even distributional ones. One must guard only against introducing bias into the choice of characters and against meaningless characters (Section 3.7).

From our assertion of the equal taxonomic value of every character (see Sections 3.2 and 3.3 for our definition of a character) it is only a small step to the practice of assigning equal weight for every character when using it to evaluate taxonomic relationships. This is a point in direct conflict with traditional taxonomic practice, over which much controversy has raged. We discuss this issue in some detail in Section 3.9. The impossibility of developing criteria for a priori weighting of taxonomic characters has been quite generally conceded.

The actual computation of a measure of resemblance can be done in a variety of ways (Chapter 4). Resemblance is expressed by coefficients of similarity usually ranging between unity and zero, the former for perfect agreement, the latter for none whatever, or by coefficients of dissimilarity (distance) usually ranging between zero and an undefined positive value, the former for identity, the latter for maximal distance or disparity. The calculations are likely to be rather tedious and computer processing is needed for any but very minor studies.

Resemblance coefficients are tabulated in matrix form with one coefficient for every pair of taxonomic entities. If a symmetrical (mirror image) matrix is to be tabulated for t entities, a $t \times t$ matrix will result. Similarities among taxonomic entities can be represented geometrically by points in a space. A maximum of $t - 1$ dimensions is needed for a correct representation of the t points (taxonomic entities) in the space. The distances between the points can be regarded as taxonomic distances.

1.5   THE CONSTRUCTION OF TAXA

Classification in numerical taxonomy is generally based on a matrix of resemblances, in which taxa are constructed through various techniques designed to disclose and summarize the structure of the matrix. By some methods classification can be carried out on the original data matrix. A rough, graphical representation of the structure of the matrix can be obtained by shading the various elements of the matrix differentially (see Figure 5-16) if they have previously been roughly grouped so that supposedly similar forms are near each other. If, as is methodologically preferable, the entities are placed in the matrix without predetermined order, visual grouping is not easily accomplished without rearrangement. The various computational methods for clustering will process the data equally efficiently whether they are ordered or not. Since computational methods simultaneously provide some numerical evaluation of the taxonomic relationships, they are to be preferred.

These numerical methods are collectively called cluster analysis (see Sections 5.4 and 5.5). They are methods for establishing and defining clusters of mutually similar entities from the $t \times t$ resemblance matrix. These clusters may be likened to hills and peaks on a topographic chart, and the criteria for establishing the clusters are analogous to the contour lines of such a map. Rigid criteria correspond to high elevation lines that surround isolated high peaks—for example, species groups in a matrix of resemblances between species. As the criteria become more relaxed the clusters grow and become interrelated in the same way that isolated peaks acquire broader bases and become connected to form mountain complexes and eventually chains, with progress from higher- to lower-level contour lines. The clusters are generally based on phenetic resemblances only and have no necessary phyletic connotations. Differences in methods of clustering refer mainly to rules for forming clusters and for partitioning the organisms in taxonomic (character) space.

The important common aspect of all these methods is that they permit the delimitation of taxonomic groups in an objective manner, given a matrix of coefficients of relationship. Boundaries for taxonomic groups can be visualized as the contour lines already discussed or they can be represented as the intersections of horizontal transects with the branches of the tree-like diagrams of relationship commonly employed in numerical taxonomy. Comparable limits can be drawn for all taxonomic groups within a particular study. Boundaries or transects at progressively lower levels of resemblance would create taxa of increasingly higher taxonomic rank.

The number and position of boundaries or transects should follow some prearranged system. Clearly they will depend on the size of the similarity matrix: too many boundaries would provide too fine a classification; too few would leave much structure unrevealed. The aims of the investigator, conventions in the particular group, and questions of convenience and esthetics would all affect the placing of boundaries. We consider the number of taxa to be established at any rank a relatively unimportant and arbitrary detail. But once boundaries have been established, the structure within a taxon (and hence the number of lower ranked taxa contained therein) depends entirely on the resemblance values of entities and is not subject to the manipulations of the investigator. In other words, *the position and number of the boundaries or transects is arbitrary, but they must be based on comparable criteria in all regions of the taxonomic space under consideration.*

We view monotypic taxa or very sizeable ones with equanimity. Their occurrence does not lead us, respectively, to lumping or splitting. In taking such a position we are motivated by an effort to supply taxa with some objective and definable criteria for intragroup cohesion. This will largely depend on the method of cluster formation (Sections 5.3 and 5.4).

Biologists who use the results of taxonomic research are much concerned with the stability of a classification. The stability of a scheme based on numerical taxonomy may be affected in two ways.

1. More information (in the form of new characters) may accumulate. If the initial evaluation of resemblances has been based on a large sample of characters it is our contention that the relative similarities would change little on the addition of further characters (Section 3.8).

2. New taxonomic entities may be included in subsequent studies. Application of the previous criteria for levels and number of transects may result in new and different taxa. Agreement will have to be reached by practitioners of numerical taxonomy on whether the established system should be rearranged to suit the new results or whether the new data should be judged by standards of relationship already established—that is, whether the boundaries should be continued at the level at which they were drawn in the first study (see Section 5.10).

What level in the hierarchy are we to call a subgenus, a genus, a family? Have these terms any significance of their own other than as indications of the relative levels of the boundaries? It is generally accepted in conventional systematics that genera (and other categories) in such diverse groups as insects, birds, and flowering plants do not represent taxa of equivalent relationship. Is it possible for numerical taxonomy to set up such equivalent categories, although these would be based on entirely different groups of characters (discussion in Section 5.11)? It would appear preferable to employ a new series of terms for the hierarchic system established by numerical taxonomy, which would include in the terms a quantitative estimate of the similarity among members of the group. The phenon terminology (see Section 5.11) is an attempt at such a new system. Other, more sophisticated methods remain to be developed.

An important aspect of the construction of natural taxa is their representation. This can be done by means of tree-like diagrams indicating phenetic relationship, graphs, or three-dimensional models of taxa in phenetic hyperspace, and two-dimensional projections or stereoscopic images of such models (see Section 5.9).

The representation of taxonomic relationships in a space of two or three dimensions (ordination; see Section 5.6) leads to investigations of patterns of taxonomic structure that are of considerable ecological and evolutionary interest (see Sections 5.14 and 7.3).

1.6   THE RECOGNITION OF PHYLETIC RELATIONSHIPS

Although the major emphasis of numerical taxonomy to date has been the creation of stable phenetic groups, the methodology also offers constructive approaches for the study of phyletic lineages and evolution (Chapter 6). Measures of phenetic similarity between organisms of different geological periods will provide objective information regarding rate and direction of evolution, or may assist in the solution of some stratigraphic problems. Much useful information may also be obtained for studies of speciation by comparing phenetic differences with genetic or geographical data.

Recent years have seen renewed emphasis on operational approaches for studies that are phyletic or phylogenetic. As we explain in Section 2.3, these two words have meanings that sometimes overlap. Such analyses aim at estimating the branching sequences of the evolutionary lines represented by the taxa in a study. The basic data comprise conventional morphological characters as well as amino acid sequences in proteins. The methods employed so far attempt to find the most parsimonious (minimum length) phylogenetic trees needed to establish the phenetic pattern of the taxa. This type of reasoning has been employed by phylogeneticists for many years. What is new is the systematization of the procedure and the simultaneous consideration of many characters, assigning them equal a priori weights but sometimes weighting them during clustering on the basis of some operational criteria. By these methods, and others under development, evolutionary lineages are being estimated by means of computers.

1.7 PHENETIC AND PHYLOGENETIC TAXONOMY

Sokal and Camin (1965) have distinguished operational, empirical, and numerical approaches to taxonomy, whose importance is considered to be in that order. In *operational taxonomy* statements and hypotheses about nature must be subject to meaningful questions, i.e., those that can be tested by observation or experiment (see Section 2.1). *Empirical taxonomy* is based on many observed and recorded characters, taxa being grouped according to a majority of shared characters (Simpson, 1961; Sokal, 1962b). The method by which the number of shared characters is determined varies with the empirical school. It may, or may not, be quantitative. Numerical taxonomy as generally practiced is both operational and empirical.

The taxonomic procedures outlined in the earlier sections of this chapter are of a strictly empirical nature. As such they are related to the procedures of some typologists in the past. This is not, however, an automatic disqualification of our views, as Simpson (1961) has implied (for a defense of "statistical typology" see Sokal, 1962b). The fundamental test of the validity of empiricism in taxonomy must be whether it can be used as a consequential and consistent method for arranging organized nature. We believe that it can be so used and that it is the only reasonable approach.

A basic attitude of numerical taxonomists is the strict separation of phylogenetic speculation from taxonomic procedure. Taxonomic relationships are evaluated purely on the basis of the resemblances existing *now* in the material at hand. These phenetic relationships do not take into account the origin of the resemblance found nor the rate at which resemblances may have increased or decreased in the past.

This attitude is taken because, as is discussed in detail in Section 2.6, we do not at the moment possess a classificatory scheme, graphic or otherwise, able simultaneously to yield information on the degree of resemblance, descent, and rate of evolutionary progress. Any scheme attempting to combine these approaches would be excessively complicated. The separation of overall similarity (phenetics) from evolutionary branching sequences (cladistics) is an important advance in taxonomic thinking (see Section 2.3 for a detailed discussion). It is often difficult to make this separation, in view of long established methods of thought acquired with the early training of most practicing taxonomists of the present generation. Yet the most recent generation of taxonomists just out of graduate school has had little difficulty with this concept and recognizes the value of the separate ways of regarding taxonomic relationships.

Not only do we insist on the separation of phenetic from phylogenetic considerations in taxonomic procedure, but we also feel that only phenetic evidence can be used to establish a satisfactory classification. We hold this belief for several reasons.

1. The available fossil record is so fragmentary that the phylogeny of the vast majority of taxa is unknown. Evolutionary branching sequences must be inferred largely from phenetic relationships among existing organisms.

2. Phenetic classification is possible for all groups. By contrast, cladistic classification, based on branching sequences, requires historical inferences about the direction of evolution in a group of organisms.

3. Even when fossil evidence is available, this evidence itself must first be interpreted in a strictly phenetic manner—with the exception that a time scale is given in addition, which may restrict certain interpretations of the phylogeny—since the criteria for choosing the ancestral forms in a phylogeny are phenetic and are based on the phenetic relationship between putative ancestor and descendant.

4. From the point of view of biology in general, it is probably of more interest to describe the overall similarity of organisms than their branching sequences. If the classifications are to have predictive value, it is evident that those based on overall similarity will be most predictive.

To accept taxonomic relationships based only on phenetic criteria is difficult for many present-day biologists. For almost a century there has been an intimate conceptual association between taxonomic and phylogenetic reasoning, so that terms such as "specialized," "primitive," "analogous," and many others have assumed double meanings whose differences are rarely made distinct. Yet the effort to do so must be made to achieve clarity and to progress in the understanding of taxonomic principles.

The issues summarized in this section are discussed in greater detail in Sections 2.2 to 2.6.

1.8   THE ADVANTAGES OF NUMERICAL TAXONOMY

The methods discussed in this book have a number of advantages, which will be listed below. First, however, we should refer to their principal aims, *repeatability* and *objectivity*. If observations are repeatable within an acceptable error and if taxonomic procedures are clearly circumscribed, it is hoped that numerical methods will lead different scientists employing the same data base and working independently to obtain comparable estimates of the resemblance among any group of organisms.

Closely tied up with repeatability is the notion of objectivity. The *Random House Dictionary of the English Language* defines "objective" as "free from personal feelings or prejudice; based on facts; unbiased." This is a relative concept, seldom fully realized. By including many characters without previous arbitrary selection or elimination, and by providing explicit methods of processing the data and evaluating the results, we reduce subjective bias and increase objectivity.

Of the manifold advantages of numerical taxonomy we may briefly mention the following, which will be discussed in greater detail in appropriate sections of the text.

1. Numerical taxonomy has the power to integrate data from a variety of sources, such as morphology, physiology, chemistry, affinities between DNA strands, amino acid sequences of proteins, and more. This is very difficult to do by conventional taxonomy.

2. Through the automation of large portions of the taxonomic process, greater efficiency is promoted (Sokal and Sneath, 1966). Thus, much taxonomic work can be done by less highly skilled workers or automata.

3. The data coded in numerical form can be integrated with existing electronic data processing systems in taxonomic institutions and used for the creation of descriptions, keys, catalogs, maps, and other documents.

4. Being quantitative, the methods provide greater discrimination along the spectrum of taxonomic differences and are more sensitive in delimiting taxa. Thus they should give better classifications and keys than can be obtained by the conventional methods.

5. The creation of explicit data tables for numerical taxonomy has already forced workers in this field to use more and better-described characters. This necessarily will improve the quality of conventional taxonomy as well.

6. A fundamental advantage of numerical taxonomy has been the reexamination of the principles of taxonomy and of the purposes of classification. This has benefited taxonomy in general, and has led to the posing of some fundamental questions discussed in numerous subsequent sections of this book and in many publications in the recent systematic literature.

7. Numerical taxonomy has led to the reinterpretation of a number of biological concepts and to the posing of new biological and evolutionary questions. Thus, the method is coming into its own as a heuristic tool in biological research. In Chapters 6 and 7 and Section 10.3 we take up these novel aspects of numerical taxonomy.

1.9   IDENTIFICATION OF SPECIMENS

Once a classification has been established by some operational method, the construction by computer of appropriate keys for identifying specimens is an obvious next step. Some of the errors that are occasionally commited in setting up dichotomous keys would be avoided by logical and consistent programs. For purposes of taxonomic keys and identification of specimens, weighting of characters must be introduced to emphasize those characters that are most effective in distinguishing between previously established taxa. Weight must not only be statistical in the sense of providing efficient criteria for differentiation, but must also take into account the ease of observing or measuring a given character, the probability of its being recordable in a study specimen, the chance of structures bearing the character being damaged or confused, etc. Alternative structures must be provided for identification in case the preferred structures are not available.

Although the logic of key making may change little, the physical form of keys can be adapted to electronic data processing and a variety of forms may prove useful in different circumstances. The more sophisticated keys will usually be probabilistic; that is, they will give, on assumptions reasonable in the light of current knowledge, the likelihood that the identification of a given specimen is correct. On-line computing permits a new interaction between taxonomists and computers for the identification of specimens. Taxonomic keys can be stored in the computer and can be presented in the form of a dialogue between the taxonomist and the computer. The subject of keys and identification is discussed in some detail in Chapter 8.

1.10   NOMENCLATURE

Any change in taxonomic procedure will inevitably entail a change in nomenclature. Although in a formal way it can be, in practice the subject of nomenclature cannot be dissociated from taxonomy per se. As long as we are going to study and group organisms, by whatever criteria, we will want names or labels to refer to them. We shall see in Chapter 9 that the present system of nomenclature does not adequately fulfill the various functions expected of it. Some changes in the system will become necessary simply due to the inevitable adoption of electronic data processing for taxonomic information. Proposals for revision of the nomenclature are only in their early formative stages. They will be reviewed in Chapter 9.

If numerical taxonomy has a special effect on nomenclature, it will be mostly through revision of existing classifications. If a numerical study shows considerable changes from the relationships implied by the orthodox nomenclature, the author has no choice but to alter the classification and to make such changes of name as are then required. One must do this if one is to provide other biologists with the benefits of improved classifications.

The present overdependence of the system of nomenclature upon types has been changed by the advent of numerical taxonomy, for it is now in principle possible to determine the limit of taxonomic groups, which formerly had been a matter of individual opinion. Numerical nomenclature for intermediate forms may also prove of some use.

1.11   THE DEVELOPMENT OF NUMERICAL METHODS IN TAXONOMY

The earliest attempts to apply numerical methods to taxonomy date from the rise of biometrics in the last century. As early as 1898 Heincke used a measure of phenetic distance to distinguish between races of the herring, while in 1909 Czekanowski employed a distance coefficient in physical anthropology. It was early realized that biometrics could be applied to systematics, but the only important development was that of discriminant functions (Fisher, 1936), which is useful in only a limited area of taxonomy.

One of the earliest statistics of interest to systematists was the "Coefficient of Racial Likeness" (Pearson, 1926). It was extensively applied in physical anthropology but does not seem to have been taken up by taxonomists. The *C.R.L.* was a type of taxonomic similarity coefficient, and was subsequently developed by Mahalanobis in the form of the "Generalized Distance" statistic, which is also formally a coefficient of this kind (see Rao, 1948). Anderson and Whitaker (1934) and Anderson and Abbe (1934) employed a similar statistic, which they called the "General Index." These statistics, though mathematically adequate, did not lead to notable advances for two reasons: (1) they were developed mainly as discriminant functions to aid in the allocation of individuals to existing taxa and not as methods for creating taxa; and (2) as a consequence of (1) these workers selected principally those characters which gave the best discrimination between the taxa, and—since it is usually only necessary to employ a small number of such characters once they have been found—these methods were in practice based on few rather than many characters. Some of the characters were selected on a priori grounds, and their small number led to instability on repeating the work with other characters. The lack of adequate computing facilities in the early part of this century also limited multivariate work to few characters. These techniques, with others developed later, have been very widely and successfully used for the study of certain limited taxonomic problems. We may, for example, cite the elegant work of Blackith (1957) on sexual and phase variation in locusts, Reyment (1963) on fossils, and that of Jeffers and Black (1963).

Some studies with aims similar to those of numerical taxonomy today should be mentioned. Smirnov (1925) established types on a quantitative basis. His work has been discussed and evaluated from different points of view by Hennig (1950) and Sokal (1962b). Haltenorth (1937) in a study of similarities among eight species of large cats developed a coefficient similar to that of Cain and Harrison (1958). At about the same time Zarapkin (1939) developed a rather elaborate technique called

Divergenzanalyse, arriving at a quantity analogous to taxonomic distance. The *Affinitätsrechnung* of Schilder and Schilder (1951) is also a computation of taxonomic distance. We believe that these methods did not succeed at the time they were developed because of the entrenched nature of phylogenetic systematics and since for any substantial number of characters or taxa the methods advocated by these authors presented computational difficulties insurmountable at the time.

Other early methods, specifically intended for taxonomy, are those of Forbes (1933), Anderson and Owenbey (1939), Sturtevant (1939, 1942), Boeke (1942), James (1953), Stallings and Turner (1957), Hudson, Lanzillotti, and Edwards (1959), and Chillcot (1960), all based on variations of matching coefficients. Terentjev (1931, 1959) introduced a method of defining clusters of characters called "pleiades." These authors did not develop their methods sufficiently to meet the main needs of numerical taxonomy and hesitated to give equal weight to every feature or to employ large numbers of characters. Similar trends can also be seen in the history of bacterial classification, where the earlier reliance on a few morphological or physiological characters has given place to attempts at classification in the Adansonian tradition (see Sneath, 1962). There developed in Poland during the 1950's the Wroclaw school of taxonomy based on the coefficient of Czekanowski, whose procedures for classification were similar to numerical taxonomy (Florek et al., 1951a,b; Perkal, 1951). Instead of a formalized and exact clustering, the Polish workers use graphic approaches involving linkage diagrams and multiple contour lines. The technique apparently remained undiscovered by English-speaking numerical taxonomists whose approach relied more upon numerical and computer techniques. However, a review by Hubac (1964) drew attention to the work of the Wroclaw group and caused at least one worker to adapt the Wroclaw techniques for the graphic representation of conventional taxometric results (Moss, 1967). (See also Section 5.9.)

One of the main conceptual difficulties that retarded progress in numerical taxonomy was the problem of weighting and the liberating effect of accepting equal weighting can scarcely be overemphasized. The use of many characters is also a prerequisite of progress. In addition, the use of methods of cluster analysis in building the taxonomic hierarchy has been a major advance. It is in these three developments that numerical taxonomy chiefly differs from the earlier ideas and methods.

The modern work in numerical taxonomy was initiated by the almost simultaneous publications of Sneath (1957a,b) at the National Institute for Medical Research, London, on bacteria, and of Michener and Sokal (1957) and Sokal and Michener (1958) at the University of Kansas, on bees. These authors developed their ideas independently and, upon learning of each other's work prior to the publication of these papers, initiated a correspondence and collaboration which resulted in the development of principles and techniques (Sneath and Sokal, 1962) and culminated in the publication of a book on this subject (Sokal and Sneath, 1963). At almost the same time two other groups independently developed numerical approaches to the taxonomy problem. Cain and Harrison (1958) at Oxford proposed a mean character difference for taxonomic purposes, and Rogers and Tanimoto (1960) at the New York Botanical Garden developed a taxonomic technique based on a coefficient of association.

Since those early days, and especially since the publication of our *Principles of Numerical Taxonomy*, there has been a very rapid increase in the development of methods for numerical taxonomy and in the application of these techniques. The subject is still in a period of active growth and change and we lack the historical perspective to single out every significant landmark in its development, but the following chronological chart may give an idea of the development of the field. The number of papers published in the field has risen from 60 that appeared in the five years from 1957 to 1961 to over 200 annually in recent years.

1957–1961	Development of first methods and of theory of numerical taxonomy.
1962	First test of the nonspecificity hypothesis (Rohlf); first criterion of goodness of a classification (Sokal and Rohlf); first comprehensive publication of theory and methods (Sneath and Sokal).
1963	Publication of *Principles of Numerical Taxonomy* (Sokal and Sneath).
1964	First cladistic technique for continuous characters (Edwards and Cavalli-Sforza).
1965	First cladistic technique for discrete characters (Camin and Sokal); first critical analysis of the methodology of numerical taxonomy (Williams and Dale).
1966	First paper on taxonomy of scanned images (Sokal and Rohlf); electronic data processing linked to taxonomy (Sokal and Sneath); first generalization of hierarchical clustering methods (Lance and Williams).
1967	Numerical cladistics of proteins (Fitch and Margoliash); first attempts at definitions of operational homology (Jardine, Key, Sneath); congruence of phenetics with genome message (Heberlein, De Ley, Tijtgat); first computer-oriented identification methods (Goodall, Morse, Dybowski, Franklin).
1968	Investigation of intra-OTU variation (Crovello).
1969	First generalized methods for numerical cladistics (Kluge and Farris); reexamination of the logic of biological classifications (Jardine); attempts at statistical validation of clustering procedures (Mountford, Switzer).
1970	Reexamination of the biological species concept (Sokal and Crovello).
1971	Publication of *Mathematical Taxonomy*, stressing logically consistent criteria for taxonomy and overlapping clusters (Jardine and Sibson).

2

Taxonomic Principles

To anyone familiar with current developments in taxonomy, it is clear that taxonomy is undergoing a period of rapid conceptual and procedural change, probably unequaled since the immediate post-Darwinian era. Numerous publications in journals such as *Systematic Zoology, Evolution*, and *Taxon* reflect the effervescence of ideas that have suddenly enlivened this field. The preceding period of similarly rapid progress in biological systematics has been called the "New Systematics"; it may be said to have begun with the appearance of the book of the same name edited by Huxley (1940). Advances in genetics, cytology, and geographic variation during the period led to considerable progress in the understanding of evolutionary mechanisms at the species and infraspecies levels. However, the New Systematics contributed little to our understanding of the nature and evolution of the higher categories and of taxonomic structure in general. Books such as those by Rensch (1947), Schmalhausen (1949), Simpson (1953), and Hennig (1966) deal with the latter topics but they contain little more than descriptive generalizations. The failure of the New Systematics to provide an adequate base for animal taxonomy is also discussed by Blackwelder (1967a, p. 336).

Traditional taxonomy attempts to fulfill too many functions and as a consequence fulfills none of them well. It attempts (1) to classify, (2) to name, (3) to indicate degree of resemblance, and (4) to show relationship by descent—all at the same time. We shall show in separate sections in this chapter that it is impossible not only in practice but also in theory for the traditional system to perform these tasks adequately. In Section 2.1 we shall discuss empirical and operational approaches to the subject. Section 2.2 will concern itself with the problems of the natural system; in Section 2.3 we shall analyze taxonomic relationships. Problems of phenetic taxonomy are detailed in Section 2.4 and those of cladistic taxonomy are detailed in Section 2.5. A decision must be made on which of these methods is to serve as a basis of classification; this will be found in Section 2.6. The last two sections, 2.7 and 2.8, deal with problems of taxonomic rank and with desirable properties of a taxonomic system, respectively.

2.1   EMPIRICAL AND OPERATIONAL APPROACHES

Recent statements on taxonomic theory (Ehrlich and Holm, 1962; Davis and Heywood, 1963; Sokal and Camin, 1965; Jardine and Sibson, 1971) have stressed the *empirical approach* in taxonomic work. The main emphasis in empirical taxonomy is to base classifications and taxonomic judgment on firm observation and not upon "phylogenetic" assumptions. We think the theoretical framework of conventional phylogenetic systematics has served as a straitjacket for taxonomic concepts and ideas at all levels, from that of the infraspecific population to that of phyla; in the final analysis it is evolutionary theory that has suffered as well, since it is only as good as the taxonomic data fed into it. Thus, so long as taxonomic descriptions and judgments are made to conform to concepts such as the biological species (Mayr, 1963), or monophyly (Simpson, 1961), taxonomists tend to consider difficulties and discrepancies as embarrassing exceptions to generally accepted principles. But in recent years the emphasis has been on the description of those variational patterns that do in fact exist in nature. New principles will emerge from such studies and a body of theory whose heuristic aspects are beginning to emerge is already forming (see Chapters 6 and 7, and Section 10.3).

The emphasis on empirical analysis of taxonomic data led naturally to what has become known as the *operational approach* to taxonomy (Sokal and Camin, 1965) by analogy and extension of P. W. Bridgman's ideas. In our context operationism implies that statements and hypotheses about nature be subject to meaningful questions; that is, those that can be tested by observation and experiment. One must establish criteria for defining categories and operations; otherwise it would be impossible to engage in a meaningful scientific dialogue about them. Sokal and Camin illustrate this concept by stating that

> ...if we wished to determine whether A is more related to B than it is to C, we have to give clear definitions of what we mean by "more related," i.e., by what criteria more or less relatedness can be measured, and we must be able to issue a series of instructions by which we, our assistants, or our colleagues can determine the relationships and solve the problem originally posed. We must be certain that the data we work with are subject to definable logical operations. Secondly, we must be concerned that the operations carried out to answer questions raised about the material are such that they can be communicated unambiguously to other intelligent persons as well as to machines able to handle the logic and computation required.

Such a guiding principle seems natural and desirable to us and we are unaware of any substantial arguments advanced against this approach in taxonomic work. The term operationism, however, is something of a red flag to certain philosophers of science (Hawkins, 1964; Hull, 1967, 1968a). Hawkins (1964) points out that emphasis on operational procedures and strictly quantitative measurements may be unwarranted in the biological and social sciences, since the theory to which the body of data pertains is itself ill-defined because of the complexity of the subject matter: "For the purposes of . . . crude theory, crude measurements suffice." Hawkins believes that in biological taxonomy the conventional types of observations are sufficiently reliable to serve the needs of existing theory. We would differ on this point, since we feel that useful theory cannot be obtained in this field until more precise measurements and definable operations have become standard procedure in biological taxonomy.

Hull (1968a) points out that operationism carried to the extreme would lead to inference-free direct observations, unlikely to lead to theoretical advances. He adds that if, in contrast, empirical taxonomy itself relies on inference and theory, its proponents should not condemn phylogenetic systematists for their excursions into similar fields.

Not all concepts can be made equally operational and some not at all. Nor is it necessary, as Hull (1968a) points out, that operationism be extended to all members of a class of concepts.

We do not object to nonoperational concepts categorically, although we would always prefer more-operational concepts to less-operational ones; but when the nonoperational concepts are vague and ill-defined, and have no heuristic value, we are opposed to them. Concepts such as the biological species (sensu Mayr et al.) and the phylogenetic notion of homology were vital in the development of modern biological theory. Today they are more of a hindrance than a stepping stone to new discoveries and it is for this reason as much as their low operational value that we wish to redefine them or possibly even dispense with them.

Operational and empirical taxonomy (by which we mean operational procedures during classification, and empirical observations of taxonomic data) are not wholly congruent. Although much of empirical taxonomy is operational it is conceivable that nonoperational criteria could be used to process empirical observations (Figure 2-1). Numerical taxonomy as generally practiced is both operational and empirical. Although it is quite possible to visualize a numerical taxonomy that is not empirical, all numerical taxonomies are likely to be operational.

2.2   THE NATURAL SYSTEM

Great difficulties have always accompanied attempts at defining a natural system. Thus we have the significant comment of Linnaeus himself (mentioned by de Candolle, 1813, p. 60) that he had been unable to discover a natural method of

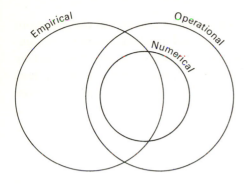

Taxonomy

FIGURE 2-1
Diagram of the relationship between operational, empirical, and numerical taxonomies. [From Sokal and Camin (1965).]

classification. Until recently, apart from the work of Gilmour mentioned later, very little was written on this. Danser (1950) realized the difficulty of defining natural groups but was not able to state any exact or scientific definition for them, ending with the hope that "... some day systematics will arrive at a more exact stage, but this does not alter the fact that already now we are entitled to face its problems, be it for the moment in a more intuitive but nevertheless scientific manner." Simpson (1961, p. 57) agrees that "in fact much of the theoretical discussion in the history of taxonomy has, beneath its impersonal language and objective facade, been an attempt to find some theoretical basis for these personal and subjective results."

The term "natural classification" has had a variety of meanings. In the early days of systematics it meant a classification that was in accord with nature, but this meaning was undefined further (though it was commonly implied to describe the opposite of artificial or arbitrary systems, but these were not well defined themselves). Later it came to mean variously a taxonomy based on maximum correlations between characters, or one based on phylogeny, and the latter sense is one in which traditional taxonomists use it today, though we explain later why we feel this is not a suitable usage. The development of the concept of the natural system is briefly discussed in the following pages.

The earliest attempts at systematics were based, as Cain (1958) has shown, on Aristotelian logic. This was the method used by early systematists such as Cesalpino (*1519–1603*) and even largely by Linnaeus (*1707–1778*). The Aristotelian system as applied to taxonomy consisted in the attempt to discover and define the *essence* of a taxonomic group (what we may somewhat loosely think of as its "real nature" or "what makes the thing what it is"). In Aristotelian logic this essence is expressed in axioms that give rise to properties that are inevitable consequences; for example, the essence of a triangle on a plane surface is expressed by its definition as a figure bounded by three straight sides, and an inevitable consequence is that any two sides together are longer than the third. Such logical systems are known as systems of *analyzed entities*, and early systematists supposed that biological classifications could be of this kind. The terms *genus* and *species* had technical meanings in logic, and these were taken over into taxonomy. These points are well discussed by Thompson (1952) and Cain (1958, 1962). Aristotelian logic does not, however, lend itself to biological taxonomy, which is a system of *unanalyzed entities*, whose properties cannot be inferred from the definitions—at least not if the taxonomy is to be a natural one in any of the usual senses.

Caspar Bauhin (*1550–1624*), John Ray (*1627?–1705*), and Pierre Magnol (*1638–1715*) had a strong intuitive sense of what natural taxa were, although they did not express themselves clearly. This is what de Candolle (1813, p. 66) aptly called "groping" (*tâtonnement*). According to de Candolle, Magnol claimed to have a clear idea of a natural family of plants even though he could not point to any one character which was diagnostic of that family.

This comment by Magnol and a similar comment by Ray (quoted by Cain, 1959c) were among the first admissions that it might not be possible to find *any single* diagnostic character for a natural taxonomic group. This is a point of the very greatest importance, which can scarcely be overemphasized. Michener (1970) supports this view for all taxa but species; others, among them Sneath (1957a) and Jardine and Sibson (1971), state that it is true of natural taxa of *any* rank. While "artificial" or "arbitrary" taxa can indeed be defined by a single character, this is not necessarily true of natural taxa. Every systematist knows of instances where a character previously considered to be diagnostic of a taxon is lacking in a newly discovered organism that clearly belongs to the taxon. A striking example is the lack in some species of fish of red blood corpuscles (Ruud, 1954), hitherto considered to be an invariable attribute of all vertebrates. Fortunately, as Michener (1957) says, natural taxa generally do possess some distinctive characters in practice although they need not do so in theory.

Biologists are indebted to Beckner (1959) for a clear enunciation of the important concept of "polytypic" natural taxa. Its implications in taxonomy have been discussed by Simpson (1961, pp. 41–57). Since this term and its converse, "monotypic," have meanings already well established in systematics, the substitute terms "polythetic" and "monothetic" suggested by Sneath (1962) have come into general use. We have assumed, in interpreting Beckner's concepts, that he intended the term "property" to be taken in the general sense of some definite value of a taxonomic character.

The idea ruling *monothetic groups* is that they are formed by rigid and successive logical divisions so that the possession of a unique set of features is both sufficient and necessary for membership in the group thus defined. They are called monothetic because the defining set of features is unique. That is, all the members of any group possess all of the features that are used to define that group. Any monothetic system (such as that of Maccacaro, 1958, or that of Williams and Lambert, 1959) will always carry the risk of serious misclassification if we wish to make natural phenetic groups. This is because an organism that happens to be aberrant in the character state (see Section 3.2) used to make the primary division will inevitably be moved to a taxon away from the required position, even if it is identical with its natural congeners in every other character state. The disadvantage of monothetic groups is that they do not yield "natural" taxa, except by a lucky choice of the character used for division. The advantage of monothetic groups is that keys and hierarchies are readily made.

By contrast, in a *polythetic group*, organisms are placed together that have the greatest number of shared character states, and no single state is either essential to group membership or is sufficient to make an organism a member of the group. This concept was stated many years ago (for example, by Jevons, 1877, pp. 682–698) and more recently by Kaplan and Schott (1951) and Gasking (1960). For its formal expression we cannot do better than to quote Beckner's definition (1959, p. 22):

> A class is ordinarily defined by reference to a set of properties which are both necessary and sufficient (by stipulation) for membership in the class. It is possible, however, to define a group K in terms of a set G of properties $f_1, f_2, \ldots, f_n$ in a different manner. Suppose we have an aggregation of individuals (we shall not as yet call them a class) such that:
>   1. Each one possesses a large (but unspecified) number of the properties in G.
>   2. Each f in G is possessed by large numbers of these individuals and
>   3. No f in G is possessed by every individual in the aggregate.
>
> By the terms of 3, no f is necessary for membership in this aggregate; and nothing has been said to either warrant or rule out the possibility that some f in G is sufficient for membership in the aggregate.

He then goes on to say that a class is polythetic if the first two conditions are fulfilled and is fully polythetic if condition 3 is also fulfilled. He points out that taxonomic groups are polythetic classes, but that polythetic concepts are by no means restricted to taxonomy or even to biology, for Wittgenstein emphasized their importance in ordinary language and especially in philosophy—polythetic ideas are implied by the concepts of "meaning," "referring," "description," and so on. There is a close parallel between Wittgenstein's "family resemblance" and taxonomic resemblance. As we have noted above, natural taxa are usually not fully polythetic, since one can usually find some character states common to all members of a taxon. It is possible that they are never fully polythetic, because there may be some character states (or alleles) that are identical in all members of a given taxon; even if there are many alleles or pseudoalleles of a gene, there may well be parts of the gene that are identical in all members. Recent work on protein sequences (discussed at greater length in Section 3.5) suggests that there are at least parts of genes that are most probably invariant in all members of a taxon, while work in population genetics (for example, that of Kimura, 1968) implies that the number of alleles at a locus in a population is limited by the mechanism of gene replacement during evolution. Both ideas convey the implication that taxa are not fully polythetic. Nevertheless, for practical purposes we must consider the possibility of a taxon being fully polythetic, since we cannot be sure that we have observed any characters that are common to all members.

Beckner also points out the importance of condition 2. If, for example, the various f's are found in only one individual of the aggregate, then each individual will possess a unique subset of the f's and will share no f's with any other individual. Such a situation does not yield a polythetic class. This is well discussed by Hull (1965). Let us illustrate such cases with the aid of two-state (presence–absence) characters: for example, individuals **a**, **b**, **c**, and **d**, do not form a polythetic class with the respective f's (presence or plus-states) of characters $\{1, 2, 3\}$; $\{4, 5, 6\}$; $\{7, 8, 9\}$; and $\{10, 11, 12\}$. If, however, as in the table shown below, individual **a** possesses $\{1, 2, 3\}$; individual **b** possesses $\{2, 3, 4, 5\}$; individual **c** possesses $\{1, 2, 4, 6\}$; and individual **d** possesses $\{1, 3, 4\}$; then the class of $\{$**a**, **b**, **c**, **d**$\}$ is polythetic (and in this instance is also fully polythetic, since no one character state is found in all the four individuals). This may be displayed in an arrangement such as this one:

Characters	Individuals					
	a	b	c	d	e	f
1	+	−	+	+	−	−
2	+	+	+	−	−	−
3	+	+	−	+	−	−
4	−	+	+	+	−	−
5	−	+	−	−	+	+
6	−	−	+	−	+	+

Individuals **e** and **f**, however, form a fully monothetic group.

One of the difficulties of Beckner's definition is that in natural taxa we do commonly have f's that are not possessed by large numbers of the class. Furthermore, we cannot test whether any given f is possessed by large numbers of the class before we have made the class, and therefore we cannot decide whether to admit this f into the set G. This difficulty can be avoided by defining class membership in terms of common (or shared) attributes. Polythetic groups can of course themselves be arranged polythetically to give higher polythetic groups, as is done in building a hierarchy in the natural system. The advantages of polythetic groups are that they are "natural," have a high content of information, and are useful for many purposes. Their disadvantages are that they may partly overlap one another (so that hierarchies and keys are less easy to make than with monothetic groups) and that they are not perfectly suited for any single purpose.

An important practical difference between "classification from below" (agglomerative clustering—the grouping of individuals into species, species into genera,

genera into tribes, tribes into families, and so on) and "classification from above" (divisive clustering—the division of the kingdoms into phyla, phyla into classes, and so on) is that the latter process is usually based on monothetic criteria. Classification from above therefore carries the risk that the divisions do not give "natural" taxa, yet it is a necessary practice in order to isolate a group of organisms of a manageable size for study. The important point is that the group classified from above may be incomplete or very heterogeneous; that is, some of its closest relatives may have been omitted, either through ignorance or because the forms have been misclassified.

A thorough early reevaluation of systematics was made by Adanson, a botanist of independent and original views. He rejected the a priori assumptions on the importance of different characters (which were a consequence of Aristotelian logic); he correctly realized that natural taxa are based on the concept of "similarity"— which is measured by taking all characters into consideration—and that the taxa are separated from each other by means of correlated features (Adanson, 1763, pp. clv, clxiv). The method he used was very cumbersome. He made a number of separate classifications, each based on one character, and examined them to find which classifications divided up the creatures in the same way. These classifications he took as indicating the most natural divisions, which were, of course, therefore based on the maximum correlations among the characters. By treating every character in the same way he was in effect giving them equal weight; it was upon this important corollary that his contemporaries attacked him (see de Candolle, 1813, pp. 70–72), without realizing that their own beliefs on the relative importance of various characters, far from being based on a priori assumptions as they imagined, were in fact a posteriori deductions from intuitive taxonomies of precisely the kind Adanson was recommending (Sneath, 1957a; Cain, 1959a,b). Adanson's earliest work in this direction was on molluscs (1757). No other workers, except perhaps Vicq-d'Azyr (1792) and Whewell (1840), seem to have followed up Adanson's ideas until recently.

We may ask why Adanson's method, though excellent in theory, was a failure in practice. Stearn (1961) considers that the material available in Adanson's day was too limited to allow of success, and we may add that such methods were quite impracticable before the advent of computing machines. Nevertheless, as de Candolle admitted, Adanson's taxa were for the most part more natural than earlier arrangements, although the superiority was not very marked. A fuller review of Adanson's contributions is given in Sokal and Sneath (1963) and by Stafleu (1963) and Sneath (1964c); for some other views see Jacobs (1966), Leroy (1967), Guédès (1967), and Burtt (1966). Burtt charges that the numerical taxonomists misinterpret Adanson's ideas and more specifically that Sneath (1964c) is wrong in proclaiming Adanson to be the father of numerical taxonomy. We prefer to let historians of science pursue this argument. For although it was—and remains —important to trace the roots of the historical origins of an idea in science, the development of numerical taxonomy has so far outpaced the early primitive ideas on this subject that to have to rely on Adanson's views for a validation of modern numerical phenetics seems as irrelevant as to rely on Mendel's writings for a validation of the findings of the molecular geneticists.

In the preevolutionary days of systematics it had been found empirically that a nested, hierarchical system gave the most satisfactory and "natural" arrangement of the data. Such a system could generally be constructed on the basis of a few characters. The art of the practice lay in finding suitable characters, to prevent the classification from creating strange bedfellows clearly incongruous when judged by their great differences in other characteristics. There was little attempt either to understand why it was possible to construct a system or to discover the rational method of choosing the "right" characters. We discuss below the development of the understanding of what it is that makes taxonomic groups "natural" and how it is possible *after creating such natural taxa* to discover characters that are suitable for discriminating between them.

Until the impact of the theory of evolution, the subsequent development of systematics took place largely in France (de Candolle, A. L. de Jussieu, Cuvier, and Lamarck) and was in the direction of greater sophistication on the theme of the coordination of characters into a harmonious whole. This was carried even to the point of implying that a whole animal could be reconstructed from one bone. One can, of course, identify a known animal from one bone, but to reconstruct from it a new animal with all its soft parts is a feat of a different order, as Simpson (1961, p. 44) points out.

The advent of the theory of evolution changed the practice of systematics very little, although the professed philosophical basis of systematics was radically altered. Natural classifications were considered to be those established on the basis of monophyletic taxa (see also Section 2.3). The theory of evolution did, however, provide a credible explanation for the nested distribution of taxa.

Little more was done on taxonomic theory until the conceptual basis of natural taxonomies was discussed from the standpoint of logic in a classic paper by Gilmour (1937) and expanded in later works (Gilmour, 1940, 1951, 1961; Gilmour and Walters, 1963). He pointed out that logicians have long realized that the central idea underlying "natural" groupings is the great usefulness of a method that can group together entities in such a way that members of a group possess many attributes in common. Indeed, we maintain that the elusive property of naturalness is simply the degree to which this principle obtains. The idea of overall similarity follows from this and is a function of the individual similarities in each of the many characters over which two entities are being compared. As Gilmour points out, natural classifications are not restricted to biological ones (see Chapter 11).

One of Gilmour's main points is that the nature of a taxonomy depends on its purpose. We could arrange living creatures in many ways, but we choose one way because we think it is best for some purpose. If the purpose is restricted, then the

classification is a special classification, often called "arbitrary." Such a classification conveys less information than a general or "natural" one. For example, we
can divide mammals into carnivores and herbivores for the purpose of ecology;
then the designation "carnivore" only tells us the kind of food they eat. We hold
the view with Gilmour that a "natural" taxonomy is a general arrangement
intended for general use by all scientists. In addition, intermediate situations can
occur between the highly natural (such as the class Mammalia) and the wholly
artificial (such as creatures whose generic names begin with the letter "A"). An
example of a partly natural group is the group that gardeners call "Alpines"—
plants that share numerous growth and physiological characteristics reflecting
their adaptation to alpine conditions. Edwards and Cavalli-Sforza (1964) have
pointed out that general or natural classifications are not too tightly defined, but
the terms have nevertheless considerable value in illustrating the basic logic that
underlies taxonomy. Mayr (1969a, p. 67) states that "the concept [of a natural
system] is so permeated with essentialist-creationist ideology that its use invariably
evokes a misconception among nontaxonomists." Since he does not believe in the
existence of the "Natural System," he prefers not to use the term at all.

We believe that natural classifications are of great usefulness because when the
members of a group share many correlated attributes, the "implied information"
or "content of information" (Sneath, 1957a) is high; this amounts to Gilmour's
dictum that a system of classification is the more natural the more propositions can
be made regarding its constituent classes. Remane (1956, p. 4) tries to show that
the predictive value of taxonomic groups is only true of natural taxa, not of
artificial ones. It is obvious that artificial groups established on a single character
are of low predictive value. Nevertheless, such groups may by chance prove to be
partly natural, since such a single character may be highly correlated with the other
characters of the taxa in question. It would be possible to devise a measure of the
extent to which this is true of any character in any given taxonomic system. Such
techniques are discussed in some detail in connection with probabilistic similarity
cofficients in Section 4.6, and with identification problems in Chapter 8, but a brief
account follows here.

The concept of information content is one that may not be easy to visualize
without an example. We therefore give below a very simple illustration of the
principle involved. When using small numbers of specimens and characters one
cannot include many "atypical" characters if the groupings are to be reasonably
sharp, but a few have been included to show that the principle applies to polythetic
groups as well as monothetic ones.

Suppose we have ten specimens scored for five qualitative characters, and the
"natural" groups are **A** and **B** as shown below. We can now count, for each group
in turn, the number of characters about which we can be reasonably certain, if we
were given a randomly sampled specimen of that group. By extension we may make
similar statements about a new specimen, which has been allocated with complete

certainty to the group in question. Whether a confident statement about the expected state of any one character is possible is shown in the column headed "Confident statement possible?" for both groups. The predicted state of the character is shown in parentheses.

We need to decide on some level of confidence, and purely for illustration we have used a level that is not very stringent, having an 80 percent (or better) chance of being correct. The reader may if he wishes use other levels, such as the most stringent possible (100 percent), and satisfy himself that the general behavior of the system is the same. If the level is reduced to 50 percent, the system loses its power of prediction, which is then effectively nil. The division into groups **A** and **B** is:

Characters	Group A Specimens						Confident statement possible?	Group B Specimens				Confident statement possible?
	a	b	c	d	e	f		g	h	i	j	
1	+	+	+	+	+	+	Yes (+)	−	−	−	−	Yes (−)
2	+	−	+	+	+	+	Yes (+)	−	−	−	−	Yes (−)
3	−	−	−	−	+	−	Yes (−)	+	+	+	+	Yes (+)
4	−	−	−	−	−	−	Yes (−)	+	−	+	+	No
5	−	−	−	−	−	−	Yes (−)	+	+	+	+	Yes (+)
Numbers of confident statements possible							5					4

There are five confident statements possible about group **A** but only four for group **B** (because character *4* does not have a constancy of 80 percent in group **B**), giving a total of nine statements for both groups.

Suppose we now divide the ten specimens differently, so that **e** and **f** are transferred to group **B**, yielding the new groups **A'** and **B'**. The table now becomes:

Characters	Group A' Specimens				Confident statement possible?	Group B' Specimens						Confident statement possible?
	a	b	c	d		e	f	g	h	i	j	
1	+	+	+	+	Yes (+)	+	+	−	−	−	−	No
2	+	−	+	+	No	+	+	−	−	−	−	No
3	−	−	−	−	Yes (−)	+	−	+	+	+	+	Yes (+)
4	−	−	−	−	Yes (−)	−	−	+	−	+	+	No
5	−	−	−	−	Yes (−)	−	−	+	+	+	+	No
Numbers of confident statements possible					4							1

It is now seen that the numbers of confident statements possible have decreased for both groups, and the total is now only five. In this example the transfer of only one specimen from **A** to **B** will not reduce the number of confident statements at the 80 percent level (unless specimen **e** is transferred), but it will at the 90 percent level. This kind of measure of naturalness is very close to what Gilmour was suggesting, but in practice it is more satisfactory to use other measures of information (see Section 4.6 and Chapter 8).

A natural classification can be used for a great variety of purposes, while an artificial one serves only the limited purpose for which it was constructed. As Sneath (1958) has emphasized, natural or "general" classifications can never be perfect for all purposes, since this is a consequence of the way natural groupings are made. By putting together entities with the highest proportion of shared attributes, taxonomists refrain from insisting that the taxa shall share any particular attribute, as a very simple trial would show. This is the reason for emphasizing the historical importance of the realization that natural taxa do not necessarily possess any single specified feature. This spelled the doom of the Aristotelian concept of an essence of a taxon, for natural groups are in logic unanalyzed entities. Simpson (1961) rejects as illogical the contention by Gilmour (1951) that a classification serving a large number of purposes will be more natural than one which is more specialized and that the most useful and generally applicable classification will be the most natural one. We feel that Gilmour's usage corresponds to the intuitive sense of naturalness which taxonomists have possessed since even before Darwin. This usage is now becoming more widely accepted in taxonomy (e.g., Davis and Heywood, 1963). Gilmour's dictum—that a system of classification is the more natural the more propositions can be made regarding its constituent classes—admits of objective measurement and testing, in contradistinction to Simpson's natural system. Furthermore, Gilmour's system has powerful predictive properties; it is therefore the one we recommend.

2.3 TAXONOMIC RELATIONSHIPS

There has been much recent progress in understanding the nature and kinds of taxonomic relationships. Before we can enter upon a detailed analysis of this matter, we must clarify the meaning of the term *taxonomic relationship* in the recent literature and especially our employment of this term in this book. The variety of meanings in which "taxonomic relationship" has been employed in the literature has led to confusion and misunderstanding among taxonomists. The meanings attributed to it seem to fall into two major classes. One use of the term is in a general sense in which the relationship may be phenetic, phyletic, cladistic, genetic, and, for that matter, on any other conceivable basis. The second meaning is more restrictive, including among "relationships" or "taxonomic relationships" only those usually called phylogenetic.

The first, and wider, defined use of the term is more prevalent in the literature and is the one adopted here. Among others, Simpson (1961, pp. 58, 62, 129, 130) and Michener (1957, p. 160; 1963, p. 154) both use taxonomic relationships broadly for phyletic as well as phenetic relationships. Acceptance of the broad definition for taxonomic relationships necessitates the use of a qualifying adjective to make the intended meaning more precise. Thus we speak of phenetic relationships, cladistic relationships, and so forth.

The confusion over meaning stems in part from the fact that in English the single term relationship symbolizes at least two separate concepts. *Webster's New Collegiate Dictionary* (1959) lists two definitions. The first is "the state of being related"; that is, "connected by reason of an established or discoverable relation." Thus there may be spatial, morphological, ecological, temporal, as well as genealogical relationships. The second is "Kinship; consanguinity, or affinity." In other languages these two concepts are identified by separate terms, hence their meanings cannot be confused. Thus, in German, the first meaning of relationship is *Verhältnis* or *Beziehung*, while the second is *Verwandtschaft*.

Our decision to retain the wider meaning of the word, accompanied by qualifying adjectives, is based on our reluctance to remove the term from the general scientific vocabulary. There appears no suitable synonym in the English language for the term in its first meaning and we can ill do without the general scientific and philosophic concept of "relationship."

In the past we have used "affinity" essentially as synonymous to relationship, qualifying it by adjectives to clarify the meaning intended (Sokal and Sneath, 1963). Cain and Harrison (1960b) have referred to phenetic affinity and phyletic affinity as synonymous with phenetic and phyletic relationship. It would appear that affinity carries the connotation of kinship, hence relationship by descent, in the minds of a great many biologists, for example, Haas (1962). We have therefore avoided using this term in the present volume.

Phenetic relationship has been defined as "... arrangement by overall similarity, based on all available characters without any weighting..." (Cain and Harrison, 1960b, p. 3). It is important to add that this measure of similarity does not carry with it any necessary implication as to relationship by ancestry, but does imply exhaustive estimates of similarity among phenotypes. Ghiselin (1966, 1969) has criticized the concept of overall similarity, believing it to be philosophically unsound. In a formal sense he may be correct: in fact such measures as actually used in phenetic taxonomy are measures "over some" rather than "over all," meaning measures over some sets of characters rather than over all available characters. It seems to us that the term phenetic similarity should be extended to mean similarity over any set and number of phenotypic characters. Whether the number of characters in an organism is finite and whether a measure of overall similarity, in the strict sense, is possible is discussed in Section 3.8.

Phenetic similarity based on restricted sets of characters may be of value in special evolutionary studies. Phenetic similarity of organs may reveal functional or adaptational similarity (as shown in a study of wings of terns, gulls, and skimmers by Schnell, 1969). The resemblance of mouth parts of nectar-gathering insects may be related to the phenetic similarity of floral structures; similarity among larval forms in a group of organisms may be contrasted with that among adults and related to differences and similarities among their habitats (as illustrated in bees of the genus *Allodape*, C. D. Michener, personal communication).

Burtt (1964) has raised the issue whether the term phenetic similarity should be restricted to similarity computed from equally weighted characters only (as in the original definition of Cain and Harrison). Regardless of the merits of equal weighting of characters (which we continue to maintain and shall discuss in Section 3.9) it would seem that phenetic similarity can be based on equally or unequally weighted characters as long as the operation for obtaining the similarity has been defined explicitly by the investigator.

Stemming from these considerations we may redefine *phenetic relationship* as *similarity (resemblance) based on a set of phenotypic characteristics of the objects or organisms under study.* While phenetic similarity may be an indicator of cladistic relationship it is not necessarily congruent with the latter. The magnitude of phenetic relationships between pairs of any set of objects depends on the kinds of characters and similarity coefficients employed.

The term phenetic relationship has given rise to other similar terms and combinations. *Phenetics* is that aspect of taxonomic relationship concerned with phenetic relations but it is also used as synonymous with the study of phenetic relationships. Combinations such as *phenograms* and *cophenetic value* will be encountered later.

Cladistic relationship was defined by Cain and Harrison (1960b) as relationship expressing recency of common ancestry. Cain and Harrison also speak of cladistic affinity to connote the degree of recency of ancestry. The term *cladistics* is used here to mean a study of the pathways of evolution; that is, how many branches are there, which branch came off from which other branch, and in what sequence? Cladistic relationship can be defined as—and represented by—*a branching (and occasionally anastomosing) network of ancestor-descendant relationships.* These treelike networks expressing cladistic relationships are called *cladograms* (Mayr, 1965; Camin and Sokal, 1965).

Any attempt to measure degree of cladistic relationships runs into several difficulties. One approach is that of Hennig (1966, p. 74) that "A species x is more closely related to another species y than it is to a third species z if, and only if, it has at least one stem species in common with species y that is not also a stem species of z." This concept is illustrated in Figure 2-2. Hennig points out that by his definition species **B** would be phylogenetically closer to species **C** than it would to species **A**. Such a definition is strictly cladistic. Hennig adds that some phylogenetic systematists would object to such a definition because **B** is at a lesser distance from **A** than it is from **C** or **D** (in current terminology this is a phenetic relationship), while others would point out that the time (T_2) of the origin of **B** is closer to that of **A** (T_1) than to that of **C** and **D** (T_3). The latter is a chronistic concept of phylogenetic

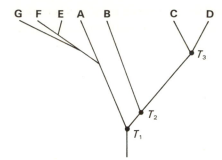

FIGURE 2-2
Cladogram (dendrogram of cladistic relationships) to illustrate the idea of cladistic affinity. Boldface capital letters represent species, T_i represents time. [Modified from Hennig (1957, Figure 5) and Sokal and Camin (1965).]

relationship. If the cladistic sequence of a group of organisms is known, Hennig's definition is operational. However, by the nature of the definition many cladistic relationships are undefined. Thus **B** is equally closely related (sensu Hennig) to **C** or to **D**. Since all species in this cladogram share a common ancestor at the first branch at time T_1, species **B**, **C**, and **D** are equally related to **A**, **E**, **F**, and **G**.

Another plan would be to count the number of furcations that one would have to pass in order to go from one OTU to another in a cladogram. This is the cladistic difference discussed by Farris (1969b). By application of this criterion **D** in Figure 2-2 would be closer to **C** than it would be to **B**; **D** would be closer to **A** that it would be to **G**. However, this concept has some serious problems in practice, since an unknown number of branches would come off each stem, one for each extinct line.

Chronistic relationship refers to relationship between pairs of OTU's on the time scale of evolution. Frequently this is given as the vertical axis in a cladogram. Colbert (1963) has presented an extensive discussion of the time dimension in phylogeny. The relationship between phenetics and the time dimension is discussed in Section 6.1. Chronistic relationships of themselves are of limited usefulness. When combined with phenetic relationships they are of great interest for finding evolutionary trends and rates.

Phyletic or *phylogenetic relationship* are terms that have been employed in the past to include, usually in some undefined combination, the three types of relationships discussed above. Some authors, such as Davis and Heywood (1963), discuss phenetic and cladistic aspects of phyletic relationship separately. Others (Mayr, 1965; Simpson, 1961) consider all three elements (phenetics, cladistics, and chronistics) for establishing phylogenetic classifications. Yet others (following Hennig, 1966) would prefer phylogenetic relationships to be entirely cladistic. Thus the reader must be on his guard for different interpretations of the terms phyletic or phylogenetic.

Since phyletic relationship can have different meanings, it is desirable to replace it by the more precise terms defined above when precision is of the essence. However, there will frequently be instances when a more general, inclusive term will be useful, and we propose to employ the terms phyletic or phylogenetic relationship in this more general way.

2.4 PROBLEMS OF ESTIMATING PHENETIC RELATIONSHIPS

Numerous problems beset the taxonomist wishing to estimate relationships between taxonomic units. This is true whether phenetic or cladistic criteria are used. There are four major *problems of phenetic classifications* requiring discussion. They are (1) incongruence between classifications based on different parts of the body or different life history stages, (2) differences in estimates of relationships produced by different similarity coefficients, (3) differences in interpreting relationships produced by different clustering methods, and (4) the possible effects of parallelism and convergence on taxonomic judgments based on estimates of phenetic relationships.

Incongruence and Methodology

Incongruence between Classifications Based on Different Organs or Different Life History Stages. Incongruence between such classifications is related to the general problem of choice of characters discussed in Section 3.6. Phenetic similarity between pairs of taxonomic units will depend on the sets of characters chosen. It has therefore been realized that general classifications should be based on as broad a phenetic spectrum as possible, and that those classifications based on more restricted sets of characters such as from single organs and life history stages may have heuristic value for special problems.

Different Estimates of Relative Phenetic Similarity Depending on the Similarity Coefficient Employed for a Given Study. The general subject of such estimates, and similarity coefficients, will be taken up in detail in several sections of Chapter 4. It is not at all clear at this point that a unique measure of similarity between pairs of OTU's is possible or even desirable.

Different Classifications Resulting from Different Clustering Methods. Unless the sole aim of taxonomy is to reconstruct cladistic relationships (an approach we cannot recommend; see detailed discussion in Sections 2.5 and 2.6), there will be a variety of criteria for optimal classifications and it will be possible to reclassify the same similarity matrix in a number of ways so as to bring out different desirable classificatory aspects. Details of the problems of comparing different clustering methods and approaches to evaluating these are discussed in Section 5.10.

For purposes of the present discussion we might summarize our attitude as follows: while differences in sets of characters, similarity coefficients, and clustering methods will lead to different results in phenetic taxonomy, there are realistic expectations of finding commonly accepted solutions for frequently occurring classes of data and situations. Furthermore, some types of differences in results may themselves be of great interest leading to new insights into the nature of the organisms or of the relationships being studied.

Convergence and Parallel Evolution

The issue is often raised that the results of phenetic taxonomy are likely to be erroneous (by the criteria of phylogenetic taxonomy) because of the possibility of serious discrepancies between the relationships based on similarity and those based on descent. It is almost a truism that an intimate relation must exist between phenetic evidence and the degree of relation by ancestry. It should also be obvious that, while the two kinds of taxonomy are equally valid for their own purposes, they stand in a peculiar relationship to each other: if knowledge about phyletic relations is required, it must be inferred from phenetic evidence; phenetic relations, however, are inferred not from phyletic hypotheses but from the specimens themselves. A ready analogy offers itself here: we may estimate the similarity between geometric objects on the basis of the nature and size of their dimensions, without any implication as to their past history or how they were developed or constructed by geometers. Yet under some conditions we may *infer* from their geometric form certain probabilities about their past history. We may, for example, suggest that a regular octahedron was developed by geometers subsequent to the development of the square.

Convergence and parallelism are terms over which there is considerable confusion (see Haas and Simpson, 1946, for a full discussion), since they may mean convergence or parallelism in one organ (or in one character complex) or of the entire phenotype. Many authors do not specify which they mean. In this book we use the terms with reference to phenetic similarity, whether the changes are contemporary (isochronous) or not (heterochronous). Convergence restricted to a small part of the phenotype would not, of course, produce convergence in overall similarity; for example, the bats are convergent on the birds with respect to flight, but in the remainder of their phenotype they are divergent from birds, when compared with the reptilian ancestor both have in common.

Convergence.   This is no problem at all when considered strictly by the criteria of phenetics. So long as we are concerned with phenetic relationships the similarity value obtained by numerical methods will be representative. It is only when we wish to draw *phylogenetic* conclusions—rather than phenetic ones—that convergence may confuse the issue.

It is necessary to specify in what respects lines are convergent. It is quite possible for two lines to converge in respect to one organ and to diverge in respect to others. The only kind of convergence that is pertinent to our present argument is that where the lines converge in so many respects (that is, characters) that it causes an increase in the overall similarity of the two lines. This, which can be called "overall convergence" to distinguish it from convergence in a few respects, might cause serious discrepancies between the taxonomy yielded by phenetic methods and a classification based on a cladogram. Convergence in a few respects ("organ convergence") will not cause discrepancies, since these few respects will have little effect on the resemblance of the many nonconvergent character states included in the analysis. There is of course no sharp line between overall convergence and organ convergence.

The pertinent question, then, is: Does marked overall convergence ever occur? And it is one which urgently needs study. There are many examples in which numerical taxonomy could be readily employed to test this question: *Canis* and the thylacine marsupial wolf, the marsupial and eutherian moles, the seals and sirenians, and some xerophytic or parasitic plants, could all be compared. We believe the overall similarity of pairs of this kind is not high. Indeed, if this were not so, it is uncertain how they were recognized as "convergent pairs" and not just close relatives. The very obvious and striking similarities in appearance will, we believe, account for very few of the total features analyzed. Any reasonably random and unbiased selection would, we think, include far more features which did not show convergence. And even in those systems whose adaptations seem primarily responsible for the observed convergence one finds many differences due to different modes of achieving the same function with diverse anatomical parts. We know of no cases of striking overall convergence where the phylogeny is thoroughly known. Sneath (1961) has pointed out the ridiculous implications of total overall convergence in higher animals, which may not be so ridiculous in viruses (see Section 2.6). It is perhaps worth noting that when considering many characters there is every expectation that evolutionary processes will *overall* be divergent. This is a consequence of variation, which has a strong element of randomness. In order to obtain convergence, the possible kinds of variation must be restricted; that is, there must be more change in different features than in common features. We have no reason to believe that, in general, natural selection will have this effect (except possibly in situations like that described below), since it will act largely upon random mutations and in many different directions.

What we have said above applies to the higher levels of taxonomic rank. At lower levels, at the genus and below, there is a possibility of some degree of overall convergence. For example, the introduction of a new food plant into an archipelago possessing several island races of a fruit-eating bird (which had diverged slightly from one another over the course of time) might produce such a strong selection pressure—in the same direction and in all the islands simultaneously—that these races would rapidly evolve toward adaptation to feeding on the new plant, and this might outweigh the slow accumulation of genetic differences that had been continuously occurring in each race. The overall similarity between these races might then increase somewhat, resulting in some degree of overall convergence. In the absence of knowledge of the past, it is difficult to see how any systematic procedure would elucidate the case, and numerical taxonomy is in no worse position than others. It is possible that convergence on this level and of this degree may be frequent but undetectable, at least by any of the procedures known today.

Finally, we may emphasize an obvious but often forgotten point. If we do indeed wish to study convergence, we can only do this by comparing a cladogram with the phenetic arrangement through time. In no other way can we detect the process of convergence, and any attempt to restrict taxonomy to cladograms would defeat its own ends.

Sokal and Camin (1965) have pointed out that most cases termed convergence in conventional taxonomy would not be so considered in numerical taxonomy. When characters are recognized to be only *superficially* similar they would have to be coded as different characters and therefore would not contribute to the measure of similarity, but will in fact reinforce divergence once the characters are coded differently. For example, a detailed comparison of the wings of birds and bats may show them to be less similar to each other than either is to be a generalized tetrapod forelimb. As soon as we go into sufficient structural or physiological detail, convergent characters or processes would resolve themselves into different ones because of their diverse origins. Examples in point would be similar structures produced in different phylogenetic lines (the eyes of vertebrates and cephalopods are similar, but important embryological, histological, and structural differences would cause their characters to be coded differently). Without any knowledge of the underlying physiological mechanisms we would code known instances of DDT-resistance in various insects uniformly, but we would possibly code them quite differently if the physiological mechanisms conferring resistance on the insects were known. If similar characters or their states are not recognized to be different in taxa of diverse phylogenetic origins (that is, the convergence misleads the taxon-omist), or if in fact the characters are identically convergent, the characters would be considered operationally homologous (see Section 3.4) and the states scored identically. In the first of these cases the characters would be miscoded and would erroneously contribute to phenetic similarity. In blue insects whose color is due to different causes (pigment and optical interference), we may—in ignorance of the true nature of the colors—code the insects identically as "blue" by operational homology. The similarity due to this common coding is the error in evaluating their phenetic similarity and is removed as soon as the state "blue" is recoded appropriately.

Sokal and Camin (1965) emphasize that for certain *special* classifications it may be desirable to leave such miscoded characters identically coded, since it is neces-sary to study nature at different levels of organization and complexity. An under-standing of life processes at the finest level does not necessarily contribute to an elucidation of processes at a higher level of organization. Different phenetic relationships may therefore be appropriate for different levels of organization. Thus, when discussing problems of population or ecological genetics, the resistance of insects to an insecticide may be the significant feature, although it may be erroneously coded in terms of the physiological mechanism of the resistance. In a study of their physiology, the mechanisms by which these insects are resistant become the feature to be coded and correct homologies may be established. In a pharmacognostic classification of plants the presence of a specific alkaloid may be coded identically, although it may be known to arise through different metabolic pathways.

Parallel Evolution.   This occurs when two genetically isolated stocks evolve so as to keep constant the difference in those attributes which are under consideration. Parallel evolution seems to us to be similar to convergence and subject to analogous reservations of definition, but to a lesser degree. The same problems and dilemmas arise, and again there is very little clear evidence in favor of extensive parallel evolution if all the features are included in the taxonomic analysis. Possibly the best instances of what may be *overall* parallel evolution (the inclusion of the word "overall" implies a constancy in the overall similarity, as with convergence discussed above) occur in certain ferns (Holttum, 1949) and also in certain ammonites. The apparently parallel trends in the degree of convolution and ornamentation of the shells of ammonites represent very few characters out of the many present during life. In addition to this, these characters are likely to be selected by the environment in the same way; for example, it is possible that certain forms of the shell may have protected many different species of ammonite from a particular predator. Another example of this may be the repeated evolution of increased curvature of the shell in lineages of oysters, leading to the *Gryphea* phenotype, which has been interpreted as a recurrent adaptation to a muddy sea bottom (discussed by Joysey, 1959). This may well have involved only a few characters, since we do not know what changes occurred in the soft parts of these molluscs.

The reviews of Trueman (1930), Swinnerton (1932), George (1933), and Joysey (1959) may be consulted for some of the better-known instances of this phenomenon; these authors all emphasize the difficulty in deciding whether the apparently parallel lineages are indeed independent phyletic lines, or whether the forms found in any given stratum should be grouped together in a monophyletic taxon. If the latter procedure is correct, then parallel evolution is simulated by similar adaptive radiations in successive taxa (stages) of an evolutionary line.

As with convergence, most apparent examples of parallel evolution are due to parallel trends in a few characters. Again, quantitative studies of this problem are urgently needed. Even if overall parallel evolution does occur, it will be no easy matter to prove convincingly the validity of the phyletic lineages concerned, and the phenomenon will probably be of small degree.

Readers of the above account may feel at variance with our ideas because our definitions of convergence and parallelism, while not alien to the usage of sytematists, are probably not the most common ones; hence our comments may be thought not relevant to the central ideas of convergence and parallelism as customarily conceived. While the terms convergence and parallelism permeate evolutionary literature, it is difficult to find definitions for them. Remane (1956), for example, uses but does not define convergence and parallelism. Rensch (1947) considers convergence to be simply nonparallel evolution. Simpson (1961, p. 78) has attempted to coin precise definitions for these terms. He defines parallelism as "the development of similar characters separately in two or more lineages of common ancestry and on the basis of, or channelled by, characteristics of that ancestry." Convergence he defines as "the development of similar characters separately in two or more lineages without a common ancestry pertinent to the similarity, but involving adaptations to similar ecological status." The inferences that are customarily made from such a definition, and that are also discussed by Simpson (1961, pp. 103 ff.), are that convergence occurs between forms that are relatively far apart, while parallelism occurs only among lines that are relatively closely related. These also are the meanings ascribed to these terms by Mayr (1969a). These definitions are useful in a general way for describing evolutionary phenomena. It is useful to have a term to describe the superficial similarity of fishes and whales, for example, and to distinguish it from the independent acquisition of similar coloration in several closely related species of insects living in deserts. However, these definitions would have little operational value even if the complete evolutionary history and genetic structure of the taxa in question were known. They are of even less value in the more realistic case where such data are not available. Simpson mentions the occasional literal interpretation of convergence as the narrowing of differences between lines and the parallel change of these lines as constituting parallelism, but he does not consider such an interpretation to be a particularly useful taxonomic concept.

It seems to us, on the contrary, that in any taxonomy based on a phenetic system (in fact, any taxonomy one wishes to base on measurable quantities), the definition of parallelism and convergence should be entirely based on the parallel or convergent nature of the differences between the lines. We hold these beliefs for several reasons.

1. No fundamental and useful distinction can be established between convergence and parallelism sensu Simpson and Mayr. To say that convergence takes place only between distantly related forms prejudges the issue of relationship completely. We would no longer be able to use cases of convergence or parallelism in our classificatory schemes because relationships would have been predetermined before any decision could be taken on whether a convergence or parallelism is at hand.

2. The definitions of Simpson (1961) and Mayr (1969a) and their subsequent discussions lead one to believe that by convergence is meant the construction of similar structures based on different genetic systems, and that by parallelism is meant the construction of similar structures based on similar genetic systems. However, these assertions are based on inferences from the systematic positions of the organisms involved and lead to the same difficulties noted under 1.

3. In the absence of knowledge about genetic homologies, we are faced in the main with phenetic changes that may be divergent, parallel, or convergent. The only useful distinction is whether changes are parallel or convergent, or better still

whether there is relatively more or less divergence, since often less divergence between some members of two taxa than between the majority of the members of the taxa may be considered as convergence or at least parallelism. Of course, sensu strictu, this can only be done if fossil series of organisms are available. Where only living end points of evolutionary change are known, parallelism or convergence cannot be shown but must be inferred with much uncertainty from recent evidence.

4. It must also be stressed that at least at the molecular level true phenetic convergence can indeed take place. If the following is the true cladogeny of a protein position in two taxa

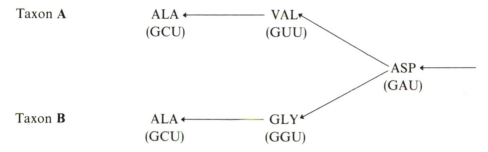

then the two alanines are identical regardless of their different origin from aspartic acid via valine and glycine, respectively. With respect to this protein position the two taxa are in fact identical; and the DNA representation of these two alanines might also be identical, as shown above by the nucleotide triplets in parentheses.

Components of Phenetic Similarity

We shall conclude our discussion of phenetic relationship by reviewing two components that combine to yield phenetic similarity: similarity due to common ancestry is called *patristic similarity* (Cain and Harrison, 1960b) and that due to parallelism and convergence (true or miscoded) is *homoplastic similarity* (Simpson, 1961). Patristic similarity is in general less than total phenetic similarity, since it consists of the latter minus the homoplastic component, as pointed out by Cain and Harrison.

It is evident that cladistic relationship is not directly related to the above two types of relationship. They are phenetic, and measure similarities between the forms. Cladistics is concerned with the way in which the lineages branched and not with the degree of difference. A genealogy or pedigree is an example of a cladistic scheme.

Patristic similarity can be further subdivided; the present analysis differs in some respects from an earlier one by Sokal and Camin (1965). One component of patristic similarity may represent *primitive similarity*, due to character states that are identical in two descendent taxa and identical with the states of the characters in their most immediate common ancestor. In Figure 2-3 both lineages **C** and **E** have diverged from a common ancestor **A**. The states of the two characters *1* and

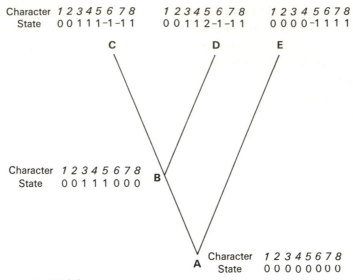

Character *1 2 3 4 5 6 7 8*   *1 2 3 4 5 6 7 8*   *1 2 3 4 5 6 7 8*
State   0 0 1 1 1 –1 –1 1   0 0 1 1 2 –1 –1 1   0 0 0 0 –1 1 1 1

C   D   E

Character *1 2 3 4 5 6 7 8* B
State   0 0 1 1 1 0 0 0

A Character *1 2 3 4 5 6 7 8*
   State   0 0 0 0 0 0 0 0

FIGURE 2-3
Cladogram showing three extant organisms, **C**, **D**, and **E**, and two
ancestors, **A** and **B**, with eight characters, to illustrate patristic and
homoplastic components of phenetic similarity (see text).

2 have not changed in the two descendants from the states that occurred in the
ancestor. This represents an element of primitive similarity with respect to homologous character states between the descendants and also between either descendant and the ancestor. Hennig (1966) calls this component of patristic similarity
symplesiomorphy. Our term is simpler and of greater mnemonic value.

The other component of patristic similarity is *derived similarity.* This results
from character states that are the same in the descendants but not identical with
those in a more remote common ancestor, as illustrated by characters *3* and *4* in
lineages **C** and **D** in Figure 2-3 with respect to the ancestor **A**. This concept is only
applicable when the common ancestor is more remote than the most immediate
one, because **C** and **D** show primitive similarity for characters *3* and *4* with respect
to their immediate ancestor **B**. These characters became modified to state 1 in intermediate ancestor **B**. The contribution of the two characters *3* and *4* to the similarity
between **C** and **D** is the same as that for characters *1* and *2* in lineages **C** and **E**, but
3 and *4* contribute less to the similarity of **C** or **D** with **A** than to that of **E** with **A**.
Hennig (1966) calls this relationship *synapomorphy.* Derived similarity includes
cases where the states in the two descendants are not identical because of further
evolution in one of the lineages. An example is character *5* in lineages **C** and **D**.
In the latter the character evolved to state 2 from the ancestral state 1 in **B**. But **C**
and **D** are relatively more similar to each other than either is to **E**, which has
evolved state −1 for character *5*. Derived similarity should be differentiated conceptually from primitive similarity, the other component of patristic similarity. If

the phenetics of the ancestors are known, the primitive similarity can be separated from the derived similarity. Character states that exhibit derived similarity are generally considered to be homologous in the classical sense of the term.

Another evolutionary phenomenon is illustrated by characters 6 and 7, which evolved independently in lineages C and D, from the ancestral state 0 in A and B to the derived state -1. The evolutionary phenomenon involved is clearly parallelism as generally understood. Another term for similarity due to parallelism is *homoiology* (L. Plate, cited by Hennig, 1966). Character 8 evolved three times to the derived state 1 in lineages C, D, and E, and in this case it illustrates convergence of C and D on E with respect to A and B.

With respect to characters 3 and 4, taxa C and D in Figure 2-3 are both divergent from A, although C and D are identical for these characters (showing derived similarity with respect to A, and primitive similarity with respect to B). With respect to character 5, taxa C and D are again divergent from A (D is also divergent from B), but if we assume that the character must evolve from state 0 through state 1 before reaching state 2, then taxa C and D show derived similarity *as well as* divergence with respect to each other and A.

Statements about amounts of phenetic similarity in Figure 2-3 are only relative. Thus we suppose that the phenetic similarity between taxa C and D will be greater than between C and E. The amount by which this will be true depends on the measure of similarity employed.

Finally, we should point out that Figure 2-3 does not necessarily imply that the evolutionary step from state 0 to state -1 in characters 6 and 7 and to state 1 in character 8 occurred simultaneously in both lineages. If it occurred early in the line B to C and late in the line B to D (heterochronic parallelism, Maslin, 1952), we could at the appropriate points in time speak of a divergence of the line BC from BD, followed by a convergence of C upon D. Divergence and convergence are here employed as phenetic concepts. This meaning of convergence differs from the one used conventionally, although some taxonomists employ the term phenetically without clearly defining it. Heterochronic parallelism would seem to be a preferable term. Since primitive and derived similarity result from common ancestry their sum yields the overall patristic similarity.

The homoplastic component of phenetic similarity represents those components that show parallelism and those that are convergent. Homoplastic similarity is thus due to homoiology as well as to truly convergent characters; that is, those that are identical but of different ancestry, like the protein position discussed above. It also expresses similarity due to characters or their states, or both, that have been misinterpreted or miscoded in a convergent manner. Recognition of homoplasy is important for two reasons. Once we recognize that a character has been miscoded we are led to revise and correct our coding system. Also, homoplasy suggests that similar evolutionary forces are at work in the lines under study, which tend to make some of their characters resemble each other. This is a phenomenon of general biological interest. The analysis of phenetic similarity between two taxa with respect to a given ancestor more remote than the most immediate one can be summarized as follows:

Phenetic similarity = Patristic similarity + Homoplastic similarity

Patristic
similarity = Primitive similarity + Derived similarity (together these represent classical homology)

Homoplastic
similarity = Similarity due to parallelisms (homoiology)
+ Similarity due to convergence (identical characters or states, or both, derived from different phyletic lines, or different characters or character states erroneously coded as homologous).

Hennig (1966, p. 146) has independently arrived at a similar partitioning of similarity into symplesiomorphy (= primitive similarity), synapomorphy (= derived similarity), and convergence (= homoplastic similarity).

We should point out that when taxonomic relations are expressed as dissimilarities, phenetic dissimilarity equals patristic dissimilarity decreased by the similarity due to homoplasy.

There are other ways of partitioning phenetic similarity, discussed in Section 4.11. One of these is the distinction between size and shape.

In conclusion it may be noted that many of the difficulties discussed in this section are due to our uncertainty of what kind of similarity is most appropriate in taxonomy, and this is more of a problem of the taxonomist than of the statistician.

2.5   PROBLEMS OF ESTIMATING CLADISTIC RELATIONSHIPS

With the advent of the Darwinian theory of evolution the natural classifications of earlier taxonomists became explicable in terms of descent from a common ancestor: a taxon was now interpreted as a monophyletic array of related forms. It has, however, been frequently pointed out (as by Bather, 1927, and Remane, 1956) that this change of philosophy did not bring with it a change in method. Taxonomy proceeded as before; only its terminology had changed. Remane (1956) quotes Naef (1919, pp. 35–36):

> ... und was Haeckel und die Phylogenetiker zunächst getan haben, war nichts anderes als die Übersetzungen der speziellen Einsichten, die sich an diese Lehre früher geknüpft hatten, in eine Sprache durch Anwendung einer neuen Terminologie, ohne doch die Lehre selbst einer Vertiefung zuzuführen oder einer kritischen Betrachtung zu unterwerfen. Auch die—wenig abgeklärten—Grundbegriffe der alten Morphologie wurden

von Haeckel einfach in die neue Sprache übersetzt, die dem Wesen nach eine genealogische war. Dabei wurde dann

aus Systematik	Phylogenetik,
aus Formverwandtschaft	Blutsverwandtschaft,
aus Metamorphose	Stammesentwicklung,
aus systematischen Stufenreihen	Ahnenreihen,
aus Typus	Stammform,
aus typischen Zuständen	ursprüngliche,
aus atypischen	abgeänderte,
aus niederen Tieren	primitive,
aus atypischer Ähnlichkeit	Konvergenz,
aus Ableitung	Abstammung usw. usw.

Freely translated, this reads:

... and what Haeckel and the phylogeneticists did at first was nothing more than to translate the special insights which had been gained in this field of study into another language by applying a new terminology without any deeper analysis or a more critical examination of the concepts. The poorly defined fundamental concepts of the old morphology were also translated by Haeckel into this new, essentially genealogical language. Thus

Systematics	became phylogenetics
Relationship by form	relationship by blood
Metamorphosis	evolution of phyletic lines
Systematic series	ancestral series
Types	archetypes
Typical states	primitive states
Atypical states	derived states
Lower animals	primitive ones
Atypical similarity	convergence
Derivation	descent, and so on and so forth.

In view of the conclusions of the previous sections it is obvious that phylogenetic relationships need to be subdivided and refined into phenetic and cladistic relationships. Classifications based on "phylogenetic relationships" without this clear separation and definition will be ambiguous and of little value to either phenetic or cladistic taxonomists. Hennig (1966, p. 77) states this position well: "... combining different systems in a syncretistic system robs the combinations of any scientific value, since it could never be used as the basis for investigations that presuppose knowledge of a particular kind of relationship between the organisms." The necessity for separating phenetic and cladistic aspects of taxonomic relationships and of basing classifications on only one or the other aspect are recognized by most modern taxonomists who have given thought to this matter.

If cladistic relationships are to be the basis for erecting classifications, we must inquire into the possibility of estimating such relationships with a reasonable degree of reliability. Supposedly cladistic arrangements have been published innumerable times in the systematic literature ever since the publication of *The Origin of Species*. There are few biological topics on which more misinformation has been published. Even in paleontology the ratio of fact to speculation is rather low. "Works which refer to the fossil evidence of evolution usually cite a few of the well-known cases of evolutionary series as if they were merely representatives of a host that might have been quoted, instead of stressing the fact that records of such cases are rare" (Challinor, 1959).

The critical question to be answered is: Can cladistic lineages be estimated from information available to systematists? Hennig (1966), for whom the determination of cladistic sequences is the crucial problem in taxonomy because of his advocacy of strictly cladistic methods of classification, presents by far the most detailed and carefully reasoned discussion of this issue in the modern literature. The criteria considered by Hennig include consideration of problems of homology, character phylogeny, reversibility, convergence, and parallelism, as well as a discussion of chorological and chronological character variation. While Hennig prefers to maintain a phylogenetic definition of homology, he realizes correctly that the actual recognition of homologies has to be through character correspondences and his ideas for determining homologies in practice are thus very close to the concept of homology espoused by us and discussed in Section 3.4. However, as is shown there, this concept is generally phenetic and cannot be employed for rigorous cladistic analysis.

Hennig's section on character phylogeny (1966, pp. 95 ff.) is the heart of his discussion of the use of morphological evidence for cladistic analysis. These considerations are fundamental in cladistic work, whether it is done by the more traditional methods, by Hennig's system, or by the operational techniques of numerical cladistics discussed in Sections 6.3 and 6.4. To satisfy his criterion of monophyly, Hennig feels that only synapomorphous relationships (derived similarity) between species can assist in the establishment of monophyletic lineages. This means that one must distinguish primitive from derived character states. Among the criteria cited by Hennig that would aid in this undertaking is geological evidence. Character states found in earlier geological strata can be considered primitive when compared to character states found in later geological strata. Although such considerations may be useful in an approximate general way, they do not sustain critical scrutiny. The only reason we compare forms in these two strata is because they are similar to each other. Thus we use the phenetic similarity of putatively related forms to make judgments on whether a character in question should even be compared. Consequently, a decision from geological evidence, on whether a character state is primitive or derived is fundamentally based on the phenetic similarity of the forms studied. It might be argued that with extensive fossil series the ordering of the character states would be conclusively evident, but again it should be pointed out that the only way to establish these series is by the phenetic similarity of the forms under consideration. We are left with what is essentially a character analysis in a phenetic numerical taxonomy of these organisms (see also Colless, 1967b, and Section 3.4 for a discussion of this point).

An identical argument holds for chorological progression series. Such series are formed by species that displace each other in a geographical progression and exhibit primitive characters in the area of origin and derived characters in the most recently invaded territory. Yet again the decision on whether one should even compare these related species, or the very fact that these species are related, is based on phenetic considerations.

A third criterion for ordering character states is based on Haeckel's well-known law of recapitulation by which the successive embryonic stages of an animal are said to mirror its phylogeny. It is now realized that the many exceptions from this law can lead to serious errors of interpretation. Hennig (1966, p. 96) concludes that "the transformation stages that led to a character condition during phylogeny, ...cannot be read off with certainty from the ontogeny of the species," although "the 'criterion of ontogenetic character precedence' remains an important aid in phylogenetic systematics, provided it is not uncritically evaluated more highly than other aids."

Hennig's fourth criterion is his strongest. He claims that under certain circumstances knowledge of the evolutionary trends of the states of one character may yield information on the evolutionary order in another character. Figure 2-4 illustrates this concept. Character h has evolved from primitive state h_0, which is retained in species **A**, to state h_1 in species **B** and state h_2 in species **C**. Paraphrased into the terminology of this book, Hennig states that if the evolutionary order of the states of character h is known, the conclusion that the species **B** and **C** are most closely related is compatible only with the assumption that in the second character i with three character states i_0, i_1, and i_2, the primitive condition is represented by i_0

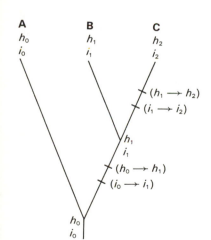

A   **B**   **C**
h_0 h_1 h_2
i_0 i_1 i_2

$(h_1 \longrightarrow h_2)$
$(i_1 \longrightarrow i_2)$

h_1
i_1
$(h_0 \longrightarrow h_1)$
$(i_0 \longrightarrow i_1)$

h_0
i_0

FIGURE 2-4
The three species **A**, **B**, and **C** have evolved from a common ancestor which possessed the character states h_0 and i_0 for characters h and i respectively. At points in evolution shown by the cross bars the states evolved into other states (as shown by the transitions indicated by arrows and given in parentheses). [Modified from Hennig (1966).]

and the most strongly derived condition by i_2. However, this statement does not meet the real issue. In any given case we are unlikely to know the true cladistics of the species in the manner in which it is shown in Figure 2-4. All that we would have is an hypothesis about the order of the states of character h and from this we can reconstruct the most parsimonious cladogram (Section 6.4) assuming that our coding of character h is correct. If in fact we were to know the cladistics of the three species, then the whole argument is beside the point. We would be little interested in evaluating the character states of i since *in Hennig's taxonomy* (as distinct from evolutionary studies) the main purpose of knowing which are primitive and which are derived is in being able to reconstruct the cladistic relationships of the three species.

To make the situation more realistic we should assume that we do not know the cladistics of the three species. Then what one needs to do is to try to fit various interpretations of the order of the character states i_0, i_1, and i_2 to possible cladistic arrangements compatible with our interpretation of the character states of h. In Figure 2-5 we can see that there are three possible cladistic arrangements of the three species (assuming that they have to be connected and do not issue in a single trifurcation from a common ancestor). To decide which of these cladograms is more likely we must make certain ancillary assumptions. The most obvious assumption is parsimony of evolutionary steps (changes in character states). Since in a realistic situation one is hardly ever certain about the evolutionary order of any one character, such as character h in this example, one must jointly determine the parsimony of the two characters together. On examining Figure 2-5 we find that, depending on how we code character i, either cladogram b (the one indicated as correct in Hennig's original figure) is the most parsimonious, or e and f are equally parsimonious. And these two are not necessarily the same, when different assumptions are made about the order in which the character states are permitted to change. Considerations of this sort led Camin and Sokal (1965) to their method of cladistic analysis, which jointly considers numerous characters coded according to assumed evolutionary sequences. The details of this method and philosophy will be taken up in Sections 6.3 and 6.4.

The point to be made here is that these methods, when done with many characters and when the possible alternatives are evaluated by computer, may lead to reasonable reconstructions of the cladistic sequences of the species under study; yet they cannot invariably lead to precisely constructed cladograms for the following reasons: (1) With more than three species and more numerous character states the number of possible solutions to be tested becomes very large. It is not obvious from Hennig's writings whether he is aware of the very large number of possible cladistic solutions in any phylogenetic scheme. It would be totally impractical to test all of these even on a computer, let alone by traditional procedures. (2) The methods chosen are more in the nature of overall compatibility tests of characters. It is conceivable at least in theory that uniformly miscoded characters could be

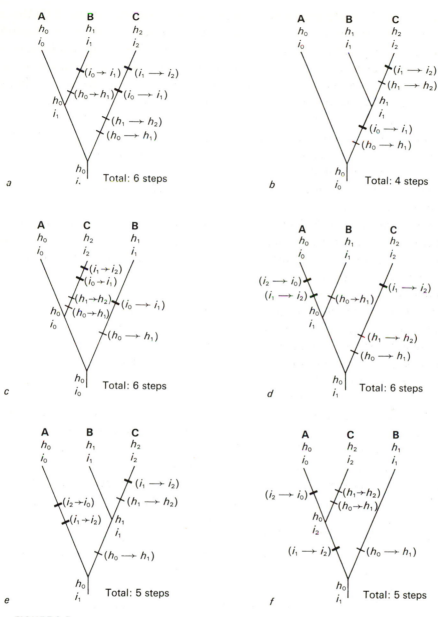

FIGURE 2-5

The three possible cladistic arrangements of the species **A**, **B** and **C** from Figure 2-4 are shown in *a*, *b*, and *c*, and in *d*, *e*, and *f* (omitting the case of divergence of all three from a single point). In *a*, *b*, and *c* the assumptions about the evolution of the characters *h* and *i* are that they followed the order of the subscripts 0, 1, and 2 in both cases. The most parsimonious solution is then cladogram *b* with four evolutationary steps.

In *d*, *e*, and *f* it is assumed that the order i_0, i_1, i_2, postulated above, is incorrect, and using new assumptions that character *i* evolved $i_1 \rightarrow i_2 \rightarrow i_0$ it is seen that cladograms *e* and *f* are equally parsimonious with 5 evolutionary steps each.

mutually totally compatible and yet be entirely incorrectly interpreted. (3) Hennig correctly asserts that certain interpretations of a character i would be incompatible with other previous interpretations of character h. These considerations, which have been independently developed by Wilson (1965) and termed a consistency test by him, serve mainly to exclude impossible or improbable interpretations but cannot serve to confirm assumed interpretations of character h or character i. These views are stated more fully and firmly by Colless (1966, 1967b).

Among other criteria for evaluating the phylogenetic order of character states, Hennig mentions the supposed correlation between the character states of hosts and parasites. The pitfalls in this field are obvious and even greater than those encountered by matching the states of one character with cladograms produced from another character.

We can conclude with Hennig (1966, p. 146) that "there are criteria and rules for determining whether the characters of different species belong to a transformation series, and for deciding whether they are to be evaluated as plesiomorphous [that is, primitive patristic] or apomorphous [or derived patristic], *but these do not have absolute validity*" (our italics and bracketed synonyms). Colless (1967b) shows that without resort to phenetic principles Hennig's attempts to recognize plesiomorphous from apomorphous character states are unsuccessful. He concludes "... that the 'Hennig System' is simply an intuitive, prototypical form of statistico-phenetic taxonomy, and that its advocates simply misunderstand the nature of the latter process." This claim has been disputed by Schlee (1969) and defended in a rebuttal by Colless (1969a,b).

Other authors are even less explicit in furnishing reliable criteria for the recognition of the evolutionary interpretation of sequences of character states. A single quotation from Simpson (1961, p. 102) may suffice: "Perhaps the most conclusive evidence as to primitive (and hence ancestral) characters is provided when one condition in a group or one end of a sequence has a homologue in another group of more remote common ancestry." It is obvious that without knowledge of the evolutionary history (presumably as evidenced by the fossil record) we could make little progress here. Remane (1956) suggests recognizing cladistic relationships through the recognition of homologues, for which the main criteria are morphological correspondences. Remane's methods therefore cannot separate phenetic and cladistic relationships.

If it is difficult or impossible to estimate cladistic relationships from primary evidence (characters of the taxonomic units), we might ask whether cladistic deductions can be made from overall phenetic similarities between organisms. Let us take a closer look at the reasons for wishing to develop taxonomies that are in accord with phylogeny. This ideal is expressed mainly by the dictum that taxa should be monophyletic groups. There will therefore be difficulty only when the taxa given by phenetic taxonomy are not monophyletic but polyphyletic. We believe that numerical phenetics will in general give monophyletic taxa because we believe that phenetic groups are usually monophyletic. To contradict this belief we must have evidence that phenetic groups, created by adequate and acceptable numerical techniques, are not monophyletic. A clear example would be where convergence has occurred to an extent that causes confusion. But we must first show that this degree of convergence has indeed occurred. Figure 2-6 shows that there are several interpretations of what is at first sight a simple problem. There may indeed have been convergence so that organisms **A** and **B** are more similar phenetically than their ancestors **A'** and **B'**, and **B** is convergent on **A**, though by

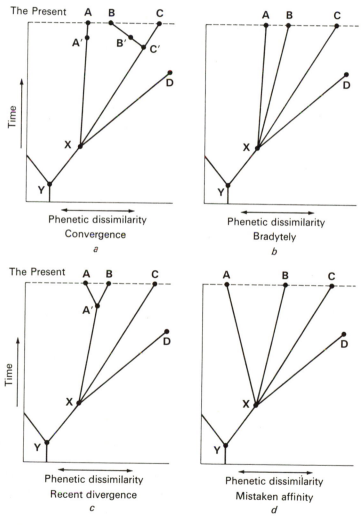

FIGURE 2-6
Alternative interpretations of apparent convergence. For explanation, see text.

ancestry related more closely to organism **C**; see Figure 2-6,*a*. But this conclusion may be uncertain. The evolution may have occurred as in Figure 2-6,*b*, where there is no convergence but divergence at different rates, with the phyletic line **B** evolving slowly (bradytely). Or, **B** may in fact have descended from **A'** so that **A**, **A'**, and **B** are monophyletic; see Figure 2-6,*c*. Finally we may have been mistaken in thinking that **B** was convergent on **A**, for a careful estimate of resemblance may show, taking all their attributes into consideration, that **B** is more similar to **C** than to **A**; here we had been misled by some striking but superficial or restricted set of characters. As soon as the homoplastic similarity due to these characters is removed the greater phenetic relationships of **B** to **C** becomes evident, as in Figure 2-6,*d*. It is clear that from a consideration of the organisms **A**, **B**, and **C** alone (without the evidence of the fossil forms **A'**, **B'**, **C'**, and **X**), we cannot distinguish between these alternatives except to recognize the last of them. Indeed, we have as yet no acceptable evidence that convergence of this kind—that is, overall convergence in phenetic resemblance—does take place to any marked extent. If it did, it would be exceedingly difficult to prove, for we would have to have an excellent series of fossils to be certain that we had not made any mistake in reconstructing the cladogram. The known examples of convergence are all open to the objection that relatively very few characters are affected.

Discrepancies between phenetic and phyletic taxonomies can also occur without convergence—for example by divergence at different rates, as in Figure 2-6,*b*, where **A** and **B** are more similar phenetically than either is to **C**, although cladistically all three are equally related.

Even if we grant that overall convergence can occur (as, for example, has been suggested for some groups of birds), we must ask ourselves why we should wish to make taxonomies based on monophyletic groups. Suppose the convergence had become so absurdly extreme that the two forms are almost indistinguishable and can readily and successfully hybridize: what is the purpose of separating them on grounds of ancestry when in all other attributes they are virtually the same? The purpose cannot be to emphasize minor dissimilarities, nor to serve as a guide to the behavior of these forms with respect to their genetic properties, or any other class of properties. The purpose is presumably, therefore, to show that this convergence had occurred, a fact which could be expressed in simple terms without any need for setting up the whole apparatus of formal systematics, and, as is discussed above, the classification would thus be for this special purpose, and not a general classification. Jardine (1969b) points out that in a natural classification monophyly can be regarded only as a diagnostic criterion, not as an absolute criterion for establishing the system. Where independent phyletic lines fuse into one, the whole problem becomes thoroughly confused, whether we know to expect confusion or not, since there are so many alternative ways of dividing the network of phyletic lines. This is a common occurrence in plants, through the mechanism of allopolyploidy.

Simpson (1961, p. 120) has pointed out correctly that most definitions of monophyly are nonoperational because "they are so vague that they provide insufficient criteria for separating one from the other [monophyly from polyphyly] by analysis of evidence." We can ignore the naive statements of earlier authors defining monophyly as descent from a single pair of progenitors, statements made in ignorance of contributions of modern evolutionary theory. Hennig (1966, p. 73) defines a monophyletic group as "... a group of species descended from a single ('stem') species, and which includes all species descended from this stem species. Briefly, a monophyletic group comprises all descendants of a group of individuals that at their time belonged to a (potential) reproductive community, i.e., to a single species." He points out the sometimes neglected fact that the monophyletic group has to include *all* species derived from this ancestral species, not only those which the taxonomist wishes to classify. As Simpson correctly states, the problem with a definition of this sort is that it is difficult to know how far back one has to trace separate stems in order to arrive at the common stem form. For example, by Hennig's definition it might well be that the mammals could be made monophyletic only by tracing them back to an unknown early reptile stem.

A useful analysis of Hennig's concept of monophyly has been undertaken by Ashlock (1971). He distinguishes between two kinds of monophyletic groups— those that are *holophyletic*, wherein all the descendants of the most recent common ancestor are contained in the taxon (this is a monophyletic group sensu Hennig), and those that are *paraphyletic* and do not contain all of the descendants of the most recent common ancestor of that group. A *polyphyletic* group according to Ashlock is one whose most recent common ancestor is not cladistically a member of that group. Because most if not all taxa presumably have common ancestors when one is willing to trace back their ancestry far enough, polyphyly in Hennig's and Ashlock's sense therefore means that the recognized taxon of descendants no longer shares derived patristic (apomorphous) characters. In the final analysis, these concepts again are based on phenetic similarity of descendants and ancestors, as further detailed in our critique of Simpson's definition of monophyly below. Yet another redefinition of these terms has been furnished by Nelson (1971).

Remane (1956) solves the monophyly-polyphyly problem more drastically by not accepting as natural any groups with characters which do not conform to his criteria of homology. Thus, by not recognizing polythetic taxa, he decreases the probability of a taxonomist's having to recognize polyphyletic groups. To admit the existence of polyphyly would be fatal to Remane's system, since he relies upon a closed circle of reasoning from monophyly to a natural system, to homologous structures, and back to monophyly.

Bigelow (1956) has pointed out that in all supposedly monophyletic classifications overall similarities and differences are usually not disregarded. Even in those cases where the ideal of a monophyletic classification could be attained, it often is disregarded in favor of a phenetic classification by supposedly phylogenetic taxonomists. Bigelow feels that "if classification is to correspond with evolution, it must be based on the extent of overall difference, not on time."

The redefinition of monophyly (Simpson, 1960, 1961) is not free from ambiguity in its practical application: "Monophyly is the derivation of a taxon through one or more lineages (temporal succession of ancestral-descendant populations) from one immediately ancestral taxon of the same or lower rank" (Simpson, 1961, p. 124). We therefore need to know what is meant by the term taxon in each instance, and we need to decide the relative rank of the taxa. The difficulties were pointed out in *Principles of Numerical Taxonomy*. Simpson does not give clear criteria for deciding this rank. Mayr's (1969a, p. 75) brief discussion of the problem relies on distinctions between the phenotype and genotype of supposedly polyphyletic taxa. Since these distinctions can rarely be made it is hardly an operational criterion. In his criticism of Hennig's definition Mayr considers a rigidly cladistic interpretation of monophyly "contradicted by common sense" and implicitly modifies it by degree of phenetic diversity.

Let us return to the question of whether phylogenetic deductions can be made from phenetic similarities. If the resemblances are based on living organisms alone, we can only speculate on the phylogeny; to check our speculations we must have fossil evidence. Yet there are some conclusions that are more probable than others. We believe that these are as follows. (1) Phenetic clusters based on living organisms are more likely than not to be monophyletic sensu Hennig. Thus phenetically adjacent taxa represent phyletic "twigs," which usually originate from the same branch; in other words, overall convergence is unlikely. (2) In the absence of direct evidence our best estimate of the attributes of a common ancestor of a cluster must be derived from the properties of the cluster itself. In short, if we have no fossil evidence, the existing pattern is our best guide to the past history—though this may often be wrong. An argument similar to the first argument in favor of equal weighting of characters (Section 3.9) applies here: if we have no evidence that evolutionary rates differed, it is difficult to proceed further without assuming these to have been constant and equal in all the phyletic lines studied. If the reader thinks of a cross section through the top of a shrub with the vertical dimension representing time and the horizontal representing phenetic dissimilarity, he will have a ready, though somewhat inadequate, simile for the situation.

The two points mentioned can then be illustrated as follows. (1) Adjacent twigs will generally arise from the same branch. Admittedly it will be very difficult to detect substantial homoplastic similarity (overall convergence) near the tips of the branches, and this may have occurred quite commonly, together with some reticulate evolution due to the fusing of phyletic lines; yet gross degrees of overall convergence between the tips of the main branches will be very much less likely, and its improbability will increase with the taxonomic difference between the branches below the plane of section. (2) In the main, the branches from which the twigs arise will lie more or less directly below the twigs; but we will have no way of telling whether the twigs arose almost vertically or whether they came off at a pronounced slant, for we can have no confidence that the twigs will fill the phenetic space evenly in the way in which the branches of an actual shrub do in order to obtain adequate sunlight.

The above deductions on the phylogeny (which are made from organisms belonging to one point in time) cannot give any estimate of the rates of evolutionary change, which may have differed in different phyletic lines. To estimate rates we must have data from several points in time.

These guide lines will hold only a simple form where one or two geological strata are represented. Where there are more than two, we will commonly be faced with the question: have rates of evolution been constant over the whole period? For example (see Figure 2-7), we may have an extant form **a** which is more similar to a slightly earlier form **b** than it is to a much earlier form **c**, though **b** and **c** are more similar. Did **a** then evolve recently from **b** (the more similar), or more slowly and directly from **c** (the less similar)? In the absence of other evidence, such as geographical isolation or a fuller fossil record, no certain decision can be taken even in such a simple case. It is very easy to find cases that are more complex than the example given above. One lineage, for example, may evolve rapidly and another slowly, and it may be impossible or implausible to draw a cladogram in which all the lineages evolve at the same rate throughout the period. A diagram of the phenetic relations (expressed in one or two dimensions) versus time will usually make clear the degree to which we can safely reconstruct the phyletic tree, and where we must indicate by dotted and queried lines our uncertainty as to the course of the descent.

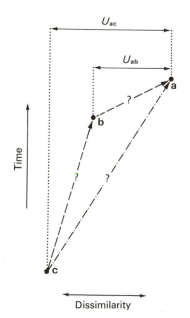

FIGURE 2-7

The time-rate problem. Organism **c** is the ancestor of both **a** and **b**. The phenetic dissimilarities between **a** and **b** and between **a** and **c** are indicated, respectively, by U_{ab} and U_{ac}. With only the data shown one cannot tell whether **a** arose directly from **c** or via **b**.

We have seen that one cannot derive evolutionary rates from similarity coefficients among recent forms. This is shown by the "pregroup-exgroup problem" discussed by Michener and Sokal (1957). Is it possible to distinguish whether an aberrant member of a cluster of forms was derived phyletically from one of the members of the cluster, or from the ancestral stem below the point at which the rest of the cluster arose (see Figure 2-8)? Michener and Sokal suggested that if the aberrant member, **x**, showed approximately the same resemblance to all the members of the cluster it was most likely "pregroup," or derived from the common stem; see Figure 2-8,*a*. If, however, the similarity with one member of the cluster was much greater than the mean similarity with the cluster, then the aberrant member, **y**, was likely to be "exgroup," or derived from the same stock as the member it most closely resembles—**d** in Figure 2-8,*b*. It is nevertheless possible to account for the observed resemblances in the figure by means of a number of cladistic schemes, if the evolutionary rates differ in the lineages.

Although the discussion of this section leads to the conclusion that cladistic relationships are difficult to estimate we should point out that the numerical cladists (e.g., Farris, Kluge, and Eckardt, 1970) have made considerable progress toward developing lines of reasoning which will determine cladograms with a fair degree of certainty (see Sections 6.3 and 6.4). These authors point out that even a weak correlation between phenetic similarity and cladistic relationship may provide useful information. In cases where the evolution of character states can be "safely" established, cladistic inferences are relatively simple. In other cases it is often necessary to undertake various types of phenetic analysis that are consistent with the assumptions of the cladogeny, using the evidence so obtained to develop hypotheses about character state evolution. These hypotheses in turn can be used to develop consistent cladograms based on a larger number of characters.

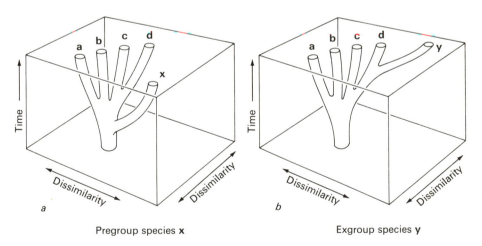

Pregroup species **x**   Exgroup species **y**

FIGURE 2-8
The pregroup-exgroup problem. For explanation, see text.

2.6 CHOICE OF A BASIS FOR CLASSIFICATION

We have seen that there are problems in both phenetic and cladistic approaches to the study of taxonomic relationships; which of these systems should be used by taxonomists in classifying nature? We may make our decisions at two levels of reasoning. (1) For which type of relationship are the problems of estimation less serious, hence more likely to be overcome? (2) Which type of relationship is inherently of greater interest and usefulness to systematists and biologists in general?

The present development of systematic theory and practice is such that the problems of estimating phenetic relationships (discussed in Section 2.4), though occasionally serious, are largely technical challenges that will be overcome before very long. In fact, discrepancies in taxonomic results attributable to these problems frequently yield useful information. By contrast, we do not have equally sanguine prospects for solving the problems of estimating cladistic sequences. Numerical cladistics may only provide a measure of compatibilities of characters, but not necessarily cladistic sequences. Numerous assumptions about ordering of characters and about evolutionary processes must be made to arrive at an estimate of cladistic relationships. Incisive mathematical techniques may be developed, which will solve some of these problems, yet the multiplicity of pathways for cladogenies may be too great to permit unequivocal decisions among them. More fundamentally, the question has recently been raised of the degree to which purportedly cladistic evidence is in fact phenetic evidence (e.g., Colless, 1967b). This point is discussed in greater detail in Section 6.3.

Turning to a discussion of the second question, we note that the difficulty with the use of the phylogenetic approach in systematics is that we cannot make use of phylogeny for classification, since in the vast majority of cases phylogenies are unknown and possibly unknowable. The theoretical principle of descent with modification—phylogenetics—is clearly responsible for the existence and structure of a natural system of classification; we may agree with Tschulok (1922) that the natural system can be considered as proof of the theory of evolution. However, since we have only an infinitesimal portion of phylogenetic history in the fossil record, it is almost impossible to establish natural taxa on a phylogenetic basis. Conversely, it is unsound to derive a definitive phylogeny from a tentative natural classification.

Again, we turn to Hennig for the most incisive analysis of the issues. Unfortunately, however, he derives erroneous conclusions from these. Aware of the limitations of the phylogenetic approach to taxonomy, he justifies such a procedure on the basis of four arguments (Hennig, 1966, pp. 22 ff.). The first is that the phylogenetic system is the most meaningful of all possible systems because all other types of classifications, such as ecological, zoogeographic, or morphological, can be derived and explained through the phylogenetic system. Indeed, none of the special classifications could occupy such a central and all-explanatory position as does a phylogenetic system. The theory of evolution is the most adequate, most unitary, and indeed simplest hypothesis to which a great variety of biological phenomena—geographic distribution, physiological adaptation, morphological similarity, or biocoenotic complexity—can be related. Phylogeny can thus be seen as the central cause of much biology, yet it cannot be used for an explanatory concept, as phylogenies are not known in the vast majority of instances. Hence a phenetic classification, although it may not be able to explain the above-mentioned biological phenomena, is at least a self-sufficient, factual procedure and will in most cases be the best classification we can get.

Hennig's second reason for preferring a phylogenetic taxonomy has been negated by the development of numerical taxonomy. He states that phylogenetic relationships are at least in principle measurable, but that similarities are not. But the development and success of numerical phenetics has invalidated that argument. Hennig's third point is that although phenetic and cladistic classifications will be identical in many groups, to the extent that they differ, phenetic systems are less suitable as general reference systems for biological systematics. This supposed primacy of phylogenetic classifications is questionable and we discuss the issue at length below. Hennig's fourth point is that the problem of incongruence will make measurement of morphological similarity between taxa impossible. However, by adding characters of various life history stages and organ systems we can obtain reliable similarities between genomes in all stages of expression.

Even if we could make use of phylogeny to create classifications, we may still ask whether this is necessarily desirable. To do so would discard much important and interesting phenetic information. An allopolyploid might originate repeatedly, giving rise to phenetically identical new species each time. In some groups phyletic classifications might prove chaotic—for example in viruses and especially in bacteriophages. Work by Jacob and Wollman (1959), Lederberg (1960), and Cowie (1967) suggests that bacteriophages are being constantly derived from bacteria as genetic entities that acquire the properties of autonomy and parasitism. This work not only implies that bacteriophages which are identical (or almost so) may be polyphyletic, but, even more disconcerting, it suggests that they may be of composite origin. They are able to transfer genes from one bacterial form to another and to incorporate such genes into their own genomes; it is therefore possible that they derive some of their own genes from one host and some from another. Other viruses may be similar in these respects to bacteriophages, and this raises the disturbing possibility that, for example, an arthropod-borne virus may be part insect, part mammal, and part bird. Recent work by Subak-Sharpe et al. (1966) has shown that the pattern of small DNA viruses is very close to that of mammalian DNA, suggesting that these viruses may have arisen from mammalian DNA sequences. The only way to bring order into such a system is by a phenetic classification, such as those of Andrewes and Sneath (1958), Sneath (1962), Bellett (1969), or Gibbs (1969).

Even more disturbing to our notions of phylogenetic taxonomy is accumulating evidence (Loening and Ingle, 1967; Stutz and Noll, 1967) that chloroplasts of higher plants have evolved from symbiotic blue green algae. Thus there may be groups where the validity of a phylogenetic classification, even if possible, may be in serious doubt.

Cain and Harrison (1960b) pointed out that the principal disadvantage of phenetic classification is that some convergence or parallelisms may go unrecognized and that polyphyletic groups might be mistaken for monophyletic ones. However, the recognition of monophyly can scarcely be thought to be the only worthwhile or even the preeminent activity of systematists. In fact, the analysis of phenetic similarity and phenetic structure may be of as great or greater evolutionary interest in studies of adaptations of organisms to their environments, correlation with breeding structure, evolutionary rates (where time factors are known), analyses of ecological niches, and other concepts.

Sokal and Camin (1965) have discussed the issue of phenetic versus cladistic classifications at length and much of what follows is directly based on their account. Since phenetic, cladistic, and chronistic aspects are necessary to understand the systematics of organisms, Sokal and Camin feel that systematics as a whole must be based on all of these considerations. Simpson (1961) has called phenetic taxonomy "shallow and incomplete." From the viewpoint of the entire field of systematics this is correct and would be equally true of a phyletic taxonomy based only on the tracing of cladistic pathways. Systematists taking an extreme position on either phenetics or cladistics are like the blind men trying to describe an elephant in the legendary folk tale (made memorable for Americans through a poem by J. G. Saxe), each man describing the whole elephant on the basis of the part he can touch. The difficulty with a composite approach to phylogenetic taxonomy, as espoused by Simpson, is that no operational methods exist for a synthesis of the various approaches. Basing taxonomy on all three approaches requires art or compromise, both of which are inadmissible as bases for a precise science. The acrimonious debate as to which system is better is thus of little profit. The question is for what purpose a classification is to be established.

An important consideration in any discussion of problems of systematics and classifications is recognition of the duality of systematics (Sokal, 1964a). One branch of systematics, classification, has the relatively unexciting but supremely important function of ordering nature into a practical and generally useful system. This can best be done on the basis of a phenetic classification, as will be shown below. Systematics in the wide sense also has the more challenging task of understanding the mechanisms which have brought about this order. These two aspects of systematics are separable and should be separated. The reluctance of conventional phylogenetic taxonomists to recognize this separation may be based on a conscious or subconscious feeling that classification by itself is an unimportant enterprise of low caliber and not worth doing. We think otherwise and are supported in this view by many eminent systematists (e.g., Mayr, 1968). In fact, taxonomists often justify their activities by claiming that other biologists would be unable to work unless they knew what organisms they were working with. For this entirely legitimate reason a stable and consistent system of classification and nomenclature must be developed and maintained. By not allowing for the separation between classification and what Michener (1963) has called the "explanatory element," conventional systematists perpetuate a system which is inherently unstable and hypothetical by the very nature of its operations.

Because phenetic classifications require only description, they are possible for all groups and are more likely to be obtained as a first stage in the taxonomic process. This trend is likely to continue and be reinforced by the continuing development of automatic data-recording equipment of various types. On the other hand, since cladistic classifications require more knowledge about the organisms and the direction evolution is taking, they must necessarily follow phenetic clustering. If a uniform system of classification is to be established, it might therefore be argued that it should be done on a phenetic basis, with a complementary cladistic classification added whenever possible.

From the viewpoint of evolutionary study, both phenetic and cladistic classifications are of great interest, leading to an understanding of evolutionary principles. From the point of view of biology in general, however, it is probably of more interest to describe the existing overall similarity of the organisms, rather than their cladistic affiliations. This argument may be denied by those whose interests in biology largely concern evolutionary rates and lineages. However, not all biologists have such an orientation. The outlook of many physiologists, biochemists, behaviorists, and others is purely functional. The primary use of a system of classification for these biologists is to group organisms sharing as many properties as possible so that predictions can be made from one member of a taxon to others. Thus a biochemist finding a new substance in the blood of a species of *Rana* would be tempted to look at other members of this genus for further occurrences of this substance. Although such a phenomenon undoubtedly has an evolutionary basis (patristic similarity, see Section 2.4), as far as this application is concerned the process of prediction is based on the purely phenetic properties of the taxon. Similarly we use taxonomy as an efficient device for summarizing detailed information on many taxonomic units and for predicting properties of members of taxa. Thus, when told that a new species of the genus *Rana* has been described, we can immediately make many predictions about its appearance without ever having seen it. This follows from the naturalness of the genus *Rana*, in the sense in which we use the concept in this book.

In most cases where primitive patristic elements constitute an appreciable portion of the phenetic similarity, the cladistic and phenetic classifications will largely coincide. However, in the possibly rare cases where this is not so, the general superiority of a phenetic over a cladistic classification becomes even more evident.

This is illustrated by the admittedly implausible case of Figure 2-9 in which the evolution of line **x** of cluster **A** into the phenetic region of cluster **B** has made **x** resemble the members of cluster **B** to such a degree that a phenetic classification would include it in **B**, but somewhat closer to cluster **A** than most members of **B**. Although the divergence of **x** from cluster **A** is of great evolutionary interest, the overall similarity of **x** to the members of cluster **B** is more generally useful. Predictions of the characters of **x** will be more successful when based on characters known for taxon **B** than for those based on taxon **A**, because of **x**'s greater overall similarity with **B**. But it is also true that additional knowledge about the cladistic relationship of **x** to taxon **A** will further aid predictions about characters of **x**. Thus we can see, even from this simple example, that phenetic as well as cladistic relations are important.

In this connection we should mention an often misunderstood point. Natural classifications in Gilmour's sense are not necessarily cladistic (and vice versa). In the example in Section 2.2 organisms **e** and **f** might be phylogenetically convergent on organisms **a–d**, giving **A′** and **B′** as the cladistic groups. Yet groups **A′** and **B′** would not contain the most information in the sense meant by Gilmour, even if the additional information about their origin was added as a sixth character, that is, "belonging to clade **A′** versus clade **B′**." For then Groups **A** and **B** would have respectively five and five statements at the 80 percent level, but groups **A′** and **B′** would have only five and two statements, respectively. Colless (1969d) points out that if particular cladistic groups are preferred on the grounds that they are maximally predictive, this means that they must also be phenetic groups.

Decisions on which system of classification to adopt are affected by the method of representation of the taxonomic structure that is employed. Verbal or written descriptions of relations among organisms have proved quite inadequate. For this

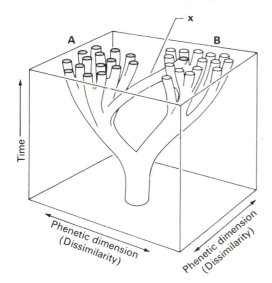

FIGURE 2-9
Phylogenetic tree to illustrate possible incongruence between phenetic and cladistic relationships. Phenetic relationships are shown in two dimensions, the third dimension is time. Cladistically, **x** belongs to the monophyletic taxon **A**, but phenetically it would be considered part of taxon **B** because of extensive parallelism. [Modified from Sokal and Camin (1965).]

reason a variety of mnemonic and didactic aids have been developed, most of them graphic. These are largely different forms of trees of relationships (or phylogenetic trees). Mayr, Linsley, and Usinger (1953) have called these drawings *dendrograms,* which seems a suitable term without any implication about the nature of the relationship. The terms *phenogram* and *cladogram* (Mayr, 1965; Camin and Sokal, 1965) have come into general use to define dendrograms representing phenetic and cladistic relationships, respectively.

A dendritic description of the taxonomic system has much to recommend it and seems in many ways to be the "natural" way of illustrating relationships and descent. A vague, general agreement on the interpretation of diagrams of relationship exists among taxonomists, yet when a given diagram is subjected to detailed, critical scrutiny we rarely find consistency of meaning within it. The interpretation of the basic facts that a diagram offers is likely to be based on varying degrees of certainty in different parts of the tree. No generally accepted conventions for constructing such diagrams exist; hence, seemingly similar diagrams may have quite different meanings, which are often not clearly enunciated by the author of the tree. In an illuminating discussion on the different components that can be included in phylogenetic relationships, Cain and Harrison (1960b) have shown that often an author himself has not a clear idea of the meaning of a diagram of relationships he presents.

The following symbolisms have been used most frequently.

1. The vertical axis (or radius in circular dendrograms)—to indicate time, either in absolute units or in relative evolutionary ones (most frequently unspecified).

2. Furcations—to indicate branches in the phyletic sequence, and so to show the relationships between the forms based on the lineages alone (not considering their phenetic similarities); that is, the cladistic relations.

3. Location and relative position of tips of branches with respect to each other—to indicate phenetic relationships.

4. Location of furcation along a vertical axis (which now designates resemblance)—to indicate closeness of relationships between taxa represented by stems issuing from the furcation. Symbolisms 3 and 4 are often used in combination to indicate what part of the relationships is due to convergence and what part is due to patristic similarity.

5. Levels of tips along a vertical axis (or along the radius of a circular dendrogram)—to indicate whether the forms are extant or extinct, and also to give some estimate of the time scale of the extinction.

6. Levels of tips along a vertical axis (or along the radius of a circular dendrogram)—to indicate degree of perfection or complexity of form. This convention, related to the *scala naturae* of earlier science, is largely out of fashion, although some of its ideas and its vocabulary are still employed occasionally. Thus Rensch (1947) uses the term *Höherentwicklung* (anagenesis).

7. The angle between stems—to represent velocity of phenetic differentiation.

8. Thickness of stems—to represent abundance at a given point in time. Abundance is usually measured by the number of species or taxa contained within the stem, but occasionally represents the number of organisms supposedly extant.

It is easily seen that 1, 2, and 5 can be combined into a single diagram. Unless the rate of evolution has been constant, 4 cannot also be included. It is generally impossible to represent phenetic relationships on a two-dimensional graph; hence 3 is bound to be a distorted representation. The use of the angle to indicate velocity of evolutionary change, 7, is never very successful except in the simplest diagrams. Abundance, 8, can usually be added to most diagrams, although the results are often not very esthetic. The basic difficulty is the graphic representation of phenetic resemblances and phenetic change. These are multidimensional relationships and cannot satisfactorily be compressed into a two-dimensional diagram. The difficulties of representation have led to other schemes of presenting the relationships, but these have not been too successful. Increasing familiarity with multivariate procedures has resulted in new attacks on problems of representing phenetic similarity among taxonomic units (see Section 5.9 for an extended discussion), and it has been realized that earlier attempts to force these into hierarchic relationships by means of phenograms have not always resulted in the most satisfying classifications.

Hennig (1966) states that for similarities among organisms to be recognized as valid expressions of taxonomic relationships, they must permit the taxa to be arranged in a hierarchic system. We disagree with this point of view for the following reasons. Hybridization may well have occurred more often than we realize in animals (and is of course well known and frequent in plants). It may therefore be impossible to make hierarchic relations in some groups because of extreme reticulate evolution. Possibly in consequence, morphological similarity relationships can sometimes be represented better in nonhierarchic ways (see Section 5.3) than by the traditional hierarchic system. Hennig (1966, p. 21) is therefore incorrect in stating that relationships of morphological similarity between species can best be represented in a hierarchic system. It should, however, be noted that reticulate phenetic relationships can often be approximated satisfactorily by a hierarchic, dendritic relation. Thus, for example, the reticulate relationships among six species of the genus *Cyrnus* illustrated by Hennig (1966, Figure 20) can be shown to yield a satisfactorily hierarchic structure.

Hennig feels that it is fallacious to assume similarity relationships between living organisms to be primary, but that genealogical relationships are primary. This seems to us to be similar to the argument that either the male or the female is primary in reproduction. The critical issue is not which is the more important relationship, but which is the relationship that we *can* investigate and from which we may derive useful principles. On these counts phenetics clearly qualifies. The only possible way of transmitting the various types of information listed in points

1 through 8 is by three separate graphs for (a) time and branches (cladistic relationships), possibly combining symbolisms 1, 2, 5, 7, and 8; (b) phenetic relationships between junctions of stems only, as customarily employed in dendrograms in numerical taxonomy—symbolism 4; and (c) complexity of form or organization—symbolism 6. Phenetic relationships among tips of branches (symbolism 3) usually cannot be represented adequately even in a two- or three-dimensional space. Proper representation would require two-dimensional cross sections through the hyperspace that is necessary to represent such relations. We shall consider the mechanics of such a presentation in greater detail in Section 5.9.

As regards aspect 2—phyletic sequence—it is obvious that a diagram can be constructed only if phylogenetic evidence can be obtained from fossils or in some other reliable fashion. We have already pointed out the dangers inherent in inferring phylogenies from Recent organisms. The sequences in phyletic lines are often much more uncertain than authors wish to admit. Some authors indicate probable descent by dotted lines. If there are many such dotted sections, the chances of the diagram being substantially and seriously misleading may be very high indeed. Unfortunately there seems to be no study on this point to tell how misleading earlier phylogenies have been when compared with later detailed and convincing fossil evidence. Such a study might be illuminating. It is true that many authors, quite properly, disavow any phylogenetic significance for their diagrams and caution readers against considering them to be in any way reflections of evolutionary history. We ourselves frequently follow conventional practice in arranging taxa by a system of hierarchic, nested categories which roughly give an indication of point 4. This and other forms of representation are taken up in Section 5.9.

In summary, for systematics to be thoroughly understood, phenetics, cladistics, and chronistics must all be considered. For classification, that is, for the formal description and cataloging of organized nature, phenetics appears preferable.

2.7   TAXONOMIC RANK

Criteria for Taxonomic Rank.   Phenetic as well as genetic criteria are commonly used; at and below the level of biological species these criteria may be in conflict, however. In the absence of data on breeding and in all apomictic groups (which together include the great majority of practical applications in systematics), the criteria for species are based on the phenetic similarity between the individuals and on phenotypic gaps (Michener, 1970; Sokal and Crovello, 1970). The rank of higher categories must perforce depend on phenetic criteria. The intrusion of an entirely different criterion for taxonomic rank in those few situations where genetic or phyletic relations are known with certainty, seems to us to be a needless source of confusion. Alternative terminologies have been suggested for genetic and also ecological entities (especially in botany, where these problems are most acute; compare Gilmour and Gregor, 1939; Camp and Gilly, 1943; Gilmour and Heslop-

Harrison, 1954); these terms have not been widely used. For reasons of clarity it is desirable that the meaning in which taxonomic rank is used should be specified. In this book it will be used in the sense of phenetic rank, unless otherwise indicated. A detailed discussion of the species concept is deferred to Section 7.1. Jardine (1969b), in erecting his logical system for taxonomy (see Section 2.8), finds little justification for considering any one category in a natural biological classification more or less real than any other category except in the sense that some diagnostic criteria may be more or less reliable than others, as for example gene flow in studies of putatively specific populations. This position is retained in his subsequent work (Jardine and Sibson, 1971).

It is undesirable for the rank of a group to be affected by the number of contained subgroups. There is a modern tendency to make each family contain only a few genera and each genus only a few species; in some works most genera are monotypic. The rank should be based on relative degree of similarity alone. In our view it is better to introduce new rank categories (such as subfamily or superclass) than to use the number of contained subgroups as an arbiter of rank.

There are as yet no criteria for any absolute measure of taxonomic rank. We do not know how to decide whether a family of birds, for example, is equivalent phenetically to a family of insects. The hope has been expressed by one of us (Sneath in Heywood and McNeill, 1964, p. 159) that protein sequence studies might lead to such criteria, but it is evident that this was overly sanguine; different types of protein evidently show widely different phenetic differences at the same taxonomic ranks. Recent work on DNA (e.g., Park and De Ley, 1967; Dessauer, 1969) seems also fraught with problems. The difficulties in using the age of clades as a measure of taxonomic rank are well discussed by Hennig (1966, p. 183).

Limits of Taxa. The limits of taxa have sometimes been suggested as criteria for taxonomic rank. Limits can be considered from two points of view. One can trace the change of taxa with time, looking at the phylogenetic tree in its entirety. This is what Simpson (1945) has called vertical classification. One can also look at a cross section of the tree and obtain the relationships among taxa at a given point in time (horizontal classification, Simpson, 1945). With vertical classification (which can only be practiced on fossil material) it is obvious that when one phyletic lineage evolves into a new form there can be no sharp division between the ancestral and the descendant species, other than an arbitrary one, except in the case of allopolyploids and other forms of hybrid origins. The accidents of discovery of fossil forms inevitably affect classificatory decisions, since the divisions will at first be placed where there are gaps in the fossil record. As the gaps are closed by new discoveries, the most common practice is to choose for the dividing line some prominent, but commonly arbitrary, evolutionary step—for example, the change in jaw structure in the evolution of reptiles into mammals. So long as the arbitrariness is clearly realized, these methods are unobjectionable and are matters of convenience. The choice of such an arbitrary step is not without some danger, however, for it may lead to incongruous situations.

A better plan, commonly advocated in paleontology when a relatively full fossil record is known, is to place the divisions at places where abrupt changes in the rate or direction of evolution make for rational and convenient groups or where phyletic lines branch. The demerit of this course is that the divisions then come at those very parts of the lineages which are of special interest for students of evolution. Nevertheless, the bulk of the total material will be grouped in a convenient way.

To turn to horizontal classification, difficulties will arise if a phenetic group is not identical with a phyletic group. For example, the appearance of a sterility barrier will at once divide a normal genetic species into two sibling species (from the point of view of the "biological" species definition). Yet for many generations (until the two sibling species have accumulated sufficient genetic differences in the course of their independent evolution) they may remain one single phenetic group, because the differences that cause the sterility barrier (plus the few other accumulated differences) will be insignificant in comparison with the many variable attributes of other kinds that the two sibling species will share. The sibling species will therefore constitute a single natural taxon, in the sense in which we use the term. Indeed, our point has apparently been misunderstood by Mayr (1965) who criticizes numerical taxonomy because it would in such a situation group together the sibling species. We must emphasize that this is precisely what it is intended to do; a phenetic method should give phenetic groups.

A point that has had little attention, although Simpson (1961) and Michener (1963, 1970, p. 23) refer to it, is to what extent taxonomic rank is represented by variability within a taxon, the distances between the centers of taxa, or by the size of the gap between them. The last may, of course, be quite small if the taxa almost overlap. Nor is it clear whether some compromise is desirable. This problem appears to underlie the difficulties of Gregg's Paradox (see Section 2.8) where taxonomic ranks are defined in terms of set theory (Gregg, 1954). It is clear that if the rank of a taxon is measured by its internal variability one may have the paradoxical situation where the only species of an order, and consequently the order itself, has very little variability. It also appears undesirable to base rank primarily on the size of gaps between taxa; rare intermediate forms may then become critical in recognizing rank. Because of the logical relationships between the phenetic distances involved in these criteria, the question is best discussed in detail in connection with the topic of patterns in phenetic space in Section 5.11.

Taxa of low rank may be difficult to define and to arrange hierarchically. Such groups would appear in a numerical taxonomic study as contiguous and indistinct clusters of individuals. Methods for studying this are discussed in Chapter 7.

The Hierarchy of Characters.   This refers to the claim—fallacious, and fortunately on the way out—that one can lay down a priori rules as to which kinds of characters separate species, which kinds separate genera, which kinds separate families, etc. This is the antithesis of the a posteriori method of discovery, which *finds* those features that do in fact separate the previously recognized natural taxa, and is, we maintain, the correct procedure. The concept of a hierarchy of characters is now mainly of historical interest, being specially connected with the French systematists (the de Jussieus, Cuvier, and de Candolle), and is often termed the principle of subordination of characters. The rules were avowedly based on the fancied importance of different organ systems. Yet they were qualified by so many exceptions, as well as by appeals to the constancy of characters within taxa (a presupposition that the taxa are already established), that it is clear they were never workable rules (Cain, 1958). We believe that no such hierarchies can be made a priori. It would indeed be curious if evolution, which is responsible for the natural hierarchy, should be so obliging as to operate only on certain classes of characters at specified taxonomic tanks. Some taxonomists prefer to base their classification on what they suppose to be nonadaptive characters. Population geneticists are currently debating vigorously whether any characters can be considered nonadaptive. Even if such characters exist, it would be almost impossible for taxonomists to recognize them as such. The converse view—that taxa should be based on adaptive characters (Inger, 1958)—is quite impracticable, as has been shown by Sokal (1959).

2.8   DESIRABLE PROPERTIES OF A TAXONOMIC SYSTEM

A taxonomic system will reflect the purposes for which it is constructed, and we will therefore need to specify these in some detail. It is a measure of the current re-evaluation of taxonomy, with its consequent uncertainties, that different taxonomic textbooks (e.g., Simpson, 1961; Davis and Heywood, 1963; Hennig, 1966; Blackwelder, 1967a; Mayr, 1969a; Jardine and Sibson, 1971) all give different emphasis to the details of these purposes. It may be that certain purposes—and hence, properties—are incompatible, and the taxonomist must then decide which to prefer, or whether to look for alternative purposes that represent some sort of compromise. Deutsch (1966) has made a first attempt at providing operational and quantitative definitions of some ways for evaluating taxonomic schemes. He lists the following as characteristics of a taxonomic scheme:

> organizing power
> predictive range
> structural complexity
> functional complexity
> performance
> novelty of performance
> structural novelty
> functional novelty

predictive accuracy
social costs (inverse of "economy")
net change in accuracy
net change in cost
originality
fruitfulness
self-transcendence

For operational definitions of these characteristics, some of which are clearly applicable to biological classifications, see the original reference and a similar discussion by Fisher (1969).

In the past, numerical taxonomists have given much attention to imitating existing taxonomic practices. This may, however, not be the most profitable approach. We may be able in the future to develop quite new methods that are not intended for the traditional purposes of taxonomy, but which have value for biology in general as well as for systematics in particular. This is still in the future; the present is a time of rapid change of taxonomic ideas, and it is difficult to foresee the detailed needs of even the immediate future of systematics. The following discussion is therefore written largely from the traditional point of view; that is, we discuss the properties of classificatory systems that taxonomists consider desirable today.

Desirable properties and purposes of classifications are linked together in a way that makes them difficult to discuss individually; most have repercussions on many others. We shall divide them roughly into three sections: (1) those concerned with "naturalness" of taxa; (2) those concerned with ease of manipulation; and (3) those that facilitate retrieval of past information.

Naturalness.   The basis of natural classifications has been discussed in Section 2.2, and it is clear that natural classifications (in some sense of the word classification) are what are principally required. There is of course a place for special classifications for special and restricted purposes; these pose their own problems, but these are often peculiar to the immediate study. In general, consistency and accuracy are the main needs.

In Section 2.2 reasons were given for preferring a natural system in Gilmour's sense, and in Section 2.6 we have explained why we believe that phenetic rather than phyletic groups are desirable.

Natural and polythetic phenetic groups are thus desirable, and we have seen that these possess high information content, and also have a high predictive value for characters of new organisms. For example, Watson, Williams, and Lance (1966) have emphasized the importance of this, saying "those who need a new source of an unusual plant product turn first to the taxonomic relatives of species known to produce it." The requirement of naturalness also implies that some criterion of optimality is used in the class-making steps; this may be at a simple level, as in the example given in Section 2.2, or may be more sophisticated, depending largely upon what information is to be assumed about the system—whether, for example, size factors are to be ignored so that only information about shape is included. A more extensive discussion of optimality criteria is given in Section 5.10. In view of the proliferation (both of statistics and also of concepts) relating to different components of phenetic resemblance, it is particularly important that the maker of a classification should have a clear idea of what he wants, and he should indicate what components are used as the basis for his classification.

Since the time of Linnaeus taxonomists have thought of the system of nature as nested and mutually exclusive. Early attempts to define nonoverlapping, taxonomic hierarchies in terms of symbolic logic were made by Woodger (1937, 1951, 1952) and continued in an important book by Gregg (1954). In Gregg's study, a taxonomic system was defined as a sequential partition of a set of taxonomic units into disjoint subsets, each representing a taxon with its rank given by the number of partitioning steps it has undergone. The definition of taxa in this context led to a difficulty that has become known in the literature as "Gregg's Paradox." This applies to monotypic taxa that conventionally differ in rank, yet would not do so under Gregg's model. Several authors (Sklar, 1964; Van Valen, 1964; Buck and Hull, 1966; Farris, 1967b; Gregg, 1967, 1968) have attempted to overcome this difficulty.

The most recent contribution has been a new logical model for taxonomy developed by Jardine, Jardine, and Sibson (1967), extended by Jardine and Sibson (1968a), and reviewed in semitechnical form in Jardine (1969b). Using set-theoretical definitions, Jardine arrived at a definition of a taxon as an equivalence class, that is, a set of objects related by an appropriately defined equivalence relation. Each taxon Y in a nonoverlapping taxonomic hierarchy is defined as an ordered pair of terms, the first member being *Ext* Y. This is short for *extension* Y and implies the entire set of objects or taxa that share the particular relationship implied by the equivalence relation at a specified rank in the hierarchy. The second member in the definition of a taxon is its *rank*, defined as $J(\mathbf{a}, \mathbf{b})$, the lowest ordinal rank of the taxon of whose extension $\mathbf{a}$, $\mathbf{b}$ are members. Since by Jardine's definition one can distinguish taxa with identical extensions but different ranks, monotypic taxa do not encounter the difficulty of Gregg's Paradox. The basic units in Jardine's system are what he calls basic taxa and what we would call OTU's (see Section 3.1). These may be individuals but are more likely descriptive vectors or matrices representing populations. We shall not enter upon a detailed explanation and derivation of the system established by Jardine and his associates here, but refer the reader to the sources already quoted. It is important to note, however, that Jardine's nonoverlapping hierarchic system leads to six theorems, which can be derived from his model and are generally accepted statements about taxonomic hierarchies. Fundamentally they concern the "nestedness" of the hierarchy, except for his last theorem, which shows that a nonoverlapping hierarchy must obey the ultrametric inequality (see Section 4.2).

Given a natural nonoverlapping taxonomic hierarchy, it can be represented in the form of a dendrogram because of its ultrametric property (see Section 4.2). However, as pointed out by numerous authors (Jardine, 1969b; Farris, 1969a; Hartigan, 1967) dissimilarity matrices are generally not ultrametrics and hence cannot be mapped by a one-to-one relationship into a dendrogram. But when the similarity or dissimilarity coefficient is not even a metric, serious distortion of the taxonomic relationships and "reversals" frequently occur when the results are represented in dendritic form (see Sections 4.2, 5.5). The degree to which such dendrograms and their corresponding classifications distort the similarity relationships among the assemblage of objects represented by the classification is estimated by various measures of stress and distortion discussed in Section 5.10.

An important, if somewhat poorly defined, criterion is concordance between a classification and other facts about the organisms. This applies at two levels, the logical and the technical. At the logical level it is an invariable test of goodness in all branches of statistics because this is our only way of checking the validity of any hypothesis; if challenged we must resort to additional evidence. There is, then, a good reason for saying that an important test of a classification is whether it agrees with another classification of the organisms based on additional data, or in other words, exhibits stability upon addition of new information. There are some problems in setting up completely logical and self-consistent tests of this kind (Sneath, 1967b), notably problems about the validity of the new characters and organisms to be employed. Also, it is true that lack of concordance does not necessarily condemn a classification, because the incongruence between classifications based on different stages of the life cycle may have a biological explanation. This is also to some extent a side issue, for if our aim (as it generally is) is to classify organisms (life cycles) rather than simply specimens (life stages) we should incorporate as much information as possible from all life stages and all organs and character-systems. The amount of information from each life stage (or organ) will thus depend on the amount of information known about it, and this would seem to be the best way of obtaining a representative set of information about the organism *as we know it*.

At a more technical level, we would rightly be suspicious of a classification that violently disagreed with certain kinds of technical evidence, with serology or DNA pairing studies, for example (provided that the evidence was beyond question). These external criteria must be viewed with common sense; they must be both biologically sound and also highly informative in the sense of containing much information. Thus there are general grounds for believing that serology, DNA studies, and protein sequences reflect a great deal of genetic information. In contradistinction to this, the presence of a simple chemical compound would be a poor court to appeal to; one such substance would be quite unacceptable, although a hundred might provide a good check upon a given classification.

It may be noted that both of the last two paragraphs are tests of badness of taxonomies rather than of their goodness. Concordance does not ensure that the classification is good, but discordance implies that it is bad. Nevertheless a number of such external criteria which all agree will make for a near certainty that the classification is valid. We do not see that there is much difference here between testing the validity of taxonomies and any other scientific phenomena.

As much information as possible is desirable for constructing taxonomies, for three reasons: to allow enough power to the method (e.g., a single presence–absence feature would obviously only permit the recognition of two taxa, the "haves" and the "have-nots"), to give a representative cross-section of attributes of the organisms, and to reduce chance and sampling error to a minimum.

Ease of Manipulation.   Most if not all of the above points are to a large extent compatible. We now turn to some that may frequently not be readily compatible with the above requirements. Criteria that relate to the working ease of a taxonomy include a hierarchic arrangement if it is possible (it may not always be), and the ability to produce good taxonomic keys (i.e., discrimination between taxa should be as easy as possible). These criteria may be in conflict with those of naturalness described above. Thus it may be unwise to force a hierarchic arrangement onto material that does not possess the structure necessary for a hierarchy; on the other hand the memory-saving properties of hierarchic systems are of considerable practical importance. Discriminatory requirements may also conflict. It is too readily assumed that taxonomies should be based on measurements in what may be loosely termed "units of discriminatory power," at the expense of a faithful representation of phenetic resemblances and of the information content of the taxa. We would hold that discrimination is a rather special purpose of taxonomies, and as a general rule phenetic resemblances are better suited to most taxonomic work.

Williams and Dale (1965) have pointed out that classifications can either be probabilistic or nonprobabilistic. The former require a null hypothesis, and frequently no suitable test of such a hypothesis may be available. There are, of course, advantages in probabilistic models when these are feasible, but this is by no means always the case. Again, we would view this as a desirable additional property when possible, but not an overriding one.

Information Retrieval.   We believe that the retrieval of information, although often placed high on the list of priorities (for example, one often hears the dictum that "a plant's name is the key to its literature"), is in fact subsidiary to questions of naturalness. If groups are indeed natural, the main requirement for easy information retrieval is a suitable nomenclature. This therefore becomes a subsidiary desideratum of a taxonomic system, important, but not as important as natural taxa, and we believe that in the future ready adaptability to electronic data processing will become an important element in any nomenclature.

3

Taxonomic Evidence

We now proceed to a detailed discussion of the data necessary for obtaining estimates of resemblance between taxa. This leads us first into a consideration of the fundamental taxonomic units employed in numerical taxonomy, then to a discussion of the nature of taxonomic characters, questions of homology, and problems in character sampling and evaluation, such as choice and weighting of characters.

3.1 OPERATIONAL TAXONOMIC UNITS

In order to restrict one's study to a manageable taxonomic group, one must first make a preliminary selection of specimens. No sharp distinction is made between the selection of specimens, discussed in some detail here, and the selection of characters treated in Section 3.6, since these generally proceed *pari passu*. Inasmuch as a taxonomic group is selected by "classification from above," selection is therefore necessarily based on rather few characters.

A point of some importance is to guard against the exclusion of pertinent material because it does not strictly fulfill all the criteria for the working definition of the taxon to be studied. The danger is that aberrant forms or descriptions of aberrant taxa in the literature may be excluded because they do not possess some character of the taxon that the systematist considers essential or diagonostic. The inclusion of a small amount of possibly atypical or unsatisfactory material, which may have to be excluded in the final analysis, is a worthwhile insurance against an unrepresentative study. Very similar considerations will apply to material that has been extracted from the literature, though the danger of mistakes is naturally greater.

There is no need to belabor the point, now very well understood, of choosing adequate numbers of specimens for establishing taxa. These could be individuals, for establishing species, or species, for erecting higher taxa. To some degree the exemplar method proposed in the next chapter (Section 4.13) is in contradiction to this admonition, since by that method only single or a few representatives of given taxa are used in the studies. However, it is clear that such exemplars are only reference points and do not indicate the limits of the taxa which they represent. The special problems attendant on material of different ages, different stages of the life cycle, and fragmentary material are discussed in Sections 4.9 and 4.12.

What taxonomic units can be classified by numerical taxonomy? The logical fundamental unit in a large majority of instances is the individual organism. Hennig (1966, p. 6) points out that the fundamental element of biological systematics is an individual at a particular point of time. He calls this unit the "character-bearing semaphoront." For purposes of systematics one must study the multiplicity of characteristics of a sequence of semaphoronts, yielding a description of the phenetics of a life cycle, the holomorph in Hennig's terminology. The individual (or the individual holomorph) is usually an unambiguous entity.

In some studies individuals are the basic units employed. Such studies would throw light on resemblances among intraspecific variants, but would not be likely to offer much scope for comparisons at the subgeneric, generic, and higher levels. Also the comparison of numerous individuals would lead to similarity matrices of excessive size and attendant difficulties of computer processing. Consequently, at higher taxonomic levels representative specimens (the exemplar method, Section 4.13) or averages of several individuals are employed for a taxon. In many studies the unit taxon will therefore be an individual (or an average) standing for a race, a species, a genus or even a higher ranked taxon. The most common unit in classificatory studies will be the species (strictly speaking, the taxonomic unit with a binomial name which is believed to correspond to one or other of the biological units which are given the name of species). Thus we shall employ taxonomic units of different categorical ranks as the entities to be grouped into more inclusive aggregates during classification. We cannot therefore speak of fundamental taxonomic units (these would mostly be individuals), but shall refer to *operational taxonomic units (OTU's)*, which are the lowest ranking taxa employed in a given study. From study to study OTU's can therefore differ in rank; for example, they may be individuals, exemplars of genera, or averages representing species.

It is important to avoid prejudice in choosing the OTU's and, if need be, to explore, by preliminary analyses, the phenetic relations of the specimens that are to form them. In both plants and animals there may be a choice between stages in the life cycle, or there may be a choice between life forms, e.g., castes in social insects. The special problems raised by these possibilities will be discussed in Section 3.6.

Should numerical taxonomy rely on the validity of prior classification for its choice of OTU's? To be totally consistent and rigorous, a taxonomic study at any level should be based on individuals. But since not all individuals of every included population can be studied and analyzed, sampling of individuals is necessary. Williams and Lance (1965) have discussed at some length the general problem of choosing a sample of individuals, and they point out that while it is desirable to have a representative sample (so that the study can be probabilistic) this is not always possible, and nonprobabilistic studies are perfectly valid within their limitations. Also, if we wish to reexamine the relationships of numerous genera in a family we cannot reassert the validity of every genus from a study of its species (and the validity of the species from a study of their individuals). If validation at every level were attempted, preliminary studies would take so much time that the originally intended classification would never be accomplished. Operationally, we have to take on faith the validity of OTU's above the rank of individuals, to "lift ourselves by our bootstraps" into a position from which it is possible to carry out classification at the intended level.

Of the supraindividual categories, the species is probably the most frequently encountered. Although this is claimed to be a nonarbitrary category by adherents of the New Systematics (Mayr, 1963, 1969a,b), it can be shown that the definition of the so-called biological species is nonoperational and that effectively all species are so considered on phenetic grounds (see Section 7.1, and Sokal and Crovello, 1970). The employment of species as OTU's in a phenetic classification is therefore appropriate.

Problems may arise if a taxon used as an OTU proves to be variable for one or more characters. This brings up the question of whether, as OTU's, we can use higher taxa (such as genera, families, and orders), in which the majority of characters within a taxon will, of course, vary. Such taxa can be used, in principle, for the reasons discussed in Section 4.13. A second equally serious problem is the low degree of relevance of most lists of characters (see Section 4.12).

One practical consideration is what to do when the number of OTU's is unmanageably large. A solution that is frequently successful (Sneath, 1964a) is the following: run a random sample of the OTU's and from each well-defined cluster pick three OTU's as reference points for that cluster. Run a second sample including these reference OTU's, and repeat until all the OTU's have been run. Many OTU's will be seen to belong to clusters previously recognized. All the remaining "solitary" OTU's should preferably be rerun with the reference OTU's to detect smaller clusters. Experience with the exemplar method (Section 4.13) gives confidence in this procedure. Three reference OTU's per cluster also provide an internal check on the study, in that they should always cluster together closely. It may be noted that although the exact composition of the set of OTU's has some effect on the phenetic relationships that are observed, numerical taxonomies are quite robust to small changes in the composition, as shown by Crovello (1968f). This is taken up further in several sections of Chapters 4 and 5.

3.2 DEFINITION OF TAXONOMIC CHARACTERS

Procedure in taxonomy, orthodox or numerical, is based on taxonomic characters. The term character has been employed in at least two distinct senses by systematists. Its commonest usage is as a distinguishing feature of taxa—a characteristic (or feature) of one kind of organism that will distinguish it from another kind. Thus, serrated leaves may distinguish one species of plant from another and hence are called a character; similarly, punctate elytra may be used to differentiate between two species of beetles, or resistance to phenol may separate two strains of bacteria. This appears to be the sense in which Mayr (1969a, p. 413) defines a taxonomic character, as "any attribute of a member of a taxon by which it differs or may differ from a member of a different taxon." Unfortunately, such definitions point out again the dilemma of conventional taxonomic procedure: characters are restricted to differences between members of taxa, but the taxa cannot be recognized without the characters themselves being first known.

Another frequent meaning of the term character, which has been espoused by numerical taxonomists as being the more useful one in their work, is that a character is a property or "feature which varies from one kind of organism to another" (Michener and Sokal, 1957) or "anything that can be considered as a variable independent of any other thing considered at the same time" (Cain and Harrison, 1958; we assume the independence referred to is logical rather than functional or mathematical). Thus, referring to the previous examples, the nature of the margins of the leaves becomes the character, while entire, serrated, undulating, or any other types of margin become different *states* of the character (Michener and Sokal, 1957). The word "state" may imply qualitative rather than quantitative subdivision, but in the absence of a more suitable term we employ it to cover both. In the sense in which these terms are used in this book, the condition with respect to punctation of the elytral surface and the property of phenol resistance would be characters, while smooth or punctate and resistant or susceptible, respectively, would be character states.

A note of caution: Blackwelder (1967a) would refer to our "states" as characters and Colless (1967a) prefers to call them attributes, using character in the same sense as we do. Readers of N. Jardine's numerous important contributions to taxonomic theory should be aware that these same terms are used in yet a different sense by him (see Jardine, 1969b). The descriptive terms applied to individual organisms are called *attribute states*, while sets of such descriptive terms are called *attributes*. The term *character state* is used by Jardine to designate the probability distributions over the states of an attribute, while sets of such distributions are called *characters*. His terminology is unfortunate inasmuch as the meanings of the terms character and character state are by now well established in numerical taxonomy; furthermore there is nothing in these terms to connote to the casual reader the idea of a probability distribution.

No objection should be raised to defining a character as a feature which varies from one organism to another. However, if we say that it varies between kinds of organisms (or species) then we are ourselves in the same sort of dilemma as a systematist attempting to apply the definition of character in the first paragraph of this section—that is, defining characters on the basis of predefined taxa. Thus we would first have to define our species before we could define the characters. To be extremely critical, therefore, we would have to define characters entirely on the basis of the differences between individuals. Specific characters are, of course, nothing but summaries or abstractions of the characteristics of a large number of individuals.

The general definition of characters established above cannot, however, give much aid to the practicing taxonomist in the process of recognizing and describing individual characters. We shall go into this problem in the next section, from both theoretical and practical points of view.

3.3   UNIT CHARACTERS

Theoretical Considerations.   Those embarking on work in numerical taxonomy may be puzzled by the task of recognizing the basic units of information for the study, what we call the *unit characters*. In trying to define these characters we must first ask ourselves what properties we wish to recognize. Do we wish to recognize genes, or a unit element in selection, or a logical construct? If, as is now clear, genes are themselves complex entities, shall we subdivide them? And if so, to what extent?

Most definitions of a unit character have been too restricted; defining a unit character in terms of morphology, chemistry, genetics, or evolution does not allow the broad treatment needed for a general theory of systematics. For this we need to define unit characters in terms of information theory, for in every instance it is information that the characters convey to the taxonomist; this idea is closely linked to the concept of natural taxa as groups with high content of information (see Section 2.2). Sneath (1957b), has made an attempt to introduce the concept of information, and a unit character (there called a "feature") was defined as an attribute possessed by an organism about which one statement can be made, thus yielding a single piece of information. These attributes are formally logical constructs, since they will change if the technique of observation changes; the definition is therefore an operational one. If a character can vary continuously, such as with the length of an organ, the character of length can be broken down into as many steps as the observational method will allow with good reliability. Either each step is counted as a feature, or at least the minimum number of features necessary to account for the existing variation is postulated (see Section 4.8). But as is also pointed out below, these steps will not be independent, since they are not mutually exclusive.

This approach may be carried to its logical conclusion, where each unit character or feature represents an alternative that can be answered as "Yes" or "No," "Possessed" or "Not Possessed"; the information content can then be measured as "bits," as is usual in information theory.

Clearly even the most simple organisms contain a great many bits of information. We may plausibly interpret this information in terms of modern genetics. The genome consists of a series of nucleotides, which are paired one-to-one in a double helix of deoxyribonucleic acid (DNA). The genetic information resides in the sequence of the different nucleotides and may be thought of as a code message written in an alphabet of four letters, each letter representing one of the four alternative nucleotides—those containing thymine, cytosine, adenine, or guanine. The genetic code message is translated into other codes determining the amino acid sequence of proteins and the structure of other molecules in the cell, and these in turn determine the physiological and morphological properties of the organism. Although evidence is still scanty, there are two lines that confirm the close correspondence of the fine genetic structure with orthodox and numerical taxonomy. One is the data from protein sequence studies and the other is the congruence between phenetic data and DNA pairing (Section 3.6). There are, however, some difficulties in relating the fine structure of the genome, even if it were completely known, to taxonomy. Thus, if there has been a gene reduplication, should this be scored as many independent characters, or as one change, the reduplication itself? This is clearly analogous to estimating the common information (in the conventional sense of the term) in two pieces of written text, such as two editions of a book, where insertions, repetitions, and changes of order may occur.

We may tentatively identify our taxonomic bits with the genetic code and in *Principles of Numerical Taxonomy* we noted that the number of bits in the genome ranges from around 10^4 for some viruses to around 10^{11} for many higher animals and plants, assuming that the bulk of the DNA is functional. The number of functional genes is of course much smaller, since each gene is made up of many bits, probably about 1,000 as a rule. These figures are only speculative, and do not include any nonchromosomal genetic information or environmental effects, but it is clear that the content of information (in the information-theoretical sense) is much smaller in microorganisms than in higher organisms and is related to histological complexity (Sneath, 1964d). The scale of the potential store of information in a nucleus may be judged from the estimate that the Library of Congress contains between 10^{13} and 10^{14} bits (see Good, 1958). It is interesting to note that Elsasser (1958, pp. 100–104) estimated from morphological considerations that the information in man was at least 10^7 bits. It should be noted that the bits are a measure of the potential information content, not of the number of alternative permutations of the information. The latter is 2^x when the number of bits is x.

Working Definition.   Except in those few and simple organisms whose fine chemical structure is gradually being unraveled, the above considerations are premature, so a working definition, which the practicing taxonomist can use, is needed. We may define a unit character as *a taxonomic character of two or more states, which within the study at hand cannot be subdivided logically, except for subdivision brought about by the method of coding.*

Since we cannot in most cases make genetic inferences from phenetic studies of characters, we shall generally have to use phenotypic characters as our basic information, defining these as narrowly as possible. Our failure to make *logical* subdivisions may rest on ignorance of the finer structure or the causation of a character. Thus, presence or absence of a bristle in an insect may be a unit character, if we know nothing of its finer structure and have no way of subdividing it, or do not care to do so. Even if the general morphogenesis of the structure is known from a representative form, unless morphogenetic differences can be established that can serve as taxonomic characters within the group studied, the presence or absence of the bristle remains the unit character. To consider another character, the same insect may possess or lack DDT-dehydrogenase, which character we are again unable to subdivide further in view of our present knowledge. Thus, the organizational levels of unit characters may differ considerably from character to character and with advances in our knowledge. The ruling idea is that each character state should contribute one new item of information.

We anticipate in the near future that automated methods will be introduced for extracting information directly from specimens and converting it into unit characters. Such automation is now being introduced in some fields of biochemistry and microbiology, but its application to morphology may prove more difficult. A step in this direction has been taken by Sokal and Rohlf (1966) and Rohlf and Sokal (1967) who showed that remarkably good numerical taxonomic results could be obtained by overlaying biological images with a perforated mask. If a line in an image showed through a given hole this was scored as a character with value 1, and if not, with value 0. The two sets of characters were then used in a numerical study. As the method stands at present, it must be assumed that two images have been appropriately oriented and scaled for size. To generalize the technique these factors will require attention, but they do not appear insuperable problems, although some points regarding operational homology require further development (see Section 3.4). Recently Meltzer, Searle, and Brown (1967) described a method of defining large and small scale detail of shapes by means of rectangular reference patterns, which may offer another approach toward automatic scanning. There are also potential developments based on holography (Gabor, 1965).

When corresponding (operationally homologous) points are used for the method of matching images (Sneath, 1967a) the definition of unit characters as such is circumvented, and the positions of the points effectively become the unit characters. The dissimilarity, d_h, between two images is then obtained as a form of average distance between the pairs of corresponding points (one from each image) at the orientation of the images that gives the best possible match.

3.4   HOMOLOGY

A Critique of Traditional Definitions

The concept of homology is central to any taxonomy. All nonarbitrary classifications are based upon comparisons of sets of characters, evaluation of the similarities and differences between them, and, in the case of phylogenetic taxonomy, inferences about the evolutionary history of the taxa derived from these comparisons. Yet such comparisons require corresponding or homologous characters for making the taxonomic judgments, regardless of the school of taxonomy followed. If we are to compare "apples and oranges," we must compare them over a set of characteristics applicable to both of them. In view of the crucial importance of the concept of homology in taxonomy it is indeed "...a serious flaw that the terms homology and analogy are so difficult to define, or at least their results are so difficult to distinguish" (Davis and Heywood, 1963, p. 40).

The concept of homology dates back to pre-Darwinian times. Owen defined a homologue as "the same organ in different animals under every variety of form and function," and he defined special homology (this is usually simply referred to as homology nowadays) as "the correspondency of a part or organ determined by its relative position and connections, with a part or organ in a different animal." He defined an analogue as "a part or an organ in one animal which has the same function as a part or organ in a different animal." Thus these terms had no necessary evolutionary connotations. To Darwin the existence of homology was another link in the chain of evidence for his theory of evolution. However, as soon as purported homologies were called upon to serve as evidence for phylogenetic relationships, any evidence concerning these homologies derived from the phylogenetic inference immediately became suspect. Taxonomists have wrestled with this problem to this day and critical examination has always shown a phylogenetic definition of homology to be wanting. Thus Simpson's definition that "homology is resemblance due to inheritance from a common ancestry" (1961, p. 78) is self-defeating, since it is a primary purpose of the phylogenetic school of systematics to work out phylogenies, and for this they need homologies that are not defined in terms of the conclusions they wish to reach. All of the criteria used by Simpson (1961, pp. 87–92) for recognizing homologies are subject to serious objections (the interested reader may find these in our *Principles of Numerical Taxonomy*). Hennig (1966, p. 93) defines homologous characters as those "... that are to be regarded as transformation stages of the same original character... 'Transformation' naturally refers to real historical processes of evolution..." Hennig realizes that this definition can only be of theoretical value, admitting "the impossibility of determining directly the essential criterion of homologous characters—their phylogenetic derivation..." He is forced to turn to Woodger's concept of correspondences (see ensuing discussion) in order to approach the problem of homology operationally.

Among the reactions to the impracticality of a phylogenetic concept of homology has been the school of idealistic morphology (for review and entry to the pertinent literature see Zangerl, 1948, Sokal, 1962b, and Jardine, 1967). The metaphysical form of this approach is rejected by most modern biologists, but empirical and statistical forms coincide with Woodger's proposals for a homology based on a theory of correspondences. This, as we shall see below, is an essentially phenetic concept. It is also noteworthy that chemists are increasingly using the term isology (introduced by Florkin, 1962), for chemical correspondence (for example the occurrence of cyanides in two organisms) rather than homology, when little is known about their evolutionary origins.

Before we turn to a detailed discussion of various operational means of determining homologies, we need to remove a misconception still current among many taxonomists: namely, that if only the fossil record were complete and known, phylogenetic classification and hence phylogenetic definitions of homologous characters would be possible. It has been pointed out by several recent workers that even with a fairly complete fossil record, the only way of associating specimens at the various time levels is by overall resemblance of organisms and their parts. Colless (1967b) has examined this question more fully than others. He postulates a completely competent and adequately long-lived observer who could continuously record the evolutionary history of a group of taxa as it unfolded before his eyes. He agrees that the hypothetical observer would clearly know the cladistic relationships among the various branches but would still have to measure overall resemblance to get an estimate of phenetic relationships among any points along the branches. Now suppose the observer examines the evolving group of taxa at regular time intervals. This makes the model slightly more realistic. The question now becomes whether at successive time intervals T_1 and T_2 he could perceive if a form A_1 was the same or had changed. He would recognize A_1 at time T_2 by its possession of the same complex of attributes it possessed at time T_1. If the interval from T_1 to T_2 is substantial there might no longer be a form A_1 at time T_2, but a hitherto unrecorded $A_{1.1}$, closely resembling A_1, so that the observer unhesitatingly accepts it as a changed A_1, that is, as an evolutionary descendant of A_1. Colless (1967b) concludes: "The trend of this argument should by now be obvious.... the point I am making is that *any* observer making surveys at $T_1, T_2, \ldots, T_n$, can infer the course of evolution only from a detailed and direct comparison of the complexes of attributes possessed by taxa at each and every one of those times; i.e., by a survey of 'overall resemblances'." Clearly then, even under impossibly optimal conditions the determination of homologies in phylogenetic taxonomy is made essentially on a phenetic basis.

In the rest of this section we shall therefore concern ourselves with a phenetic and empirical concept of homology that we hope will become increasingly operational and quantifiable.

Homology Redefined

As we have seen, a precise and universally applicable definition of homology has not so far been proposed. Various definitions applicable to special cases will be featured below. By abstraction from them, homology may be loosely described as compositional and structural correspondence. By *compositional correspondence* we mean a qualitative resemblance in terms of biological or chemical constituents; by *structural correspondence* we refer to similarity in terms of (spatial or temporal) arrangement of parts, or in structure of biochemical pathways or in sequential arrangement of substances or organized structures.

In the past, homology has generally been a categorical concept, that is, structures either were, or were not, homologous. Attempts at making homology operational have generally led to a quantitative concept of homology that permits degrees of homology (e.g., Sattler, 1966). Thus, structures may be more homologous with one structure and less so with another. If the categorical definition of homology is preferred, then a character or a structure may be considered homologous with the character or structure with which it shares the greatest degree of similarity or correspondence, an approach Jardine (1967, 1969c) prefers. Compositional correspondence can be regarded as a special case of homology decided upon an external criterion, and structural correspondence as a special case of homology decided by internal criteria, as pointed out by Sneath (1969c). Jardine and Jardine (1969) prefer the terms "attributive matching" and "relational matching" respectively. In most biological situations the criteria can be adequately described as compositional or structural, which are the headings we employ for the present discussion, but sometimes this would unduly stretch the usual meaning of these terms. Thus the ability of a bacterium to ferment lactose would scarcely be called compositional (though it may depend ultimately on some unknown detail of composition), and it would be clearer to say that the character "lactose fermentation" is based on an external criterion involving a chemical test procedure. Indeed, in bacteriology the bulk of the characters employ external criteria. Sneath also points out that compositional criteria may introduce two additional aspects of homology. The correspondence may be quantitative, as when, in considering the correspondence based on the compositional criterion of calcification, we may find that some structures are more heavily calcified than others. Either the degree of calcification is admitted as a relevant quantitative variable or some level is used to separate calcified structures from noncalcified ones. The correspondence may also be diffused, as when a bone has a correspondence to all other bones with respect to the criterion of calcification.

All cases of structural homology contain some external element in order to distinguish the parts under study from other parts (e.g., bones from soft tissues). However, the use of external criteria does not avoid all problems; thus we may be faced with a bone in one animal that is represented by a tendon in another, and Jardine and Jardine (1969) note that most cases of homology involve compromises between compositional and structural criteria.

Operational Homology

The first approach to homology in phenetic and numerical taxonomy was to call two character states the "same" whenever they are indistinguishable. Similarly, if the abstraction or idea of two characters cannot be distinguished in the taxa, then again we would consider them to be the "same." In practice the worker will divide his organisms into major structures or other such divisions (for example, head, limbs, leaves), and he must first decide whether these are the "same." Then he can proceed to look for differences in their properties that may be used as the bases for setting up the characters and their states. Within each major structure he again looks for subsidiary structures and repeats these procedures. Characters in this sense are synthesized from the states.

For example, consider two species of insects that are both black, while others in the same genus are red. If we had no way of distinguishing the two kinds of black, we would consider them to have the same character state, "black." Similarly, we would consider "redness" and "blackness" to be states of the same character, "body color," unless we had reason to believe that this color was of a different nature in some of the insects than in others. If, for instance, we found colors due to pigments as well as colors due to optical interference phenomena (such as iridescence) occurring in the group of insects under consideration, we would then subdivide our former character "body color" into two: "pigmental body color" and "structural (interference) body color."

By way of another example, leaves on a given series of plants may be long or short. We first have to decide what a leaf is and whether the structures seen on the separate specimens are in fact leaves—that is, are the "same"—or perhaps are other structures such as modified stems. Having decided that they are leaves, we also have to agree on what we shall call a short leaf—perhaps one of less than 3 cm; if so, we shall call a long leaf one that is longer than 3 cm. In this sense all leaves shorter than 3 cm are homologously short leaves, those longer than that are homologously long. The character will now be called "length of leaf," with two states, "short" and "long." Clearly it could have had more states had one wished to construct it in such a manner. In each leaf other characters, such as the venation, pubescence, and similar properties, could be coded as characters.

We call this approach *operational homology*. Our position is largely that of
common sense; when we say that two characters are operationally homologous,
we imply that they are very much alike in general and in particular. If the characters
are "not quite the same," then more than one character is involved, and they
should be broken into several independent ones; some of these independent
characters will then be indistinguishable and will be scored as "the same" character
in the two organisms.

A difficulty of the operational definition of homology is that some homology
statements are made about unit characters, others about sets of unit characters.
Some unit characters such as color, weight, length, or pH can be defined unambiguously in terms of operations necessary to ascertain their magnitude. These
are usually properties of the entire organism, definable on external criteria. However, other unit characters are properties of parts of an organism, each part being
defined by a set of unit characters. But these characters are in turn defined by the
parts to which they belong and need reference to structural correspondences. Thus,
for example, an eye or a forelimb can each be subdivided into numerous unit
characters, yet each of these unit characters cannot be defined without reference
to the fact that it is on the eye or the forelimb. For example, color of eye or color
of appendage must be related to eye or forelimb. A circularity of comparison (Inglis,
1966) makes it difficult, therefore, to define an eye or a forelimb in any rigorous way.
In practical work in numerical taxonomy such decisions are usually carried out
on a common sense basis. However, for a rigorous definition of homology this is
not sufficient. In fact even the entire organism needs to be defined unambiguously.

To establish that two eyes in different organisms are homologous, we can use
two approaches. We can define them purely structurally, that is, consider an eye
a unit character and describe its relation to other parts of the body (cranium,
muscles, nerves) in such a way that it will become apparent that the characteristics
in the two organisms are homologous; that is, both are an eye. Although this
approach employs criteria external to the eye (e.g., a large nerve connected to the
brain), it requires criteria internal to the organism and presupposes knowledge of
the identity of these other parts of the body. Hence it depends on previous homologies established for them, finally resulting in the conclusion that the greatest
number of identical relations is obtained if both are identified as eyes. The other
approach is to use a criterion of similarity based on numerous finer unit characters,
that is, statements about the components of the eye. By showing that the structure
in question in one organism is most similar to the structure known as an eye in
another organism, we can establish their homologies and identify the structure in
question as an eye. The difficulty here is that in order to do so by the similarity
approach, we must define unit characters in both structures, and such unit characters in theory cannot be defined unless we know that the structure is an eye. With
both approaches, therefore, we are caught in a circularity of reasoning. Fortunately
we can often make a start with compositional criteria such as presence of lens

protein, which in principle could be established on chemical grounds. A rigorous approach would be to consider the eye a unit character first, determine homology provisionally on a compositional and structural basis, and then proceed to the more detailed analysis involving sets of characters per organ or complex structure.

In calling this concept operational, we did not so much have the philosophic idea of operationism in mind (see Section 2.1). Rather, we felt that we could not rely on phylogenetic inferences; we had to have some procedure for getting started or "becoming operational." As we shall show later, when the approach of operational homology is taken to its logical foundations, it becomes a numerical homology and, while not fully operational in the sense of operationism, it is certainly considerably more so than the concept of phylogenetic homology. Hull (1968a) points out correctly that in determining homologies, pheneticists do not restrict themselves to direct observations of shape and relative position. In determining what is the same state and character in different organisms, some previous knowledge of the biology of the structures and organisms concerned is clearly employed. None of the procedures for finding homologies are entirely inference-free, and in order to bootstrap ourselves into a position from which we can make some statements we must initially assume certain relationships as being obvious. We may, however, then turn around and test these relationships. As the concepts of numerical homology are developed these inferences become less necessary, or at least more clearly defined. We cannot agree with Hull, however, when he equates these non-observational inferences in phenetic taxonomy with those in phylogenetic taxonomy, which he considers to be certain law statements concerning trends in evolutionary development, the relations between embryological and evolutionary development, the correlation between morphological and cladistic affinity, and the like. But here Hull has put his finger on the very weaknesses of this line of reasoning. There are few, if any, general "laws" that can be derived from any of the considerations mentioned.

The concept of operational homology as used by us is close to the original definition of homology as employed by Owen and discussed earlier. Woodger (1945) has discussed the problem of homology from the point of view of morphological correspondences, deriving the concept in what would now be called a phenetic sense, with structural and compositional considerations both contributing to the correspondences. He points out that in making such morphological correspondences we pair the different parts of the structure in two organisms, with the aim of obtaining the greatest number of one-to-one pairings. For example, if we pair head of cat with head of dog we find numerous subsidiary pairings within this major pairing—such as eye with eye, and brain with brain; further pairings occur within the latter structures—lens with lens, retina with retina, cerebellum with cerebellum, and so on, down to histological levels and farther. A similar comparison between head and leg would show few such correspondences. Woodger illustrates his arguments by the example of the pentadactyl limb. He shows that the pairing of the "same" bone in two forms—humerus with humerus, radius with radius, and so on —depends on the spatial relations of the bones. For example, the humerus is proximal to the other bones, the radius and ulna are both immediately distal to the humerus, the ulna being postaxial to the radius, and so on. In a newly studied creature we call a bone the radius if we find that it bears these relations to the other bones, and if they in turn bear their own proper relations to each other. Such matching sets of bones Woodger calls *isomorphic*. He also discusses the difficulties that arise if some bones are atrophied or if their articulations are abnormal, so that it may be difficult to recognize which bone corresponds to which. Woodger correctly says that morphologists pair off organs one with the other intuitively so as to make the greatest number of one-to-one correspondences.

As we attempt to apply the ideas of operational homology in actual practice we encounter considerable difficulty. First, how are we to determine the homology of the large number of characters that are employed in numerical taxonomic studies? It is unrealistic to expect that in the near future methods will be developed for the rapid determination of the homologies of large numbers of characteristics. In almost all numerical taxonomic studies carried out to date, the judgment of the taxonomist regarding homologies is implicitly accepted. The homologies are deliberately phenetic, and evolutionary considerations in evaluating these homologies are intentionally avoided, although we cannot always be certain that this is done. However, since we believe that much purported phylogenetic homology is in fact a measure of phenetic and relational resemblances, some phylogenetic coloring of the judgments of the phenetic taxonomists is probably harmless. Operational homology, when carried out at this level, clearly does not conform to the dicta of operationism. Only as the concept of homology becomes more quantitative (see later discussion) will it be more operational in the strict sense.

Problems of homology have not been serious in numerical taxonomic work in practice. Fisher and Rohlf (1969) and Moss (1968b) found that numerical taxonomies are quite robust to small errors in homology. Fisher and Rohlf found with correlation coefficients that changes were minimal when less than 10 percent of characters were wrongly homologized (there was also an effect on distance coefficients, but this was smaller probably because the permutation process they used preserved the general size factor, see Section 4.11). It may be noted that in constructing phenetic classifications in bacteriology it is possible to formulate a list of characters and their states that does not presuppose a prior knowledge of homologies. Nevertheless these problems must be watched; Jardine and Sibson (1971, p. 171), for example, believe that a numerical taxonomic study by Hamann (1961) may be misleading because of difficulties in homology with families of monocotyledons.

To date, the various proposals for quantifying the concept of homology have sought to analyze the mental processes of systematists and comparative anatomists. These approaches can be classified under two main headings—the structural approach in which correspondences in positional relations between parts are the primary criterion (but which also can include relations such as are exemplified in biochemical pathways), and the phenetic approaches, in which homology is based on overall similarities of organs and complex structures in terms of their constituent unit characters. In the various approaches below, the definitional approach applies to unit characters only. The other approaches all are based on sets of characters. *Structural correspondence* comprises (1) a multidimensional approach and (2) a one-dimensional approach. The *phenetic approaches* to homology include (1) the definitional approach, (2) similarity of complex structures based on defined unit characters, (3) similarity based on undefined characters, and (4) geometric similarity.

Structural Correspondence

Multidimensional Structural Correspondences.   Much of the classical concept of homology as derived from the work of the comparative anatomists and especially as applied in vertebrate anatomy focuses on the positions of various parts relative to each other. The pioneer work in the rigorous formulation of this notion of homology is that of Woodger (1945), whose concept of homology involved $1:1$ correspondences between morphological parts maintaining their positions relative to each other. These ideas have recently been further developed by Withers (1964). The idea of using evidence for homology from the position of structures is well established. Inglis (1966) states that in making judgments on homology from structural relations one must assume the known homologies of other structures in the organisms. ". . . Thus a statement of absolute resemblance involves a system of comparison, using as reference points as yet uncompared organs or features, the validity of whose use in this comparison is only justified by later reference back to the first organs or features compared . . . Comparison leading to the recognition of absolute resemblance is therefore circular and involves whole organisms." Extending the line of argument, Key (1967) visualizes homology as being that state of arrangement or matching of characters that will result in the greatest estimated similarity between OTU's being compared. His definition of homology is: "Feature a_1 of organism A is said to be *homologous* with feature b_1 of organism B if comparison of a_1 and b_1 with each other, rather than with any third feature, is a necessary condition for minimising the overal difference between A and B." The difficulty with making such an approach practical is that in order to measure the overall difference between any two OTU's and minimize it by rearranging the characters, one must already have an acceptable system for coding the characters and to do so one needs to have at least a tentative concept of homology.

A significant breakthrough in quantifying the concept of homology has come from Jardine (1967). He follows Woodger in using the criterion of agreement in relative position and connections with each other. For the determination of homologies, a set of relations is established of the type "anterior to," "ventral to," or "right of." A set of parts of an organism is said to be connected by these relationships, which must be nonreflexive, that is, if part 3 is anterior to part 1, part 1 cannot also be anterior to part 3. By Jardine's definition a *similarity* is a function rearranging the parts of one organism with respect to another, in order to obtain 1:1 correspondences with respect to the relations between the parts. That function which produces a unique maximal correspondence (in Jardine's term, a similarity having as members the greatest number of ordered pairs of parts) is called h, a *homology*.

These definitions lead to an algorithm that permits the computation of the unique largest similarity. This is the largest number of correspondences between some or all of the parts from each organism in which the spatial relations between parts are preserved. Jardine's program permits "undecidable cases" and the program has the option of allowing for two or more parts to be considered together as a single part. This permits a part in one organism to correspond to an aggregate of several parts in another. Jardine and Jardine (1967) have applied this technique to a number of simple cases such as bones of human, cat, rat, and dog skulls, where it yielded satisfactory results. In investigating a more challenging case, the homology of the skull bones of the holostean *Amia calva* and of the teleost *Clupea finta*, the method did not yield any unique largest correspondence but several equally large correspondences. These differed in the way in which the bones of the suborbital series were paired but gave consistent correspondences between other bones. The results largely supported W. K. Gregory's interpretation of these homologies.

Despite the success of his approach, Jardine (1969a,c) has shown that different early assumptions—expressed in the list of relations admitted as relevant—are critical. Thus, in computing the homology of the skulls of the cat and rat, the generally accepted homology is found if the relations "anterior to," "ventral to," and "distal to" are employed. Moreover, if the relation "adjacent to" is added, several equally large correspondences are found. This is because the frontal and squamosal bones in the rat are adjacent, unlike the cat. In a detailed study of skulls of fossil fishes Jardine showed how different assumptions of this kind would give maximal correspondences that agreed with the homologies proposed by various authors, each of whom had apparently made different assumptions of this type. This was particularly noticeable with decisions on whether the lateral lines (sensory canals) could move from one bone to another during evolution. Difficult problems are also posed in distinguishing between the fusion of two bones and the loss of one of them.

Such observations reveal our inadequate knowledge of how to reach appropriate assumptions. Although Jardine is undoubtedly correct in saying that a background knowledge from fossil studies plays a part in this (for example, bones rarely change their antero-posterior relations, but commonly change in adjacency) we would not overemphasize this dependence on paleontology. If this dependence were critical it would prevent homologization in groups without fossils, and a child who knew nothing of phylogeny could not homologize the hand of a man with that of an ape.

Jardine's approach is clearly a numerical homology, but as contrasted with those efforts based on phenetic approaches, his measure of similarity is maximum correspondence within a network of relationships. Jardine's approach is a considerable advance over earlier more or less intuitive assessments of relational homology. As formulated, it clearly determines the homology of unit characters (the parts). The extent to which it can be applied for characters in general is still in question. First, though it can be applied to subunits of parts in terms of its logical and mathematical formulation, can we make relational statements about ever-smaller parts or bones in the same way we can about discrete bones or sclerites? Here is where the compositional aspect of homology (cartilage, connective tissue, muscular tissue, etc.) could usefully be combined with the structural relation. Also, more complex relational statements will be made as we extend our purview beyond strict classical morphology. Can such a system be employed for homologizing biochemical pathways or biochemical structures? What of behavioral homologies? Possibly a "hybrid" system employing phenetic similarities between complex structures as well as the greatest numbers of relational correspondences may be developed.

One-dimensional Structural Correspondences.   In biology this approach to homology has mainly concerned amino acid sequences in proteins. Whether approaches of this sort might be useful in other sequentially organized structures or properties, such as metameric organs or organisms, remains to be investigated. One of the ways of finding homologous positions in protein chains is to align the chains side by side in the manner that gives maximum agreement between pairs of amino acids. This is the method for homologizing amino acid sequences by cross-association (sliding matches) as described by Sackin, Sneath, and Merriam (1965). A second method suggested by Fitch (1966a) finds the minimum number of mutational changes in the nucleic acid sequence that would convert one protein into another. Thus, if at a given position in the chain one amino acid is glycine and the other phenylalanine, the minimum change would be two nucleotide mutations, from glycine codons GGU or GGC to UUU or UUC (the only codons for phenylalanine), and not from the glycine codons GGA or GGG, which would require three mutations to give phenylalanine. By arranging protein sequences according to their best matches, homologous positions may be established and evolution at these sites investigated. The complexities and difficulties in homologizing protein sequence positions when there are deletions or gaps in one chain (see Cantor, 1968; Sackin, 1969, 1971; Fitch, 1969, 1970a,b) are such that even the simpler sequential correspondences raise unsuspected problems.

Phenetic Approaches

The Definitional Phenetic Approach.   This is especially suited to unit characters that are not part of an organized structural tissue. Identically defined characters can be described in different organisms; these are homologous by definition. This definition may be in terms of general appearance, structure, or composition. Thus, weight or maximum length, if rigorously defined, is homologous by definition in two OTU's. Similarly, the presence of hemocyanin in organisms would be homologous by definition. Note that these statements are not related to any organ or tissue and thus need no inferences about previous homologies of such structures.

Similarity of Complex Structures Based on Defined Unit Characters.   One way to establish homology between organs, organ systems, and other complex structures is by their overall similarity. This process is comparable to the classification of the organisms themselves. It attempts to define structures in terms of unit characters (Section 3.3) and applies to them the concept of operational homology discussed above. Some intuitive judgment enters in such definitions of characters. The aim is to compare these organs, character by character, and ultimately to group structures so that there is the greatest number of common properties in the "organ taxa" set up. We may call those structures homologous that exhibit maximal similarity with each other among a set of potential structures derived from the two organisms being compared. This is evident from a consideration of why we pair an eye with another eye and not with an ear. In both alternatives there are some shared character states (both organs are carried on the cranium, both are special sense organs), and some differences (no eye is identical with any other eye), but there is no single property which we can satisfactorily postulate a priori as being essential to the definition of an eye or an ear (since we may be sure that some morphologists will soon find an exception to our rule). Therefore, the concept of natural organ taxa is polythetic (see Section 2.2), just as natural taxonomic groups are; hence natural organ taxa can, like natural taxa, only be defined as arrangements by which the groups so formed possess the greatest possible number of common properties. Such reasoning leads us to a *numerical homology*—that is, a numerical taxonomy of organic structure. As pointed out by Colless (1969a) this would also lead further to the homologizing of the whole organism, and numerical homology would then be intimately related to some aspects of phenetic resemblance. In contrast, the definitional approach is analogous to monothetic grouping methods.

  By basing homology on maximal overall similarity and basing the latter on correspondences between unit characters, we find ourselves with the realization that increasing knowledge and refinement of techniques will lead to ever finer unit characters in terms of structural complexity. This in turn leads to clusters and structures at increasingly finer levels. Thus not only do we find within the concept of natural taxonomic groups similar concepts of natural organ groups, but also within these there may be yet other like concepts (such as natural cell groups or natural gene groups) hierarchically arranged like a nest of Chinese boxes. There must, however, be some limit to this process, even if the limit lies at the fine structure of the genes.

However, even if we proceed to this level, the problem of homology remains with us, now commonly called *genetic homology*. Eventually one may visualize genetic homology as correspondences in the nucleotide base sequence of the double helix of DNA. At this stage in our understanding of molecular biology it may be more useful to think of amino acid sequences of proteins. It is possible to identify regions of the protein molecule in which corresponding sequences can be found even in quite distantly related organisms. For example, cytochrome *c* seems to have regions that can be traced back to a common ancestor of yeast and man.

There are still many problems in any approach to genetic homology. The sporadic occurrence of enzymes unusual in animals, such as the carbohydrate-splitting enzymes of the snail, is presumably based on evolution de novo, for it would be difficult to believe that these enzymes in snails are derived from their remote protistan ancestry. It seems very likely, however, that when such re-evolution does occur it is due to reduplication of a gene controlling the production of a similar enzyme, followed by change in the enzymatic properties of one of the duplicate enzymes. We assume that the actual sequence of amino acids is duplicated and leads to separate evolutionary structural divergence. The classical instance of such a phenomenon is the reduplication of hemoglobin genes that have undergone independent evolution in mammals (Ingram, 1961), possibly with internal duplication as well (Zuckerkandl and Pauling, 1965b; Fitch, 1966b). A further possibility is the transfer of genes from very dissimilar organisms, by various mechanisms. For example, viruses might transfer genes to distantly related plants or animals (Sokal and Sneath, 1963, p. 73; Anderson, 1970).

An important consideration with reference to tracing a homology to the levels of fine genetic structure is whether we should, in fact, do so. This problem is discussed in greater detail in Section 3.6 and here we might simply point out that for many types of evolutionary and taxonomic problems it would not be desirable to proceed to the molecular level, but to evaluate homologies at a higher structural level.

Considering homology as maximal overall similarity between complex structures also has implications for the definition of *analogy*. We have not so far paid much attention to definition of this concept. If by analogy we mean functional but not morphological similarity, one must be able to separate logically the definition of structure and function, which may not be easy. It seems to us that function is also an aspect of the phenotype and that statements about function can be unit characters in the same way that statements about morphology can. If that point of view is accepted, then it follows that functional identity or similarity will lead to phenetic similarity just as morphological identity does. Furthermore, if overall resemblance is based only on functional characters, it may well be that structures will be considered to be highly similar that, in fact, might have considerable morphological disparity. When estimating overall similarity we should not separate structure from function; we would not wish to base similarity simply on function (unless we intend to make a special classification such as one for functional anatomy or for ecology). We do not believe that false overall similarity (due to analogies) would enter very prominently or frequently into a phenetic measure of homology since analogous organs (wings of birds and butterflies) would be dissimilar in so many structural and functional character states compared to their few common functional and morphological character states. So considered, homology and analogy are not categorically distinct concepts, but are near one end point and at an intermediate point, respectively, of a continuum of relationships in which total identity is at one extreme and total dissimilarity at the other. The general region of high similarity is homology. Cases of analogy would be found at intermediate levels of similarity, where it could be shown that the subset of characters dealing with functional relationships expresses similarity though a subset of morphological characters does not. Rather than make a categorical contrast between homology and analogy, therefore, we prefer to speak of greater or lesser resemblance among organs, organ systems, and other levels of complexity.

Similarity Based on Undefined Characters. One way to deal with problems of homology is to ignore details of structure, given certain preliminary assumptions about the spatial orientation of specimens with respect to the observer. This is the approach undertaken experimentally in a study by Rohlf and Sokal (1967). Basically, the technique consists of recording agreement in visible structures over selected minute areas of the images of two organisms. This was accomplished by preparing random masks from 25 punch cards, each perforated with 25 randomly chosen holes. These masks were placed over black and white drawings of 29 "species" of Caminalcules (a group of imaginary animals created by Joseph H. Camin; see Section 6.4) and 32 species of culicine mosquitos. Black lines appearing through a hole were scored 1, and empty holes were scored 0 (see Figure 3-1). Similarity matrices were constructed on the basis of matching scores for corresponding masks and holes (625 pseudocharacters). The taxonomic structure obtained from scanning images in this way compared favorably with the structure obtained by deliberately looking for and scoring characters. The surprising results would suggest that precise homologies may not be necessary for expressing phenetic similarity between organisms. In fact, however, correspondence between scans does rest on general morphological similarity both in structure and in relationship of the images compared. The "characters" in this study are thus not biologically (morphologically or relationally) defined but represent given positions in the scan. It is quite possible that in complicated images parts of the image might be demonstrated homologous

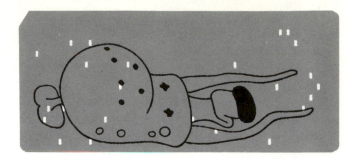

FIGURE 3-1

Automatic scanning of Caminalcules. At the top, one of the 25 masks, containing 25 randomly chosen holes, is shown superimposed over a drawing of a Caminalcule. In making the composite shown below, each of the 625 holes was scored 1 when a black line appeared through it and a 0 when no black showed. (Actually, fewer than 625 holes are visible here, since many holes on different cards coincided). [From "Numerical Taxonomy" by Robert R. Sokal. Copyright © 1966 by Scientific American, Inc. All rights reserved.]

by the scanning method. Complete scans of the entire image (all positions on the grid or mask as contrasted with the random scans reported above) also gave satisfactory results, but classifications were not necessarily better than those achieved by the random scanning method.

This approach has clear limitations. Great disparity in size of organisms would lead to unjustified dissimilarity, so that adjustment to a common scale would be essential. In their work with the Caminalcules, Rohlf and Sokal made no attempt to allow for differences in size. Thus a large proportion of the phenetic differences recorded by the random scanning method were simply differences in size. In spite of this introduction of obvious error (unless size should be weighted more heavily than is usually done by biologists), the taxonomic structure obtained was quite satisfactory. In aligning the organisms before scanning, a primitive type of homology is used. Decision must be taken on what is anterior and posterior in the animal, the location of the "shoulder," bilateral symmetry, and so forth. The orientation of the images is clearly important, and consistent criteria must be chosen.

Differences in the orientation of appendages, such as wings or legs, would make related organisms appear far more dissimilar than they otherwise would be. Such adjustments could be made by operators on graphic display consoles or automatically by rather sophisticated programming (as described below).

Geometric Similarity.   Rather than estimate the similarity between samples of images as just discussed, one could try to fit entire images to each other. This would require enlargement, diminution, rotation, translation, and disproportionate stretching of some images to a standard or pattern image. Such an approach could in theory be undertaken entirely "blindly." By this we mean that an attempt could be made to fit the outline of image B to that of image A in the best possible manner. Even if these images are not at all phenetically related, as for example the outline of a *Paramecium* and that of a human skull, there will nevertheless be some position of best fit. The goodness of this fit should be of a low order. When homologous images are compared, these differences should be much less and it should be possible to transform one image essentially into the other.

If one employs the method of matching diagrams described by Sneath (1967a), it is necessary to mark homologous (i.e., corresponding) points on the diagrams. This can be done by eye, in most cases no doubt without difficulty. Thus we might mark the tip of the nasal bone on two skulls. This involves decisions on homologies, for example, what is the nasal bone and what is its tip. Sneath notes that one can envisage an automatic method, though it has difficulties that have not been explored, which would represent the lines of the diagrams as series of closely-spaced points and scale the figures to the same size, overlapping them at their centroids. One diagram, B, of a pair under comparison (A and B) is then rotated, so that the sum of distances (or squared distances) between every point on B and the *nearest* point on A is minimized. There is no analytical solution to this, because radially symmetrical objects would show several positions of low misfit (local minima) and the lowest of these would have to be found (the global minimum). This would require a succession of small angular rotations to find the global minimum, which should, as a rule, lead to the correct general orientation of the two figures (both skulls for example, facing the same way). There then remains the problem of fitting smaller areas of the diagrams to obtain homologies between the finer details, which might be carried out successively for smaller and smaller areas. In theory the end result of this process would give progressive distortions of the parts of the diagrams until every point on diagram B lay on top of the homologous point in A. The mathematics of performing the distortions would pose considerable difficulties in computation, though trend surfacing of the displacements between the points of B and the nearest points of A might offer a solution.

It is not yet clear which of the above approaches will be the most fruitful in having the greatest general application for taxonomy. Quite possibly an integrated approach employing several of these concepts will be developed. In the meantime, the numerical taxonomist wishing to proceed expeditiously in classifying groups of organisms may need to rely almost entirely on his intuitive appreciation of homologies when defining and coding characters. Other studies undoubtedly will be carried out whose main aim will be establishing homologies between characters or organized sets of characters. All such studies, regardless of the approach employed, if they are eventually to yield data matrices for numerical taxonomy, must employ some form of maximal criterion of similarity or correspondence to establish the homologies; in effect one must use a categorical concept of homology.

3.5   KINDS OF CHARACTERS

The philosophical may argue that it is not possible to make absolute measures of resemblance, because such measures would involve an arbitrary selection among the endless array of attributes which could in some sense be called characters of the organisms. Nevertheless, meaningful estimates of resemblance can be made once there is agreement on what characters are to be admitted as relevant in taxonomy.

It is, of course, quite impossible to give an adequate catalog of all the taxonomic characters that can be used in various groups. Such a catalog would comprise nothing less than a description of organized nature. Only specialists in the various groups will be in a position to define and describe unit taxonomic characters in the organisms they are studying. In their search for characters they ought to follow two guide lines, one of which at least is not included in customary taxonomic practice. First, use all kinds of characters from all parts of the body and from all the stages of the life cycle. Second, use all characters varying within the group studied, not merely conventional diagnostic characters. The latter are likely to be constant within the members of a given taxon. The exclusive use of such characters would prejudge the very issue—the establishment of taxa free from subjective bias—which numerical taxonomy wishes to solve. If the studies are based to a very large degree on characters previously described in the literature, there is some danger that diagnostic characters will be favored inordinately, since there is a historical weight in favor of diagnostic characters in the published literature. There is at least the suggestion of such a bias (Sokal and Rohlf, 1970) in the original set of characters describing the *Hoplitis* complex (Michener and Sokal, 1957), chosen by Michener from his published work and his recollection of the taxonomic relationships of this group. Reanalysis of the same data by various techniques has shown that the Michener and Sokal data matrix yields tighter clusters with wider gaps and fewer intermediate species than comparable taxometric studies of other organisms.

Taxonomic characters can be grouped roughly into

   a, morphological characters (external, internal, microscopic, including cytological and developmental characters),
   b, physiological and chemical characters,
   c, behavioral characters,

  d, ecological and distributional characters (habitats, food, hosts, parasites,
    population dynamics, geographical distribution).

This list is far from being exhaustive. Readers interested in greater detail may
wish to consult Mayr (1969b, p. 127), who has prepared a more complete classification and discusses each category in detail. Davis and Heywood (1963) may be
consulted on characters in botany and Skerman (1967) and Lockhart and Liston
(1970) for bacteriology.

From the above list of characters we single out chemical characters in group *b*
and ecological and distributional characters (group *d*) as requiring special
discussion.

In recent years there has been a great increase in work on chemical constituents
as taxonomic characters. Comprehensive references to the literature are given in
Blackwelder (1967a, p. 624) for animals, and by Hegenauer (1962–69) and B. L.
Turner (1969) for plants, and a recent symposium (Sibley, 1969) covers all these
groups. There are possible applications, too, to paleontological material (Abelson,
1957; Florkin, 1966). Reference was made in *Principles of Numerical Taxonomy* to
techniques that yield many characters in a single technical procedure, called poly-
phenic methods. These include chromatography, electrophoresis, spectroscopy,
and other techniques; they are now widely used and are reviewed in the references
cited above.

Although they may pose certain unusual problems in coding, *chemical characters*
are just as valid for taxonomy as any others. Despite a few claims to the contrary,
they seem on the whole no better and no worse than any other kind of character
for constructing natural taxa, discriminating between species, or other systematic
work. There are, however, two points about them that are apt to be misunderstood.
First, there is a prevalent tendency to assess the chemical resemblances intuitively,
and to compare chromatograms and the like by eye. Even when chemical resemblances are estimated directly, there is often no awareness that the next, and necessary, step is to search for taxonomic structure (for example by cluster analysis). We
discuss numerical taxonomic methods for chemical data in Section 5.12.

Second, there are two main classes of chemical information which, although
not always sharply distinct, have very different implications in taxonomy. Simple
chemical substances, enzymatic reactions, and some other sources of chemical
data yield relatively little information about the organism. They may be dependent
on a few metabolic properties and are commonly recorded as present or absent,
or given in quantitative terms. These are what Zuckerkandl and Pauling (1965a)
call "episemantic molecules." Other classes of biochemical compounds are quite
different in that their structures are highly complex, and, if known, yield a great deal
of information. The nucleic acid coding of the genome is one example, and protein
sequences is another: these are "semantic molecules" or "informational macromolecules." For this reason, methods that rely on the high information content of
a protein, for example, are far more informative taxonomically than the presence

or absence of a single simple chemical compound. These methods comprise protein sequence studies, serology, and nucleic acid pairing, and they are discussed below.

Crick (1958) foretold the advent of *protein taxonomy*, which is now developing swiftly. A compendium of present knowledge is the *Atlas* of Dayhoff (1969a), and a good simple account is given by Dayhoff (1969b). Functional proteins commonly consist of two or more different sorts of polypeptides held together by various forms of chemical bonding. Thus adult human hemoglobin consists of two Hb_α chains and two Hb_β chains, forming a tetramer. Most interest for our purpose lies in the sequence of amino acids in a given polypeptide. The human Hb_α chain contains 141 amino acids, and the sequence starts, Valine-Leucine-Serine..., and ends,.... Phenylalanine-Leucine-Alanine. There could be an enormous number of different polypeptides of length 141. In general there will be 20^n different polypeptides of length n.

The sequence of amino acids differs in the same functional protein of different animals. Thus human and horse cytochrome c differ in 12 out of the 104 amino acids. Protein sequence studies also have a unique power not possessed by any other taxonomic method; they allow estimation of resemblance between organisms that are exceedingly diverse. It would, for example, be hard to think of any reliable characters to use to compare man with yeast (morphology, simple chemicals, cytology, etc.). Yet the protein sequences of the cytochrome c of man and yeast are identical in 62 out of the 99 comparable positions, while the expected number for random sequences of amino acids is only about 7; indeed, it is even possible to detect some resemblance between cytochrome c of bacteria and man (Sackin, 1969). The number of identities and nonidentities between amino acids is therefore of great potential interest in many areas of systematics. The problem of character selection does not arise in its ordinary form (Section 3.6), for the sequence contains all the characters the protein possesses, at least if the primary structure alone is considered.

For a given kind of protein, such as hemoglobin, it is already evident that relatively small parts of the molecule have a fixed structure (i.e., in all organisms), presumably because of functional requirements. The folded structure of a protein appears to be determined principally if not entirely by the sequence of amino acids in the polypeptide chains. This sequence in turn is determined by the nucleotide sequence in the messenger RNA, and that is determined by the nucleotide sequence of the DNA.

It must, of course, be pointed out that the structure of one protein represents the fine structure of only one cistron. Whether or not the differences in any one protein are representative of the differences in the whole genome is a question that is taken up in more detail on the next pages and in Section 5.12, but the concordance is sufficient to make protein sequences a particularly interesting new field.

None of the chemical techniques discussed so far yields measures of taxonomic resemblance directly, though by counting similarities and differences in chemical substances they can be made to do so. The next two techniques discussed in this section are different and they give resemblance measures directly.

A technique rather different from those considered so far is *comparative serology*. It yields quantitative measures of taxonomic relationship, but does so through the production of antibodies in experimental animals (usually rabbits). The basic principle is of wide application: a protein of one organism will react strongly with antibodies against it, but the same protein from a different organism will react less strongly. For example, antibodies to horse albumin, made by injecting the albumin into a rabbit, will react strongly with horse albumin, but weakly if at all with albumin of cow or pig. These reactions can readily be quantified, and are well known to be as a rule congruent with phenetic similarity. The characters that determine these reactions are usually not known, for they consist of details (as yet poorly understood) of the structure of biological macromolecules, usually proteins. General considerations of the nature of these characters are taken up here, and the kind of relationship that is revealed by serological techniques is discussed in Section 5.12. General reviews of this field are given in Leone (1964) and Hawkes (1968).

The many small differences and resemblances in proteins can be thought of as a large sample of the features of the organism. The fine structure of the genes in the form of the nucleotide sequence is translated into the sequence of amino acids in the proteins, as has been discussed. This, in turn, is expressed in the serological reactions and is one of the major factors determining these (e.g., Arnheim, Prager, and Wilson, 1969). On the other hand it is clear that the sequence of one protein cannot be taken as a random sample of all the genetic features of the organism. There is, therefore, a danger that if we study serologically a single protein (this is desirable for technical reasons), we may in effect be studying the fine structure of only one gene, which may not be representative of all the genes. It may be noted in passing that the value of serology and protein sequence studies does not depend on the "conservatism" of the proteins, for this is not what is required either for phenetic or cladistic studies; if all birds had retained the same serum proteins as their reptilian ancestor, then avian serology would be uniformly uninteresting. In fact, the information we now have shows that evolution in protein sequences is quite congruent with evolution in other respects, and also with phenetic taxonomic relations. This is supported by other evidence; the well-known correspondence between orthodox taxonomy and serology is one piece of evidence; another is the congruence between serological resemblances based on different kinds of proteins (albumins, globulins) and between proteins of different stages of the life cycle (Boyden, DeFalco, and Gemeroy, 1951; Wilhelmi, 1940; Spiegel, 1960; Marable and Glenn, 1964; Mohagheghpour and Leone, 1969).

At present we still do not have very much evidence on whether the many protein sequence differences have approximately equal weight in determining the serological results, but it is likely that this is true to a certain extent, although some physicochemical features are essential to antigenicity. Therefore there is good reason to suppose that comparative serology should generally yield the same conclusions as numerical taxonomy, provided the sequence differences of the proteins used for serology are fairly representative of the whole genome. One might speak of it as a method for estimating resemblance in which the immunized animal acts as the computing machine when it produces the specific antibodies.

It may sometimes be possible to break down serological data to give antigenic formulas for the different antigenic factors. Where this can be done, these factors can be included like other characters in numerical taxonomic analyses (Lockhart and Holt, 1964), but more often the serological similarities must be treated as resemblance coefficients (see Section 5.12).

There seems no likelihood that any of the methods mentioned above will prove to be an adequate *sole* basis for taxonomy. To qualify as such, a method would have to reflect accurately the entire phenome of a given level of complexity (see Section 3.6). The last of the chemical methods discussed here—that using nucleic acid pairing—attempts to do just that; its future development will therefore be watched with great interest.

The *nucleic acid pairing technique* (often called hybridization) was pioneered by Doty, Marmur, Eigner, and Schildkraut (1960), and technical advances were made particularly by McCarthy and Bolton (1963). The "characters" are the nucleotide bases of the nucleic acids (most often the DNA of the genome) in their specific sequences. The resemblances thus obtained are discussed in Section 5.12.

Biochemists have sometimes argued that their characters are superior to morphological ones because they are closer to the genotype, and that chemical data can be arrayed in the order: DNA pairing, protein sequences, serology, and simple chemicals as gene products; each is increasingly removed from the nucleotide sequence of the genome. This argument is attractive on theoretical grounds, but we do not believe it should be emphasized much in practice. We still know little about the fidelity of expression of the genotype in chemical or morphological systems; technical problems (including the choice of resemblance measures) can vitiate the results and the adequacy of the techniques can only be assessed against other phenetic criteria; the biochemist may often only sample very small parts of the genome. Also, as we note in Section 3.6, classifications of genomes may not necessarily be the ideal of all systematics.

The factors of *ecology* and *distribution*, which have become much emphasized in recent years, require some comment in connection with their use for numerical taxonomy. While they are regularly reported, when known, they are not too frequently employed for classificatory purposes. Some difficulty may be encountered in coding them for numerical taxonomy. When an ecological character expresses some sort of gradient, such as life zones in mountainous areas, depth in soil, temperature maxima, and other gradients, multistate coding is straightforward. However, how are we to code phytogeographic or zoogeographic distributional characters, host plant preferences, or parasitic fauna? With distributional data, a two-dimensional breakdown into two characters is sometimes possible. In the other cases the information may have to be partitioned into a number of two-state characters. Thus, where several species or genera of hosts occur for a group of parasites under study, each of the former may have to be a single two-state character marked "present" when parasitized, and "absent" when not. We may, however, wish to express systematic relations among host plants by appropriate coding. If a group of parasites lives on four hosts, species **A** and **B** of genus **X** and species **C** and **D** of genus **Y**, we could have one character for genus **X** or **Y** and one for species (**A** or **B** and **C** or **D**) for each of the two genera.

Geographic distributions are characters that may need to be used with caution. In most cases it is not possible to be sure that they represent any character in the genotype. Similar care is needed with many ecological, behavioral, and parasitological observations. For example, parasitological characters may sometimes depend on chance infestation and not on the genotype of the host or of the parasite. Thus, attempts to use the Mallophaga in classifying birds, for example, are made difficult by doubts as to whether some mallophagans are stragglers from other birds or have quite recently become established on their hosts through cross-infestation (for example, see Clay, 1949, and the discussion of Hennig, 1966 pp. 107, 175). Sometimes, however, there is remarkable congruence between phenetics of hosts and parasites, and Kistner and Pasteels (1970) describe an instance. Characters such as host specificity also pose the difficulty of a decision about how they should be scored. Thus some viruses are restricted to one species of mammal, while others attack both birds and mammals. One might decide to score class specificity, ordinal specificity, familial specificity, generic specificity, and so on, and to consider that higher categories should contribute more weight than lower ones. Since we have little detailed information on this problem, and hence no satisfactory method of allocating weights, we would suggest that only a few such characters be used and that each should be given equal weight. Although this may reduce the information, it will also avoid introducing bias. Ibrahim and Threlfall (1966a) give an example of such a study on fungi. Rather similar work has been done by 't Mannetje (1967b, 1969), and by Colwell, Moffett, and Sutton (1968), where susceptibility to the root nodule bacterium *Rhizobium* and to bacteriophages, respectively, were used as characters for numerical taxonomy. As we note in the next section the question of whether characters are genetically determined may not be so important as has been thought.

Little need be said about avoidance of bias in choosing the characters. It is clear that when we use only a set of characters known to show resemblance between certain groups, the similarity coefficients that will result from this study will reflect that choice. In an extreme case, by choosing only those characters that were the same in two organisms, one would obtain perfect but spurious resemblance between them. A systematic survey of all known characters, or the inclusion of all characters the investigator has been able to observe, should prevent bias of this sort. Almost every new technique in biology gives new characters that can be employed in systematics. These new characters must be incorporated into the existing body of taxonomic data, and it is our belief that only numerical taxonomy can adequately do this.

3.6   CHOICE OF CHARACTERS

Having defined unit characters in taxonomy in Sections 3.2 and 3.3, and having surveyed the kinds of taxonomic characters in Section 3.5, we now proceed to the difficult question of which characters should be chosen as a basis for estimating the similarity between OTU's. In attempting to answer this question, we encounter two relevant problems: (1) What biological factors do the characters represent? (2) Are all characters of equal value and information in providing evidence on phenetic similarity?

Implied in the first problem was the hope during the early work on numerical taxonomy that taxonomic characters would provide information on the genetic factors differentiating OTU's. Although numerical taxonomy measures similarity between the *phenomes* of OTU's (Soulé, 1967b; defined by us as *the total phenetic manifestations of the genome of an organism or a taxon*), it was hoped that this would also lead to estimates of similarity between the genomes of these OTU's. In this connection we developed the *nexus hypothesis* which assumes that every phenetic character is likely to be affected by more than one genetic factor and that, conversely, most genes affect more than one character. The result is a complicated nexus of cause and effect. Any character should give information about several genes and it should be possible in general to pick up the effect of a given gene through any one of several characters.

While the nexus hypothesis is undoubtedly true in a general way (for detailed evidence see *Principles of Numerical Taxonomy* or general references to physiological and developmental genetics), we have recently felt that it has lost some of its relevance for phenetic taxonomy. The distinction between phenotype and genotype has become rather vague in view of recent insights into the nature of fine genetic structure. We have already suggested in Section 3.4 that the complexities—at many levels—involved in translating the genetic code into the final elaborations of external and internal morphology, physiology, and behavior are of such an order of magnitude that it is probably futile to hope that these can be understood in the near future. Not only are these interactions complex, but they also involve switch-mechanisms so that part of the genome of an organism is not functional at any one life history stage. What one wishes to measure in phenetic taxonomy is the *expression of the genome* of the organism through its life history—its phenome, in fact. Realizing this, it becomes less important from our point of view to know to what degree the phenome reflects the genome. We remain saddled with the problem of having to investigate different levels of organized complexity, but for a general taxonomy all known levels should be utilized. For special classifications, external characters or biochemical ones or others might be preferred.

In discussing which characters to employ in a given taxonomic study, we earlier postulated the *hypothesis of nonspecificity*, which stated that there are no distinct large *classes* of genes affecting exclusively one class of characters, such as morphological, physiological, or ethological characters, or affecting special regions of the organisms, such as head, skeleton, or leaves. If this assumption were warranted, then obtaining a disproportionately large number of characters from one body region or of one special kind would not restrict our information to a special class of genes. Furthermore, there would be no a priori grounds for favoring one character over another.

The nonspecificity hypothesis has been shown to be only partially correct. Identical classifications are not produced from different sets of characters for the same OTU's. Measurement of the agreement between two classifications can most simply be done by correlating the matrix of similarity coefficients based on one set of characters with the matrix based on the other set of characters. This is one version of the method of cophenetic correlations, which we now prefer to call matrix correlations (Section 5.10). The resulting coefficient $r_{S_1 S_2}$ between two similarity matrices (or, occasionally, $r_{C_1 C_2}$ between two phenograms) estimates the congruence between classifications implied by the similarity matrices (or phenograms). By *congruence* we mean the *degree of correspondence between arrangements of OTU's in a classification*. Identical classifications are perfectly congruent. Tests of congruence may be made at a variety of organizational levels. Thus, we may look for the congruence between classifications based on different organs or regions of the body. Examples would be classifications based on characters of the brain contrasted with those based on characters of the intestinal tract; or characters of the epithelium contrasted with those of the connective tissue; or head versus body or wing characters. Other measures of congruence could be for dimorphic or polymorphic manifestations of an OTU, such as between classifications based on females or males, or those based on diverse life history stages, such as larvae, pupae, and adults in insects, or different castes in social insects, or different adult forms in cyclomorphic organisms such as aphids.

On the whole, classifications based on separate sets of characters—be these from organs or life history stages—agree partially. In vertebrate paleontology and systematics of recent vertebrates, classifications based on skeletons—or even portions of skeletons—are frequently congruent with those based on other parts of the organism's anatomy. It is well known that morphological evidence from a newly investigated organ system frequently confirms previous classifications.

Although the problems of incongruence have been fully appreciated by taxonomists for a good many years (Hennig, 1950, and Remane, 1956 give good reviews of this), it has only been through numerical taxonomy that quantitative estimates of congruence have become possible. Table 3-1 summarizes a selection of measures of congruence obtained to date, listing the organisms, the numbers and types of characters, and the measures of congruence expressed as matrix correlation coefficients. It will be noted that there is a considerable range of these co-efficients, from quite low values (classifications essentially independent) to fairly high ones.

Rohlf (1965) has applied a randomization test, such as are described in Sokal and Rohlf (1969), to test the significance of matrix correlations for three sets of data reported in Table 3-1. Rohlf's results indicate that correlation coefficients between matrices based on subsets of characters from different body regions or life history stages (with the exception of that between distance matrices in the butterfly data by Ehrlich and Ehrlich, 1967) are significantly lower than a random allocation of characters to each of two subsets would produce. Similar findings were obtained in a preliminary study of this point (Sokal and Sneath, 1963, p. 89; and by others). It is therefore unlikely that the subsets of characters that have been compared in these measures of congruence described identical taxonomic relationships. Readers will notice that the estimate of the congruence is in part dependent on the kind of similarity coefficient (correlation, distance, or other coefficient) on which the matrix correlation is based. This aspect of the matter does not relate directly to the problem of congruence but to the different nature of these coefficients, which will be discussed in Section 4.7.

On theoretical grounds, as well as from the results obtained so far, it would appear that congruence will be greater the higher the rank. Thus, for example, there is little doubt that the orders of insects would be as faithfully reflected in their larvae as in their adults, or that the classes of vertebrates would be recognizable from their skulls, their pelvic girdles, or their circulatory systems, and that these systems would yield roughly the same classifications. However, at lower taxonomic ranks this may no longer follow, and phenetic classifications may differ when based on sets of characters responding to differing environmental challenges. In a quantitative study of two cyclomorphic adult forms, Sokal and Thomas (1965) found that alates and stem mothers of the aphid *Pemphigus populi-transversus* share only one common factor out of five contributing to interlocality covariation (among local populations within a portion of the range of the species). Thus these organisms show little congruence. Larvae and adults of the same species will also frequently not cluster congruently.

The reasons for the lack of congruence may be twofold. Certain genes appear to be active only at specific times in the life history of an organism (Beerman and Clever, 1964). The lack of congruence may also reflect different adaptational patterns and evolutionary rates for those genes active during the development of a particular life history stage.

Factor analysis (for a detailed account, see Section 5.6) has important implica-tions for choice of characters. Factor analysis of a matrix of correlations among a suite of characters aims to extract the k common factors affecting the suite. Thus the covariational pattern of n characters can be expressed in terms of k factors, where $n > k$. It might therefore be argued that rather than choose n characters for an estimate of phenetic similarity, phenetic similarity should be based upon the k factors, using such characters as are necessary to evaluate the factor endowment of each OTU. There are two problems with this approach, one practical, the other theoretical. The practical problem is that in order to carry out a factor analysis one needs to have scores and correlations of the characters. Factor analysis therefore cannot really save effort, because one must record and analyze the characters before a factor analysis. Similar considerations also apply to finding clusters of highly correlated characters (which also have a high information content in a special sense; see Bisby, 1970b) with the aim of using only one character from each cluster in a subsequent analysis, and this approach carries additional dangers not possessed by factor analysis (Sneath, 1967b). Thus, unless there are theoretical advantages in estimating phenetic similarity from factor scores, it might be best to stick to the whole set of characters.

Theoretically, factor analysis seems attractive inasmuch as it reduces the number of phenetic dimensions necessary for the visualization of the organisms. But if we found that from a set of n (say 100) characters, k (say five) important factors have been extracted, should these factors be equally weighted in measuring the similarity between OTU's? There is no simple answer to this question. Should factors be weighted in terms of the number of characters they affect? This would be very difficult to evaluate. Should they be weighted in terms of the amount of variation they engender? This might be measured by the sum of squared factor loadings. Or should each factor be equally weighted regardless of its "strength" in affecting the characters? The first two approaches might yield estimates of similarity more or less approaching the intuitive appreciation of resemblances by conventional taxonomists. The third solution might yield estimates quite different from the intuitive measures, but possibly closer to the "refined similarity" produced by systematists who carefully analyze their impressions and discount the overwhelming effects of one or the other factor, such as general size, hairiness, calcification, or similar general adaptational trends. We are not prepared to say that the best estimate of similarity is the one that most closely approximates the concept as interpreted by conventional taxonomists. Rather, it should be justifiable on its own terms. More research is needed to develop an independent criterion of weighting factors.

Finally, it may be possible to use factor analysis economically to determine whether more characters should be measured. Techniques could be developed that would test additional characters to find out whether they belong to factors already known and measured or whether these new characters represent factors that have not as yet been evaluated and employed for estimating overall similarities. Sokal and Rohlf (1970) have made some experiments on these lines, and one of

TABLE 3-1

Results of studies of congruence between different sets of characters

Organisms	Comparison			Correlation coefficient between		Source of data
	Sets of characters	Number of characters in the respective sets	Similarity coefficient	Listed comparison	Randomized sets of characters	
Different body parts or systems						
Species of *Hoplitis* complex (bees)	Head vs. nonhead	60:62	*r* *d*	0.61 0.33	0.81* 0.64*	Michener and Sokal (1966), and Rohlf (1965)
Genera of butterflies	Head vs. thorax	38:105	*r,* *d*	0.79 0.68		Ehrlich and Ehrlich (1967), and Rohlf (1965)
	Head vs. abdomen	38:27	*r* *d*	0.37 0.28		
	Thorax vs. abdomen	105:27	*r* *d*	0.35 0.37		
	External vs. internal	100:96	*r* *d*	0.69 0.76	0.79* 0.78*	
	External vs. musculature	100:75	*r* *d*	0.70 0.76		
Species in two drosophilid genera	Head vs. thorax	39:51	*d*	0.45		Bächli (1971)
	Head vs. abdomen	39:62	*d*	0.25		
	Thorax vs. abdomen	51:62	*d*	0.30		

Species of mite genera *Dermanyssus* and *Liponyssoides*	Gnathosome vs. nongnathosome	15:120	r d	0.29 0.82		Moss (1968b)
	Dorsal idiosomal vs. nondorsal idiosomal	45:90	r d	0.35 0.80		
	Ventral idiosomal vs. nonventral idiosomal	40:95	r d	0.52 0.85		
	Leg vs. nonleg	35:100	r d	0.53 0.71		
Species of mosquito genera *Aedes* and *Psorophora*	Leg vs. nonleg	67:91	r d	0.42 0.29		Hendrickson and Sokal (1968)
Genera of gallinaceous birds	Wing muscles vs. leg muscles	82:93	difference scores assigned by investigators	0.62 0.67		P. H. A. Sneath (unpublished), Sokal and Sneath (1963) based on Hudson, Lanzillotti, and Edwards (1959), and on Hudson et al. (1966)
	Leg muscles vs. foot muscles	77:56	S_{SM}	0.72	0.83†	
	Extensor muscles vs. flexor muscles	52:81	S_{SM}	0.79	0.82†	
Genera and species of Lari (birds)	External vs. skeletal	72:51	r d	0.73 0.43		Schnell (1970a,b)
Species of angiosperm genus *Sarcostemma*	Floral vs. vegetative	61:32	r d	0.17 0.23		Johnson and Holm (1968)

(*continued*)

TABLE 3-1 (continued)

Organisms	Comparison		Similarity coefficient	Correlation coefficient between		Source of data
	Sets of characters	Number of characters in the respective sets		Listed comparison	Randomized sets of characters	
Species of angiosperm genus *Salix*	Vegetative vs. sexual	72:125	$1 - d$	0.29	0.86†	Crovello (1969)
Strains of bacterial genus *Chromobacterium*	Morphological vs. physiological and chemical	29:76	S_J	0.61	0.75	Sneath (1972)
Different life stages — Species of *Hoplitis* complex (bees)	Males vs. females	53:69	r / d	0.71 / 0.35	0.81* / 0.67*	Michener and Sokal (1966), and Rohlf (1965)
Species of mosquito genus *Aedes*	Adults vs. larvae	77:71	r / d	0.29 / 0.59	0.60* / 0.75*	Rohlf (1963a, 1965)
Species of angiosperm genus *Salix*	Male sexual vs. female sexual	60:65	$1 - d$	0.80	0.82†	Crovello (1969)

NOTE: The table shows values for the matrix correlation coefficient $r_{S_1 S_2}$ between the similarity values in matrices S_1 and S_2 derived from pairs of character sets. Different types of similarity or dissimilarity coefficients were employed. The $r_{S_1 S_2}$ value is dependent on the numbers of characters; therefore its approximate value for random sets of characters is shown where published (or, if the value is marked with a dagger, it has been estimated by interpolation from values for slightly different numbers of characters). This dependence is revealed in the tendency for comparisons involving small numbers of characters to give low correlations; the difference between two values of $r_{S_1 S_2}$ is a better indication of the magnitude of incongruence than is the value of $r_{S_1 S_2}$ for the listed comparison alone. One would not expect higher correlations than those found for the randomized sets.

*Means computed from Table 2 in Rohlf (1965).

†Values estimated from the assumption that the value of $1 - r^2$ from randomized sets of n_1 and n_2 characters is approximately the square root of the sum of components proportional to $1/n_1$ and $1/n_2$ respectively.

their findings was that of three persons who chose a set of characters, one had made a selection that left three important factors unrepresented (see also Section 3.8).

In view of the partial lack of congruence it is important that characters should be distributed as widely and evenly as possible over the organisms studied. Colless (1969c) advocates a deliberate policy of stratified sampling from various organs to ensure this. He believes (personal communication) that in this way a more "switched-on" genome is sampled and that it is possible to minimize the functional, coadapted correlation in this manner, increasing the content of independent information. A somewhat different issue is whether to use published data from standard monographs. Crovello (1968g) believes that they are more reliable than is usually thought, but in common with most taxonomists, we would view such data with suspicion unless they had been collected with great attention to their comparableness, accuracy, and completeness. We draw the reader's attention to a recent review of the subject by Farris (1971).

3.7 INADMISSIBLE CHARACTERS

The proper selection of characters is clearly a critical point in the application of numerical taxonomy, as it is in other taxonomies, and misunderstandings have arisen on this score. There are, however, certain kinds of characters whose nature clearly disqualifies them from employment in a numerical taxonomic study. These are listed in the present section as inadmissible characters.

Meaningless Characters.   It is undesirable to use attributes that are not a reflection of the inherent nature of the organisms themselves. For this reason taxonomists do not include the names or numbers given to specimens, nor do they employ characters whose response to the environment is so variable that it is not possible to decide what is environmentally and what is genetically determined. The number of leaves on a branch of a tree may be an example of the latter, though if acceptable evidence is forthcoming that this number is relatively constant in a species it might be admissible. However, characters affected strongly by environmental influences would be appropriate if numerical analysis were to be made of the effect of environment on phenetic relationship (Section 7.4). Similar cases may occur where other special kinds of investigation are planned. This is a matter of scientific judgment, not simply of taxonomic method, and each case must be treated on its merits.

Logically Correlated Characters.   We must exclude as redundant any property that is a logical consequence of another. We cannot use both presence of hemoglobin and redness of blood if the latter is defined as possession of hemoglobin. Mathematical manipulations that constitute logical consequences should be avoided : for example, we could not employ both the length and half the length of an organ, or the radius and the circumference of a circular structure. Similarly, characters that are tautological—those that are true by definition as well as those that are based on properties known to be obligatory—should not be included. An example of tautology is to score both tallness and height of a man. An example of a character that is true by definition is to score "presence of calcium in raphides" after having scored "raphides composed of calcium oxalate," a substance that by definition contains calcium. To score in this instance "raphides insoluble in acetic acid" would also be scoring a character that reveals a known and invariable property of calcium oxalate, though it is not part of the definition of that compound ; if this is known, the property must be omitted. In making these qualifications for admissibility we are fully aware that many or most of the "inadmissible" characters would be inadmissible on more than one count. Thus, if we use two tautological characters, we would find on examining our data after they had been prepared for machine computation that the two characters are perfectly correlated. According to our rules on empirical correlations (see below) there would be much suspicion about using both of these characters. It is quite likely that we would therefore reject one of them by the empirical correlation criterion.

Partial Logical Correlations.   Many cases will arise where the dependence of one character upon another is not total but only partial. Cain and Harrison (1958, p. 89) illustrate this by an example:

> Degree of melanization of the skin in mg/sq cm must not be used together with skin colour estimated by some colorimetric method if the melanin is making a contribution, which is some function of its own density, to the skin colour, unless this contribution can be subtracted from the measurement of skin colour.

We would recommend the following procedure in cases of partial logical correlations. When a character *2* depends in part upon another character *1*, the decision whether to employ *2* as well as *1* will depend on the nature of the factors other than *1* that affect *2*. If, to the best of our knowledge, these factors reflect heritable variations, we would include *2*. But if these factors represent experimental or technical error or are otherwise unaccountable, we would not use character *2*. If it is possible to partition the variation into heritable and other components it may be justifiable to use the heritable component as a character. Hall (1969b) suggests a method for reducing correlation of this kind if some estimate of this can be made. If, for example, a series of n' measurements are made of leaf widths at different places along the leaf it is obvious that these n' characters are partially correlated. Hall suggests that the contribution of character i to a resemblance coefficient should be multiplied by a weight w_i, which equals $[1 + h(n' - 1)]/n'$. Here h is an estimate of "homological indistinctness" (of the independence of the characters), varying from 0 (completely correlated) to 1 (quite independent).

Invariant Characters. We would exclude characters that are invariant over the entire sample of OTU's. To include them would not add any information about resemblances among the OTU's. Employment of invariant characters would either not affect similarity coefficients, or else would induce simple transformations of them. However, there may be instances, notably in bacteriology, where established techniques prescribe a list of tests to be performed, which thus yields invariant characters, and with protein sequences the sequence comprises the entire character set. It may be argued that from the point of view of obtaining standardized results, all characters should be included in the computation; however, since all coefficients obtained are only relative quantities, we would recommend that for most work invariant characters not be included.

Empirical Correlations. How should we decide if two characters not logically related, but highly correlated empirically, are to be counted as separate unit characters? It is possible to give extreme examples that are absurd. Thus it is observationally true to say that certain avian characters are invariably associated, and likewise certain mammalian characters. Should we attribute this effect to a single character in which birds and mammals differ—a gene, perhaps, which if mutated would turn a bird into a mammal at one jump? Clearly, we would here prefer to postulate many independent genes, and we would treat these features as independent despite the strictest correlation. In still other instances we would not assume independence so easily. The close correlation between white skin and pink eyes of total albinos in most vertebrates would be counted as a single character, since the total absence of pigment implies lack of retinal pigment.

Yet it remains true that we often need to postulate independent characters even in cases such as the albinos, for occasional albinos do have some retinal pigment. The same is true of most other apparently dependent associations. Any exception will suffice to prove that more than one character is involved. Even strictly functional associations are not as dependent as they seem at first sight: the need for the teeth to meet is only true for a species as a whole; aberrant individuals can and do occur. The fact that selection keeps two characters (the position of the top teeth and the position of the bottom teeth) in close correspondence does not necessarily imply unitary causation of these characters. In coding such a species for analysis we would employ two characters in spite of their stringent empirical correlation in the material at hand.

In serially homologous structures such as segments of an annelid or appendages of an arthropod, or in generally homologous structures such as hairs on the body surface, a character affecting equally all the members of the series could be subdivided into separate characters for each member. However, no new information would be brought about by such a procedure. In such a case we would employ only one character.

In summary, when we have evidence that more than one factor affects two correlated characters within a study, regardless of whether this evidence comes from within the study or from outside, we would include both characters; otherwise we would employ only one. Our position is that we assume at least some independent sources of variation in any empirical correlation, unless we have reason to believe otherwise. This would err in the direction of redundancy, but it would be counterbalanced by the likelihood of obtaining new information.

3.8   THE REQUISITE NUMBER OF CHARACTERS

An important question in any numerical taxonomic study is what number of characters is required to obtain stable classifications. In spite of the large amount of work in numerical taxonomy, we still are unable to provide generally valid answers to this question. An early recommendation that no less than 60 characters should be used, whenever possible, and that if at all feasible, considerably more characters should be employed still seems reasonable, although as shall be seen, we cannot justify this requirement on either empirical or theoretical grounds. What is the correct number of characters is related to the problem of the congruence of classifications based on sets of characters from different body parts or life history stages (discussion in Section 3.6). Had the hypothesis of nonspecificity been fully valid, any set of characters would lead to sample estimates of a parametric similarity value and the question of number of characters would simply be a statistical one: what is the requisite number of characters to be sampled to obtain an estimate of similarity with confidence bands of desired width and at a desired probability level? But since it appears obvious that different sets of characters will yield somewhat different phenetic information, we cannot dispose of the problem so simply.

An early consideration has been shown to be of little utility. This was the hypothesis of the factor asymptote in which certain assumptions were made about the number of genetic factors underlying the expression of a taxonomic character and estimates were made of the number of genetic factors that could be sampled with a sample of taxonomic characters. On any reasonable assumption it was found that relatively little of the genotype would be sampled, even if a substantial number of characters were recorded. The details of the argument are furnished in *Principles of Numerical Taxonomy*. Since we are now more concerned with the phenome rather than the genome (see Section 3.6), the hypothesis has lost much of its theoretical relevance. In view of our discussion in Section 3.6 of phenetic factors as these relate to congruence, the hypothesis of the factor asymptote might be resuscitated in a new guise. Is there an asymptote for the number of phenetic factors discovered as one increases the number of characters in a study, and especially as one proceeds from one class of characters to another? Attempts to answer this question are fraught with considerable experimental and computational difficulties, but sooner or later we should be obtaining evidence on this point.

We earlier postulated the matches asymptote, which assumed that as the number of characters sampled increases, the value of the similarity coefficient becomes more stable. This was thought to be so because one can express the resemblance between OTU's as a proportion of characters agreeing (matching), out of the total number being compared, and assume that the similarity between the two operational taxonomic units is an estimate of the parametric proportion of character matches. Such a single, definite proportion of matches (if we were able to sample all the characters) might, for example, be the matches in the nucleotide sequence of the DNA of these OTU's. If such an hypothesis were tenable, then the requisite number of characters could be simply computed from ordinary sampling theory in statistics and would need no special defense. However, it seems doubtful to us at this time that there is a single parametric measure of similarity between OTU's. It might be argued that a measure of the parametric similarity between the DNA's of the OTU's or between protein sequences could be developed. In fact, the coefficient of cross-association (sliding matches, see Section 4.4) attempts to do just that. Clearly, difficulties will be encountered by problems of nonhomologous regions, duplications, deletions, and the like. But at least theoretically we can conceive of such a measure. However, in view of our considerations about the complexity of the interactions that lead from the genetic code to the manifestations of the phenome, we doubt whether the parametric similarity between the DNA (even if it can be estimated) is a useful measure of the similarity of the organisms at the various levels of complexity we may wish to study. Yet, if phenetic similarity is not a single quantity but a shifting concept depending on the method of measurement as well as the character base, is there any hope of arriving even at an educated guess for the requisite number of characters?

Fortunately, an empirical fact may help us here. When large numbers of characters are measured, the estimate of similarity obeys what might be described as a principle of inertia. As more and more characters are added, it takes an increasingly large number of characters with quite different phenetic information to alter appreciably a given estimate of phenetic similarity. Thus, while classifications of the same OTU's based on different sets of characters might start out as different constellations in phenetic hyperspace, they would eventually converge toward the same general region, though they would not necessarily be identical as more characters were being added to the system.

The problem with an experimental test of any of this work is that it is quite difficult to obtain randomly chosen additional sets of characters. This is certainly true when the same investigator wishes to obtain additional sets of characters. His past experience in obtaining the first set of characters would undoubtedly bias him either toward avoiding characters of the sort that he had previously recorded or, on the contrary, might influence him into finding more characters like them. One way to carry out such an experiment without bias would be to have a second investigator find a number $n' > n$ characters in the hope that at least $n' - n$ new characters would be described (and hopefully more, since it is unlikely that the same n characters would be found by the second investigator). An added difficulty is that different investigators will define identical and similar characters in different ways, and then the use of some analyical technique is required to define common "pure" characters in the two character matrices produced by the two investigators. This is a task that factor analysis might usefully perform. In the only such study to date, Sokal and Rohlf (1970) found that of 17 common factors in a study of 25 species of bees of the *Hoplitis* complex, 16 had been found by the experienced taxonomist who first classified them on the basis of 119 characters. The "least traditional" taxonomist found 14 common factors based on 53 characters, while a third inexperienced person found all 17 common factors based on 62 characters. Thus it is obvious that the results would have been largely the same regardless of the order in which the characters had been added. If the addition of further characters followed the pattern illustrated here, we would not expect the taxonomic structure to change appreciably. An interesting verification of the asymptotic approach to taxonomic stability as the number of characters increases is shown in an analysis of human populations (Jardine, 1971).

The problem could be investigated by a simulation experiment. A variable number k of character classes could be generated each containing n_i characters ($i = 1, 2, \ldots, k$; and $n_1, n_2, \ldots, n_k$ not necessarily equal). Each suite of n_i characters could have its own distinct correlational pattern, with some of the characters of each class i being correlated with characters in the other classes. The distribution by which this "spillover" into other character suites could be governed might be described by two distinct models, one a J-shaped curve similar to a Poisson distribution with a low mean, the other a uniform distribution. One could then investigate to what degree the classification of t OTU's changes as one moves from one suite of characters to another. Thus, one might try the first suite of characters n_1, compute a similarity matrix among OTU's, and then compute a similarity matrix based on the n_2 characters, as well as one on the $n_1 + n_2$ characters. Will the matrix correlation of successive similarity matrices due to additional character sets become stabilized and very high? What would be the effect of the number of characters and the distribution of characters over the character sets?

Until we have answers to these questions, and also indications of whether this model has relevance to the actual conditions occurring in nature, an answer to the question of how many characters to use is quite difficult. Caution would dictate that we consider the similarity to be open ended, hoping that its inertia will stabilize overall similarity values in a bounded region of phenetic space. The practical advice that can be given at this time is to take as many characters as is feasible and distribute them as widely as is possible over the various body regions, life history stages, tissues, and levels of organization of the organisms. Since congruence is always less than is expected from random samples of characters, the number of characters used will set a lower limit to the confidence levels of the similarity coefficients. The investigator should therefore employ at the very least as many characters as will give the confidence limits he wishes (see Section 4.10). Since we believe that there will be correlation between suites of successively taken characters and that the phenomenon of inertia will exert a stabilizing effect on similarities, we do not feel as strongly as Ehrlich and Ehrlich (1967) that overall similarity cannot be dealt with in practice.

3.9   THE PROBLEM OF CHARACTER WEIGHTING

The problem of weighting characters is apt to give rise to misunderstanding, and we have therefore attempted to clarify this. Numerical taxonomists are generally in agreement in giving each character equal weight when creating taxonomic groups, although Burtt (1964) has pointed out that numerical taxonomy need not necessarily be based on equally weighted characters: he calls such equally weighted classifications *isocratic*. We ourselves believe that equal weighting is desirable, and discuss here its justification.

We should first emphasize that we are not here discussing the use of characters in identification. After a manner of speaking, "weighting" is used in such a procedure and properly so. However, the construction of taxonomic keys and the identification of specimens belong to a later stage of taxonomic procedure subsequent to the formation of the taxa concerned. We are therefore discussing *a priori weighting*, before a classification is commenced, and what we feel is objectionable is to presuppose knowledge that is not yet available, either about the classification of the organisms, or about the presumed significance of their characters.

We also emphasize that we are not advocating equal weight for character complexes, such as flowers or leaves. These are broken down into their unit characters, and hence effectively receive weights in proportion to their complexity or information content. Unit characters should quite appropriately receive unit weight. Expressed thus, the problem does not appear so illogical.

Kendrick (1964, 1965) has suggested that characters should be weighted depending on the "rank" of the organs which they describe. Thus, for example, if two OTU's **a** and **b** have leaves with several "subcharacters" such as "smooth–hairy, simple–compound," etc., while OTU **c** has no leaves, a situation could arise wherein **a** and **b** would differ in the entire set of subcharacters, agreeing only in the character "possession of leaves." By contrast either **a** or **b** could be compared with **c** only in the character "possession of leaves" for which they would differ. Thus the similarity between **a** and **b** with respect to leaves might be less than that between either of the two and **c**, a situation Kendrick feels is undesirable. OTU's **a** and **b** would appear to be more dissimilar than either is from **c** simply because the dissimilarities between **a** and **c** cannot be specified and the subcharacters for **c** would have to be left unscored. However, we believe that if we were to take a representative sample of characters from the entire organism, it is unlikely that we would get so distorted a picture of the dissimilarities. In fact by some methods of measuring resemblances, **a** and **b** would be no more dissimilar than either is to **c**.

To support this point of view let us take the absurdly extreme case wherein **a** and **b** are two flowering plants with different leaves and **c** is a fungus. It seems disturbing to find that **a** and **b** are more dissimilar with respect to leaf morphology than **a** and **c**, which differ simply in their possession of leaves. We are disturbed by this disproportionate dissimilarity because we recognize the fungus as being so very different from the flowering plants. But if we include the entire repertoire of morphological, histological, and physiological characteristics possessed by the angiosperm OTU's and the fungus it is obvious that the great similarity of the former to each other will emerge in contrast with their marked dissimilarity to the fungus. By taking only the characters of the leaves we are selecting a very specific fraction of the phenome. This point also relates to how one should treat inapplicable comparisons, and is a point returned to in Section 4.12.

In addition, it should also be made clear that we are not here discussing the mathematical weighting attendant on different coding and scaling procedures (Section 4.8). "Weighting" produced that way is directed to the specific purpose of the classification and the estimation of different components of phenetic resemblance, such as the removal of a general size factor or the kind of taxonomic structure we wish to evaluate. It is of course true that characters not employed are automatically given zero weight, but this is inevitable by any procedure. It does not justify arbitrary or irrational weights for those that are employed.

That every character should be given equal weight is implicit in the work of Adanson and the writings of Gilmour (1937, 1940, 1951) and Cain and Harrison (1958); and it was stated explicitly by Sneath (1957a) and Michener and Sokal (1957). Sneath reached this conclusion on considerations stemming from Gilmour's work on epistemology. Since natural taxa ideally contain the greatest possible content of implied information, this can only be measured in the number of statements that can be made about its members, which is independent of how important we may think any statement is. This argument has been developed at some length in connection with the "general" nature of natural classifications (see Sneath, 1958). Michener and Sokal (1957) concluded that even if desirable, there is no rational way of allocating weight to characters and therefore one must in practice give them all equal weight. In addition, when many characters are employed, the statistical analysis of similarity is only slightly affected by weighting some characters (unless this weighting is extreme). Thus Moss (1968b) noted that an accidental error, which weighted 61 character states of 18 characters nearly a thousandfold, nevertheless had a negligible effect on the resemblance coefficients in a study based on 135 characters and 17 OTU's. It is perhaps worth reemphasizing this point, for if in practice the measures of overall similarity do yield substantially the same results when based on many characters, whatever the weighting (within reason) of individual characters, it would seem unnecessary to argue the point further. It would be useful if further studies could be made of the extent to which arbitrary weights affect taxonomy, in order to provide some guidelines for avoiding excessive distortion.

Some other suggestions have been made for weighting. Thus Williams, Dale, and Macnaughton-Smith (1963) have proposed for ecological applicatons that characters be weighted in proportion to their average correlation with all other characters. This is likely to have much the same effect as the suggestion by Mayr (1964) that one should first make a classification on unweighted characters and from this obtain the character weights (presumably according to their relative value in separating the taxa found) and second, repeat the classification using these *a posteriori weights*. Jizba (1964) has tried this with geological material, but his second, third, and even fourth classifications differed little from his first. Similar experiments in biological taxonomy are needed. By such a procedure the taxa might perhaps be somewhat more sharply demarcated, but even this is in doubt. One would have to ask whether iterative computer runs using reiterated series of weights would be desirable, and what the criterion for stopping iterations would be.

It is possible to use the rarity of features, or alternatively their commonness, in the whole set of OTU's to weight the characters, but the results are likely to be little different (except perhaps by a scaling factor, e.g., the example given by Sneath, 1965). Smirnov (1968) has suggested a similar kind of weighting, which is discussed more fully in Section 4.4 though we feel it has little to commend it.

Methods of numerical cladistics weight characters or character states as a result of explicit algorithms during the clustering procedure. Investigators may or may not agree with the aptness of these algorithms but their explicitness, and the fact that weight is not a priori, distinguish this procedure from weighting as practiced in conventional taxonomy.

One such method weights characters according to some function of their variability within OTU's (generally populations or species). Thus Farris (1966) suggests that characters varying little within populations are more reliable indicators of cladistic relationship than variable characters; invariable characters could therefore be weighted more heavily for cladistic studies. Eades (1970) makes similar proposals. However, this generalization about characters and evolution seems to us to be still uncertain. More interesting is the suggestion (Farris, 1969b) that weighting of characters for cladistics could be in proportion to the degree to which they fit a parsimonious cladogram, the aim being to improve iteratively the fit and the parsimony, as discussed further in Section 6.4. Goodman (1969) has a somewhat different aim in proposing that a character difference should be weighted inversely according to the within-OTU variance, because in his studies on plant breeding the variance may be largely due to environmental effects, so that genotypic differences might be better expressed if variable characters are deemphasized. Also variance may differ markedly within different OTU's. This particular scheme has the drawback that the weight becomes indeterminately large for characters constant within OTU's, so that a few of these characters may dominate the analyses. A better suggestion, by Flake, von Rudloff, and Turner (1969), is to multiply the squared difference in character i by a weight w_i^2, where w_i is $1 - (s_{i,\text{within}}/s_{i,\text{total}})$, where $s_{i,\text{within}}$ is the average within-OTU standard deviation for character i, and $s_{i,\text{total}}$ is the total standard deviation. If w_i is negative it is taken to be zero, that is, the character is excluded as being too variable within OTU's for safe use.

All the methods mentioned in the last paragraph give most weight to those characters that vary least within OTU's. Such weighting increases the ease with which taxa can be discriminated, and thus leads towards generalized distance, further taken up in Sections 4.3 and 8.5. We would consider these as methods for special purposes rather than for general taxonomic analyses. It may be noted too that Crovello's character state difference (Crovello, 1968e; Section 4.13) gives less weight to characters that have the same degree of variation within all OTU's, and more to those that are highly variable within some OTU's and invariant within others. However, this effect is unlikely to be very pronounced in practice.

The arguments in favor of equal weighting fall under seven headings.

1. If it cannot be decided how to weight the features, one must give them equal weight—unless it is proposed to allocate weight on irrational grounds.

2. To create taxonomic groups, one must first decide how to weight the features that are to be employed for classifications. Therefore, one can use no criterion that presupposes the existence of these taxa. For example, one cannot choose the constant features—to know if they *are* constant one must first set up taxonomic groups, and these have not yet been established.

3. The concept of taxonomic importance has no exact meaning. If "importance" means "importance to me because I am interested in it," this is only special pleading. If "importance" means basic or fundamental, this can only mean that it sums up a number of other characters: if they are unknown, they are hypothetical; if known, the character is not single but multiple. If "importance" means essential to survival, the taxonomy can estimate viability but not resemblance. If "importance" means "correlation with other features," then the added weight is due to these other features; where we observe the correlation breaking down, we do not regard the feature as important.

4. If differential weighting is admitted, exact rules must be given for estimating it. One must know whether the weight to be given to the possession of feathers is twice or twenty or two hundred times that given to possession of claws, and why. We do not know of any method for estimating this, and even if such a method were to be developed we doubt if any systematist would have the patience to use it because of the hundreds of characters he would need.

5. The nature of a taxonomy depends upon its purpose: a systematist could arrange living creatures in many ways but chooses one way because he thinks it is the best for some purpose. We hold the view that a "natural" or "orthodox" taxonomy is a general arrangement intended for general use by all kinds of

scientists (Gilmour, 1937; Sneath, 1958). It cannot therefore give greater weight to features of one sort, or it ceases to be a general arrangement. Being general, it is best for general purposes but is perfect for none.

6. The property of "naturalness" is, we believe, due to the high content of implied information that is possessed by a natural group. A group such as the Mammalia at once tells us much about its members with a high degree of certainty. A group such as "black animals" tells us nothing more than that they are all black. The content of information is measured by the number of statements that can be made about its members: each statement has unit value, and whether we think them important or not is irrelevant.

7. The use of many characters greatly evens out the effective weight that each character contributes to the similarity coefficient. Unless highly unequal weights are given to some characters, the very employment of many characters tends to make the taxonomy equally weighted.

Equal weighting can therefore be defended on several independent grounds: it is the only practical solution, it and only it can give the sort of natural taxonomy that we want, and it will appear automatically during the mathematical manipulations. Singly, these arguments are cogent; taken together, we think they are overwhelming.

4

The Estimation of
Taxonomic Resemblance

This chapter presents a detailed exposition and evaluation of various numerical methods that have been advocated for estimating the resemblance between OTU's. We would like to remind the reader before proceeding that the terms "resemblance" and "similarity" are used interchangeably throughout this book, and that unless specifically qualified they imply solely a phenetic relationship.

4.1 THE DATA MATRIX

To estimate the resemblance between pairs of OTU's, we adopt the convention of arranging data for numerical taxonomy in the form of an $n \times t$ matrix, whose t columns represent the t OTU's to be grouped on the basis of resemblances and whose n rows are n unit characters. Each entry X_{ij} in such a matrix is the score of OTU j for character i.

Characters	OTU's		
	1	2 $\quad \cdots \quad$ t	
1	X_{11}	X_{12} $\quad \cdots$	X_{1t}
2	X_{21}	X_{22} $\quad \cdots$	X_{2t}
$\vdots$	$\vdots$	$\vdots$	$\vdots$
n	X_{n1}	X_{n2} $\quad \cdots$	X_{nt}

As far as possible a consistent system of symbols has been adopted throughout the text, and formulae from other authors have been rewritten to conform with the symbolism of this book. We should also point out that many multivariate statistics texts use n for sample size (here the number of OTU's, which we denote as t) and use p for variable number (here denoted as n). That usage is not standard and inasmuch as our notation has been well established in numerical taxonomy, we retain it here.

It is important to note that the values in the matrix may represent the original or crude character scores, or they may be the transformed scores after such procedures as standardization (Section 4.8). Because resemblances can be usually calculated from either, it is not convenient to employ consistently a different symbol for transformed scores, but where appropriate they will given as X'. In most practical applications the transformations will be calculated automatically by the computer before the calculation of resemblances. When two OTU's are involved in an expression we shall designate them by j and k. Two characters will be symbolized by h and i.

The scores in a data matrix may be expressed in various ways depending in part on the type of character. This is considered at length in Section 4.8, but we will briefly mention here the main types. Two-state characters (all-or-none characters) such as "present" or "absent" may be symbolized by $+$ and $-$, but the use of 1 and 0 facilitates the numerical treatment of such data. Multistate characters may be of two kinds, quantitative and qualitative. The former have states that may be ordered in magnitude, and they can be scored as continuous variables, rank orders, percentages of the maximum expression, and so on. The score is therefore numerical. Qualitative multistate characters cannot be ordered and may be illustrated by a series *, □, and §. Of such a series one can only say that any two are the same or different. There is no implication that the difference between * and □ is greater or less than between * and §, for example. These may be scored as numbers or letters depending on the computer program, and some programs employ alphabetic symbols to avoid confusion with quantitative multistate characters.

It is also necessary to have a symbol for missing entries in the data matrix. These are unknown or inapplicable values (the two are usually not distinguished); NC for "no comparison with this entry" is commonly used.

Cattell (1952) has pointed out that most data matrices of the sort discussed in this section can be examined from at least two points of view. The association of pairs of characters (rows) can be examined over all OTU's (columns). This is called *R technique*. The converse practice, the association of pairs of OTU's (columns) over all characters (rows), has been called *Q technique*. Cattell (1966b, p. 294) in his most extensive recent discussion of the subject maintains that Q and R techniques should be restricted to factor analysis, and that Q' technique should be the term applied for cluster analysis of OTU's as customarily practiced in numerical taxonomy. However, numerical taxonomists have adapted Q and R in the general sense for cluster as well as factor analysis rather than restrict them to the latter field only. In taxonomy (as in other sciences) both Q and R techniques have been employed, and their relations are discussed in Section 5.8. Our main emphasis in numerical taxonomy is on Q studies. They refer to the quantification of relations between organisms with the aim of producing classifications of organisms. Criticisms of the Q technique applied to problems in psychology (Cattell, 1952) do not apply to work in taxonomy (Sokal and Michener, 1958).

Williams and Dale (1965) point out that there may be some confusion over the use of Q and R technique, because a factor analysis may compare characters but finally lead to a grouping of OTU's. They therefore suggest that Q and R should refer to the purpose of the analysis, the usage employed here. R technique leads to a classification of characters, Q technique to one of OTU's. The main mathematical steps are formally the same, and an R study is made by transposing the data matrix so that the characters (rows) become the individuals comparable to the former OTU's, and the actual OTU's or taxa (columns) become the characteristics (attributes) over which association is computed.

To clarify this further, Williams and Dale also suggest that when the relationships are represented in a hyperspace, the kind of space that is operated upon should not be called R-space or Q-space, but A-space and I-space, as follows:

A-space (*attribute space*) has formally n dimensions, one for each attribute or character, in which there are t points that represent the OTU's.

I-space (*individual space*) has formally t dimensions, one for each OTU, in which are n points representing the attributes or characters.

As will be seen in the next chapter, we frequently attempt to reduce the dimensionality of both A- and I-space. The above definitions will be considered to cover, as well, less than n or t dimensions, respectively.

In either space the character values are the coordinates of the points in the hyperspace. Most numerical taxonomic methods operate by Q technique on an A-space, but some cluster and factor analyses are Q techniques operating on an I-space. These distinctions are particularly important in fields like ecology, in which one is apt to become confused between attributes and individuals (see Section 11.1).

4.2   AN INTRODUCTION TO SIMILARITY COEFFICIENTS

Similarity between two OTU's is generally estimated by means of a *similarity coefficient*, which is a quantification of the resemblance between the elements in the two columns of the data matrix representing the character states of the two OTU's in question. Although the term "coefficient of resemblance" would be more appropriate, "similarity coefficient" is the commonest term, and we have retained it to cover both coefficients of similarity in the strict sense and those of dissimilarity.

Similarity coefficients have been employed for various classificatory tasks in biological taxonomy and other fields since the last century (Goodman and Kruskal, 1959). The intensive development of numerical taxonomy since the middle of the 1950's has resulted in the rapid elaboration of such coefficients, frequently on an empirical basis without adequate theoretical justification. Furthermore, lack of communication among diverse disciplines has led to considerable duplication of the same coefficient in different fields. Several studies have attempted to summarize and compare these coefficients (Goodman and Kruskal, 1954, 1959, 1963; Dagnelie, 1960; Sokal and Sneath, 1963; Cole, 1949, 1957; Moore and Russell, 1967; Cheetham and Hazel, 1969). The major attempt to evaluate these coefficients on the basis of their mathematical properties is that of Williams and Dale (1965). Many of their valuable contributions to the subject are incorporated in the discussion that follows.

In *Principles of Numerical Taxonomy* we attempted to include at least all those coefficients that had been suggested for biological taxonomy and as many of those employed in other fields as was feasible. After the recent extensive development of numerical taxonomy it would be impractical to continue this aim. We shall, therefore, limit ourselves to a discussion of the major classes of coefficients, citing in detail those few that have been extensively employed or those with especially interesting and potentially fruitful properties, adding brief references to some others.

The choice of coefficients is limited by the scale of measurement of the character states in the data matrix. Thus, for example, from qualitative multistate characters it would be improper to compute any of the similarity coefficients that assume continuous variables. As in ordinary statistics, the number of choices of algorithms decreases with changes in the data scale from continuous to ranked to presence–absence characters. It becomes most limited when dealing with qualitative multistate characters. Details on which methods are compatible with any given type of scaling are furnished below.

A fundamental question when measuring similarity between any two OTU's is whether or not one hopes to estimate some parametric value of the similarity. This is both a statistical and a biological question. In statistics there are various methods of correlation and association (depending on the scale in which the data are presented), all of which estimate the same value, the parametric correlation coefficient ρ, *given certain assumptions*. It is tempting to look for a similar parametric similarity measure in phenetics, but it is questionable whether such a measure of overall similarity exists, as has already been discussed in Sections 2.4, 3.6, and 3.8. In view of our discussion in those sections all that we can hope for at this time is to obtain a satisfactory measure of the similarity of organisms over the set of characters chosen for the taxonomic comparison or over the set of characters that could be added by the sampling method used for the chosen ones. We have already noted that the inability to estimate a parameter may not be too serious in view of the robustness of similarity measures based on increasingly large character sets.

It is unrealistic to try to compare estimates of similarity in pairs of OTU's from different studies involving different organisms, ranks, and data sets. Thus, a correlation coefficient of 0.63 between two species of bees in a study of a family of bees cannot easily be compared with a correlation coefficient of 0.85 between two families of mammals in a study of the families of mammals. By contrast we can compare the relative magnitudes of a given coefficient within any one study, where similarities between pairs of OTU's are based on the same characters and are taken from the same data matrix.

A desirable property of similarity coefficients, which, regrettably, does not obtain for many of them, is the property of *joint monotonicity* with other similarity coefficients. This implies that when the pairwise similarities of t OTU's are ranked by the magnitude of their $t(t-1)/2$ similarity coefficients of a type 1, they will be in the same order as when they are ranked by magnitude of a similarity coefficient of a type 2. Joint monotonicity would indicate a degree of robustness of the estimation of taxonomic resemblance, although it will not ensure identical classifications. Absence of joint monotonicity means that clumps or clusters formed by the same clustering algorithm will have different elements depending on the nature of the similarity coefficients. If the similarity coefficients are to represent the same set of taxonomic resemblances, lack of joint monotonicity is undesirable, because it will lead to different classifications. It might be argued by some, however, that different coefficients should lead to different taxonomic structures, for it can be shown that different coefficients estimate different aspects of the taxonomic relationship.

Another important aspect of similarity coefficients is whether they are probabilistic or just descriptive. Williams and Dale (1965) have faced this issue with respect to the classifications that result from numerical taxonomy, dividing classification into probabilistic and nonprobabilistic ones, a distinction we shall return to in the next chapter. However, considering only similarity coefficients at this juncture, we must still ask ourselves the following: are the numerical values obtained as the result of a mathematical operation on pairs of columns of data matrices estimates of a parametric value, about which we may have various degrees of confidence, based on our assumptions about the underlying distribution of the variables? Or, are they simply quantitatively descriptive numbers whose significance cannot be evaluated by conventional statistical techniques, either because we lack knowledge of the underlying distribution of the variables on which they are based or because no reasonable null hypothesis can be constructed? In the present state of knowledge the answer appears to be generally somewhere between the two. In fact, the nature of these distributions must be the subject of much numerical taxonomic research. As long as the similarity coefficients are based on large numbers of characters, differences in magnitude of coefficients within a matrix can be interpreted with some confidence. Attempts have been made from early in numerical taxonomy to develop independent methods of testing the significance of similarity coefficients and we shall discuss these efforts in subsequent sections, especially in Section 4.10. Except for the special case of those coefficients related to information theory, the most profitable approach so far has been that of comparing actual similarity matrices with those based on random data sets; this is also discussed in Section 4.10.

The question might be asked: are similarity coefficients needed at all? Can we not operate upon a data matrix in such a way as to obtain taxonomic structure directly without the intervening step of obtaining a similarity matrix? Several of the techniques of partitioning sets of OTU's by various optimality criteria (see Section 5.10) have bypassed the similarity matrix. The data vectors representing each OTU are grouped into sets possessing some desirable internal cohesion and external differences from other related sets. However, once such criteria of optimality are established, they become similarity measures almost by definition, since sets or parts comprising two OTU's must be permitted in such a scheme (although sets or parts containing two members need not occur in any optimal classification as judged by this criterion). Thus, although the intermediate step of a conventional similarity matrix is not always necessary for erecting a classification, the by-product of the classification will almost always yield, or at least imply, a measure of similarity. One should also remember that bypassing the stage of formal similarity coefficients does not reduce any uncertainty about the significance of the overall classification, a matter also discussed in Section 5.10.

For practical purposes the types of similarity coefficients can be divided roughly into four groups, whose boundaries, however, are often quite fuzzy.

Distance coefficients, discussed in Section 4.3, measure the distance between OTU's in a space defined in various ways. The most familiar measure of distance is simple Euclidean distance in a character space of one or more dimensions. However, there are other distance measures whose metric is not necessarily Euclidean, as we shall see. Distance measures have inherently the greatest intellectual appeal to taxonomists as they are in many ways the easiest to visualize. It must be pointed out that distance coefficients are the converse of similarity coefficients. They are, in fact, measures of *dissimilarity*. In an analogous way, complementary functions of association or correlation coefficients can be thought of as measures of dissimilarity and, in fact, many of the association coefficients discussed in Section 4.4 can be expressed as distance measures when they are stated as complements of their maximum value, or in terms of a special metric.

Association coefficients, discussed in Section 4.4, are based on various algorithms involving qualitative data (multistate characters or two-state characters). Basically they are measures of the agreement of the states in the two data columns representing the OTU's, although they may sometimes be special cases of distance or angular coefficients. They can also be applied to ranked and continuous data when one is prepared to sacrifice information in the characters by transformation of the more informative character scale, as for example by recoding multistate meristic characters into two states.

Correlation coefficients and other angular coefficients, discussed in Section 4.5, measure proportionality and independence between pairs of OTU vectors. Most commonly this is the product-moment correlation coefficient applied to continuous variables, but other correlation coefficients for ranks or two-state characters have also been employed in numerical taxonomy. Correlation coefficients may be viewed as special cases of angular coefficients, a wider class that is also discussed in Section 4.11. Relations between distance and correlation coefficients will be illustrated in Section 4.5.

Probabilistic similarity coefficients are among the most recently developed measures of similarity. They include the information statistics which measure the homogeneity of the system by partitioning or subpartitioning sets of OTU's. The technique relates these coefficients to the idea expressed earlier: similarity coefficients can be avoided and one can go straight from the data matrix into a classification by using only a criterion of goodness of the classification as a guide in establishing partitions. Information statistics generally are such optimality criteria. They have important properties in that they are usually additive, are distributed as chi-square and are probabilistic; that is, in those restricted cases when we can establish a null hypothesis about them, it can be tested by conventional statistical means. Again, in special cases these coefficients can be represented as distance coefficients. We will discuss these coefficients in Section 4.6.

Williams and Dale (1965) stress the importance of being able to visualize the relationships between taxonomic units in a geometric manner. Ideally such relationships are those of distances in a three-dimensional Euclidean space, familiar to us from everyday experience. Since only in exceptional cases can we truly represent relationships among OTU's in three dimensions, we can either approximate the true relationships as distances in a three-dimensional Euclidean space, or we can consider them in a multidimensional Euclidean space which, although we cannot visualize it, retains many of the familiar properties of 3-space. In fact, as Williams and Dale (1965) and L. A. S. Johnson (1968) point out, the space need not even be Euclidean in the strict sense, but it should if possible be metric, which means that its topology is determined by a metric function. Therefore, it is desirable that measures of dissimilarity (or functions that convert similarity coefficients into measures of dissimilarity, such as complements from the maximum value of association coefficients or the arc cosine of the correlation coefficient) observe the property of a *metric*. By this is meant that these measures or functions satisfy four axioms over the entire set of OTU's. If the function for pairs of OTU's **a, b, c,** . . . is defined as φ, where φ is a real nonnegative number, these axioms are:

1. $\varphi(\mathbf{a}, \mathbf{b}) \geq 0$, and $\varphi(\mathbf{a}, \mathbf{a}) = \varphi(\mathbf{b}, \mathbf{b}) = 0$
2. $\varphi(\mathbf{a}, \mathbf{b}) = \varphi(\mathbf{b}, \mathbf{a})$
3. $\varphi(\mathbf{a}, \mathbf{c}) \leq \varphi(\mathbf{a}, \mathbf{b}) + \varphi(\mathbf{b}, \mathbf{c})$
4.   If $\mathbf{a} \neq \mathbf{b}$, then $\varphi(\mathbf{a}, \mathbf{b}) > 0$

Axiom 1 states that identical OTU's are indistinguishable while nonidentical ones may or may not be distinguishable by the dissimilarity function. Axiom 2 is the important symmetry relationship. The value of the function φ is the same from **a** to **b** as it is from **b** to **a**. The third axiom is the well-known triangle inequality stating that the function between **a** and **c** cannot be greater than the sum of the functions between **a** and **b** and **b** and **c**. The fourth axiom states that if **a** and **b** differ (in their character states) then the function between them must be greater than zero. For the familiar Euclidean distances these four axioms are true, and thus they are metrics. Euclidean distances possess the additional property of obeying Pythagoras' theorem. Not all measures of dissimilarity are metrics. Some may not fulfill axiom 4 for all OTU's in the taxon; that is, despite the known difference in character states between two OTU's **a** and **b**, the function may still be zero. Such systems are called *pseudometric* or *semimetric*. Of necessity some useful measures are semimetric. We shall encounter some of these in subsequent sections. Whenever possible, however, for ease of comprehension of the relationships, we shall confine our attention to coefficients whose properties lead us to metric spaces.

When the third axiom is relaxed so that

3.* $\varphi(\mathbf{a}, \mathbf{c}) \leq \max [\varphi(\mathbf{a}, \mathbf{b}), \varphi(\mathbf{b}, \mathbf{c})]$

the pair-function is called an *ultrametric*. A phenogram (see Section 5.9) is an example of an ultrametric pair-function for any pair of OTU's. The ultrametric inequality (axiom 3*) insures that the pair-function implied by the dendogram is monotonically increasing or decreasing, depending on whether it is a dissimilarity or a similarity function.

4.3   DISTANCE COEFFICIENTS

Because of their intuitive appeal, we commence our discussion of similarity (and dissimilarity) coefficients with a description of various measures of distance, although historically, and in terms of frequency of application, association coefficients can lay claim to primacy. We have already stated that both association coefficients and correlation coefficients can be related to distances. The general concept of distance can be imparted by means of the following models.

Let us assume we have t OTU's for which $n = 2$ characters were studied. We now draw a conventional pair of rectangular coordinates representing A-space, letting the abscissa X_1 represent character 1 and the ordinate X_2 character 2. Next we plot the position of the t OTU's with respect to these coordinate axes. A hypothetical case is shown in Figure 4-1. If any two OTU's are identical in terms of the two characters under consideration, their positions will coincide and the distance between them will be zero. The greater the disparity between them, the greater will be their distance. Thus, distance is seen to be the complement of similarity.

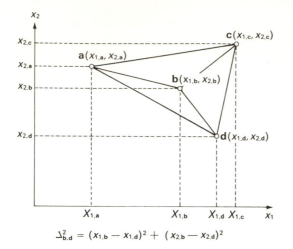

$$\Delta_{b,d}^2 = (x_{1,b} - x_{1,d})^2 + (x_{2,b} - x_{2,d})^2$$

FIGURE 4-1
Representation of four OTU's (**a**, **b**, **c**, and **d**) as points on a plane determined by their character states for two characters *1* and *2*. Each character is represented by a dimension—in this case two, X_1 and X_2. We have arbitrarily assigned the order $a < b < d < c$ for the states of character *1*, and $d < b < a < c$ for the states of *2*. To obtain the taxonomic distance, $d_{b,d}$, between OTU's **b** and **d** as defined in the text, we must divide $\Delta_{b,d}^2$ by *n*, the number of characters, and take the square root of the quotient.

When we wish to estimate taxonomic distance on the basis of three characters, we must add a third coordinate, X_3, to our diagram. On paper such a three-dimensional model can only be shown as a two-dimensional projection (Figure 4-2). We cannot visualize the geometry of adding a fourth and subsequent characters. The requirements of each new coordinate axis are that it be at right angles to all previous ones. Although we cannot depict such an axis graphically, we can postulate its existence and demonstrate algebraically that most of the geometric theorems of conventional three-dimensional space can be extended to *n* dimensions in Euclidean hyperspace. Thus we are at liberty to postulate *n* dimensions for *n* characters and we can compute the distance between any two OTU's in hyperspace.

In the formulations that follow, scaling of the characters may vary as discussed in Section 4.8 and the symbols would then refer to the scaled values. The symbolism for the elements of the data matrix is as in Section 4.1.

Average Differences. The expression

$$\frac{1}{n} \sum_{i=1}^{n} (X_{ij} - X_{ik})$$

is not particularly suitable for measuring the distance between OTU's **j** and **k**, since the differences could be negative as well as positive (i.e., it violates axiom 2 above). In an uncorrelated multivariate normal distribution the expected value of this expression would be zero. The obvious correction would be to use

$$\frac{1}{n} \sum_{i=1}^{n} |X_{ij} - X_{ik}|$$

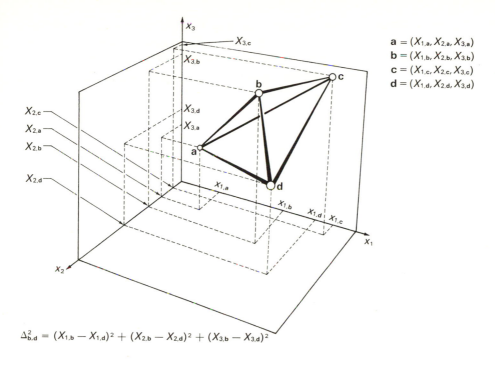

$$a = (X_{1,a}, X_{2,a}, X_{3,a})$$
$$b = (X_{1,b}, X_{2,b}, X_{3,b})$$
$$c = (X_{1,c}, X_{2,c}, X_{3,c})$$
$$d = (X_{1,d}, X_{2,d}, X_{3,d})$$

$$\Delta^2_{b,d} = (X_{1,b} - X_{1,d})^2 + (X_{2,b} - X_{2,d})^2 + (X_{3,b} - X_{3,d})^2$$

FIGURE 4-2

Representation of four OTU's in a three dimensional space obtained from Figure 4-1 by adding a third dimension, X_3. The order of character states is now $a < b < d < c$, $c < a < b < d$, and $a < d < b < c$, for X_1, X_2, and X_3, respectively.

the absolute (positive) values of the differences between the OTU's for each character. This is the *mean character difference (M.C.D.)*, which has been proposed by Cain and Harrison (1958) as a measure of taxonomic resemblance. It had previously been used in anthropology by Czekanowski (1909, 1932), who called it *durchschnittliche Differenz*, and by Haltenorth (1937), who employed the coefficient in an extensive study of 86 characters of eight species of the large cats, each character being a mean based on a large number of specimens. A reanalysis of Haltenorth's original data by Sokal—employing d_{jk}, the coefficient of taxonomic distance described below—resembled Haltenorth's results closely.

The simplicity of the *M.C.D.* statistic is in its favor. It is a metric. However, it does suffer some disadvantages. It will always underestimate the Euclidean distance between the taxa in space, which may or may not be of consequence. It also lacks some of the desirable attributes of the alternative measure, the taxonomic distance or its square described below. In general, the *M.C.D.* stands in the same relation to taxonomic distance as the average deviation stands in relation to the standard deviation, and it suffers from similar disabilities as the average deviation.

Taxonomic Distance.   The Euclidean distance between two OTU's in two- and three-dimensional spaces has been illustrated in Figures 4-1 and 4-2. We can generalize this concept of the Euclidean distance between two points in an *n*-dimensional space. The formula for such a distance, Δ_{jk}, between OTU's **j** and **k**, is

$$\Delta_{jk} = \left[\sum_{i=1}^{n} (X_{ij} - X_{ik})^2 \right]^{1/2}$$

This quantity was first defined for numerical taxonomy by Sokal (1961, as δ_{jk}), by Sokal and Sneath (1963), and by Rohlf and Sokal (1965). Since Δ_{jk} increases with the number of characters used in the comparison, an average distance is commonly computed. This is

$$d_{jk} = \sqrt{\Delta_{jk}^2/n}$$

After standardizing each of the *n* characters, assuming that the characters were independent and normally distributed with a mean of zero and a variance of unity, Rohlf (1962) computed the expected value of *d* for even values of *n* as

$$\mathscr{E}(d) = \frac{(n - 1)! \, (\pi/n)^{1/2}}{2^{n-2} \left[\left(\frac{n}{2} - 1 \right)! \right]^2}$$

Using Stirling's formula this reduces to

$$\mathscr{E}(d) \approx \sqrt{2} \left(1 - \frac{1}{n} \right)^{1/2} \left(1 + \frac{1}{n - 2} \right)^{n-1} \frac{1}{e}$$

Thus for the larger values of *n* generally employed in numerical taxonomy the expected value of *d* approaches $\sqrt{2}$ very closely. The expected variance of *d* is

$$\mathscr{E}(\sigma_d^2) = 2 - [\mathscr{E}(d)]^2 \approx 1/n$$

which approaches zero as *n* tends to infinity. Figure 4-3 gives the 95 percent confidence limits to the expected value of *d* at various sample sizes, *n*. After $n \approx 75$, the variance decreases very slowly.

The distance coefficient based on standardized characters can be easily transformed into the form of a correlation coefficient, $r_{p(jk)}$, Cattell's coefficient of

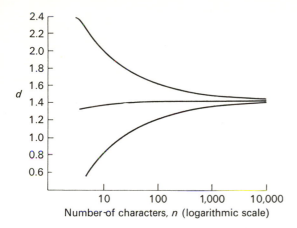

FIGURE 4-3
Expected values of d and 95 percent confidence limits of the taxonomic distance on the asumption that the observations on which the OTU's are based (that is, the n standardized characters used) are independent and normally distributed with a mean of zero and unit variance.

pattern similarity (Cattell, 1949; Rohlf and Sokal, 1965; Cattell, Coulter, and Tsujioka, 1966). This coefficient is computed as follows:

$$r_{p(jk)} = \frac{2\chi^2_{.5[n]} - nd^2_{jk}}{2\chi^2_{.5[n]} + nd^2_{jk}}$$

where $\chi^2_{.5[n]}$ is the median chi-square value for n degrees of freedom. Within the range of distance coefficients observed in numerical taxonomic studies, Rohlf and Sokal (1965) find that there is practically a linear relationship between d and r_p. Thus classifications based on these two coefficients would be nearly identical.

It should be pointed out that M.C.D. and d_{jk} are special cases of a class of metric distance functions (the Minkowski metrics) whose general form can be stated as

$$d_r(j, k) = \left(\sum_{i=1}^{n} |X_{ij} - X_{ik}|^r \right)^{1/r}$$

In this system $d_1(j, k)$ is known as the *Manhattan* or *city-block metric*. There has been renewed interest in its properties in recent years in connection with developments in numerical cladistics (see Section 6.4). When adjusted for the number of characters the Manhattan distance becomes Cain and Harrison's mean character difference, $d_1(j, k)/n = M.C.D.$ These metrics, unlike Euclidean distances, are not invariant under rotation of the character space as performed in ordination (Section 5.6). Lance and Williams (1967a) give several variations of Manhattan metric distance coefficients. We need mention here only their *Canberra metric*, defined as

$$d_{CANB.}(j, k) = \sum_{i=1}^{n} \left(\frac{|X_{ij} - X_{ik}|}{(X_{ij} + X_{ik})} \right)$$

For positive character states it has attractive properties as it is a property solely of the individuals or groups being compared and is not affected by the range of the entire characters. It is also sensitive to proportional rather than absolute differences. On the other hand it is restricted to positive character states. Hodson (1969) uses a Euclidean form of the Canberra metric suggested by R. J. Hartzig, which can accommodate negative states. An alternative solution (J. C. Gower, personal communication) would be to replace the denominator by $|X_{ij}| + |X_{ik}|$. The complement of similarity coefficient S_G defined by Gower (1971a, see our Section 4.4) is also a Manhattan metric when applied to quantitative characters or to two-state characters coded zero and one. It depends on the range of all characters for its scaling factor. Obviously, $d_2(\mathbf{j}, \mathbf{k}) = \Delta_{\mathbf{jk}}$ and $d_2(\mathbf{j}, \mathbf{k})/n^{1/2} = d_{\mathbf{jk}}$.

Other Formulations. The idea for a distance coefficient has come to many people. So far as we can learn, a measure of distance was first employed by Heincke as early as 1898. Schilder and Schilder (1951) demonstrated such a coefficient without standardizing characters. In comparing several populations of snakes, Clark (1952) computed distances for each character as a ratio varying between zero and unity. This is called the *coefficient of divergence,*

$$CD_{\mathbf{jk}} = \left[\frac{1}{n} \sum_{i=1}^{n} \left(\frac{X_{ij} - X_{ik}}{X_{ij} + X_{ik}} \right)^2 \right]^{1/2}$$

obviously related to the Canberra metric. It has also been used by Rhodes et al. (1969) in botany.

Bielicki (1962) describes a coefficient of distance for use in anthropology. Based upon an earlier statistic of Wanke (1953), it is similar to the coefficient of Clark (1952).

Sokal (1961) brought taxonomic distance to the attention of numerical taxonomists, employing the formulation of $d_{\mathbf{jk}}^2$ and suggesting standardization of character state codes for each character.

Zarapkin (1934, 1939, 1943) employed a statistical approach related to distance. He employed a so-called standard population for which he would compute the mean and standard deviation of each character considered. Zarapkin computed the standard deviations of all standardized deviations between each OTU and the standard population. This quantity, computed over all the characters, he called $\mathfrak{S}$. To the extent that the mean of the deviations between any OTU and the standard population approaches zero, the quantity is nothing but the distance between the OTU and the standard population expressed in the standardized units of the latter. The difference between $\mathfrak{S}_{\mathbf{jk}}^2$ and $d_{\mathbf{jk}}^2$ is the size coefficient of Penrose, C_Q^2 (see Section 4.8). We would not recommend this method for reasons detailed in Sokal and Sneath (1963).

Related to taxonomic distance is the *coefficient of racial likeness (C.R.L.)*, developed by Karl Pearson (1926) for measuring resemblances between samples of skulls of various origins. The reader will note that because the OTU's are samples they are analogous to clusters or taxa, that is they have explicit intra-OTU variability, and to emphasize this we use the capital letters **J** and **K** for the OTU's being compared. Means and variances of continuous characters (lengths, ratios, and others) are computed for each sample and the coefficient is defined as

$$C.R.L. = \left[\frac{1}{n} \sum_{i=1}^{n} \left(\frac{(\bar{X}_{i\mathbf{J}} - \bar{X}_{i\mathbf{K}})^2}{(s_{i\mathbf{J}}^2/t_{\mathbf{J}}) + (s_{i\mathbf{K}}^2/t_{\mathbf{K}})} \right) \right]^{1/2} - \frac{2}{n}$$

where $\bar{X}_{i\mathbf{J}}$ stands for the sample mean of the ith character for sample **J**, $s_{i\mathbf{J}}^2$ for the variance of the same, and $t_{\mathbf{J}}$ for the sample size of **J**.

When we wrote our *Principles of Numerical Taxonomy* in 1963 we did not consider the *generalized distance* developed by Mahalanobis (1936) and Rao (1948) to be of particular usefulness in numerical taxonomy. We took this position for several reasons. During the early work on numerical taxonomy very few studies were carried out at the lower taxonomic levels and consequently the OTU's were typically individuals or exemplars of samples rather than samples from populations. Second, the computational loads involved in computing generalized distances for large numbers of characters and many samples seemed beyond the capabilities of the computational machinery then available. A third objection was that character states were usually discretely coded in numerical taxonomy, with generalized distance requiring continuous measurement variables. Fourth, we had serious doubts about the validity in most taxonomic samples of the assumptions underlying the generalized distance model. Now that computers have become more powerful and versatile and more emphasis has been placed on numerical taxonomic studies at the specific and infraspecific levels, the first three objections are becoming increasingly irrelevant. Multivariate normality and, particularly, homoscedasticity of sample dispersion matrices continue to be questionable in most multivariate taxonomic studies. Nevertheless, considerable robustness to violations of these assumptions has been demonstrated and the computation of generalized distance has become a useful tool in the toolchest of numerical taxonomists. This trend is in conformity with the general breakdown of a sharp distinction between multivariate and other numerical methods in taxonomy.

Generalized distances are computed by maximizing the difference between pairs of means for those linear combinations of characters that have maximal variance between pairs of groups relative to the pooled variance within groups for the same linear combinations. Generalized distances can be computed by the formula $D_{\mathbf{JK}}^2 = \delta_{\mathbf{JK}}' \mathbf{W}^{-1} \delta_{\mathbf{JK}}$, where $\mathbf{W}^{-1}$ is the inverse of the pooled variance-covariance (dispersion) matrix within samples (dimension $n \times n$) and $\mathbf{\delta_{JK}}$ is a vector of differences between means of samples $\mathbf{J}$ and $\mathbf{K}$ for all characters. A separate $D^2_{\mathbf{JK}}$ must be computed for each sample pair $\mathbf{JK}$ ($\mathbf{J} \neq \mathbf{K}$). Further details are given in Section 8.5. For computational purposes the above formula is awkward and readers wishing to use this method are urged to turn to one of several available programs, such as the MULDIS routine of the NT-SYS system of multivariate computer programs developed by Rohlf, Kishpaugh, and Kirk (1971).

These computations are of necessity carried out by computer. It should be pointed out that the generalized distance is equivalent to distances between mean discriminant values in a generalized discriminant function; thus there is an intimate relationship between discriminant functions and generalized distance. Therefore bear in mind that generalized distance measures the distance as a function of the overlap between pairs of populations and transforms the original distances so as to maximize the power of discrimination between individual specimens in previously constructed groups (the OTU's in the above discussion; see Section 8.5). In doing so it may distort the original distances considerably, and the taxonomist should consider whether this is what he requires. It should be noted that when a large number of characters is employed there is a danger that small eigenvalues, whose exact magnitude may be affected by rounding errors, may make a large haphazard contribution to the distance. $D^2_{\mathbf{JK}}$ is also heavily dependent on assumptions of multivariate normal distributions within the OTU's, which will usually not hold above the population or species level. Several papers in the recent literature have employed this method to estimate the relative distances between population samples and to obtain taxonomic structure from such a matrix (Fisher, 1969; Goodman, 1967a; Jewsbury, 1968; Miller and Kahn, 1962 p. 259 ff; Reyment and Naidin, 1962). If correlations between characters are not marked, d^2 from characters standardized within groups and D^2 will be almost the same except for a scaling factor (e.g., Huizinga, 1962, 1965). This is also true when all correlations are equal (Penrose, 1954).

Bartels et al. (1970) have introduced a novel "topographic" distance measure originated by D. W. Calhoun, which uses only rank orders of the OTU's on each character axis. For any pair of OTU's $\mathbf{j}$ and $\mathbf{k}$ the Calhoun distance is the proportion of the entire set of OTU's (excluding $\mathbf{j}$ and $\mathbf{k}$ themselves) that have character states intermediate between that for $\mathbf{j}$ and that for $\mathbf{k}$ for one or more of the characters. When n is large the distance will become maximal (1.0) if $\mathbf{j}$ and $\mathbf{k}$ happen to possess the largest and smallest values for any one of the characters; it may then be more appropriate to use instead the mean proportions of intermediate OTU's over the n characters.

The coefficient of mismatch between diagrams, d_h (Section 3.3) is also a distance measure. It differs from all those mentioned above in that the characters are now the coordinates of marker points, thus effectively locking the characters into a rigid spatial framework.

4.4 ASSOCIATION COEFFICIENTS

There are so many association coefficients in the biological and nonbiological literature that any attempt at an exhaustive catalog of them would require many pages, and it is doubtful whether such an exhaustive list would be of special value to the aspiring student of taxonomy unless each coefficient were thoroughly described, and its properties evaluated and compared with those of other coefficients. Although some comparative studies of association coefficients have been published (Cheetham and Hazel, 1969; Cole, 1949, 1957; Dagnelie, 1960; Goodman and Kruskal, 1954, 1959, 1963; Sokal and Sneath, 1963), we still lack an overall evaluation of the many coefficients; most of them have been applied no more than once and thus cannot even be empirically validated. We have, therefore, taken the position that only those coefficients that have been extensively used in numerical taxonomy should be presented here, and even among those we have chosen a few types representative of more diverse approaches. Fortunately, as Williams and Dale (1965) point out, many of the published coefficients are mathematically related to one another. It is our position that of the various classes of association coefficients preference should be given to the most simple ones.

So far, we have not defined *association coefficient*. Of the various similarity measures these are the most difficult to define. They are pair-functions that measure the agreement between pairs of OTU's over an array of two-state or multi-state characters. Many of these coefficients measure the number of actual agreements as compared with the number of theoretically possible ones. Characters coded in two or a few states are especially suitable for the computation of association coefficients, although even continuous characters can be coded (albeit usually with loss of information) to yield association coefficients. We should point out that the terms employed here for these coefficients (such as association, similarity, relationship) have been used in a variety of meanings in English and other languages. The designations adopted here are, therefore, arbitrary.

In the most common model, association coefficients are computed with two-state characters, which are for convenience coded 0 or 1. The 0,1 code can represent the presence of absence of a characteristic or property such as a bristle or a pigment; it may stand for the success of failure of a biochemical reaction or an ethological test; or it may be an arbitrary designation as in a structure having only two shapes, either rounded or pointed, where 0 might designate rounded, and 1 pointed. When character states are compared over pairs of columns in a conventional data matrix the outcome can be summarized in a conventional 2×2 frequency table such as the one shown in the text figure at the top of the next page. In the left upper quadrant of the figure, we would place the number of characters coded 1 in both OTU's, while in the right lower quadrant we would write the number of characters coded 0 for both OTU's. The other two quadrants register the number of characters in which the two OTU's disagree, being coded 1 for OTU **j** and 0 for

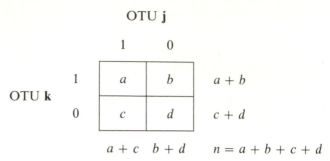

OTU k (or the converse). Earlier we proposed a more elaborate symbolism for these frequencies involving subscripts to sample size n (Sokal and Sneath, 1963), which Williams and Dale (1965) point out is more informative but also more clumsy in algebraic expressions and hence likely to be misread. Therefore, we now return to a, b, c, d, and n, the traditional symbolism for 2×2 tables in the statistical literature, and shall interpret these symbols as defined above. The marginal totals are the sums of these frequencies, with n being reserved for the sum of the four frequencies, which equals the number of characters in the study (unless there are missing observations). We shall find it convenient to define m as the number of matches or agreements $(m = a + d)$, and to let u be the number of mismatches $(u = b + c)$, whence $m + u = n$.

By basing a pair-function in numerical taxonomy on a 2×2 frequency table we are not necessarily implying that these data are appropriate for analysis in a 2×2 contingency table as conventionally employed in statistical work. Sokal and Rohlf (1969, p. 588) distinguish three models for the analyses of 2×2 tables. None of these models is appropriate for testing such a table as it is used in numerical taxonomy. The reason is that we are not dealing with a random sample of measurements of the same property and testing the independence of the "positiveness" of the characters shared by the two OTU's. Since we know that the characters are correlated we clearly cannot use any of the conventional models of 2×2 tables. So far we have only considered Q analyses. But even when we consider R analyses in which the degree of association between two characters is measured over a set of OTU's, we cannot consider the 2×2 table to be a contingency table between the two characters because we do not have a random sample of OTU's, but usually have instead a selected set of OTU's representing a given taxon.

We can sum the cells of each diagonal of the 2×2 table, yielding two frequencies, that of matches and that of mismatches. This is desirable when the meaning of "positive" and "negative" can vary from the presence or absence of a structure or a chemical reaction to two alternative states of a binary character without the implication of absence carried by the term "negative." Thus in the latter instance the choice of which of the two states is to be called "positive" is quite arbitrary. When a strain of micro-organisms becomes drug resistant should we call the resistant or the susceptible strain "positive"? We cannot imply that there is a parametric value of positiveness that is estimated by the proportion in our sample. In such instances the number of matches, that is, agreements in character states measured by $m = a + d$, or the number of mismatches $u = b + c$, are simpler and possibly more meaningful quantities.

The Coefficient of Jaccard (Sneath):

$$S_J = \frac{a}{a + u} = \frac{a}{a + b + c}$$

Sneath (1957a) used a coefficient he called the *similarity*, which has had a considerable history of application in R and Q studies in ecology. The earliest record of its employment we have found is by Jaccard (1908), and we shall therefore refer to it as the coefficient of Jaccard, S_J. It is clear that $S_J \to 0$ as $a/u \to 0$, and that as $u \to 0$, $S_J \to 1$. The coefficient of Jaccard omits consideration of negative matches. In its class it is the simplest of the coefficients. It is monotonic with the coefficient of Dice (1945), also used by Sørensen (1948). The formula for this coefficient, $S_D = 2a/(2a + u) = 2a/(2a + b + c)$, gives more weight to the matches than to mismatches. It varies between 0 and 1, as does the coefficient of Jaccard. A related dissimilarity coefficient, $(b + c)/(2a + b + c)$, by Williams and his colleagues (e.g., Watson, Williams, and Lance, 1966, 1967) is called by them the nonmetric coefficient. Ising and Fröst (1969) have introduced a similar coefficient with the interesting property of giving a fractional value to negative matches in proportion to the expectation from the overall frequency of the characters in the $n \times t$ matrix. The complements of S_J, S_D and the "nonmetric coefficients" are nonmetric; in fact, they are pseudometric and also do not necessarily obey the triangle inequality. Furthermore, we have philosophical objections to the employment of S_J and S_D in most instances, inasmuch as they do not consider matches in negative character states.

Whether negative matches should be incorporated into a coefficient of association may occasion serious doubts. It may be argued that basing similarity between two species on the mutual absence of a certain character is improper. The absence of wings, when observed among a group of distantly related organisms (such as camel, louse, and nematode), would surely be an absurd indication of similarity. Sneath (1957b) excluded negative matches from consideration in his similarity coefficient, but most of the applications of association coefficients since 1957 (largely in the field of microbiology) have included negative matches in their coefficients. In bacteriological work the problem may be slightly different than in higher organisms because of the practice of applying a standard series of tests to a group of bacteria. Exclusion of negative matches from the computation of a coefficient of association may be the safe procedure here, especially since a large group of negatives may on occasion be due to an unrecognized metabolic block preventing the expression of many other characters. This case is similar to that of missing organs, discussed later (Section 4.12). The difficulty particularly with microorganisms is to know what characters are missing. In morphological characters this is determined by position—an insect, for example, cannot have wing veins in an absent wing. Thus we must code the veins NC (see Section 4.8). But our knowledge of metabolic characters is more limited. It may be impossible to decide if enzymes $A, B, C, D, \ldots$ are present but not expressed because of lack of Z, which is necessary for activity. This is analogous in a morphological context to being unable to decide how to score subsidiary characters because we do not know whether the organ is present. The coefficient of Jaccard is appropriate when negative matches are to be excluded.

The Simple Matching Coefficient:

$$S_{SM} = \frac{m}{m + u} = \frac{m}{n} = \frac{a + d}{a + b + c + d}$$

This is one of the oldest and simplest coefficients, introduced to numerical taxonomy by Sokal and Michener (1958) and used repeatedly since. Sneath (1962) called it S_S. It is the affinity index used by Brisbane and Rovira (1961). We shall not trace its early history here. From the formula it follows that $S_{SM} \to 0$ as $m/u \to 0$, and that $S_{SM} \to 1$ as $u/m \to 0$. In its complementary form, $1 - S_{SM}$, the simple matching coefficient is equal to the squared Euclidean distance based on unstandardized character states, which can take the value of 0 or 1, that is, $\sqrt{1 - S_{SM}} = d$. The square root of the complement of the simple matching coefficient is therefore a metric pair-function that can be imbedded in a Euclidean space where all OTU's occupy corners of a unit hypercube. The coefficient of Rogers and Tanimoto (1960), $S_{RT} = m/(m + 2u) = (a + d)/(a + 2b + 2c + d)$, first suggested by these authors for numerical taxonomy, is monotonic with the simple matching coefficient and has similar boundary values and properties. It has also been used as a distance (see Section 4.6). We prefer the simpler S_{SM}.

Ignoring the heterogeneity of column vectors of the data matrix, Sokal and Sneath (1963) suggested that the variance of a simple matching coefficient would approximate that of a binomial distribution, that is, $S_{SM}(1 - S_{SM})/n$. Goodall (1967) has examined the distribution of the matching coefficient in considerable detail, allowing for the fact that the probabilities of two OTU's agreeing for any one character will differ from character to character. Since the distribution of positive character states in taxonomic populations is not known, he assumed a variety of plausible models ranging from lognormal to uniform and two-valued distributions. He also examined several real data sets from taxonomy and ecology. Goodall found that for 20 characters the binomial variance consistently over-estimated the calculated variance of S_{SM}, usually by a small percentage (between

10 and 20 percent), and even in the case of extremely unequal frequencies of positive states among the characters, by no more than 50 percent. Thus the use of the binomial variance to set confidence limits to a similarity coefficient (or to test the difference between two similarity coefficients) is a conservative procedure. Monte Carlo tests of a larger number of characters confirmed these conclusions.

On the basis of these findings Goodall (1967) recommends setting confidence limits to expected values of the simple matching coefficient computed as $\mu = (\Sigma_{i=1}^{n} p_i)/n$, where p_i is the proportion of character i that is of state 1 in the set of OTU's under consideration. The confidence limits are set assuming a normal distribution of $\arcsin \sqrt{S_{SM}}$ around the mean $\arcsin \sqrt{\mu}$ with variance $[\mu(1 - \mu) - (\Sigma_{i=1}^{n} p_i^2 - n\mu^2)]/n[4\mu(1 - \mu)]$. For greater precision in a given example, Monte Carlo techniques may be preferred. The question of significance of similarity coefficients will be discussed in greater detail in Section 4.10.

An additional problem: because S_{SM} approximates a binomial distribution, high values of this coefficient have very low variances. This can be somewhat overcome by the angular transformation as employed above by Goodall.

The Yule Coefficient. This coefficient for data arranged in a 2×2 table is computed as

$$S_Y = (ad - bc)/(ad + bc)$$

Its numerator is the determinant of the 2×2 table and the limits of S_Y are from -1 to $+1$. In the former case there are no matches at all, in the latter, matches are perfect. Except by Brisbane and Rovira (1961) this coefficient has not been employed in numerical taxonomy. Others related to it, which are seldom used but described in greater detail by Sokal and Sneath (1963), include the well-known coefficient, $S_\varphi = (ad - bc)/[(a + b)(a + c)(c + d)(b + d)]^{1/2}$, which is the product moment correlation coefficient r for data coded 0,1, and the coefficient of Hamann, $S_H = (m - u)/n = (a + d - b - c)/(a + b + c + d)$, employed by that author in an early numerical taxonomic study (Hamann, 1961). All these coefficients balance matches against mismatches, a concept that does not appear of special utility in the estimation of similarity.

Multistate Characters. When multistate characters are ordered they can be processed the way quantitative characters can, and comparisons for any character should reflect the degree of disagreement rather than simply state it as a match or mismatch. However, when multistate characters are qualitative (i.e., when we are unable to array the states into a logical order), a measure of agreement as summarized in an association coefficient can only be based on one of two considerations. We may consider a match between two OTU's in a multistate character the equal of that in a two-state character (i.e., no allowance is made for the possibility that a match in a five-state character might be less likely than in a two-state character). This approach is used when the simple matching coefficient, S_{SM}, is extended to multistate characters. The related coefficient of Rogers and Tanimoto (1960), S_{RT}, has been explicitly employed in this manner, and for multistate as well as two-state characters it is simply the number of matches divided by the sum of the number of matches plus twice the number of mismatches.

The other approach is to consider the probability of a given match taking place. It stems from the not unfamiliar reasoning that similarities representing agreements in rare character states should count more heavily in the establishment of a classification than those based on commoner characteristics. We must usually resort to probability statements based on the observed distribution of the character states in the selected sample of OTU's that is to be classified, since we rarely if ever have any knowledge of the distribution of these character states among OTU's in nature. At best this concept would be hard to define: is it distribution among all OTU's? Clearly some taxonomic limits must be drawn, but where? Who draws them? And is there an objective method for doing so? Do we consider the distribution of character states among all known OTU's, all extant OTU's, or all OTU's belonging to the taxon that have *ever* existed? We are obviously faced by serious problems here.

Smirnov (1969) has been the foremost advocate of a probabilistic weighting of comparisons based on multistate characters. The reference cited (published in Russian) is the most comprehensive and up-to-date version of his philosophy and methodology. For the English-speaking reader an introduction to his similarity coefficient is furnished in Sokal and Sneath (1963) and a more extensive English account is found in Smirnov (1968). Briefly, his coefficient, $t_{1,2}$, can be described as the mean over all character states of a series of weights ranging from a minimum weight $1/(s-1)$ to a maximum weight of $(s/2)-1$, with -1 the weight for a mismatch. The symbol s stands for the number of taxa in the study (t in this book). The weights are computed as a function of the ratio of the number of OTU's in the study exhibiting a certain character state over those not exhibiting it. The details of the method can be consulted in the references cited. Smirnov's philosophy for establishing his method is defended in his 1968 paper. He believes that a rare character is more important for establishing relationship than a common one. If phenetic similarity is desired this clearly is an untenable assumption, even if consideration is given to the random and often haphazard selection of characters for a given numerical taxonomic study. From a careful perusal of Smirnov (1968) it would appear that he is looking for a unique natural system that presumably, although this is not explicitly stated, is a phylogenetic one. There may be some justification for expecting rare characteristics to be more indicative of phylogenetic lines than common ones, yet clearly rarity in itself is no guarantee of patristic similarity. There are also practical difficulties in that the value of a similarity coefficient will necessarily change as the taxa considered are made more or less inclusive. This is somewhat true of other measures of phenetic similarity as well, but would be especially noticeable in Smirnov's system. Gower (1970) has discussed these problems in relation to a similar proposal by Burnaby (1970). For all these and other reasons we are not convinced of the value of Smirnov's coefficient in phenetic taxonomy and believe its validation in phylogenetic taxonomy needs further justification. Smirnov's system has been extensively applied in numerical taxonomic studies in Russia. References to such studies can be found in Smirnov (1968, 1969). We doubt whether the results of using Smirnov's system are likely in practice to be appreciably different from those obtained from the more usual coefficients. Katz and Torres (1965) found that Smirnov's method gave results similar to that of Rogers and Tanimoto (1960). Sneath (1965) noted that weighting character states according to their rarity made little difference in the resulting classification. It is also possible to take the view that the commoner matches should receive more weight than rare ones. For example Gambaryan (1964, 1965) employs for 0,1 characters a matching coefficient in which each of the m matches is first multiplied by a weight w_i that equals $-[p_i \log_2 p_i + (1 - p_i) \log_2 (1 - p_i)]$, where p_i is the proportion of 1 states among the t OTU's. This weight is greatest for characters with $p = 0.5$. This approach is related to the information statistics discussed in Section 4.6. Shmidt (1970) found that Gambaryan's coefficient and Smirnov's gave very congruent results.

The General Similarity Coefficient of Gower. Gower (1971a) has proposed a general coefficient of similarity that is applicable to all three types of characters, two-state, multistate (ordered and qualitative), and quantitative. For each of these types the coefficient relates to one of the established association or distance coefficients. One version of Gower's coefficient, is obtained for two individuals $\mathbf{j}$ and $\mathbf{k}$ by assigning a score $0 \le s_{ijk} \le 1$ and a weight w_{ijk} for character i. The coefficient is defined as

$$S_G = \frac{\left(\sum_{i=1}^{n} w_{ijk} s_{ijk} \right)}{\sum_{i=1}^{n} w_{ijk}}$$

The weight w_{ijk} is set to 1 when a comparison is considered valid for character i and to 0 when the value of the state for character i is unknown for one or both OTU's. For two-state characters s_{ijk} is 1 for matches and 0 for mismatches. Gower typically sets $w_{ijk} = 0$ in two-state characters when two OTU's match for the negative state of a two-state character. In this respect he follows the philosophy of the Jaccard and Dice coefficients. However, the option of permitting negative matches is furnished. In a data matrix consisting of two-state characters only, Gower's general coefficient therefore becomes S_J, the coefficient of Jaccard. For multistate characters (ordered or qualitative) s_{ijk} is 1 for matches between states for that character and is 0 for a mismatch. The number of states in the multistate character is not taken into consideration and in this detail the coefficient resembles a simple matching coefficient applied to a data matrix involving multistate characters. For quantitative characters Gower sets $s_{ijk} = 1 - (|X_{ij} - X_{ik}|/R_i)$, where X_{ij} and X_{ik} are as customarily defined in the data matrix (Section 4.1), and R_i is the range of character i in the sample or over the entire known population. This may be assigned from a knowledge of the total range of the character in the population or from the range as observed in the sample. This formulation results in $s_{ijk} = 1$ when character states are identical and $s_{ijk} = 0$ when the two character states in OTU's **j** and **k** span the extremes of the range of the character. This formulation is the complement of the mean character difference of Cain and Harrison (1958).

A similar formulation has been independently developed by Colless (1967a), by Carmichael, Julius, and Martin (1965) and also by Anderson (1971). All these coefficients permit the mixing of different types of characters.

Gower points out that similarity matrices based on his coefficient make it possible to represent the t OTU's as a set of points in Euclidean space (i.e., the matrices are positive semi-definite). A convenient representation can be obtained by taking the distance between the **j**th and **k**th OTU's proportional to $(1 - S_{jk})^{1/2}$ (Gower, 1966a). It can be shown that not only are similarity matrices of Gower's coefficient based on data matrices containing each type of character positive semi-definite, but also that any combination of the three types treated by the general formula will retain this property. NC's may cause the similarity matrix to lose its positive semi-definite property. This is also true of R-type (or character-) correlation matrices involving substantial numbers of NC's.

When Gower's coefficient is modified to allow negative matches in two-state characters it appears to be a very attractive index for expressing phenetic similarity between two OTU's based on mixed types of characters. It has been used in several numerical taxonomic studies of organisms (e.g., Eddy and Carpenter, 1964; Sheals, 1964; and Sims, 1966) as well as two studies of soils (Rayner, 1965, 1966). Gower (1971a) has also developed his coefficient to permit arbitrary weighting of characters and weighting of primary and secondary characters in the sense of Kendrick and Proctor (1964).

Among the special coefficients of association is one developed by Sackin, Sneath, and Merriam (1965) for cross association of nonnumeric sequences. This procedure is useful in matching amino acid sequences in proteins or stratigraphic sequences in geology (Sackin and Merriam, 1969). The similarity coefficient employed is the Gower coefficient as applied to qualitative multistate characters. These authors compute the similarity coefficients between two sequences in different positions of mutual alignment, studying the magnitude of the similarity coefficient as a function of its alignment position. Specially high peaks of alignment indicate homologies between corresponding positions. There are special techniques for inversions, deletions, and similar abnormalities of the linear order of the structures being studied.

A general discussion of ecological similarity coefficients is furnished in the valuable book by Greig-Smith (1964). In ecological classification, which generally measures similarity between stands expressed in terms of species composition, the frequency of the number of individuals of each species is often incorporated in the measure of similarity. Representative of this formulation is Morisita's coefficient, $2\Sigma n_{1i}n_{2i}/(\lambda_1 + \lambda_2)N_1 N_2$, where summation is over the species contained in stands 1 and 2, with n_{1i} being the number of individuals of species i in stand 1, N_1 the total number of individuals in stand 1, and λ_1 is a measure of diversity devised by E. H. Simpson, estimated as $\Sigma n_{1i}(n_{1i} - 1)/N_1(N_1 - 1)$, again summed over the species occurring in each stand.

Another ecological index that accounts for sample size was proposed by Mountford (1962). The argument behind his coefficient is that indices in general increase with sample size because large samples are more likely to include rare species. This problem, of more moment in ecology than in taxonomy, led Mountford to the development of a new index more independent of sample size than S_J or S_D, based on the logarithmic series distribution following C. B. Williams. Based on a 2 × 2 (presence–absence) arrangement of the data, his coefficient I is obtained as the positive root to the equation

$$e^{(a+b)I} + e^{(a+c)I} = 1 + e^{(a+b+c)I}$$

A good approximation for I is the expression $2a/(ab + ac + bc)$.

4.5   CORRELATION COEFFICIENTS

Among the most frequently employed coefficients of similarity in numerical taxonomy is the Pearson product-moment correlation coefficient calculated between pairs of OTU's. This coefficient was employed in psychometric work as long ago as Stephenson (1936), who originated the Q technique under the name of inverted factor technique. It was first employed in numerical taxonomy by Michener and Sokal (1957), and by Sokal and Michener (1958), and has been employed numerous times since in plants (Soria and Heiser, 1961; Morishima, 1969b), animals (Moss, 1967; Hendrickson and Sokal, 1968; Boyce, 1964), in immunotaxonomy (Basford et al., 1968), and ecology (Fujii, 1969), to cite just a few instances. The application of the correlation coefficient has been quite frequent also in the classification of plant associations (see Greig-Smith, 1964 for a review of the literature) and in animal distributions (Fisher, 1968). It is also used in soil classification (Moore and Russell, 1967).

In general the coefficient has been used on data where most if not all of the characters were present in more than two states. This coefficient, computed between OTU's **j** and **k**, is

$$
r_{jk} = \frac{\sum\limits_{i=1}^{n}\left(X_{ij} - \bar{X}_j\right)\left(X_{ik} - \bar{X}_k\right)}{\sqrt{\sum\limits_{i=1}^{n}\left(X_{ij} - \bar{X}_j\right)^2 \sum\limits_{i=1}^{n}\left(X_{ik} - \bar{X}_k\right)^2}}
$$

where X_{ij} stands for the character state value of character i in OTU j, $\bar{X}_j$ is the mean of all state values for OTU j, and n is the number of characters sampled. Since this formula is based on moments around the mean, it takes into account the magnitudes of mismatches between taxa for characters with more than two states. In this respect correlation coefficients are superior to some coefficients of association described in Section 4.4. Correlation coefficients range from -1 to $+1$. Although high negative correlations between OTU's are possible, they are unlikely in actual data, since pairs of OTU's antithetical for a sizeable number of characters are improbable if there are more than a few OTU's in the study. This is due to functional restrictions on characters, which would make it difficult for an organism occupying a region in A-space to be antipodal to the region occupied by another organism. For example, it is improbable that one OTU would score low on every measurement scale on which another OTU scored high, and at the same time score high on every scale on which the other scored low.

In statistical theory the Pearson product-moment correlation coefficient estimates a parameter, ρ, of the bivariate normal frequency distribution. However, in numerical taxonomic data such a distribution is unlikely in view of the heterogeneity of the column vectors. The characters (rows of the data matrix) are not independent of each other, and therefore we cannot base significance tests of correlation coefficients in numerical taxonomy on the usual assumptions. If different characters are measured in widely varying scales this would impair the validity of the correlation coefficient even further. For this reason psychometricians had long advocated standardization of the rows of the data matrix (equivalent to standardization of characters over all OTU's). This practice has been generally followed by numerical taxonomists. In such cases, the mean of the column vectors frequently approaches 0 and permits a representation of inter-OTU relations by a relatively simple geometrical model, as Rohlf and Sokal (1965) have pointed out. This is illustrated in Figure 4-4, where three OTU's are shown in a three-dimensional A-space. The angles between the vectors bearing the OTU's at their tips are measures of the similarity between the OTU's. The smaller the angles the greater the similarity between the OTU's. In those instances in which the vectors carrying the OTU's are of unit length and the column means are zero the cosine of the angle

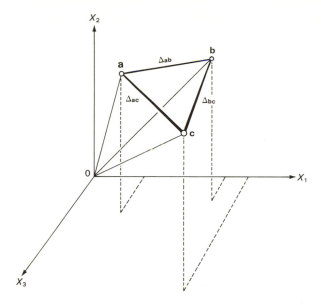

FIGURE 4-4
Two dimensional representation of three OTU's plotted in a
three-dimensional A-space. The OTU's **a**, **b** and **c** are plotted
with respect to three character axes X_1, X_2, and X_3 defining
a three-dimensional space. Distances between OTU's are
identified by Δ. Correlations between OTU's are functions of
the angle between the lines connecting them to the origin, O.
For further explanation see text. [Modified from Rohlf and
Sokal (1965).]

between any two vectors is an exact measure of the correlation coefficient. Standardization of the characters will approximately normalize the column vectors of the
OTU's and correlation coefficients based on such data will therefore approximate
the cosines of the angles of a geometrical model in A-space for these OTU's. The
sampling error of r may be obtained by transforming into Fisher's z, whose variance
is $1/(n-3)$. Rohlf and Sokal (1965) believe the expected value of the average
correlation coefficient in a matrix of correlations based on standardized characters
to be zero but have not been able to demonstrate this mathematically. Other
angular coefficients have occasionally been used in numerical taxonomy, for
example, the cosine shape coefficient (Boyce, 1965, 1969, and see Section 4.11) and
Kendall's rank order correlation τ (Daget and Hureau, 1968).

Correlation coefficients are generally nonmetric functions. When they are converted to some simple complementary form to correspond to distances they do not
generally obey the triangle inequality and it can also be shown that perfect correlation could occur between nonidenticals, such as two column vectors, one of which
is the other multiplied by a scalar. Orloci (1967a) points out that semichord

distances equal to $R \sin \frac{1}{2} \theta$ (where R, the radius of the hypersphere, can be considered as unity, and θ is the arc cos r) will yield a Euclidean metric. This transformation can be computed as $\sin \frac{1}{2}(\text{arc cos } r) = [\frac{1}{2}(1 - r)]^{1/2}$.

Eades (1965) has pointed out an undesirable property of the correlation coefficient for numerical taxonomy. In special cases with few characters the direction of the coding of the character states will affect the values of the correlation coefficient. Minkoff (1965) too has stressed that the taxonomic use of r_{jk} requires that all characters have the same directional and dimensional properties. However, we believe that in the large suites of characters (particularly morphological measurements) customarily processed in numerical taxonomy as well as in large data matrices involving many OTU's, such undesirable properties would not manifest themselves, and it has in fact been the experience of numerical taxonomists that when the interpretation of taxonomic structure is made on the basis of phenograms, correlation coefficients are usually the most suitable measure when the results are evaluated by conventional taxonomists. Boyce (1969) points out that Minkoff's conditions are fulfilled in phenetic studies based on morphology when the forms are characterized by linear dimensions. In such studies problems of size and shape (Section 4.11) arise especially often, explaining the success of the correlation coefficient as a measure of resemblance; if such problems must be faced then directional and dimensional properties of characters must also be suitably handled.

4.6   PROBABILISTIC SIMILARITY COEFFICIENTS

Various workers have repeatedly attempted to develop similarity coefficients that would take into account the distribution of the frequencies of the character states over the set of OTU's. The philosophy here is that agreement among rare character states is a less probable event than agreement for frequent character states and should therefore be weighted more heavily. We have already encountered this philosophy in Section 4.4 when discussing Smirnov's coefficient, which we might well have included in the present section, except that it did not calculate the exact probabilities of given matches.

A probabilistic similarity index utilizing two-state, ordered multistate, and measurement characters has been developed by Goodall (1964, 1966c), who computes the cumulative probability that a given pair of OTU's **j** and **k** will be as similar or more similar than can be empirically ascertained for each character on the basis of the observed distribution of its states in the set of OTU's under study. The probabilities for the entire suite of characters are combined, using Fisher's method for combining probabilities (see Sokal and Rohlf, 1969, p. 623). The details of Goodall's formulation can be consulted in the appendix to his paper. The computational load required for the application of this method is considerable, and for large studies it may be prohibitive. Goodall (1966a) reanalyzed earlier bacterial studies by Sneath and collaborators by this method and found that his probabilistic similarity coefficient in general gave results congruent with the earlier study. However, his index was more sensitive when used to find smaller groupings. Another study is by Clifford and Goodall (1967). We have some questions about whether probability levels could be validly used to develop a phenogram as Goodall suggests. Once a cluster has been proven present, the null hypothesis, on the basis of which the index was originally calculated, is violated; hence probabilities would have to be recalculated for each cluster. Common probability levels cannot therefore be used as criteria of rank. Our most serious reservation, however, is the fundamental one already raised: whether rarer character states should be given enhanced weight. Goodall (1966c) claims that the working taxonomist prefers to enhance the weights of rarer characteristics, but this does not seem to us to be a generally accepted dictum of taxonomic practice.

Probabilistic similarity indices may be of more relevance in ecological classification, where the occurrence or abundance of a species in a stand or sampling unit is a stochastic function. Species in such studies are the equivalent of characters in ordinary numerical taxonomy, and stands correspond to OTU's. Even here strict criteria of phenetic similarity would preclude the unequal weighting of rare species.

Given the probabilistic nature of phytoecological sampling, there has been some tendency by plant ecologists to introduce information-theoretical concepts, based on the frequencies or probabilities of the occurrence of the various species over the collection of stands, into numerical taxonomic work (e.g., Watson, Williams, and Lance, 1966). If the probabilistic assumptions are valid, these concepts have certain built-in advantages. They are based on probability theory, hence amenable to statistical testing using the well known relationship between the information statistic and the chi-square distribution. The information statistics are additive, which permits the partitioning of indices for larger groups into information statistics for subgroups. Furthermore, measures of information in ecological classifications tie in well with measures of diversity developed for faunistic and floristic work in recent ecological and biogeographic theory. Given a probability distribution for the occurrence of various character states over the t available OTU's ("the universe of discourse"; Goodall, 1966a), a measure of information or entropy can be constructed as follows. It should be noted that "information" in the technical sense is not a measure of the amount of knowledge about a group, but rather is a measure of disorder, variance, or confusion, or as Orloci (1969b, who may be consulted for further details) puts it, "surprisal." We define the measure of disorder for character i as

$$H(i) = - \sum_{g=1}^{m_i} p_{ig} \ln p_{ig}$$

where m_i is the number of different states in character i, and where p_{ig} is the observed proportion of the t OTU's exhibiting state g for character i. Consequently

$\sum_{g=1}^{m_i} p_{ig} = 1$. If the n characters are not correlated, one can sum the separate values of $H(i)$ to yield the total information of the group: $I = t \sum_{i=1}^{n} H(i)$. Remember that in an ecological setting characters (rows) are species and OTU's (columns) are stands or quadrats, and this formula is therefore applied to data matrices where the frequencies of any one species are recorded over a series of stands. These frequencies have to be grouped into classes to provide the m_i "states" of a species (or character) i; for example, into classes designated "absent," "rare," "common," "abundant." It is sometimes easier to compute from the frequencies, f_{ig}, directly

$$I = nt \ln t - \sum_{i=1}^{n} \sum_{g=1}^{m_i} f_{ig} \ln f_{ig}$$

which is equivalent to the formula for I given previously.

Often the frequency data can be recoded into two-state characters, presence or absence of the species in a given stand. In such a case

$$H(i) = - [p_i \ln p_i + (1 - p_i) \ln (1 - p_i)]$$

and the total information of the data matrix is then written as

$$I = - t \sum_{i=1}^{n} [p_i \ln p_i + (1 - p_i) \ln (1 - p_i)]$$

It is simpler to rewrite this formula in terms of frequencies for taxon **J** containing t_J OTU's.

$$I_J = nt_J \ln t_J - \sum_{i=1}^{n} [a_{iJ} \ln a_{iJ} + (t_J - a_{iJ}) \ln (t_J - a_{iJ})]$$

where a_{iJ} is the number of OTU's in taxon **J** possessing the + (present) state. The quantity $2I$ is approximately distributed as χ^2 with $2n$ degrees of freedom. The quantity I can be considered as a measure of the total information (entropy, surprisal) in the entire sample. The more homogeneous a taxon, the lower its value of I will be. One aim of clustering using information statistics is therefore to obtain taxa with low values of I, which rise steeply when these taxa are joined.

Regardless of whether two or more character states are employed in a study, a data matrix can be represented as a contingency table, where each dimension of the table is one of the characters and the number of rows and columns of the table is that of the character states. For example, if the two characters h and i have the character states shown at top left, next page, for OTU's **1** through **4**, we can record the frequencies of pluses and minuses in the contingency table in the center.

Characters	OTU's			
	1	2	3	4
h	+	+	−	−
i	+	−	−	−

		i +	i −	
h	+	1	1	2
h	−	0	2	2
		1	3	4

	+	−	
1	2	0	2
2	1	1	2
3	0	2	2
4	0	2	2

Such a table permits an alternative method for computing I over the two characters for the taxon comprising OTU's **1** through **4**. Only the cells representing combinations $+ -$ and $- +$ give surprisal, or information, since $+ +$ and $- -$ are homogeneous. This can be confirmed by rewriting the data in the form shown at the right and computing I_{hi} as in the formula for I_J, above.

$$I_{hi} = (2 \ln 2 - 2 \ln 2 - 0 \ln 0) + (2 \ln 2 - 1 \ln 1 - 1 \ln 1)$$

$$+ (2 \ln 2 - 0 \ln 0 - 2 \ln 2) + (2 \ln 2 - 0 \ln 0 - 2 \ln 2)$$

$$= 1.386$$

One can also compute the information for each OTU based on the marginal frequencies of the 2×2 table. Thus $I_h = (4 \ln 4 - 2 \ln 2 - 2 \ln 2) = 2.773$ and $I_i = (4 \ln 4 - 1 \ln 1 - 3 \ln 3) = 2.249$. The total information will be equal to, or less than, the sum of the information contributed by the marginal totals. This is because of the correlation of the characters over the two OTU's.

The correlation of characters leads to two further concepts stressed by Estabrook (1967) and Orloci (1969b). *Joint information* $I(h, i)$ is the union of the information content of two characters over a set of OTU's; *mutual information* $I(h; i)$ is their intersection. The relation between these quantities is the following:

$$I(h) + I(i) - I(h, i) = I(h; i)$$

$I(h, i)$ is computed as $n \ln n - a \ln a - b \ln b - c \ln c - d \ln d$, or $n \ln n - \Sigma f \ln f$ summing over the cells of the 2×2 table whose frequencies are symbolized conventionally as a, b, c, and d. In the example shown in the 2×2 table above

$$I(h, i) = 4 \ln 4 - 1 \ln 1 - 1 \ln 1 - 0 \ln 0 - 2 \ln 2 = 4.159$$

We can obtain the intersection of the information space, between the two characters: $I(h; i) = I(h) + I(i) - I(h, i) = 2.773 + 2.249 - 4.159 = 0.863$. This is the mutual information that measures the surprisal due to the states shared by the two characters in the four OTU's. It can also be independently computed as the minimum discrimination information statistic (Kullback, 1968)

$$2I = \left[\sum^{m_h} \sum^{m_i} f_{ij} \ln f_{ij} - \sum^{m_h} \left(\sum^{m_i} f_{ij} \right) \ln \left(\sum^{m_i} f_{ij} \right) \right.$$

$$\left. - \sum^{m_i} \left(\sum^{m_h} f_{ij} \right) \ln \left(\sum^{m_h} f_{ij} \right) + \left(\sum^{m_h} \sum^{m_i} f_{ij} \right) \ln \left(\sum^{m_h} \sum^{m_i} f_{ij} \right) \right]$$

where m_h and m_i are the number states of characters h and i, respectively, and f_{ij} stands for the frequency or count of the ith state of character h with the jth state of character i. The data matrix could then be treated as though it were a contingency table (Sokal and Rohlf, 1969, p. 599) and can be analyzed this way. The computation of $I(h;i)$ using the $2I$ formula of Kullback is as follows for the 2×2 table given above:

$$2I = 2[(1 \ln 1 + 1 \ln 1 + 0 \ln 0 + 2 \ln 2) - (2 \ln 2 + 2 \ln 2) - (1 \ln 1 + 3 \ln 3)$$

$$+ 4 \ln 4] = 2(1.386 - 2.772 - 3.296 + 5.545) = 2(0.863)$$

Therefore, $I = I(h;i) = 0.863$. All the above information statistics were applied to characters over sets of OTU's, i.e., they are R approaches. One could, however, equally well measure information content of OTU's over characters and set up contingency tables of OTU's against other OTU's (the Q approach).

Most applications of information analysis in numerical taxonomy have employed the total information using the R approach. The most usual procedure is to consider all t OTU's and first find the two whose fusion leads to the smallest gain of information ΔI, and successively fuse OTU's or groups on this principle, thus producing clusters of maximum informational homogeneity. This general process is taken up again in several sections of the next chapter but it may be noted here that the information statistics can be applied to *sets* of OTU's, whereas most resemblance coefficients only apply to pairs of *single* OTU's or pairs of taxa.

An index based on mutual information has been discussed by Estabrook (1967) and Orloci (1969b), and some work with it is reported by Hawksworth, Estabrook, and Rogers (1968). It is the ratio of the sums of the information content exclusive to characters h and i divided by the total information possessed jointly by h and i. Under some special conditions all of these coefficients of disorder can be shown to be metrics (Rajski, 1961), and can be put in the form of a similarity coefficient (Rajski's coherence coefficient) in the conventional manner as $S(h, i) = [1 - d^2(h, i)]^{1/2}$, which would be a measure of the similarity between characters h and i based on a measure of disorder between them; $d(h, i)$, Rajski's metric, is equal to $1 - [I(h;i)/I(h, i)]$. As first specified by Estabrook (1967) this coefficient measured similarity between two characters, but it could as easily be used for OTU's.

There have been few applications of information statistic resemblance coefficients to biological systematics, most of the applications having been in ecological classification. However, the few published studies (El-Gazzar et al., 1968; Watson, Williams, and Lance, 1966, 1967; Ivimey-Cook, 1969a; McNeill, Parker, and Heywood, 1969a; El-Gazzar and Watson, 1970a) report findings based on information statistics programs that are not too different from those making use of other types of similarity coefficients. Reference to the application of information statistics in ecological classification can be found in Pielou (1969b). Considerable research into other measures of homogeneity and heterogeneity has been carried out by Hall (1967a, 1968, 1969c), whose findings are reviewed in Hall (1969a). He employs a measure which, when restricted to a pair of OTU's, is similar to $M.C.D.$, but can be applied to pairs of sets of OTU's, just as the information statistics can.

Information statistics inevitably lead to measurements of order and disorder of entire classifications and hence fit suitably into the subsequent chapter. However, as we have noted, all such measurements can be decomposed into pair functions between any two OTU's and to this degree they are similarity functions as has been shown above. It has proved difficult to adapt them to continuous, quantitative characters (see Lance and Williams, 1967a; Wallace and Boulton, 1968, Boulton and Wallace, 1970; Orloci, 1970; and Jardine and Sibson, 1971). Where this has been done they often are almost proportional to squared Euclidean distances based on standardized characters, or in the case of clusters, sums of squares. We need more experience with them before we can make firm recommendations about them. A fairly complete discussion of the role of information statistics in phytosociology is given by Orloci (1968c) and in numerical taxonomy in general by Orloci (1969b).

An early formulation of what is in essence a probabilistic and information theoretic similarity coefficient is the distance measure defined by Rogers and Tanimoto (1960). This quantity was defined by them as $d_{jk} = -\log_2 S_{jk}$, where d_{jk} is the distance between any OTU's j and k, and S_{jk} is a similarity coefficient between OTU's j and k—either their own coefficient (see Section 4.4) or some other coefficient, such as the simple matching coefficient. From this definition of distance and similarity these authors define

$$H_j = \sum_{k=1}^{t} d_{jk} = \sum_{k=1}^{t} (-\log_2 S_{jk}) \qquad \text{where } k \neq j, S_{jk} > 0$$

which is a measure of the information content (in bits) of OTU j. From this similarity coefficient, which defines a semimetric space, the authors proceed to an index of inhomogeneity useful in their clustering procedure, best described in the subsequent chapter. Hyvärinen (1962) and Joly (1969) have developed similar coefficients.

4.7   COMPARISON OF COEFFICIENTS

Having made the acquaintance of a variety of coefficients, the reader will un-doubtedly ask himself which ones he should choose for a given study, and what general recommendations can be made about various coefficients. During the early work in numerical taxonomy the hope was expressed that later comparative studies would lead to a clear resolution of such questions. Although a number of comparative studies have been made, we are not really much closer to resolution. Recent studies include Boyce (1969), Sokal and Michener (1967), and Williams, Lambert, and Lance (1966); for a review of the earlier ones see page 166 of *Principles of Numerical Taxonomy*.

We have already seen that the choice of a coefficient will frequently be guided by the scale of the data matrix for which a pairwise similarity function must be com-puted. When the measurement scale is such that several possible coefficients may be employed, the choice among coefficients is often based on the worker's prefer-ence in terms of conceptualization of the similarity measure. Thus distances are preferred by some, and association coefficients are preferred by others. As we have seen in Section 4.2, abstract considerations of metric (see Williams and Dale, 1965) may often determine the desirability of a similarity coefficient. Even more im-portant is the question of exactly what it is that taxonomists wish to estimate, in particular what components of phenetic similarity are desired. This is discussed at some length in Section 4.11, and it is clear that there are numerous conceptual problems to be resolved by taxonomists.

But when all is said and done, the validation of a similarity measure by the scientists working in a given field has so far been primarily empirical, a type of intuitive assessment of similarity based on complex phenomena of human sensory physiology. However, nonmorphological similarities, such as are computed from biochemical data matrices, are far more difficult to verify empirically. Taxonomists have begun to ask themselves questions about the meaning of similarity and about individual differences among taxonomists in recognizing and defining it. When and if some consensus can be established on the definitions for similarity, it should in principle be possible to develop an optimal similarity coefficient. However, studies to date have employed a variety of similarity coefficients, usually in com-bination with several clustering methods (Boyce, 1969; Williams, Lambert, and Lance, 1966; Sokal and Michener, 1967), and often judgments have been based on robustness of similarity measures and on the relative frequency with which various taxonomic structures tended to recur under a variety of clustering schemes. Pitfalls in such procedures are obvious. Williams, Lambert, and Lance (1966) point out the interaction observed in final classifications between different similarity coefficients and clustering techniques. Although all the following authors make some tentative recommendations (Williams, Lambert, and Lance, 1966, prefer an information statistic coefficient with centroid sorting cluster analysis in ecological data, and

Boyce, 1969, in hominoid data and Sokal and Michener, 1967, in bee data prefer correlations and UPGMA), their conclusions are based on past experience with their specific groups and not on conformity with externally defined criteria.

Perhaps the only recommendation that we would care to make at this stage in the development of the field is that, of each type of coefficient considered, the simplest one should be chosen out of consideration for ease of interpretation. Frequently more elaborate coefficients are jointly monotonic with their simpler analogs. We favor more strongly now than we did when we wrote *Principles of Numerical Taxonomy* a similarity coefficient based on binary coding of the data not only because of its simplicity and possible relationship to information theory but also because, if the coding is done correctly, there is the hope that similarity between fundamental units of variation is being estimated. There are also the obvious relations of such similarity measures to natural measures of similarity or distance between fundamental genetic units (amino acid or nucleotide sequences).

During the early work in numerical taxonomy, ease of computation was also important in the choice of coefficients. This has largely been superseded by the development of high speed computers and need not be considered except for the very largest and most elaborate studies. However, certain methods, such as the probabilistic coefficients developed by Goodall (1966c), still require an inordinate amount of computation for a large study.

A useful reference giving a variety of similarity measures from different fields is Ball (1965). These include measures suggested for information retrieval, as well as similarity coefficients in the social sciences.

4.8 CODING AND SCALING CHARACTERS

Coding and scaling are steps that convert the crude data into a form suitable for computation and also preserve the kind of information the taxonomist wishes to consider in making a classification. A good discussion is given by Burr (1968). What kind of information is most often required is discussed at greater length on the following pages, and in Section 4.11, but a simple example is the expression of characters as proportions to avoid effects of gross size. The problem of coding characters that are variable within OTU's is discussed in Section 4.13.

Coding and scaling overlap to some extent, but coding is largely concerned with logical decisions and divisions, and scaling is most often some form of mathematical transformation.

Coding

Characters are subdivided below as follows: (1) Two-state characters (all-or-none characters, binary characters, presence–absence characters); (2) quantitative multistate characters (sometimes called just quantitative characters) including both continuous and meristic ones; and (3) qualitative multistate characters.

Two-state Characters.   These may be recorded as + and − or as 1 and 0. This straightforward method of coding is referred to by Sneath (1957b) as Method A and by Beers and Lockhart (1962) and Beers, Fisher, Megraw, and Lockhart (1962) as Method 1.

It is usual to record positive characters (or attributes marked as present) as + (or as 1) and negative characters (or attributes marked as absent) as − (or as 0), but in most applications it is immaterial whether characters are scored as + or −. For ease of comprehension, however, it is usual to follow the convention above, in particular, where an organ is missing, inasmuch as the negative sign is a clearer indication that attributes belonging to that organ must be scored NC.

Quantitative Multistate Characters.   The states of quantitative multistate characters can each be expressed by a single numerical value; that is, they can be arranged in order of magnitude along a one-dimensional axis. Examples are the amount of a chemical produced by a bacterial strain, length of an animal, or amount of pubescence on a leaf. Quantitative multistate characters may be turned into several two-state characters if association coefficients, most of which require two-state characters, are to be computed. The worker has the choice of recoding the character in some fashion into several two-state characters or arbitrarily dividing the scale into two (not necessarily equally long) parts. The first course has much to recommend it. Quantitative multistate characters are very likely to be caused by more than one genetic factor and several two-state characters may thus be more appropriate. The recoding can be done in a number of ways, but it presents some logical difficulties, since one must decide whether the attributes comprising the multistate character should be treated as additive or nonadditive (see below). But since we do not in fact know whether one or several factors (genes?) are behind the expression of even two-state characters, use of the second option may be more conservative, though undoubtedly losing some information (but this loss is not great; Lance and Williams, 1967a).

Continuous quantitative characters may be divided into a small number of intervals, each representing one of the character states in the data matrix. Earlier this was done to simplify computations. Now that computers are readily available this is no longer a compelling reason.

In any case it is well known in statistics that "grouping" the observations into classes has only minor effects on the value of a statistic (unless the class intervals are very few and very unevenly spaced) for almost all kinds of frequency distributions. Thornton and Wong (1967) give an illustration in numerical taxonomy. If class intervals are set up, they should be evenly spaced unless there is reason to do otherwise (for example if a logarithmic scaling is intended). The number should not be so few that much pertinent information is lost. Certain points deserve further investigation. One of these has been noted by Sneath (1968a); it is the distorting effect of dichotomizing a continuous character into only two states when taxonomic distances are employed. Another point is whether clearer classifications are obtained from two-state characters than from quantitative ones, and if so whether it may be useful to dichotomize characters for this reason.

Qualitative Multistate Characters. The several states in qualitative multistate characters cannot be arrayed in some obvious order but still refer to a unit character on logical grounds. These characters are therefore often called unordered multi-state characters. An example would be sculpture patterns on the surface of an organism or alternative color patterns of a given structure. One way of coding these is to use a separate symbol for each state; for example:

Color of structure	State
Red	A
Yellow	B
Blue	C

A match is scored if the same symbol occurs in two OTU's, otherwise a mismatch is recorded. Another method is to convert the qualitative multistate character into several new characters. The character might then be coded:

Color of structure	Two-state characters		
	1	*2*	*3*
Red	1	0	0
Yellow	0	1	0
Blue	0	0	1

This is not an easy task inasmuch as the recoding has to be done in such a way that a positive score on one of the new characters does not automatically bring about negative scores on all other such characters derived from the same qualitative character. The reader will recognize this as the problem of avoiding logically correlated characters, which was discussed from a more general point of view in Section 3.7. It will also be noted that the comparison of different colors will yield two mismatches, though the number of matches obtained will vary with the number of two-state characters employed and will depend on whether the similarity coefficient takes negative matches into account. This type of recoding is therefore best reserved for situations where it is evident that the new characters are logically independent: for example, the states "stem spiny," "stem hairy," and "stem red," can appropriately be coded as three separate characters, because any combination of the three can theoretically occur. In practice it is commonly found that most qualitative multistate characters can be converted into several independent characters if a little thought is given to the problem.

Additive Coding. By this method, also known as additive binary coding, the multiple character states are coded as follows, illustrated by the quantity of a given chemical substance:

OTU	Multistate character	Two-state characters		
		1	*2*	*3*
a	State 0 (character undetectable)	0	0	0
b	State 1 (weak positive)	1	0	0
c	State 2 (moderate positive)	1	1	0
d	State 3 (strong positive)	1	1	1

In this way a multistate character i of m_i states is turned into $m_i - 1$ two-state characters. The scoring is termed additive because the state 3, for instance, is expressed as the sum of the effects of the states of the two-state characters *1, 2*, and *3*.

Additive scoring may exaggerate dissimilarities due to differences in overall size, and in bacteriology, growth rate and metabolic vigor, although this effect is in general not pronounced, as might be thought (Sneath, 1968d). It has the merit of retaining the information on the magnitude of difference in the characters. It does, however, involve some logical redundancy. If, for example, an organism is marked 0 on character *1*, it is by definition also 0 on characters *2, 3*, and *4*, and if it is marked 1 on character *4*, it is by definition also 1 on characters *1, 2*, and *3*. Also the contribution made by the difference on a given character toward a distance coefficient (such as the complement of S_{SM}—see Section 4.4) is effectively the absolute difference, not the squared difference. In this respect additive coding yields results like the *M.C.D.* (Section 4.3), since $1 - S_{SM} = M.C.D.$ for two-state characters.

Nonadditive Coding. Suppose we do not wish to assume that the effect on the phenotype of several small genetic changes is additive. We may then set up the following model. Two OTU's **b** and **c**, sharing a multistate character but differing in state, may be regarded as being similar in one way (i.e., in possessing a detectable value for the character), but also different in one way (i.e., in having a different value for the character). The magnitude of this difference is not considered. Their similarity on this character is therefore 1/2. If they possess the same state they are similar in both ways, with similarity 2/2. However, an OTU **a** in which the character is not detectable (has the value zero) differs in the first way from **b** and **c**, but since it could not by definition manifest any positive value of the character, it is not comparable to the second character or any further subdivision of it. All states other than the first must therefore be scored NC.

The coding scheme will therefore be as follows:

OTU	Multistate character	Two-state characters			
		1	*2*	*3*	*4*
a	State 0 (character undetectable)	0	NC	NC	NC
b	State 1 (weak positive)	1	1	0	0
c	State 2 (moderate positive)	1	NC	1	0
d	State 3 (strong positive)	1	NC	NC	1

Note that in one sense the four characters can be divided into two groups: the first character, *1*, and the second character, *2*, with characters *2*, *3*, and *4* being merely subdivisions of the entire character *2*. This scheme is that referred to by Sneath (1957b) as Method C and by Beers and Lockhart (1962) and Beers, et al. (1962) as Method 2. It has been used chiefly in bacteriology.

When many characters are employed, the different methods will usually give very similar results, and additive coding appears simple and adequate. Several practical studies in bacteria and fungi confirm these conclusions despite somewhat different views held by various workers (Lockhart, 1964; Kendrick and Proctor, 1964; Ibrahim and Threlfall, 1966a). A small practical point: one may be able to use the same computer program to obtain several different association coefficients by recoding the data in certain ways (Lockhart, 1964, 1970) but at the expense of computer storage space.

The different coding methods described so far give somewhat different average weights to each multistate character. When each state of the character is equally frequent among the OTU's and the number of such states is large, the mean similarity between all possible pairs of OTU's (expressed as matches out of the total number of comparisons, and including two minus values as a match) approaches 2/3 with additive scoring, but 1/3 on nonadditive scoring. If negative matches are not included, the mean similarities are 1/2 in both instances.

Several papers have discussed the proper treatment of characters when certain organs are missing from some of the OTU's (e.g., Proctor and Kendrick, 1963; see also Sections 3.9 and 4.12). We feel that the type of coding suggested by these authors will seldom be necessary, but a simple illustration adapted from Lockhart (1970) is shown here.

Characterization of OTU's	Characters		
	Primary (spores)	Secondary (position)	Secondary (shape)
Spores absent	A	A	A
Spores central, ovoid	B	B	B
Spores central, spherical	B	B	C
Spores terminal, ovoid	B	C	B
Spores terminal, spherical	B	C	C

It will be noted that the position and shape of spores are described as secondary characters, because they can only be expressed if the spore is present. The spore is thus called a primary character. Also, the character scores are given as letters, to indicate that they are to be considered as qualitative multistate characters.

Some special methods of coding may be called for in allometry, in cases of intra-OTU variation and in numerical cladistics, and these are discussed in Sections 4.9, 4.13, and 6.4.

Scaling

It will first be useful to consider the effect of elementary scaling operations. The simplest is *translation*; that is, the addition or subtraction of a constant from all values of a given character. In graphic terms this is the movement of the origin of the scale of measurement. Distance measures of resemblance are unaffected by this, but angular measures of the resemblance are changed if the angle is taken from the origin. A simple example will illustrate this: the distance between two buildings is the same whether we measure from the southwest or northeast corner of the town, but the angle between them will probably be different from the two view-points. The second operation is *expansion*; that is, the multiplication or division of all values by a constant. This corresponds to a change of units of measurement—from inches to centimeters, for example—or alternatively to a linear stretching or shrinking of the scale of measurement. This will in general change both distance and angular resemblance measures. In our example the distance between the houses will not be the same in inches and in centimeters. The angle between them will only be the same if we expand every dimension to the same extent. Both distances and angles will change if some characters are multiplied by one constant and other characters by other constants. These two operations (or combination of them) are simple linear transformations. The effects of nonlinear transformations, such as to logarithms or to square roots, are quite complex, and will affect all types of resemblance coefficients; they correspond to uneven stretching, skewing, and the like.

Proctor (1966) has described the geometric interpretation of some coding and scaling procedures on two-state characters, using the relation $d^2 = 1 - S_{SM}$. She points out that giving a two-state character a weight w is equivalent to representing it by a dimension running from 0 to $\sqrt{w}$ (instead of from 0 to 1). Also, additive coding of k states can be represented by a hypercube of $k - 1$ dimensions, and effectively gives a weight of $k - 1$ to the character. Nonadditive coding can be treated as a regular simplex of order $k - 1$, for the distance between any pair of states is then either zero (for agreement) or one (for any disagreement).

Problems of simple scaling will often arise. If the amount of some substance produced is 1, 10, 100, and 10,000 units, respectively, in four taxa, one can score these 0, 1, 2, and 4, using a logarithmic transformation. This indeed is desirable rather than scoring it in its original scale, since the untransformed variable would exert excessive weight in most coefficients of similarity even if the characters were ranged or standardized (see below). Transformations of this type are, of course, familiar in statistical procedure. We should emphasize here that no character must be allowed to assume excessive weight; to permit this would make nonsense of our attempt to choose a wide and numerous sample of characters. It is far better to reduce the weight on such a character by transformation than to run the risk of its swamping the measures of resemblance.

When effects of overall size are likely to interfere with the study, measurements may be converted into ratios to other characters, or to some convenient measure of overall size, such as the cube root of weight. It appears logical to express areas as square roots, and weights and volumes as cube roots to bring them into comparable units to linear measurements.

Another class of scaling methods is aimed at making the variation of the transformed characters as equal as possible. The intent is to allow each character to contribute toward the overall resemblance inversely in proportion to its variability among the entire set of OTU's. Thus a character with a small range of variation contributes as much as another character with a large variation range. Although there has been much practical work on the effects of such scaling procedures, the logic behind them has not yet been very thoroughly explored. Such procedures eliminate both the effects of size and variability of a character but one could if one wished eliminate only one of these. We may therefore consider these transformations under three headings, (1) those that equalize the gross size of each character, (2) those that equalize the variability of each character, and (3) those that do both.

The gross size of each character can be equalized in several ways. The most logical, to subtract the mean, does not appear to have been proposed. However, such a procedure would not overcome the effect of a large variance (wide range). The method advocated by Cain and Harrison (1958) is to divide each value by the largest state of that character, $X' = X/X_{max}$, thus scaling the states of the characters into the interval 0 to 1. Note, however, that although the largest state scores as 1, the smallest may not score zero, which is different from the ranging of Gower that is described next. Cain and Harrison's method leaves the relative variability (the range divided by the maximum) unchanged, and the character sizes are equalized against the standard of the largest state for each character. There appears to have been no employment so far in numerical taxonomy of transformations that equalize the variability while leaving gross size unchanged.

The simplest form of equalizing both size and variability is by *ranging*, as used by Gower (1971a) in his coefficient S_G (see Section 4.4). In ranging, the smallest value for the character is subtracted from each value and the result is divided by the range:

$$X' = (X - X_{min})/(X_{max} - X_{min})$$

The smallest state among the OTU's then has the value 0, and the largest state has the value 1. With normally distributed characters X' will approximate a monotonic function of the standard deviation (the relation of the range to the expected standard deviation for different sample size may be found in standard statistical texts such as Sokal and Rohlf, 1969).

In standardization of characters we compute the mean and standard deviation of each row (the states of each character) and express each state as a deviation from the mean in standard deviation units (Sokal, 1961). To give an example, if two states for a character are 1 and 6, its mean 4.1, and its standard deviation 0.8, then these states become $(1 - 4.1)/0.8 = -3.875$ and $(6 - 4.1)/0.8 = 2.375$, respectively. The standardization of the character states makes all character means equal to zero and character variances equal to unity. If we wish to add a new OTU we may calculate the standardized values from the previous means and standard deviations, though the resulting value will not really be correct, however, because both mean and standard deviation of the character states have changed with the addition of the new OTU. When only few new OTU's are added this is not a serious problem, as the mean and variance would be inappreciably altered. When a larger number of new OTU's is added (a case we consider unlikely if fairly exhaustive comparative study has preceded the analysis), a fresh standardization of the affected characters will be necessary.

The use of the rank order of the character state values, i.e., the jth rank of the t values of a given character, merits exploration, as it avoids certain problems associated with extreme values or very abnormal frequency distributions. Burr (1968) points out that the use of logarithms of metrical characters has an effect similar in several ways to standardization, and Schnell (1970a,b) observed that matrices based on logarithms or on standardized characters were extremely congruent. However, one should guard against characters that naturally could assume the value of zero.

If desired, a transformation to a normally distributed variable with zero mean and unit variance can be carried out by means of *rankits*. (A rankit is the average deviate of the rth largest in a sample of n observations drawn at random from a normal distribution with a mean of zero and a variance of unity; Sokal and Rohlf, 1969). This procedure has an additional effect, not given by ranging or standardization—conversion of the frequency distribution of character scores so that they approximate a normal distribution (with mean of zero and unit variance). This would only seem to be worth considering if one proposed to use a method that was highly sensitive to departures from a normal distribution. The effects of standardizing rows and columns in a data matrix are discussed in great detail by Cattell (1966b, p. 115 ff.), expressed, we regret, in a rather difficult terminology.

More complex forms of scaling may perhaps deserve consideration in future work, and the papers of Carmichael, Julius, and Martin (1965) and Talkington (1967) may be consulted. The latter paper discusses a simple method for scaling a mixed set of discrete and continuous character states.

There are some sound reasons for considering that the weight of a character should be inversely proportional to its variability. For normally distributed quantitative characters their information (in the information theory sense) is proportional to the variance, so that if the variances are made equal, then each character contributes equal information.

In a more general sense we may argue that the variation contributes most of the information, and that the gross character size and range of variation should contribute little toward phenetic resemblance, in terms of the information appropriate to taxonomy.

There are, however, some disturbing features about standardization and similar scaling procedures. Characters with small ranges of variability and those with large ranges have equal influence on the resemblance coefficients. We may not be able to distinguish these small variations from variation due to other causes. Clearly one would not wish to employ a character whose variation among the OTU's was principally due to measurement error or environmental effects. Hudson et al. (1966) emphasize this point. Also, we exclude characters that have no variation at all. Therefore one has the anomalous position that as one proceeds to increasingly less variable characters one gives the absolute degree of variation more and more weight until deciding there is no variation, and then giving it zero weight by excluding it. We may therefore need to introduce an arbitrary stopping rule.

Although the standardized state codes depend on the distribution of the states and hence on the selection of OTU's, we believe this is unlikely to prove a practical difficulty. All methods are affected to some extent by the choice of OTU's, but standardization is not too sensitive to this. The relation of standardized characters to components of phenetic similarity is discussed in Section 4.11.

Studies of the effect of standardization of characters have been carried out by Rohlf and Sokal (1965) on the 97 species of bees of the *Hoplitis* complex first analyzed by Michener and Sokal (1957) and on the 48 species of the mosquito genus *Aedes* analyzed by Rohlf (1962). These investigations have shown that standardization of characters reduces the average correlation within a matrix to approximately zero from the previous positive value. The standard deviations of the correlation coefficients based on standardized characters are larger than expected, and the coefficients are skewed to the right. Among correlations based on standardized characters, few negative correlations lower than -0.3 have been observed, while positive correlations can range almost up to unity. Underwood (1969) has explored further the effect of standardization, particularly the constraints it introduces in correlation matrices, where it always produces some negative values.

Studies by Rohlf and Sokal (1965), Sokal and Michener (1967), and Moss (1968b) show that in a number of instances standardization has had little effect on distance coefficients, but usually the effects on correlations are quite marked. The relation between r values based on standardized and unstandardized characters is not linear, being constrained by the upper limit of the unstandardized series. Standardization reduces the atypicality of aberrant OTU's, particularly when correlations are employed. It should be noted that although standardization equalizes the size of different characters (by making their means zero) it does not remove the size difference between OTU's. A small OTU has small or negative character state values after standardization.

Standardization is rarely employed with 0,1 data in taxonomy. Its effect is to give greater weight to mismatches upon characters that are rare or very common, compared to those that occur in 50 percent of the OTU's. Watson, Williams, and Lance (1966) standardized such data in a botanical study, and found it made little difference. In an ecological example, Williams, Lambert, and Lance (1966) found that standardization was if anything disadvantageous, as judged by other ecological and numerical criteria. There have been rather few studies comparing different methods of scaling (Crovello, 1968e; Pisani et al., 1969; Schnell, 1970a,b); no major differences due to different methods of scaling have been reported.

All the scaling methods discussed here were applied to rows of the data matrix; that is, each character was scaled independently of the others. There have been a few suggestions in numerical taxonomy for scaling by columns—scaling the character values of each OTU independently—or scaling both rows and columns in turn (Sokal, 1961; Williams and Dale, 1965; Sneath, 1967a). Scaling by columns will largely eliminate the gross size differences between OTU's and is discussed further in this context in Section 4.11. Double standardization (first by rows, followed by standardization by columns) will not affect correlation coefficients. The effects of this procedure on distances and other coefficients has so far not been explored in numerical taxonomy.

Coding and scaling will limit the choice of resemblance coefficients. Most obvious is that nonintegral state scores prevent the use of most coefficients of association. Distances and angular measures of resemblance can be used with integral and nonintegral scores. It is somewhat difficult to predict the effect of dividing a character into several states as compared with two states. Distances between OTU's should in general decrease when a previous two-state case is recoded into more states. However, if the previous example had many matches at zero or one, which on finer classification were shown to be short distances, the overall distance may well increase somewhat. The effect on correlations cannot be readily predicted.

A problem that has received attention in ecology is that of combining together different types of characters (two-state, quantitative, and qualitative multistate) in calculating a resemblance (e.g., Williams and Dale, 1962, 1965). In ecology there are statistical difficulties, but in biosystematic numerical taxonomy these appear less serious. Two-state and quantitative multistate characters may be readily combined in two ways: one way is to treat the two-state characters as 0 and 1 and employ the usual distance or correlation formulae; or one may convert the quantitative multistate characters into a series of two-state characters as discussed previously, and employ association coefficients. Qualitative multistate characters are probably best treated as a series of two-state characters.

Recommendations

Many of the coding and scaling methods described above have not yet been thoroughly explored in practice. A difficulty in making recommendations is that we do not yet have really good ways of checking phenetic resemblance on internal criteria.

We suggest, however, that the following methods are likely to be the most satisfactory for general work. For predominantly two-state characters (to be used with coefficients of association) they should be scored 0,1 and not standardized; where some characters in such studies are multistate we advise additive coding for these.

For studies with mixed two-state and multistate characters, Gower's method of ranging (followed by the use of his coefficient S_G) appears best (for further discussion of such studies see Lance and Williams, 1967a, 1968b). For predominantly multistate characters we advise standardization, but if size is believed to be a serious problem we suggest that the size factor be first removed by one of the methods described in Section 4.11.

4.9 GROWTH AND MORPHOLOGY

In this section we take up two related topics important in some areas of systematics —growth and morphology. Neither has received much attention in numerical taxonomy to date, but both will become increasingly important in the future. In much numerical taxonomic work, particularly with fully adult individuals, the absolute size of a quantitative character (or the ratio between it and some standard measure, such as length or weight) can be employed directly by suitable scaling and coding. Even with adult specimens, this may on occasion be an unsafe procedure, since the size of the character may be dependent on factors other than age—for example, the state of nutrition or, in bacteriology, the temperature during growth. The problem is particularly acute with fossil material, where one has no direct knowledge of the age of the individuals at death, and commonly too few specimens are available for an indirect answer to this question. In addition, it is generally found that the ratio of the size of the character to some standard character also varies with age. Therefore another way of expressing the character is desirable, and this may be done by means of one or the other of the allometry formulae. The reader is referred to Huxley (1932) and Medawar (1945) for a general treatment of this subject, and to Teissier (1960) for an account of methodology.

In most cases it is found that a straight line is obtained if the logarithm of the size of a character is plotted against the logarithm of age (or the logarithm of a standard character, such as total body length). This relation will prove adequate for most numerical taxonomic work, particularly if, as we advise, an effort is made to restrict the study to comparable stages of the life cycle. More complex relations that are discussed in the works just cited can be handled on the same principles. The usual allometry formula is

$$\log Y = \log a + b \log X$$

where X is age (or some other standard measure), Y is the size of the character under study, and a and b are constants describing, respectively, the value of Y when $X = 1$, and the slope of the line. We have adopted the customary symbolism of the regression equation rather than the converse one often applied to the allometry equation, in order to promote uniformity in statistical symbols.

The constants of this formula are normally obtained by plotting a scatter diagram of values of $\log Y$ against $\log X$ and fitting the line by the least squares method, employing the usual formulas for the regression of $\log Y$ on $\log X$. It is, of course, first necessary to be sure that the scatter diagram does approximate a straight line, failing which some other allometric transformation is required. It is also necessary to be sure that the population is homogeneous and does not, for example, consist of individuals of several different species. These points will commonly be evident from the scatter diagram on inspection, and standard statistical techniques can be used to test them. It is probable that electronic computing techniques will be essential in any large-scale work of this kind, both to obtain the regression lines and to check their significance.

Once obtained, the constants a and b can be regarded as characters themselves, since they relate the size of the character under study with the standard character. These two constants can then be scaled and coded in the usual way. Gould (1966) points out that because of the interdependence of the constants a and b it may be unwise to employ intercepts a if the values of slopes b are very unequal in the study, and that in general it is the values of b that appear to convey most information of taxonomic interest. An early, and instructive, example of allometry in taxonomy is that of Reeve (1940), who studied the heads of anteaters, in which the snout length varies greatly with age. The standard character will most usually be overall length, age, or weight, but more sophisticated parameters (such as the geometric mean of length, breadth, and thickness) may prove to be useful. An additional constant $s_{Y \cdot X}$, expressing the scatter about the regression line, may also be employed, since it may be regarded as an additional attribute of the population, but it may not always be clear how much of it is due to errors of measurement, heterogeneity of the sample, environmental effects on the phenotype, and similar factors. Caution is advised here. Further developments in allometry include its multivariate generalization (Jolicoeur, 1963; Hopkins, 1966), and a discussion in the paper of Kidwell and Chase (1967) of different methods of fitting the allometric equation if neither of the variables X and Y can be taken as the independent variable.

Allometry is a problem related to the effect of environment on characters (age, amount of available nutrition, and others) and to the problem of redundancy and empirical correlation (that is, the crude measures may depend on a small number of underlying causes). The orthodox systematist faced with this problem chooses specimens of equal age or size for comparison, intuitively judging these as equivalent. Studies may be called for, however, in which the choice of equivalent specimens is exceptionally difficult. For example, if an amphibian larva suffers a delayed metamorphosis it may not be comparable either in age or size to any stage of another amphibian; and where delayed metamorphosis is the rule, as in the axolotl, *Ambystoma mexicanum*, this may be a considerable problem. In such instances it may be necessary to restrict the characters to those that do not show pronounced allometric changes, as is of course the practice in orthodox taxonomy in these cases.

Kaesler (1967) has discussed growth in various invertebrate phyla in relation to numerical taxonomy. Fry (1964) gave an example of a problem of this kind. He was studying species of the pycnogonid genus *Ammothea*, and found that almost all characters, including meristic ones, showed strong allometric effects. The usual numerical taxonomic methods tended to group together the juveniles from several species. That is, the different juveniles resembled each other more than they resembled adults of their own species. Nevertheless, the material contained specimens of all ages, and it should in theory be possible to dissociate the growth stages of the same species. It was partly with this in mind that Sneath (1966e) suggested a computer method for finding such series or sequences. The method uses a "gravitational" model by which the points (representing the specimens in a character hyperspace) are moved iteratively into smooth curves, and the sequences of nearest neighbors are then printed out. The method can in principle discover multiple, branched, or closed sequences, but tests on Fry's data showed that a considerable number of specimens is needed to achieve satisfactory results. Shortest spanning trees (see Section 5.7) may be an alternative technique.

Numerical taxonomy may have important contributions to make to studies of growth. A step in this direction is the work of Blackith, Davies, and Moy (1963), in which canonical analysis was applied to the development of a hemipteran. The resulting growth pattern was Y-shaped; juveniles occupied the stem of the Y, and males and females respectively occupied the two tips. Phenetic methods are attractive because they can measure growth rates and directions in A-space, considering many attributes at once, and yet by ordination the salient patterns can be readily displayed. Phenetic work is being started on human growth (Tanner et al., 1967) and on functional anatomy in primates (Oxnard, 1969a,b; Oxnard and Neely, 1969). Guttman and Guttman (1965) note scattered work on this by several of the early biometricians, including the observation (referred to as the "rule of neighborhood") that bones in greater proximity in the hand have greater correlation in length. In arthropods such findings have been made by Sokal (1952) and early observations date back to Alpatov and Boschko-Stepanenko (1928).

D'Arcy Thompson (1917) pointed out that many organic shapes could be expressed as simple mathematical transformations of other shapes (further discussed by Woodger, 1945). Raup (1961, 1966) has applied this idea to the shape of shells of molluscs. He shows that the shape of many gastropod shells can be specified by such characters as the profile of the whorl at one point together with a function expressing the rate at which the whorl increases in size at each complete turn of the helix. Clearly, these characters express the morphometrics of the shell more economically than measuring a large number of dimensions of the shell at random.

The simpler of the transformation grids given by D'Arcy Thompson (see Figure 4-5) can be treated in the same way. If there is a straightforward expansion of one part of the grid, this is the only pertinent difference between the forms as represented there; all the other differences in measurements are a consequence of it. For numerical taxonomy one would prefer to use a single figure that described this expansion rather than many independent measurements, for independent measurements would indicate a much greater difference than would seem justified; moreover, many of the characters would be redundant, since their values would be logical consequences of the expansion function. The suggestion of Hall (1969b) on this point has been mentioned in Section 3.7.

Here again we meet the problem of empirical correlation. The simplest hypothesis is that mentioned above, but it is also possible that the expansion is due to a number of genetic changes, each controlling the expansion of one portion of the body. If we have no evidence bearing on this, we would prefer the simplest hypothesis.

General and simple methods need to be developed for extracting the factors responsible for such transformations. Although it may be easy to recognize that a figure such as Figure 4-5 is due to a regular expansion, it is not easy to see how many separate factors are needed to express more complicated examples, such as the insect heads and marsupial skulls shown in Figure 4-6. Not only are the grid lines deformed in several ways, the deformation is different in different parts of the head or skull, too.

A way of extracting the minimum number of factors that would account for the difference in form would be useful; then these factors could be employed as characters. A start has been made by adapting one type of trend surface analysis (power series polynomials) to the problem (Sneath, 1967a). Before this can be done it is necessary to scale both of the two diagrams under comparison, A and B, to the same size and to fit them over one another to give the least possible misfit (see Section 3.4). It is necessary also to mark operationally homologous points on the diagrams ("corresponding points"). When the diagrams are thus fitted, one can record the displacements of the points of diagram B relative to those of A.

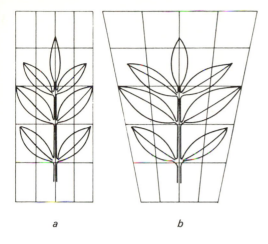

a   *b*

FIGURE 4-5

Transformation grid (after D'Arcy Thompson) applied by the authors to a hypothetical leaf, showing a regular expansion of the distal part of leaf *b* as compared with leaf *a*.

M. J. R. Healy has since pointed out (personal communication) that one can study the displacements from the midpoint of the lines that join corresponding pairs of points on *A* and *B*. These midpoints define a hypothetical diagram, *H*, intermediate in shape between *A* and *B*, and now the trend surfacing of *A* to *H* is the same as of *B* to *H* (though opposite in sign). Trend surface analysis is described briefly in Section 7.5. Further details may be found in Sneath (1967a).

The results of the trend analyses is to produce a series of coefficients of the trends, which do, in a sense, account for the difference in form, and it was found that if these were used as characters in estimating resemblance (by the *T* coefficients in Sneath, 1967a) the results were fairly close to the phenetic similarities judged on other grounds. However, one serious difficulty then arises. If one looks at some of D'Arcy Thompson's figures it becomes clear that it is not the major trends but the small scale detail that is of most taxonomic significance, and it is therefore the residual small-scale variation rather than the trends that should perhaps be used. But this variation must be distinguished from measurement error and individual variation (a problem similar to that in connection with standardization, see Section 4.8). In fact, in the example used, the size of the residual variation seemed as good a measure of taxonomic dissimilarity as any. The problem then may be stated concisely as follows: what algebraic trend surface should we compute, and should we then use for resemblances the trend coefficients or the residual misfit? This needs further exploration.

A related method is that of Smirnov (1927), who resolved the shapes of the elytra of coccinellid beetles into a number of components by a method based on Fourier series; Lu (1965) has done similar work. In this field, too, the nonlinear fitting and scaling methods of Shepard and others may be valuable (Shepard, 1962; Kruskal, 1964a,b; Shepard and Carroll, 1966).

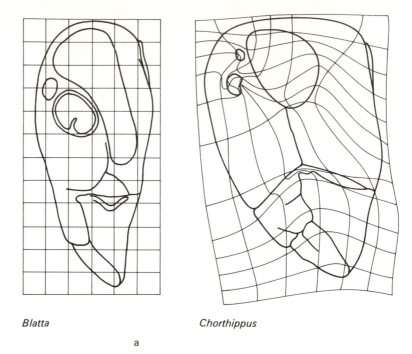

Blatta   *Chorthippus*

a

FIGURE 4-6
Transformation grids of complex forms. a, two orthopteran
heads. [Drawn from nature with grids added by A. J. Lee.]
b and c, the transformation between two marsupial skulls.
[After Parker and Haswell, and Cuvier.]

4.10 STATISTICAL SIGNIFICANCE OF SIMILARITY COEFFICIENTS

Significance testing

Some reference to standard errors and significance testing of similarity coefficients
has been made previously in connection with the presentation of the coefficients
themselves and in Section 3.8, but the general problems of significance testing in
similarity coefficients are perplexing ones that we now take up. The difficulty is
that none of the assumptions customary for developing sampling distributions for
statistics can ordinarily be justified. When we measure the similarity between
OTU's j and k over a set of n characters, the character states for any one OTU are
not samples that estimate a single mean and variance. Rather they are a heterogeneous assemblage of values, which we cannot even assume to have been randomly
sampled. This problem has been referred to in the literature as "the heterogeneity
of the column vectors of the data matrix." Not only do the character states not

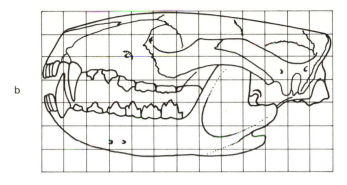

b

Dasyurus

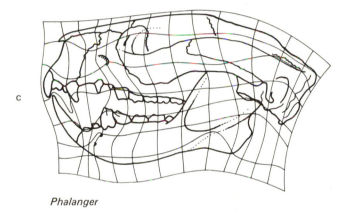

c

Phalanger

estimate a parametric population, but furthermore, the characters (rows of the data matrix) are correlated with each other and thus a similarity coefficient based on *n* characters would not reflect *n* independent dimensions of variation. Standardization of characters, which makes every character state a sample from a population with mean of zero and variance of unity, serves to some degree to alleviate the heterogeneity of column vectors.

Although we have seen that heterogeneity of column vectors makes statistical significance testing of similarity coefficients impossible in a formal sense, approximations to standard errors can be made and in a few tests (e.g., Goodall, 1967) it can be shown that the parametric significance tests approximate the exact probability tests on the data (usually on the conservative side).

In Section 4.6 we encountered similarity coefficients that are a function of the probability of occurrence of a particular combination of states in two OTU's, an approach that may be justified in ecology. In biological systematics there is probably little justification for the development of such coefficients. The data matrix is based on an undefined sample of OTU's and characters, and changes in the set of OTU's or characters chosen will result in changes in the distribution of the character states, and consequently, in changes in the values of the probabilistic similarity coefficients.

As time passes we expect that numerical taxonomy will become increasingly statistical as it considers the significance of resemblance estimates and classifications, and Bailey (1967) as well as Sneath (1967b) have pointed to this as a desirable development that will assist the proper validation of numerical taxonomies. In view of our uncertainties about the nature of sampling distributions noted above we can only outline a few general points and suggestions for future work.

We first need to ask what kinds of significance tests are appropriate in numerical taxonomy. The null hypothesis of no resemblance is probably meaningless in this context. In biological systematics there is some resemblance even between quite remote forms; in many actual studies there is considerable resemblance even between the most dissimilar OTU's. Occasionally it may be useful to ask whether two OTU's are significantly different, i.e., to test the significance of the observed resemblance value from the value for identity. Significance tests of differences between coefficients, or confidence limits on a single value, are of greater interest.

There are several kinds of comparisons in which such tests can be applied. Is the difference in resemblance of two given OTU's significant when compared on separate sets of characters (e.g., on their external versus internal morphology)? This question is difficult to formulate operationally since we cannot determine whether different sets of characters have the same parameters, and hence this question can be answered satisfactorily only in those few cases where the basic units of variation are comparable. This is the kind of question we ask in tests of the nonspecificity hypothesis, but we have so far been able to answer it with some degree of success only for entire classifications (similarity matrices), not for individual coefficients. At this time, this argument would almost exclusively limit comparisons for sets of characters in a pair of OTU's to fine biochemical structure. One might argue that similarities based on amino acid substitutions are comparable, regardless of the protein for which two OTU's are being compared. Similarly, substitutions in DNA base sequences may be comparable. For such comparisons the various approximate standard errors for association coefficients given in Section 4.4 may be employed.

Another type of comparison is more frequent and seems more useful. It is a test of the significance of the difference between two similarity coefficients in the same matrix. Such a test would provide the answer to the question: are OTU's **a** and **b** significantly more similar than **b** and **c**? The answer could affect decisions during clustering procedures. Again, heterogeneity of column vectors makes the application of conventional statistical testing rather hazardous here. Other approaches suggest themselves—nonparametric tests based on gaps, rank order of the similarity coefficients in the matrix, or randomization techniques (for a simple example in numerical taxonomy see Sokal and Rohlf, 1969, p. 634). However, such tests inevitably lead logically and directly into tests of the significance of taxonomic structure, because all significance tests on clusters must be decomposable into those on clusters containing only two OTU's. This aspect of significance testing is therefore discussed in Section 5.10.

Errors Affecting Resemblance Coefficients

Another question of interest in many numerical taxonomic studies is: what are the confidence limits of a single resemblance coefficient? This question is again of importance in interpretation of taxonomic structure, because phenons established on small differences in resemblance value may have no valid foundation.

There are three main sources of error that affect coefficients of resemblance: (1) sampling error of characters, (2) sampling error of OTU's, and (3) experimental or observational error. We shall discuss these in turn.

Sampling Error of Characters.   For binary characters, S_{SM} approaches the binomial distribution for sizable values of n (Goodall, 1967). Standardization of rows (characters) in the data matrices of quantitative multistate characters generally helps to make the similarity coefficients based on such a matrix approach the error expected on the customary statistical hypothesis of random sampling and independence of character state values. Double-standardization (by rows and columns) may improve this tendency. If resemblance coefficients followed the customary random distributions, their standard errors would be roughly proportional to $1/\sqrt{n}$. Note that if n multistate characters have been recoded as n' binary characters, the sampling error depends on n, not n', because of correlation between the states of the multistate characters. Standard errors for some coefficients are given in Sections 4.3–4.6.

However, since each column vector of character states is not a sample from a population representing a single random variable, correlation between characters will lower the true dimensionality of the A-space. Thus a resemblance coefficient would in fact be based on fewer independent dimensions, leading to a reduction in the degrees of freedom available for a significance test. Most reasonable assumptions about the correlation of variables would lead to an increase in the variance of the column vector for each OTU, since interfactor variance and covariance components would be added to a hypothesized error variance σ_j^2. There are some conceivable models that would result in a lowering of the variance, but they seem less likely. The standard error of OTU's, and hence of similarity coefficients, is therefore likely to be underestimated by the conventional standard errors based on n observations, though degrees of freedom will be overestimated. The result is the unfortunate position of being able to estimate the lower bounds of error, but not the upper bounds. This means that if in a significance test of the difference between two resemblance coefficients the null hypothesis of no difference is accepted, it can be done so quite firmly. The Type I error α is almost certainly considerably greater than our test would indicate. By contrast, rejection of the null hypothesis is fraught with danger because our estimates of the standard errors are almost certainly too low, although we do not know by what order of magnitude. The implications for confidence intervals of similarity coefficients are that by using conventional standard errors and degrees of freedom we shall make them too narrow. Clearly more work, possibly by computer simulation, employing plausible models, is required here.

Some basis for constructing such models must come from experimental results on whether sets of characters obtained by a standard technique and distributed evenly over the parts of the organisms are in practice close to random selections of what might be termed a "universe of readily available characters." If this were generally true it would give us much more confidence in estimating sampling errors, despite the present vagueness of this concept. Work such as that on the "intelligent ignoramus" (Sokal and Rohlf, 1970; see Section 5.10) will help to resolve this problem, which depends on the true dimensionality of the A-space. Our present experience is very small, but the general agreement in microbiological studies between similarity values on the same organisms based on different sets of biochemical characters would be in accord with this hypothesis. There appears to be no strong evidence to the contrary from other groups of organisms. This problem may best be explored by the technique of split character sets (see Rohlf, 1965; Lange, Stenhouse, and Offler, 1965; and Table 3-1 in Section 3.6). Clearly this question is also related to the nonspecificity hypothesis, which deals with logically definable classes of characters. The evidence on this hypothesis has already been reviewed in Section 3.6.

If conventional standard errors are employed it will be desirable to transform some coefficients to stabilize their variances over their entire range. This is standard statistical practice. Thus for correlations the z as well as the arccos transformations have sometimes been suggested or employed (e.g., Rohlf and Sokal, 1965; Rohlf, 1962, 1963a). Similarity an arcsin transformation may be used for proportions (Goodall, 1967; Sneath, 1968d), whenever binomial distributions of characters are reasonable assumptions. This may be carried even further, as suggested by C. A. B. Smith and employed by Grewal (1962) and Berry (1963), by subtracting the estimated sampling variance itself. The transformation leads to a difference measure that has an expected value of zero for two identical OTU's (populations in this instance), and which can, by chance, sometimes be negative if the arcsin term happens to be less than the correction factor. Various transformations are discussed in Sokal and Rohlf (1969).

Sampling Error of OTU's.   Quite clearly the estimate of the similarity between two taxa will depend on the reliability of the OTU's chosen to represent them. Sampling theory for OTU's is if anything more difficult to tackle than that for characters. We are generally dealing with unknown distributions of large numbers of OTU's in an as yet ill-defined phenetic hyperspace. When simplifying assumptions can be made about these distributions, such as multivariate normality in the case of some local populations of organisms (see Section 7.2), established statistical tests can be performed. However, this is the exception in numerical taxonomy. The validation of the exemplar method (Section 4.13) rests on the reliability of the OTU's as representatives of their taxa and on the assumption that at the rank level under consideration intrataxon variance is less than intertaxon variance. Moss (1968b) has made some tests of this question, finding that sampling of different OTU's from the same general phenetic structure results in phenograms that do not differ greatly. Since sampling error of OTU's is dependent on the configuration of OTU's in A-space we discuss its further implications in the next chapter (mainly in Section 5.10). The net effect of sampling error of OTU's is to increase the error of resemblance coefficients above the amount introduced by sampling error of characters.

Experimental or Observational Error.   Owing to the difficulty of making accurate and repeatable biochemical tests and chemical assays (e.g., Lockhart, 1967; Lapage et al., 1970; Tremaine and Argyle, 1970) observational error in estimating the values of character states is of more concern in microbiology than in other fields of biology. This error is clearly important in correct identification. Similar considerations apply to chemotaxonomy, e.g., Taylor and Campbell (1969). The same general principles would also be applicable to botanical studies and to those animals in which environmental effects on the phenotype might add to the variance of similarity coefficients. This problem is equally relevant in areas outside biology (such as medicine or psychology) where accurate recording of character states may not be easy. The effect of experimental error on the coefficient S_{SM} has been explored by Sneath (1971a) and Sneath and Johnson (1972). A proportion p of the nt binary test results are assumed to be erroneous (as judged against the majority result on repeated testing or by analysis of variance). The effect of these errors is to shift the value of S_{jk} that would have been obtained from error-free data to a value, S'_{jk}, that lies on the average nearer 0.5, this average being the mean, $\mathscr{E}(S'_{jk})$. The errors also introduce scatter about this mean. The following approximations were derived: $\mathscr{E}(S'_{jk}) = S_{jk}(2p - 1)^2 + 2p(1 - p)$; variance of $S'_{jk} = 2p(1 - p)[1 - 2p(1 - p)]/n$ (Sneath and Johnson, 1972). The effect of experimental error becomes serious when p is over 0.1. It is advisable in microbiological studies to replicate some of the strains so as to afford some check on test errors that might otherwise go undetected. Fortunately, in practice the error is usually less than 0.05 in bacteriology. Figures given by Weimarck (1970) suggest that it is also about 0.05 for chromatographic data in botany, though Taylor (1971) warns that there may be very large seasonal changes in plant constituents that could introduce error.

Although one could consider correcting S'_{jk} if one knew p, this requires considerable caution, because the direction of the error is not known. The effect of experimental errors for other resemblance coefficients requires exploration, but it is easy to show that for taxonomic distance based on standardized characters and using a pooled variance σ_E^2 for the experimental variances of the characters after standardization, that provided σ_E^2 is considerably less than 1, $\mathscr{E}(d') \simeq \sqrt{d^2 + 2\sigma_E^2}$ and var $(d') \simeq \sigma_E^2/n$.

Experimental errors have another consequence: they will increase the apparent sampling error. The sampling and experimental errors will usually not be correlated, so the estimated variance of a resemblance coefficient will be the sum of the sampling and experimental variances. This will give wider but more realistic confidence limits to a resemblance value. Both errors decrease with increasing n, but if characters are taken in order of their reliability, the total error at first decreases but may later increase with the addition of the more unreliable characters. This is especially true for resemblance values close to that representing identity. However, Sneath (1971a) and Sneath and Johnson (1972) found in a microbiological study that it was better to include some of the less reliable characters to obtain a reasonably high value for n than to employ only a few characters of extreme reliability. Related work is that of Jackson (1970).

Many implications of the points discussed in this section remain to be explored. They lead directly to the concept of uncertainty of the distance between OTU's in A-space, and by implication to the concept of uncertainty of the position of single OTU's in that space (although there are problems when the A-spaces are not directly comparable, as with different character sets). They would also lead toward the idea of the uncertainty of position, size, and shape of clusters, and thus link up with similar concepts in multivariate statistics and with topics discussed in Section 5.10.

4.11   THE COMPONENTS OF PHENETIC RESEMBLANCE

What is phenetic resemblance? Is there some unambiguous measure of it? The answers to these questions are complex, and a full understanding of the issues requires competence not only in biology but also in mathematics, sensory physiology, psychology, and philosophy. Early work in numerical taxonomy attempted to simulate in a crude and not too explicit way the purported thought processes of taxonomists. We shall have to know more about the psychology of assessing similarities before we can improve our techniques appreciably. Some experiments underway in the laboratory of one of us (R.R.S.) may provide partial answers. In the meantime, however, we have learned to recognize some logical and mathematical components of phenetic resemblance, which will be discussed below. As we learn more about the nature of the taxonomic process, we may learn what part these components play in it (Ehrlich, 1964; Sneath, 1967a). Different similarity coefficients emphasize these components variously, and individual taxonomists differ in their intuitive assessments of similarity. This may be responsible for observed differences in evaluations of similarity by machine or traditional methods. In fact, an understanding of these components may lead to formulae for phenetic similarity that are preferred on inherent grounds, although they depart from conventional personal assessments. Different aims will legitimately require different formulae, and we expect that new components will be employed in the systematics of the future.

Classes of Phenetic Components

The following classes of phenetic components are discussed in this order: size versus shape; large scale versus small scale detail; atypicality versus distinctiveness; vigor versus pattern; complexity; and discriminatory models. It should be noted that these need not be mutually exclusive; one can, for example, separate size and shape, and on the same data also separate atypicality from distinctiveness, giving two different sets of relationship. It will be recalled from Section 2.4 that there is yet a different way of dividing phenetic resemblance into components: into patristic and homoplastic similarity, based on evolutionary considerations.

Size and Shape.   This is the simplest of the concepts. We may, for example, say that a rat is roughly twice the size of a mouse but approximately the same shape. In classifying other rodents we would probably wish to discount size differences. Evidently overall size makes in some sense a "relatively small contribution" toward taxonomic resemblance, while the complexities of shape contribute much more. But at a deeper level of understanding we see that this is only true in one sense; the size component itself is also made up of much information about the individual characters, but we have assumed that it is desirable to consider much of this size information as the expression of a single factor, overall size.

  There are several ways in which we may measure the size of an organism, by its weight or its length, for example. But we may also think of size as a general factor that affects many characters, so that what is in our mind is a function of the magnitude of the states of many different characters. In ordinary speech we mean spatial size, and it would seem desirable to follow this convention, and to distinguish by other terms, or by qualifying expressions, size in a sense other than spatial. Shape must necessarily refer to ratios between different characters, and for clarity we again restrict this concept, unless qualified, to a spatial mode.

  Measuring size by a standard character such as weight or volume presents no special conceptual difficulty. But if we wish to express it as a function of some dimensional characters we may be in the dilemma exemplified by the question:

which is bigger, a snake or a turtle? Does the fact that a snake is much longer, even though of the same weight, make it the bigger? This relates to the two ways we might measure the size of a rectangle, by its area, or by its largest possible extension, the diagonal. The former, however, is not practicable, because its multidimensional analogue, the hypervolume, is zero if any one character state is zero.

The simplest way of removing the effect of size may well be very effective. This is to express the character values of each OTU as a ratio of a standard measure of size of the OTU (e.g., weight or volume). We would usually prefer to use the cube root of weight or of volume to convert the standard into the same units as lengths of organs and parts. Other possibilities were mentioned in Section 4.9. The resulting ratios will then be an expression of shape alone.

Penrose (1954) has suggested dividing d_{jk}^2 (C_H^2 in his symbolism) into two parts: a coefficient of "size," C_Q^2, defined as the square of the difference between character sizes in OTU's **j** and **k**

$$C_Q^2 = \frac{1}{n^2} \left[\sum_{i=1}^{n} (X_{ij} - X_{ik}) \right]^2$$

and a coefficient of "shape" (which we term C_P^2), defined as the residue after size is removed

$$C_P^2 = \frac{n-1}{n} C_Z^2 = C_H^2 - C_Q^2 = d_{jk}^2 - C_Q^2$$

$$= \frac{1}{n} \sum_{i=1}^{n} (X_{ij} - X_{ik})^2 - \frac{1}{n^2} \left[\sum_{i=1}^{n} (X_{ij} - X_{ik}) \right]^2$$

The "shape" coefficient represents the variance of differences between the character states of the OTU's being compared. It is almost the same as the square of Zarapkin's coefficient $\mathfrak{S}^2$ (Section 4.3), identified as C_Z^2 by Penrose, as seen above.

The C_Q^2 is identical to the correction term used in calculating the variance of differences between the character states. It represents the magnitude and direction of the differences. When C_Q^2 is large, the character states of the two OTU's being compared are quite different in magnitude, and the differences that exist are largely in one direction. A large C_Q^2 would appear, for example, if one OTU were very similar to another but much larger along most of the character scales. Thus, when size is an important factor, this should be revealed by the magnitude of C_Q^2. The quantities C_P^2 and C_Q^2 are distributed as chi-square with $n - 1$ and 1 degree of freedom, respectively (Rahman, 1962).

Rohlf and Sokal (1965) have pointed out that the shape coefficient is not a measure of similarity in proportions, as the name might imply. It is zero—indicating

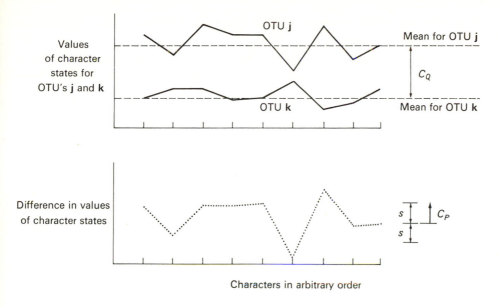

FIGURE 4-7

Representation of Penrose's coefficients of size, C_Q, and shape, C_P.

In the upper part the character state values for OTU's **j** and **k** are shown, and the distance between the dashed lines representing the means for each OTU is the coefficient C_Q, the square root of the form C_Q^2 shown by the formula in the text. This is the mean size difference between **j** and **k**.

In the lower part the differences in the states are shown. The standard deviation of these differences, s, equals the coefficient C_P (the square root of C_P^2) and represents the difference in shape.

identity in "shape"—only if the difference between two OTU's is constant for all of the characters. These authors found that the product-moment correlation coefficient, r, is a better measure of similarity in shape between two OTU's. Boyce's (1964, 1965) findings were similar. The hypothetical state values 1, 2, and 3 for three characters of an OTU **j** and 3, 4, and 5, for the same characters in OTU **k** give a Penrose shape difference of zero, because the states for corresponding characters in OTU **k** are two units greater than those of OTU **j**, although we would not say rectangular boxes of these dimensions were of the same shape. In this case the correlation coefficient suffers from the same drawback, because r between them would also be 1.0.

The Penrose coefficients can be illustrated graphically in a ratio diagram similar to that introduced by Simpson (1941), as shown in Figure 4-7. The mean distance squared between the two lines for OTU's **j** and **k** represents Penrose's size coefficient and the summed variances for each line represents his shape coefficient.

Colless (1967a) points out that Manhattan metrics can also be partitioned into a size component and a remainder. If this is required it would seem convenient to subtract Penrose's C_Q^2, representing size, from the square of the *M.C.D.*, to obtain a quantity analogous to Penrose's shape coefficient. The taxonomic properties of this procedure remain to be investigated.

In Section 4.5 we illustrated (Figure 4-4) the relation between the correlation coefficient and that of taxonomic distance. We reproduce a similar figure here (Figure 4-8) to illustrate the concepts of size and shape in a different manner related to, but not identical with the Penrose coefficients discussed above. The character values are in crude spatial measurements (e.g., all in millimeters), and are therefore all zero or positive. The size of an OTU **j** is given by the length of the vector $\mathbf{x_j}$ in the A-space. Thus for OTU **j**

$$\|\mathbf{x_j}\| = \left(\sum_{i=1}^{n} X_{ij}^2 \right)^{1/2}$$

and similarly for OTU **k**

$$\|\mathbf{x_k}\| = \left(\sum_{i=1}^{n} X_{ik}^2 \right)^{1/2}$$

The difference in shape between OTU's **j** and **k** is measured as the angle η between these two vectors which is given by

$$\cos \eta = \left(\sum_{i=1}^{n} X_{ij} X_{ik} \right) \bigg/ \left(\sum_{i=1}^{n} X_{ij}^2 \right)^{1/2} \left(\sum_{i=1}^{n} X_{ik}^2 \right)^{1/2}$$

The cosine varies from $+1$ for an angle of $0°$ (identity in shape) to -1 for an angle of $180°$ (greatest dissimilarity in shape); if X_{ij} and X_{ik} are always nonnegative, the greatest dissimilarity is represented by an angle of $90°$, yielding $\cos \eta = 0$. This angle η is zero when one OTU has exactly the same proportions as the other, which leads one to place them both on the same line-of-sight from the origin O. Boyce (1965, 1969) has explored the use of $\cos \eta$ as a shape measure and has found that it performs well in this respect. If required, vector size can be scaled by dividing by the square root of the number of characters, in the same way that Δ_{jk} is converted to d_{jk} to yield a measure independent of n. The relation of the correlation coefficient r to these three measures, which is not entirely determinate in an A-space model, has been described in Section 4.5.

Some miscellaneous comments on size and shape follow: standardization of characters by rows does not remove the size factor (though it may reduce it) because small OTU's will in general have negative standardized scores, and large ones positive scores. Moss (1968b) has given an example of this, showing that by using d_{jk} one may obtain separate clusters of large and of small OTU's. In botany the concepts of size and shape may well be less useful than in zoology, for increased size of a plant does not necessarily imply that parts like leaves and flowers are also

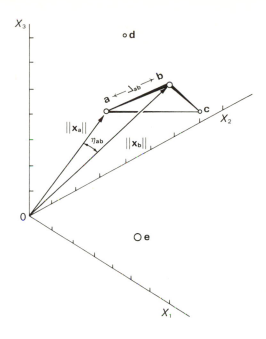

FIGURE 4-8

Representation of vector size, $\|\mathbf{x}\|$, and vector shape, the angle η.

OTU's **a–e** are shown in an A-space of three dimensions X_1, X_2, and X_3. The lengths of the arrows represent vector size and are shown for OTU's **a** and **b**, while the angle between these is η_{ab}, their vector shape difference.

The Euclidean distances between **a**, **b**, and **c** are shown, but the remaining distances and angles are omitted for clarity, as are the projections onto the axes representing the coordinates, as in Figure 4-2.

larger (see Johnson and Holm, 1968; Eshbaugh, 1970). If in an ordination (Section 5.6) the first axis has mostly positive loadings we consider the first axis to be a general size factor. This effect can be removed by subtracting the contribution of the first principal component in an A-space from the distances among a set of OTU's. Hall (1969b) has devised a special coding method for removing size effects in cases of this kind. Kendrick (1964) has suggested a method to reduce the effect of size factors, which makes the size component of taxonomic distance proportional to the size rate of two OTU's and not to the size difference. Thus, considering these hypothetical OTU's **a**, **b**, and **c**, of the same shape but of sizes 0.1, 1.0, and 10.0 units respectively, his method would give d_{ab} equal to d_{bc} equal to 0.9. Most other methods would give d_{ab} equal to 0.9 but d_{bc} equal to 9. The resemblance measure d_h^2 represents a shape dissimilarity, while the variance of a diagram is a measure of its size. Size is an asymmetric relationship (if A is bigger than B, then B is smaller than A). Boyce (1969) points out that if one adds a constant k to all character state values of one OTU, this does not affect C_p; multiplying all values by k leaves cos η the same, and r is unchanged by either procedure. Information statistics do not in general remove size factors.

Large Scale versus Small Scale Detail. It has been pointed out in Section 4.9 that one can treat transformation grids by trend surface analysis, thus isolating morphological trends and the residual misfit of fine scale detail from those trends. This partition can be made at any desired complexity of the trends, for example, separating the factor representing linear stretching of one organism compared with another from another factor not accounted for by such stretching. In the technique described in Section 4.9, however, gross size has already been removed before trend analysis, so the method makes a distinction between different factors of overall difference in shape (d_h).

Atypicality and Distinctiveness. Basically atypicality means some measure of difference from a central or typical OTU. This concept has relevance only to a set of OTU's not to a single pair. It is thus necessary to specify whether one is considering all the OTU's in a study, or some cluster of them. We discuss it here in the context of all the OTU's in a study.

To specify atypicality thus requires two things: the resemblance measure, and the point from which resemblance is measured. This point may represent an actual OTU (such as a typical specimen) or an artificial construct that may not exist as an organism (such as an average of the character values). In most applications it may be disadvantageous to use an actual OTU as the center, but the principles mentioned below can be so applied if desired.

One way of representing atypicality is by a vector model similar to that in Figure 4-8. The origin of the A-space is moved to the point O' given by the mean of each character, so that O' is the centroid, $\bar{x}$, of the set of OTU's (Figure 4-9). The distances Δ_{jk} are unchanged, since each row of the data matrix has simply been coded by subtraction of the character mean $\bar{X}_i$. The distances of the different OTU's from the new origin O' are a measure of their atypicality (Sneath, 1967a). The lengths of these vectors can be easily computed as

$$\|x'_j\| = \left[\sum_{i=1}^{n} (X'_{ij})^2 \right]^{1/2}$$

where $X'_{ij} = (X_{ij} - \bar{X}_i)$. The new angle ζ between OTU's **j** and **k** can be defined by

$$\cos \zeta = \frac{\displaystyle\sum_{i=1}^{n} X'_{ij} X'_{ik}}{\left[\displaystyle\sum_{i=1}^{n} (X'_{ij})^2 \right]^{1/2} \left[\displaystyle\sum_{i=1}^{n} (X'_{ik})^2 \right]^{1/2}}$$

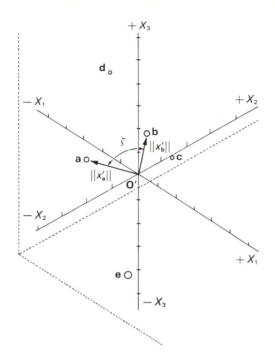

FIGURE 4-9

Representation of vector atypicality and vector distinctiveness in A-space.

The figure is obtained from Figure 4-8 by moving the origin from O to O', the centroid, $\bar{x}$, of the OTU's. The old coordinate frame is shown dotted.

The lengths of the arrows show the vector atypically for OTU's **a** and **b**; i.e., $\|x'_a\|$ and $\|x'_b\|$. The angle between them, ζ, is the vector distinctiveness between **a** and **b**. The distances between the OTU's are unchanged from those in Figure 4-8.

and this angle is termed distinctiveness by Sneath. Cos ζ approximates r_{jk} if the characters are standardized. Character standardization tends to equalize the values of atypicality and distinctiveness between OTU's. This model is also applicable to 0,1 characters: the origin is moved into the interior of the original hypercube in A-space. The $\bar{X}_i$ is simply the proportion of 1 states for character i. The origin here, of course, cannot represent an actual OTU.

The Penrose size and shape statistics are unaffected by change of origin, and so they are the same in the vector size and shape model as in the atypicality-distinctiveness model; there is no straightforward analogue of scaling by weight or volume.

There have been a number of other proposals for measuring atypicality, or its converse, typicality. The measure $-\log_2 H$ of Rogers and Tanimoto (1960) is an estimate of atypicality. Hyvärinen (1962) suggests a typicality coefficient for 0,1 characters, which he terms T_j (for OTU **j**). Hall (1965a,b) describes the *peculiarity* *index* for 0,1 characters, as follows: the more prevalent state for each character is given a weight of zero. The rarer state is given a weight of the difference between (a) half the number of OTU's in the study possessing either state and (b) the number of OTU's with the rare state. The peculiarity index for a given OTU is the sum of these weight over all characters. The OTU's with highest values are the most atypical. Goodall (1966b) describes the *deviant index*, which can also be used with quantitative multistate characters. For an OTU **j** and for each character *i* one finds the proportion of p_{ij} of OTU's that lie further from the mean of *i* than does **j**. One then sums over all characters to give $D_j = -2\Sigma \ln p_{ij}$, which has a chi-square distribution.

A main reason for introducing these measures of atypicality has been to give a criterion for excluding OTU's from clusters when they are too aberrant, and this is thus related to the logic of clustering procedures (Chapter 5).

Vigor and Pattern.   The analogs of size and shape arise with 0,1 characters in bacteriology, and have been termed *vigor* and *pattern* (Sneath, 1968d). If two bacterial strains are identical except that one is a slow-growing variant, then in physiological tests read at the standard time of incubation the slower-growing strain will show fewer positive results than the faster-growing strain. However, one would not expect any tests to be positive in the slow-growing strain that were negative in the fast one. Longer incubation of the former might give the same positive results as the latter at the standard time. Many environmental factors, often complex and hard to estimate, can raise similar problems with normal wild strains (e.g., the studies of Tsukamura, 1966, and Wayne, 1967). It was therefore suggested that vigor be introduced for the concept of metabolic ability over a *specified set* of tests, and that differences unaccounted for by this should be referred to as pattern.

Clearly such a model is only applicable to characters that can be unambiguously scored as positive, but not to those in which coding is arbitrary.

D_T^2, the square of the complement of the simple matching coefficient S_{SM}, can be partitioned into two additive parts, the square of the difference in vigor, D_V^2, and the remainder, the square of the differences in pattern, D_P^2, as shown below:

$$(1 - S_{SM})^2 = D_T^2 = D_V^2 + D_P^2$$

where, for OTU's **j** and **k** (using the symbolism for 2 × 2 tables established in Section 4.4)

$$D_T^2 = (b + c)^2/n^2$$

$$D_V^2 = (b - c)^2/n^2$$

and

$$D_P^2 = 4bc/n^2$$

The vigor of OTU **j** is $(a + c)/n$ and of OTU **k** is $(a + b)/n$.

This partitioning works well in practice in bacteriology and is useful in problems of growth rate, and it helps in elucidating problems associated with spurious resemblances between metabolically inactive strains as well. It may be noted that of the usual association coefficients, S_Y behaves most like D_P (or rather, as the complement of D_P, namely, $1 - D_P = S_P$ in Sneath's notation). It should also be noted that these coefficients are not algebraically identical to the representation of size and shape for 0,1 characters in the vector model described earlier in this section. They may be useful for problems analogous to shape and size in chemotaxonomy.

Complexity.   A special form of resemblance exists between organisms that differ greatly in complexity, that is, where the numbers of applicable characters differ greatly. Reynolds (1965) has drawn attention to this, referring to it as "comparative elaborateness." It may arise in comparing organisms of very different taxa (for example, structurally simple invertebrates, like coelenterates, with complex organisms), or in studies where some members are very degenerate in the morphological sense (such as female scale insects). It also arises in protein sequence work, and may be illustrated by a hypothetical example of short sequences of amino acids. Suppose OTU **a** has a polypeptide of sequence Alanine-Arginine-Valine, OTU **b** has the sequence Alanine-Arginine-Leucine, and OTU **c** has the sequence Alanine-Arginine-Valine-Tyrosine-Glycine. We see that **a** and **c** are identical on the three positions that are comparable in both (the first three positions) but they are not identical if one considers all five positions. OTU **b** on the other hand resembles **a** in only two of the three positions and would ordinarily be considered less like **a** than is **c**. We could add to the dissimilarity, as usually measured, an extra component of two characters that would express difference in complexity. This is related to relevance (Section 4.12) except that the missing characters are not simply unrecorded.

Discrimination Space.   A-spaces based on Mahalanobis' D^2 and similar statistics (see Sections 4.3, 8.5) are spaces in which, as F. J. Rohlf points out to us, the distances between clusters of OTU's are measured in what may be called "ease of discrimination units." This, of course, presupposes that the clusters are already given, and implies some kind of reworking of a numerical taxonomy based initially on single OTU's. Such spaces are sometimes required by taxonomists (e.g., Minkoff, 1965) but we think that their proper application is most often to problems of discrimination (Chapter 8) rather than to construction of taxonomic systems. It is, for example, possible to envisage situations where the D^2 distances between, say, two birds and a mammal were all nearly equidistant. Yet we feel that the taxonomist would usually not wish to lose the information that, by any usual phenetic method, the two birds were much more similar to each other than either was to the mammal. Burnaby (1966) has made some suggestions for partitioning D^2 into components analogous to size and shape. The relation between ordinary A-space and D-space is illustrated in Chapter 8. It corresponds to both the stretching of character dimensions and to the changing of angles between the axes from the original right angles.

Conclusions

While some of the measures described above are worthy of exploration in numerical taxonomy, we have as yet little practical knowledge of components of phenetic resemblance and the statistics best representing them. We would recommend r as the most useful similarity coefficient in that it is the purest measure of shape of the commonly used resemblance measures, and since we believe that distinctiveness is the resemblance component most useful to taxonomists, correlations on standardized characters are to be recommended for general use. Distances based on standardized characters afford important information for making clusters, and are particularly valuable for this purpose. Because of the pervasive effects of size factors we advise that workers should always consider whether they need to remove them. Even more important than recommendations on specific coefficients is the general recommendation that numerical taxonomists should consider carefully what it is they wish to measure. In fact, they should state in their publications what components of phenetic resemblance they have attempted to estimate.

4.12   UNWARRANTED COMPARISONS

Up to this point we have ignored a major problem that must have occurred to most readers: often for a given OTU, no information is available for a particular character, making a comparison between that OTU and others impossible or unwarranted. We can distinguish between the several ways in which this may occur.

Missing Data.   More frequently in some studies than in others, certain items of information may be unobtainable. The only available specimen may be damaged and have some structures missing; museum regulations may prevent dissection for the study of internal characters; distributional or ecological facts may be unknown; equipment for complex chemical or physical tests may not be available. Yet we may have information on the character in question in many OTU's of the study, and the one obvious and simplest solution—the elimination of the character from consideration—would seem deplorable. When such missing data occur, the character state should be labeled with some agreed code for "inapplicable," which should be clearly distinguishable from a minus or a zero. Commonly NC ("no comparison") is employed.

Missing Characters.   Many instances will arise when a given character present in one OTU is absent in another. Most often the zero or minus state of the character will be the appropriate code for this condition. Sometimes, however, one character may be masked by another so that we cannot score it; for example, black pigment would prevent our scoring for the character "presence of yellow pigment." The latter character would then, in a sense be a missing character, and this would be another form of missing data discussed above. When this occurs the character state should be labeled inapplicable.

Missing Organs. More frequent are instances when organs or relatively major parts of the body are absent or strongly modified in a given OTU, with the result that logically subordinate characters contained within the missing part cannot be scored. If, in a study of a group of insects, we have included "presence or absence of a wing" as one character and also five wing vein characters, we cannot score the venational characters in a wingless taxon. The only consistent procedure is to consider the wing vein characters inapplicable in the wingless taxon. Thus they cannot be compared with the corresponding states in the winged form. They also cannot be compared with the wing vein characters in another wingless taxon, since the fact that they are both not manifested does not provide a basis for comparison. Thus the wingless and the winged forms could be compared only on the basis of the single character "presence or absence of wings." We can compare the two wingless forms on the basis of the same character if we accept "negative matches" (Section 4.4). This may seem to be splitting hairs and we must admit that it is not always easy to decide when a character is inapplicable and when it is negative or absent. However, once the decision has been made the subsequent procedure is logical and consistent. A block in a metabolic pathway poses a similar problem (Section 4.4).

Kendrick and his colleagues (Proctor and Kendrick, 1963; Kendrick, 1964, 1965; Kendrick and Proctor, 1964; Kendrick and Weresub, 1966) have suggested an alternative to marking inapplicable characters NC. They distinguish between those characters that are primary (such as presence of some organ) and those that are secondary (such as color of the organ). Secondary characters qualify primary characters. A primary character is given additional weight equal to one more than the number of secondary characters that qualify it. Several slightly different scoring methods can be used, but the effect approximates treating the inapplicable characters as having zero scores, especially when there are many secondary characters. This procedure was suggested because Kendrick and his associates felt that in the taxonomy of fungi there were drawbacks to the usual method, though their arguments unfortunately rest heavily on intuitive estimates of resemblance. Ibrahim and Threlfall (1966a) found that the usual method was satisfactory with the fungi they studied, and Long (1966) notes that complex characters can usually be broken down into unit characters, so the examples of Kendrick that rest on this point seem to raise an unnecessary difficulty. Reynolds (1965) and Williams (1969, who proposes a special weighting scheme) support Kendrick, while Lockhart (1964) suggests modifications that allow a primary character to score twice. Lockhart notes that the problem is only acute when there are few characters available, because variations in scoring may then make an appreciable difference. In a study of bacteria Lockhart and Koenig (1965) compared the use of primary characters alone with the use of primary, secondary, and tertiary characters combined and treated according to the suggestions of Kendrick. The results were very similar, though the latter seemed somewhat more satisfactory, and of course contained more information. The general congruence between S_J and S_{SM} is another pointer toward the relatively small effect that scoring differences make in practice, provided a large number of characters is employed.

We confess that we are not wholly convinced by the arguments of Kendrick and his supporters, for two reasons. First, we are not sure whether one can consistently distinguish primary from secondary (and even tertiary) characters. Second, if it seemed necessary to replace an NC entry by another value, a measure of central tendency, such as the mean of that character over the remaining OTU's, would appear more reasonable than effective replacement by zero. The effect of NC entries on resemblance is quite complex (see below). Jičín and Vašíček (1969) suggest that in calculating S_{SM} it is logical to consider an NC entry as a 0.5 chance of an entry of 1, but we feel that some knowledge of the appropriate state is needed before making such a proposal.

The Estimation of Resemblance When Some Characters Are Inapplicable. No problem exists in the estimation of resemblance when coefficients of association are used. Any pair of character states including the code for "inapplicable" is simply excluded from the 2×2 table of matches and from the computation. This is an acceptable procedure unless too many of the characters are inapplicable, in which case the inclusion of the responsible OTU is inadvisable.

When the resemblance between two OTU's is calculated by means of a correlation coefficient or a distance coefficient, characters that are inapplicable for either one are omitted in the calculation of the coefficient for the pair. It is obvious that the divisor in calculations of either coefficient has to be adjusted. In the distance coefficient the number of characters has to be reduced by the number of inapplicable comparisons, but in the correlation coefficient it is often necessary to compute separate sums of squares for the denominator of each individual r, since different characters are likely to be inapplicable in different OTU pairs. Missing values often give non-Euclidean metrics when, without NC's, the metric is Euclidean. We should also add that the large statistical literature referring to missing values is not applicable in numerical taxonomy.

One final consideration: if the coefficients in a resemblance matrix are based upon different samples of characters, the resulting coefficients are subject to two sources of error. First, they may be different because the characters from which the coefficients are computed differ. A low relevance value (see below) will provide a warning of such a possibility. A second error is statistical. Confidence limits of the coefficients will vary, and a difference that is statistically significant between two coefficients in a matrix may be nonsignificant between another two coefficients in the same matrix, based upon a smaller sample of characters. We therefore hope that the number of character state codes labeled "inapplicable" can be kept at a minimum in a given study.

Relevance.   Cain and Harrison (1958) have introduced the useful concept of the relevance of a comparison. They define this as the ratio of "twice the number of applicable characters considered (since these are shown by both forms) to the number of inapplicable ones (each of which will be shown by only one of the forms)." This ratio has the undesirable property of being indeterminate at its upper limit, and it is also not clear whether Cain and Harrison included those characters inapplicable to both forms being compared. We prefer a simpler coefficient of relevance,

$$R_{jk} = \frac{a_{jk}}{n}$$

where a_{jk} is the number of characters applicable in OTU j that are also applicable in OTU k (or vice versa), and n is the number of characters employed in the study. By this formulation R_{jk} ranges from zero to unity. It should be pointed out that $\Sigma_{i=1}^{n} w_{ijk}$, the denominator of Gower's coefficient S_G, is nR_{jk}.

Crovello (1966, 1968c) has pointed out that sometimes other forms of relevance are useful, for example, in deciding whether to omit a character or an OTU on the grounds of insufficient information. He suggests that R_{jk} be called OTU by OTU relevance. In addition one can have character relevance, R_i, which is a_i/n, and also OTU relevance, R_j, which is a_j/n. In these a_i is the number of OTU's to which character i is applicable, and a_j is the number of characters applicable to OTU j. Crovello also notes that low relevance may be due either to accidental causes (e.g., broken specimens) or biological causes, and that low OTU relevance may be a useful indication of the taxa that have been inadequately studied.

We have little experience yet of the effect of low relevance. Crovello (1968g) found that the correlation coefficient between a similarity matrix based on 43 characters with almost complete entries (mean OTU relevance = 0.972) and one with about half the $n \times t$ entries missing (mean OTU relevance 0.482) was quite high ($r = 0.887$). Unpublished work by J. W. Carmichael and one of us (P.H.A.S.) on the bacterial data of Stevens (1969) showed that the correlations with the S matrix from the complete $n \times t$ table remained high as entries were replaced by NC. When one tenth, one fifth, and one third of the entries were replaced, the correlations were, respectively, 0.980, 0.953, and 0.886. Numerical taxonomies therefore seem to be fairly robust to the effects of moderately low relevance.

The effect of low relevance is to introduce uncertainty into the positions of OTU's in A-space in the dimensions of the missing characters. The effect of NC entries may sometimes be equivalent to positioning an OTU at some value near the center of the range of the character, but the effects are quite complex for many resemblance coefficients, and this requires further study. Character correlations (R studies) based on scores with a substantial number of NC's lead to irregularities when subjected to factor analysis; communalities frequently will not stabilize.

4.13   CHARACTER VARIATION WITHIN OTU's

How should we record characters if they vary within the operational taxonomic units that we employ? This would happen when the OTU represents aggregates of individuals, such as species or populations. Even biological individuals employed as OTU's may show intra-OTU variation. For example, an individual plant will have leaves of varying lengths.

We have to distinguish between characters that are (1) phenetically discrete and (2) phenetically overlapping.

Phenetically Discrete Characters.   Characters that are phenetically discrete do not vary appreciably within the OTU's; consequently, the OTU's under study can be easily grouped according to the various states for the character in question, with little or no possibility of misclassification for any given OTU. Meristic characters will often fall into this category.

Phenetically Overlapping Characters.   Characters that are phenetically overlapping differ in their means from OTU to OTU but exhibit considerable variation within OTU's and overlap between them. Characters such as those expressing size, color intensity, and ratios of body measurements would quite likely fall into this group. However, statistically discontinuous or meristic characters, such as segment or tooth number, may also be phenetically overlapping characters.

If we sample one or a few specimens from a polymorphic population, we might observe and record only one character state, although two or more would be found in the population. Or, using a local population (or subspecies) to represent an entire species might obscure the fact that some characters had different states in different local populations. Such variation can, of course, occur at higher levels, too. Some sampling error of this kind is unavoidable so long as the OTU represents a hierarchic level higher than single individuals. Thus, problems in coding the phenetic value of a character in a given taxon may arise from the following three sources (apart from observational errors): (1) unrepresentative specimens, (2) phenetically overlapping characters, and (3) mutation within clones.

Unrepresentative Specimens.   If, as is not infrequently so, a species is known and described from a single specimen, we run the danger of employing data in our computation that may not be typical or representative of it. Even in the case of phenetically discrete characters, occasional variants and mutants are bound to occur, and while it is not very likely that a single specimen taken at random from the population will show one of these, the possibility should not be neglected. With one variant per character per 1,000 individuals and a consideration of 100 characters, a specimen picked at random has almost a 10 percent chance of carrying at least one variant. However, in most studies, particularly in those applications of numerical taxonomy that may be expected in the future, one would expect that a fairly representative sample of each OTU has been examined and that aberrations have been recognized as such.

Should a higher taxon that is to be used as an OTU be represented as the weighted mean of its constituent taxa and, if so, how should the weighting be done?

One solution to this dilemma is to introduce into the study one representative of each varying constituent of each polymorphic taxon and analyze them all together. If our notions about resemblances are correct, the first clusters should represent the various polymorphic taxa; that is, the variants composing them should resemble each other much more closely than any one of them or their common taxon resembles any other taxa. However, introducing many variants adds much labor and expense to a study and may make it prohibitive.

Another solution is to use only a single representative of the polymorphic group. This would be done in the expectation that the variance of the polymorphic forms within their taxon is less than the variance among the taxa of the study. Thus, the error introduced by choosing a single representative of a taxon should not be large enough to affect seriously the estimation of the similarity among the taxa of the study. If this were not so it would raise the question of the validity of the represented taxon. We have called this approach the *exemplar method*. The single representatives of the OTU's are exemplars of the taxa they represent.

Thus, to cite an example, if we were studying the relationship of *Homo sapiens* to various anthropoids, we could use a specimen from any of the races of man. The correlation of such an individual with any given anthropoid should be independent of his race, on the average, and would therefore approximate that of some hypothetical average man with the same anthropoid.

Zarapkin (1943), in a study of hands and feet of man and three apes, compared the deviations of single specimens of the animals and found clear distinctions at the racial, specific, and generic levels that transcended individual variation. In a test of the exemplar method carried out by Sokal (1962b) on Smirnov's (1925) data on genera of syrphid flies, two taxa (genus groups), **A** and **B**, joined at a (coded) similarity level of 850. The average value of the similarities between members of **A** and **B** turned out to be 851, with a standard deviation of 20. The range of individual similarity coefficients is from 900 (upper value) to 813 (lower value). Thus we are able to estimate the amount of possible error involved by taking at random a member of taxon **A** and one of taxon **B** as exemplars of their respective groups. In this particular example, also, the magnitude of error is quite tolerable.

In recent years evidence has accumulated to show that the exemplar method is generally reliable. Da Cunha (1969), studying bees, used three specimens of each species and found that they almost always clustered together as expected. The exceptions were all minor misplacements. Similar findings were made by Funk (1964) on mites, Boyce (1965) on rodents, and Moss and Webster (1969) on nematodes. Moss (1968b) found only small differences when using different specimens of the same species in a study on mites. Sneath (1968a) using tree branches noted very close resemblance. With various groups of flowering plants Bidault (1968), Taylor and Campbell (1969), and Orloci (1970) obtained clear clustering of conspecific individuals. There is evidence that the degree of variation in clones is small enough in some plants (Balbach, 1965) and in bacteria (Liston, Weibe, and Colwell, 1963) to allow single exemplars to be used with some confidence for studies encompassing several species. In all these studies the clustering of conspecific individuals was probably as good as one could expect in view of some uncertainties about the validity of a number of the species, although occasionally the studies raised other problems, such as misidentifications (Kaesler, 1970b) and the taxonomy of polyploid series (Bidault, 1968).

It has been noted in a number of studies on congruence (see Section 3.6) that there is noticeable incongruence between sexes, and it will usually be necessary to choose exemplars from the same sex (unless both are studied). Crovello (1966, 1969) noted this incongruence in dioecious plants (*Salix*).

A more troublesome problem was found by Fry (1964) in studying pycnogonids, in which all the characters, including meristic ones, show pronounced allometric trends. He obtained some reasonable results by treating quantitative characters as two-state characters, using the following rule for dealing with intraspecific variation: if the ranges of states in a character i are separate in two OTU's then this counts as a dissimilarity; if the ranges overlap, this counts as a similarity. The resulting resemblance is in the form of the association coefficient S_{SM}. Johnston (1964) notes similar problems with mites.

A solution of the problem of intrataxon variation of characters will partly depend on what the investigator wishes to study. If he wishes to compare a typical mammal with a typical bird, he must himself decide what he means by typical—whether "central" or commonest. He must also take the consequences of his decision, for it may be that the commonest form is very eccentric. If he is in doubt, he will do well to use several forms to represent the taxon, using, when necessary, single specimens for this. We suggest that in general a combination of the two approaches will prove of most value.

Problems of Phenetically Overlapping Characters.   Whether they are qualitative (melanic forms in a group of moths), meristic (number of antennal segments in a group of grasshoppers), or continuous (size of leaves in species of elms, rate of sugar fermentation in bacterial substrains), means of phenetically overlapping characters are not very representative if they have been derived from a single reasonably homogeneous population from a limited geographical area. However, even if very complete knowledge of the variation within each taxon were available, it would still be difficult to decide how to compute a mean and its variance for every character of the taxon. One way might be weighting based on frequency in the population. If, for example, a mean for skin color in the human species is to be computed, one could multiply the various color values by the respective frequencies of these colors and thus obtain a mean for the species. A phylogenetically oriented biologist might feel, however, that the actual frequencies of the various types living at present were not really representative of the common stock from which they presumably originated. Since we do not really know the color of the ancestral stock for *Homo sapiens* and are unlikely to know ancestral character states in most instances, such considerations are not particularly useful. An unweighted mean or midpoint of the range of variation may therefore be preferable to a weighted mean. The range or standard deviation of each character could be added as a separate "character" in such studies. Lima (1965) found little difference in the resulting phenograms when this was done, and Crovello (1968e) had similar findings, but we lack specific studies on how congruent are the resemblances based separately on means and on ranges or standard deviations.

The employment of phenetically overlapping characters gives rise to problems of a statistical nature. Since the expression of a given character varies within an OTU, the mean used to describe the state of the character for a given OTU is merely an estimate subject to sampling error. No difficulty occurs in setting confidence limits to individual means, but the distribution and hence the validity of coefficients of resemblance based on such measures are difficult to evaluate. This problem arises constantly in physical anthropology. Estimates of distance (particularly Pearson's coefficient of racial likeness and Mahalanobis' generalized distance, Section 4.3) therefore take the variance of the estimates into consideration. They also permit estimates of the degree of misclassification of individuals in two populations.

Mutation Within Clones. A problem peculiar to microbiology is that many biochemical tests can select for mutations. A single strain may then give different results on two occasions, depending on whether a mutation had occurred. This is likely to affect few of the characters, and it may therefore not matter which state is scored, but if a mutation is regularly observed, this fact (and the mutation rate if measured) is a perfectly valid character of the strain. Such problems in clones of higher organisms, though rare, can be treated similarly.

Similarity Coefficients that Allow for Variation within OTU's. There are several resemblance coefficients, besides *C.R.L.* and Mahalanobis' D^2, designed to take into account the variation within OTU's as well as the difference between means.

These include those of Sanghvi (1953) and Crovello (1968e). Sanghvi gives a coefficient T^2, which is closely related to $C.R.L.$

$$T_{JK}^2 = \frac{1}{n} \frac{\sum\limits_{i=1}^{n} (\overline{X}_{iJ} - \overline{X}_{iK})^2}{(s_{iJ}^2 + s_{iK}^2)}$$

Crovello's coefficient is

$$CSD2 = \sum_{i=1}^{n} [(\overline{X}_{iJ} - \overline{X}_{iK})^2 + (s_{iJ} - s_{iK})^2]^{1/2}$$

where $\overline{X}_{iJ}$ is the mean, and s_{iJ} is the standard deviation of character i in OTU **J**. Another, more complex resemblance measure is given by Carmichael, Julius, and Martin (1965).

The increasing interest in gene frequency studies (especially stimulated in recent years through the biochemical analyses of isoenzymes) has brought about a need for the quantification of the similarity between population samples characterized by different gene frequencies. Such statistics include Sanghvi's chi-square measure (Sanghvi, 1953), and those of C. A. B. Smith (in Grewal, 1962, and in Berry, 1963), Sneath (in Sokal and Sneath, 1963, p. 157), Edwards and Cavalli-Sforza (1964), and Hiernaux (1965). With some of these statistics the resemblance of an OTU with itself need not be 1 (a lower value indicating intra-OTU heterogeneity). The resemblance measure of Edwards and Cavalli-Sforza (1964) shown below is of interest in numerical taxonomy because it is related to their model for evolutionary pathways (Section 6.4). The distance between two OTU's **J** and **K** is

$$\left[\sum_{i=1}^{n} (2 - 2 \cos \alpha_i) \right]^{1/2}$$

where there are n loci, and when for any locus i, with m alleles and frequency p_g, $\cos \alpha_i$ is

$$\sum_{g=1}^{m} (p_{gJ} p_{gK})^{1/2}$$

The cosine transformation is intended to make the sampling variance independent of the value of p.

Recently suggested coefficients include that of S. Stewart (in Selander, 1970),

$$S = \frac{1}{n} \sum_{i=1}^{n} \left(\frac{\sum\limits_{g=1}^{m} p_{giJ} p_{giK}}{\left(\sum\limits_{g=1}^{m} p_{giJ}^2 \sum\limits_{g=1}^{m} p_{giK}^2 \right)^{1/2}} \right)^2$$

where n stands for the number of loci and p_{giJ} stands for the frequency of the gth allele at the ith locus for OTU J. S. J. Rogers at the University of Texas (personal communication by R. K. Selander) has developed a distance type coefficient of the following form

$$D = \frac{1}{n} \sum_{i=1}^{n} \left[\frac{1}{2} \sum_{g=1}^{m} (p_{giJ} - p_{giK})^2 \right]^{1/2}$$

Hedrick (1971) has calculated yet a different coefficient, which he calls the probability of genetic identity. Selander (personal communication) has compared these coefficients for the six Danish populations of the house mouse described in Selander et al. (1969). Although there are distinct differences between the numerical values of the coefficients, the matrices in general correlate quite highly.

Some of the coefficients we have just given are discussed by Gower (1972), whose review we think is valuable.

5

Taxonomic Structure

In a significant way the problems discussed and the material presented in this chapter represent the crux of taxonomy, for all chapters preceding this one have prepared us for the posing of the question, what is the structure of organized nature?, and the chapters following this one (about phylogeny, populations, nomenclature, and keys) are in a sense ancillary to taxonomic structure, the primary topic considered here.

In several previous sections we have already alluded to the purposes of creating taxonomies. We might briefly review these here. Economy of reliance on memory is a paramount taxonomic goal. We can neither list nor remember all the characteristics of various organisms and higher taxa, and we therefore need a system of grouping them into a manageable number of groups whose characters are preponderantly constant. Because of high constancy and mutual intercorrelations of characters, such a grouping will carry a high predictive value. Thus, if we read of a new aphid species we can immediately predict a number of characteristics that this species is expected to possess. An aphid will with almost complete certainty be a plant feeder, possess a particular type of wing venation, be parthenogenetic in part of its life cycle, produce males by nondisjunction of the sex chromosomes, produce honeydew, secrete wax from cornicles or other glands, and so on. Since an aphid is a homopteran, we can forecast with some accuracy the general construction of its mouth parts, the texture of its wings, and other homopteran characteristics. This type of argument can, of course, be extended to the hexapod and arthropod levels of classification and even higher. It is obviously much easier for us to remember this of the group Aphididae than of each individual aphid or species of aphid. Furthermore, it is impossible to remember or appreciate the innumerable relations between the various OTU's to be classified, but this is easier when they are grouped into fewer inclusive taxa.

It is clear that such considerations lead to a second, closely related purpose of taxonomy, namely predictive power. The more natural a taxonomy is, the more predictive it will be about characters known from part of the group that have not yet been investigated in another part.

Another purpose of taxonomy, and for many biologists the paramount purpose, is the construction of phylogenies or, put in another way, the construction of classifications that reflect as faithfully as possible the putative phylogenetic history of the organisms contained therein. As we have stated in Section 2.6, this is not our primary purpose, although we consider it an important and legitimate aim of taxonomy and will consider methods of dealing with this problem in the following chapter.

A related purpose seems to us of greater interest. If we can recognize the patterns of distributions of organisms and of groups of organisms in nature, what can we learn from these patterns about the nature of the evolutionary forces that have brought about these patterns? Can we use taxonomic structure to generate evolutionary hypotheses (general hypotheses as distinct from the tracing of specific phylogenetic lineages)? Conversely, can we use taxonomic structure to test evolutionary hypotheses derived from other data or from theoretical considerations? It should be obvious from the above that the accurate description and determination of taxonomic structure is a problem of the highest importance to systematics and evolutionary biology.

It is difficult to present the material in this chapter in a self-evident logical order. We would like to know how organisms and taxa are distributed in phenetic space. From insight into the distribution of organisms one can proceed to inferences about the evolutionary and ecological origin of the patterns exhibited. The difficulty here is that it is not obvious how such patterns should be recognized. If only for the purpose of summarization one must group information on numerous characters so as to reduce the multidimensional phenetic relationships to comprehensible patterns in a few dimensions. One must also group information on numerous individuals into clusters. However, clustering in turn imposes certain constraints on the relationships among the taxonomic units and our perspectives of them. Although the end result of the clustering process is necessarily determined by the inherent structure and relations of the OTU's, it is also appreciably affected by the choice of a clustering algorithm. This algorithm in turn is chosen by the taxonomist because of certain a priori views of the distribution of the organisms in space or the type or shape of the classification that he wishes to obtain.

Since the actual constellation of OTU's in phenetic space and hypotheses about such constellations (or the desiderata of taxonomic practice) will not necessarily coincide, one must develop methods for measuring the difference between taxonomic resemblances and the desired or hypothesized classifications. Related to this issue is the matter of the goodness of a classification, regardless of its degree of conformity to a preconceived arrangement. Can criteria be developed for determining how successfully a group has been classified? It is doubtful whether universally acceptable indices of optimality can be developed. Constraints must be imposed on the type of classification wanted before meaningful criteria can be developed. Another important aspect is the necessity to represent the results of a classificatory endeavor in some graphic or at least tabular manner. The limitations of methods of representation inherent in the printed page or even three-dimensional models invariably distort the taxonomic structure obtained from the data matrix. Each of these considerations and operations rests in some sense upon decisions reached in a previous step and some of the difficulty of providing a proper rationale for taxonomy lies in breaking into this circle of reasoning.

We shall first consider the basic evidence available for finding taxonomic structure, the resemblance matrix, described in Section 5.1. Next we take up the description and definition of clusters and patterns in a totally abstract way (Section 5.2), and follow this by a discussion of appropriate systems of classifications for biological material (Section 5.3) as suggested intuitively and by traditional taxonomy. Then we shall discuss a variety of clustering methods, themselves classified by their various properties (Section 5.4). The most common type of clustering will rate a section to itself (5.5). This is followed by accounts of methods of ordination (Section 5.6) and graphs and trees (Section 5.7). The relations between Q and R techniques in classification need some mention (Section 5.8); then follows a comprehensive review of methods of representation of taxonomic structure (Section 5.9). Optimality criteria in a classification and the comparison of the results of different classificatory methods are covered in Section 5.10. Specialized aspects are the development of criteria of rank (Section 5.11) and the corroboration of a conventional phenetic classification by biochemical methods (Section 5.12). The next section (5.13) summarizes our recommendations on the proper choice of analytical methods and modes of publication of results. The actual distribution of OTU's and taxa in phenetic hyperspace and its relation to problems in evolutionary biology and ecology are taken up in the last section (5.14).

5.1 THE RESEMBLANCE MATRIX

Just as we began Chapter 4 with a consideration of the data matrix upon which one operates to obtain resemblances, so we begin this chapter with a brief description of the resemblance matrix in which we wish to find taxonomic structure. The rows and columns refer to the OTU's and the entries in the matrix are estimates of the resemblances for every OTU compared with every other OTU. The order of OTU's is the same in the rows as in the columns, and therefore a value in the principal diagonal represents an OTU compared with itself, a "self-comparison." Such a matrix is square, of order $t \times t$ where t is the number of OTU's. Using S to stand for any similarity coefficient, such a matrix is represented by:

OTU's		OTU's			
	1	2	3	...	t
1	S_{11}	S_{12}	S_{13}	...	S_{1t}
2	S_{21}	S_{22}	S_{23}	...	S_{2t}
3	S_{31}	S_{32}	S_{33}	...	S_{3t}
.	.	.	.		.
.	.	.	.		.
.	.	.	.		.
t	S_{t1}	S_{t2}	S_{t3}	...	S_{tt}

Then S_{jk} is the resemblance in the jth row and the kth column.

For most similarity coefficients an organism is considered identical with itself (exceptions are Smirnov's coefficient or some others measuring genetic resemblance between populations based on gene frequencies) and therefore the entries in the principal diagonal represent identity. This may be 1, for most association coefficients and for correlations, and the values in the matrix can then be thought of as "similarities." Or it may be 0, for distances and analogous measures (mismatching, differences), and the values in the matrix can then be thought of as "dissimilarities." Rarely one may have other values for identity (e.g., $+\infty$ for the z transformation of correlations).

Again, with almost all methods, the resemblance of **a** to **b** is the same as that of **b** to **a** (the property of symmetry discussed in Section 4.2). The top right part of the matrix is then a mirror image of the lower left part, and only one half is needed. In publications it is conventional to use the lower left triangle:

OTU's		OTU's			
	1	2	3	...	t
1	×				
2	S_{21}	×			
3	S_{31}	S_{32}	×		
.					
.					
.					
t	S_{t1}	S_{t2}	S_{t3}	...	×

This is also usually the best form for computer storage. The reader will note that the entries in the principal diagonal are marked ×. This is because these entries may or may not be recorded (they are the self-comparisons, and are usually defined by the coefficient used). Most often they are omitted.

Some procedures call for the summation of given rows (or columns) of the matrix. When this is done it is implied that the full matrix is used, not just the lower left triangle. But if the self-comparisons in the principal diagonal are to be included in the sum, this must be made clear, for there is no fixed convention. Of course, one does not need to write out the other half of the matrix (or to take up computer storage with it), because one can always obtain the same result by turning at right angles, on reaching the principal diagonal, and reading off the remaining values from the appropriate column rather than from the row (or vice versa).

5.2   PATTERNS AND CLUSTERS

Patterns

The determination of taxonomic structure in its operational formulation is but one instance of the general, widely applicable search for patterns in nature. The field of *pattern recognition* has developed in recent years into an important inter-disciplinary area between engineering, computer science, and applied mathematics. There is an ever increasing literature on the subject (e.g., Mattson and Dammann, 1965; Sebestyen, 1962; Uhr, 1966; Mendel and Fu, 1970) and there is already a journal entitled *Pattern Recognition*. Pattern recognition has come to include numerous scientific and engineering problems, some of which concern us here only peripherally. One example is the detection of sources emitting signals with high noise-to-signal ratios (as is encountered in work with radio telescopes). Another is the heightening of contrast in optical or radar scans—this is *pattern detection* as defined by Sebestyen (1962). Still another example is the recognition of alphameri-cal characters in print or handwriting (the assignment of unknown individuals to predetermined classes)—in the strictest sense of Sebestyen's definition this would be pattern recognition, though we prefer to define it as an identification problem. But these are only a few examples of the active work in pattern recognition sensu lato. The reader will realize that, though on the surface such activities seem to bear little relation to traditional biological taxonomy, the types of problems dealt with do occur in the taxonomic process as it is likely to develop in the near future; they will therefore be of concern to the generation of taxonomists now in training (for further discussion see Section 11.5 and Chapter 12).

However, the aspect of pattern recognition that will concern us most immediately will be the recognition of patterns of distribution of OTU's and groups of OTU's (taxa) in a space (commonly a hyperdimensional, phenetic A-space). The use of the term *pattern* must be qualified here. It is one of those generally descriptive and understandable terms that, when given specific meaning, often tend to limit rather than enhance comprehension and communication. It will be recalled that pattern was given a specific meaning with microbiological studies, in Section 4.11. Pattern is used by Ball (1965) to symbolize the character vector representing a single OTU. (He has abandoned this usage in Ball, 1970). Cattell, Coulter, and Tsujioka (1966), who have given this problem considerable attention, used "pattern" in another technical sense discussed below. The obvious synonyms, such as configuration or constellation, have all been used in specific technical senses by various authors repeatedly and, regrettably, our choice here is either to reuse these generally comprehensible terms or to coin new ones of less intrinsic mnemonic value, which would be less likely to be accepted.

We shall mean by pattern, therefore, any describable properties of the distribution of OTU's and groups of OTU's in an A-space. Before we proceed with an examination and discussion of possible patterns we should at least be familiar with the terminology employed by Cattell et al. (1966). In order to define types or clusters they refer to the patterns of OTU's as those properties of the individuals that imply profile as well as configuration. By *profile* is meant the vector of character states of an OTU over the suite of *n* characters. The order of the characters is of no consequence as long, of course, as homologous characters and character states are found in the same row of the original data matrix. Thus rows of the data matrix could be interchanged as long as it is done for the entire matrix and the inherent topological properties of the pattern of OTU's in A-space would not be changed thereby. By *configuration*, Cattell et al. mean the logical ordering of the character states in the vectors, as in an ordered process such as a life history, where data matrix rows representing characters cannot be meaningfully interchanged. For most of our work the idea of configuration sensu Cattell et al. will not be frequently employed, and patterns will therefore imply the distribution of the tips of the OTU vectors in the A-space without any consideration for a logical order of the characters.

It is very difficult to differentiate conceptually between detecting patterns of OTU's (and attempting to describe them) and pattern recognition in the strict sense, although we shall attempt to do so here to the extent it is possible. What causes one to recognize a string or a spheroid or a sausage-shaped cluster or a torus among OTU's dispersed in a hyperspace? As soon as any objective definition is attempted one needs operational criteria for defining the class of patterns with which one is concerned and as soon as that is done, one is back to operations that produce patterns of the sort under discussion.

Starting with individuals as OTU's (each an *n*-dimensional vector) numerical taxonomists in various fields have commonly considered two alternative null hypotheses against which to compare, if not test, the observed dispersion pattern. The two null hypotheses are regular distribution and random distribution. They are illustrated in Figure 5-1 for the two-dimensional dispersion pattern. *Regular*

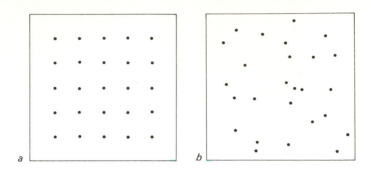

FIGURE 5-1

Two types of phenetic distribution: *a*, regular distribution, and *b*, random distribution.

distribution, which implies a deliberate spacing out as though the OTU's were placed at the intersections of a lattice, is relatively easy to detect and test for. Variances of frequencies per unit sampling space will be less than means, resulting in coefficients of dispersion $s^2/\overline{Y} < 1$ (see Sokal and Rohlf, 1969, p. 88). In ecology, where mutual interference among plants and territoriality in animals are common phenomena, such distributions are frequent. In some other sciences, such as geography, regular spacing as a result of economic or political factors is also well established (Haggett, 1965). In biological systematics no cases of regular distribution in phenetic hyperspace are known nor are they likely from any current theoretical model.

Truly *random distributions* of organisms are occasionally observed in ecological work and in various nonbiological fields. These may be based on various probability distributions, such as the Poisson, the uniform, the multivariate normal, or others. However, here again current evolutionary theory (as supported by empirical observation) makes the random distribution of organisms in phenetic A-space extremely implausible. We must therefore look for departures from randomness in the direction of condensation of points in certain regions of the A-space, leaving most of the space unoccupied. It should be noted that these patterns depend on the size scale that is being considered; e.g., a distribution may be random at one scale and clustered at another (see *Principles of Numerical Taxonomy*, p. 173).

Clusters

The relative vagueness of our definition of pattern makes it even more difficult to try and define *cluster*. We shall encounter a variety of criteria for measuring properties of clusters, and this variety opens the way to a multiplicity of different definitions of the term. For this reason we shall leave the definition of clusters conveniently vague: sets of OTU's in phenetic hyperspace that exhibit neither random nor regular distribution patterns and that meet one or more of various criteria imposed by a particular cluster definition. Various of these criteria will be discussed below.

Although most of our description of patterns will be conducted at least within a three-space and frequently mathematically within a higher-dimensional space, it will be convenient to depict the problems encountered within the familiar two-space of the printed page, reasoning from these to extensions into multidimensional Euclidean, non-Euclidean, and also nonmetric, spaces. The two-dimensional patterns we shall discuss for purposes of defining our problems have their real life analogs in the familiar areal distribution of plants and animals when the added dimensions of topography and environmental differentiation are excluded.

A sample of t points in hyperspace may be considered a partition into t parts each containing one point, which are to be grouped into k parts where $t \geq k \geq 1$ (agglomerative clustering, discussed in Section 5.4); or the sample may be conceived of as a single set of all t points, to be divided into k parts (divisive clustering, discussed in Section 5.4). Regardless of the method yielding the k groups their recognition must depend on criteria applied to these groups.

Given a subdivision of the total set into subsets, various descriptive parameters of each subset may be used in defining the structure of the points. We shall now list some of these parameters and introduce the general mathematical notions they represent. The detailed application of these parameters singly or in combination during clustering is deferred to Sections 5.4 and 5.5, in which detailed clustering methods are discussed, and to Section 5.10 on optimality criteria.

Cluster Center.   The *center* of a cluster can be represented in two general ways, as a point representing an actual organism, or as a point representing a hypothetical organism (such as the "average man" who has 0.8 wives and 2.3 children!), which is simply a useful mathematical construct. If OTU's in a study are not actual individual organisms, but are themselves hypothetical constructs, the above distinction becomes blurred, and in practice most central points will represent constructs that are to some extent artificial. For discussion, however, we shall take them separately.

There are two commonly used measures of the center of a cluster that represent abstractions.

1. The *average organism* or *centroid*, $\bar{x}$, is given by the point in phenetic space whose coordinates are the mean values of each character over the given cluster of OTU's. It is also the center of gravity of the cluster of OTU's. It is widely used in factor analytic methods, discriminant functions, and in clustering methods. If the characters are all of the 0,1 type, the centroid represents a point within the phenetic hypercube, and the coordinates are simply the observed frequencies of the various characters.

2. For 0,1 data another construct commonly used in microbiology is the *hypothetical median organism* of Liston, Weibe, and Colwell (1963). It is the

hypothetical OTU that possesses the commonest state for each character, and therefore would better have been called the hypothetical *modal* organism. For all its characters it exhibits states 0 or 1, and hence lies at a corner of the phenetic hyper-cube, at the corner that is nearest to the centroid. It may by chance represent an actual OTU, though this is unlikely. If generalized to quantitative characters, it would express the mode of each character: the "hypothetical modal organism" (see Sokal and Sneath, 1963, p. 256). Somewhat similar concepts have been pro-posed by Tsukamura and Mizuno (1968) and Hayashi (1964).

The most usual measure of an *actual* OTU is the *centrotype* (Silvestri et al., 1962). It is the OTU with the highest mean resemblance to all other OTU's of the cluster. It is the OTU nearest to the centroid for Euclidean distance models (not necessarily so for other models). Related to the centrotype is the most typical member of a cluster in a factor analysis. It is that OTU having the highest loading on a factor representing the cluster in a Q analysis (Rohlf and Sokal, 1962).

The usefulness of these concepts will be evident from their definitions: the centroid is an ideal typical form, convenient for many mathematical purposes (i.e., sum of squares of distances from it are minimum), the hypothetical median organism is the nearest to this ideal that could be obtained, whereas the centrotype is the most typical OTU actually present, and would be suitable as a type specimen, for example. Other aspects of statistical types are discussed in Sokal (1962b).

Density. The simplest characterization of a cluster is that it should be denser than other areas of the space under consideration. The notion of density can be expressed as number of OTU's per unit hypervolume or as the density/volume ratio of a convex hyperdimensional set.

Variance. The notion of density leads easily to that of the variance of points in each subset around a centroid or some other central point, such as a hypothetical median element or a mid-range (all in hyperdimensional space, of course). The generalized variance, which is the determinant of the within-group variance-covariance matrix, would seem a suitable measure here (see Anderson, 1958; Goodman, 1968a), but only when used on the same suite of characters. There are problems with this method when the matrix is not of full rank. The concept of variance leads to procedures analogous to analysis of variance (univariate and multivariate), the criterion for a satisfactory partition being that variance within the subsets is minimized relative to variance among the subsets. As we shall see, this is an important criterion of optimality in a classification. Rohlf (1970) has shown that by weighting differences in principal component scores inversely proportionally to the lengths of the principal axes of hyperellipsoid clusters, relationships can be established in which some dimensions within a cluster are of more importance for "belonging" or cohesion than are others.

Dimension. This aspect of a cluster can be measured in various ways. It is clearly related to the variance just discussed. In discriminatory methods it is more usual to employ what is effectively a radius of the cluster, so that a specified majority of OTU's lie within this radius. This of course presupposes that the cluster is acceptably close to being hyperspherical (and the points are normally distributed); if it is highly elongated the concept of a taxon radius is inappropriate unless suitable transformations can first be applied (as is one intended use of Mahalanobis' D^2 analysis; see Section 8.5).

The ellipticity of a cluster can be estimated from the standard deviations of the OTU's on the several factor axes (which are the square roots of the eigenvalues) and these are often obtainable from many ordination methods (see Rohlf, 1970 for an extended discussion of elliptic clusters). Some check, too, can be obtained from measures of straggliness mentioned below, but this area has not yet received much attention.

A taxon radius, however, is not easy to estimate from theoretical considerations. If the distribution of distances of OTU's from the centroid is plotted as a histogram, one can obtain a taxon radius empirically (see Figure 5-32 in Section 5.10). The histogram is likely to approximate a normal curve; this can be tested by standard methods, and the mean of these distances $\bar{d}_{j\bar{x}}$ and their standard deviation, s, can be found. One may then take as the taxon radius $r = \bar{d}_{j\bar{x}} + \varkappa s$. One may choose the constant $\varkappa$ according to the desired one-tailed value of the normal distribution. Thus with $\varkappa = 2.33$ one would expect 99 percent of the OTU's to lie within r. This formula is available both for the original hyperspace and for models of reduced dimensionality (e.g., three-dimensioned factor models) but r will not be the same in each case, and with few dimensions (or with extreme correlations between characters), the histogram will be skewed to the right. In fact the theoretical n-variate normal curves are shaped like the square root of the chi-square distributions with n degrees of freedom; i.e., they have mean $= \sqrt{n - (1/2)}$ and standard deviation $= \sqrt{1/2}$, when n is large.

Gyllenberg (1964, 1965b) has suggested that twice the intracluster standard deviation should be used as a radius. Thus.

$$r = 2\sqrt{\frac{\sum d_{j\bar{x}}^2}{t_J}} = \sqrt{\frac{2\sum_{j=1}^{t_J} \sum_{k=1}^{t_J} d_{jk}^2}{t_J^2}}$$

where t_J is the number of OTU's in cluster **J**. But the previous radius allows more flexibility. For multivariate normal distributions of high dimensionality, the mean distance from the centroid approximately equals $\sqrt{1/2}$ times the mean distance between all pairs of OTU's (see Section 5.11; for some studies on this latter parameter see Gilmartin, 1969a,b).

Number of Members in a Cluster. Some clustering methods allocate weights according to the number of members in a cluster, or require that a group shall contain a given minimum or maximum number of OTU's if it is to be recognized as a cluster. This is sometimes called cluster size, but should be distinguished from the other use of the term "cluster size" which refers to dimension.

Connectivity. This concept from graph theory expresses the often applied notion that in order to be members of a cluster, individuals have to be related to a certain minimal degree, i.e., connected. This connectedness may well be closeness by distance in the hyperspace within certain circumscribed limits. If all members of a subset are so close to each other that they can all be considered connected to each other (the technical expression for this relationship in graph theory is *fully connected*), then clusters composed of these points will be compact in hyperspace. Minimal connectivity means that each OTU is connected at the minimal acceptable level to every other OTU either directly or via other OTU's, but that the number of connections (edges) is $t - 1$ for t OTU's. Removal of even one connection between OTU's from a *minimally connected set* results in a disconnected set or graph. Minimally connected graphs are also known as *trees* in graph theory. Minimally connected sets are likely to give long strung-out clusters while fully connected sets must be dense and compact. Partial connectivity implies an intermediate condition between the two extremes. Useful references are Busacker and Saaty (1965) and Kansky (1963), and for application to taxonomy, Estabrook (1966).

Straggliness. This rather ill-defined criterion has been little studied. It is related to connectedness but is not identical with it. Straggliness implies much greater extension (but not necessarily a linear one) in a few of the dimensions than in the others. Examination of factor analysis plots, or calculation of the standard deviations in each factor axis, can reveal ellipticity as noted above. The histogram of distances of OTU's from the centroid will also show up highly straggly clusters, since instead of a sharp mode well away from the origin, the histogram will be flattened, falling off from the origin. More exact techniques await development. For suggestions on measuring both connectivity and straggliness the reader is referred to papers by Bonner (1964), Williams, Lambert, and Lance (1966), Wirth, Estabrook, and Rogers (1966), Johnson (1967), Needham (1967), and Gabriel and Sokal (1969).

Gaps and Moats. An important concept in defining clusters is that of the gap or moat by which the distance from each member of a cluster to the closest one in another subset is computed and converted into an index of separateness or isolation of the subsets. The idea of a moat can be absolute, i.e., the distance between closest members of different subsets, or the average of such distances for all members of both subsets; or a moat may be some ratio of the distance between the centroids of clusters and the radial vectors of the cluster envelopes to their centroids. The gap between clusters is one criterion of a new type of cluster described by Jardine, van Rijsbergen, and Jardine (1969), which they call ball clusters, and for which the greatest distance between any pair of OTU's within the cluster is less than the smallest gap from a cluster member to any OTU outside the cluster.

The criteria derived from the parameters just described far from exhaust the possibilities of cluster definitions. They are only obvious ones that come to mind as one tries to formulate the departures from randomness that define clusters.

In two-dimensional space, clusters arise from so-called contagious distributions, of which the negative binomial and the logarithmic distributions are but two examples. In plant ecology their application has been studied extensively (Greig-Smith, 1964; Pielou, 1969a; C. B. Williams, 1964). In our applications, where they would have to be extended to cover multivariate cases, their properties have not been extensively studied. However, because a number of hypothetical models can give rise to these distributions, we feel it is not particularly fruitful to test observed distributions of points against hypothetical distributions, for even conformity to the distributions will not lead to confirmation of given models in evolutionary theory. It seems more important to study the patterns of distribution as they occur in hyperspace.

Our discussion has assumed that the assemblage of OTU's is a universe within which patterns are to be recognized. A somewhat different approach is necessitated when the taxonomist assumes a priori that sampled populations consist of a mixture that he first wishes to decompose into separate populations, which are then investigated further or are used as OTU's in other, more encompassing studies. The application of such *mixture problem* techniques is usually restricted to specific or infraspecific populations in which sexes, different life history stages, genetic polymorphs, or sibling species need to be sorted out. The techniques generally employed assume that the underlying distributions are known. Among the oldest such methods is the resolution of a mixed series into two Gaussian components described by Rao (1952, p. 300). This method assumes normality of the distributions and a common variance in the two samples, and provides estimates of the means of the two samples, their variance, and the proportion in which they are mixed. A graphic method for estimating means and standard deviations of mixed normal samples has been developed by Harding (1949). Among the more recent methods is that of Bhattacharya (1967), who has developed an approximate technique for resolving a distribution into two or more Gaussian components, for each of which a separate variance can be estimated. The separation of a multivariate sample into two or more multivariate normal components is described by Wolfe (1970), who also furnishes a computer program for this procedure. When no assumptions are made about the underlying distributions, the separation of mixtures resembles cluster analysis. In the univariate case a method with which one of us (R.R.S) has had some experience is the dip-intensity statistic of Kruskal, which measures departures from unimodality (Giacomelli et al., 1971). A similar technique has been developed by Engelman and Hartigan (1969). In the multivariate case ordinary methods of numerical taxonomy furnish quite adequate separation of apparently homogeneous populations (Takade, 1971).

5.3   TAXONOMIC GOALS IN BIOLOGY

In the preceding section we considered the dispersion patterns of objects distributed in an n-dimensional phenetic hyperspace from a purely formal point of view. Any objects distributed in any A-space may be so considered. Now we must turn to the types of organization and representation of material desirable for biological purposes. Most of our discussion here will of necessity concern the classification of organisms (conventional taxonomy), but we shall also deal with ecological and biogeographic taxonomic arrangements. Our account of classification will include the conventional arranging of organisms in nested, mutually exclusive systems (hierarchies), other (nonhierarchic) systems of partitions, and ordination—the positioning of OTU's in relation to multidimensional character axes (in an A-space).

   The classical tendency in biological systematics has been hierarchic classification, and its success in representing the system of nature has led to its application in ecology, especially plant ecology and, in the past, occasionally in other fields (see for example the now embarrassingly naive classification of social aggregations of animals by Deegener, 1918). Several motivations encourage taxonomists to prefer classifications of organisms into nested, mutually exclusive taxa. It seems to be a general human tendency to stress the sharpness of distinctions between classes and to overemphasize gaps in the spectrum of phenetic variation. Mutually exclusive classes are frequently used conceptually by humans, although we are repeatedly warned against stereotyping events and individuals. Nevertheless we succumb to a natural tendency to avoid intersecting sets, which would result in some individuals being members simultaneously of more than one set. And we are so obedient to the Linnean system, which requires mutually exclusive and hierarchically ordered classes, that the process of classification has become synonymous in the minds of many biologists with a mapping of the diversity of nature into the Linnean system. This is reinforced by a third trend, the evolutionary explanation of the hierarchic arrangement of the Linnean system. If natural taxa are to be monophyletic (sensu Hennig), a dendritic pattern of organismic diversity has heuristic and intrinsic value and the aim of taxonomy would be to arrange organisms into those nested, mutually exclusive taxa that correspond most closely to the actual clades. A fourth reason for classification of organisms in nested hierarchies has been the achievement of economy of memory, which, though not necessarily restricted to such a taxonomic system, is conventionally associated with it.

There are, however, a number of reasons why workers have become disenchanted with the classical classificatory approach. The concept of the polythetic taxon (see Section 2.2) by its very nature contradicts the concept of an Aristotelian classification, although fully polythetic models can be constructed that will yield nested, mutually exclusive classifications (and undoubtedly there are many such instances in biological systematics).

In biology the move away from traditional hierarchic classifications developed in phytosociology, where even simple quantification resulted in taxa that clearly violated the relationships among observed stands. Workers in this field were among the first to note that random or uniform phenetic distribution patterns cannot give satisfactory nonarbitrary hierarchies (see Figure 7-1 in Sokal and Sneath, 1963). Only very rarely is there sharp ecotonal transition between vegetation types, and much more frequently we find broad and diffused ecotones or transitional successional stages. Special emphasis on a continuum concept in plant ecology has been placed by J. T. Curtis and his coworkers (for a review see Curtis, 1959). Further developments are due to D. W. Goodall who coined the term ordination (Goodall, 1953). This has led to the elaboration of numerous methods in which stands are plotted in two- or higher-dimensional space allowing observations of the densities of OTU's and relationships of important points of concentration. In biological systematics there has been a similar trend, growing out of the development of measures of distortion of similarity or dissimilarity matrices when these are interpreted by means of phenograms. Such measures as the cophenetic correlation coefficient (Sokal and Rohlf, 1962) and other measures of stress have led to the realization that the hierarchic classifications often are poor representations of actual phenetic relationships found in nature. Far better representations are often obtained by summarizing the data in an ordination of as few as three dimensions.

More recently, Jardine and Sibson (1968a) have shown in a more formal mathematical way that some properties of a classification they consider desirable may be violated by nested classifications, and they have suggested replacing these with overlapping hierarchic classifications. Thus there seems to be a general trend nowadays toward greater flexibility and experimentation in the types of classificatory systems acceptable to biologists.

5.4   A TAXONOMY OF CLUSTERING METHODS

Clustering Methods

The process of clustering consists of arriving at one or more partitions of a set of OTU's, the partitions to have various desirable properties as described below. A more extensive list of these is given in Wishart (1969b). Clustering may result in the grouping of OTU's, i.e., in the forming of a coarser partition from a finer one, or it may involve the breakup of an entire set of OTU's into increasingly finer partitions.

In many taxonomic procedures each data set is described by an entire series of such partitions leading to the well known taxonomic hierarchy. In other studies, a single partition exhibiting some desirable properties is considered adequate.

Not all workers define clustering in the above sense. Ball (1965) means by clustering "the finding of minimum distance (or maximum correlation) between a pattern [see our Section 5.2] and one 'single cluster center' out of a set of cluster centers." He uses the term "clumping" for agglomerative clustering as used in this book, in which the closest pair of OTU's forms the nucleus for a clump that grows around this nucleus. In his later account (Ball, 1970) he uses a definition of clustering more in line with ours: "The finding of data-derived groups on the basis of the groups being internally similar. Does not use an externally supplied label."

The few, empirically validated clustering methods of early numerical taxonomy have multiplied to form a vast and complex field nearly impossible to survey. Comprehensive reviews have been furnished by Ball (1965), Williams and Dale (1965), Spence and Taylor (1970), Wishart (1969b), and Cormack (1971). Any attempt at a complete enumeration would be foredoomed to failure and obsolescence. The difficulty of outlining the major kinds of approaches to clustering biological data is compounded by the inability of workers in the field to arrive at a logical system of classification of clustering methods. Eight aspects by which the methods differ one from another are listed in the following pages, and the pros and cons of the contrasting viewpoints, as well as their mathematical implications, will be discussed. However, there seems no way of erecting a consistent system of clustering algorithms. Binary divisions of the eight different aspects would lead to $2^8 = 256$ different types of clustering methods. Many of the possible combinations have so far not been tried, and some of them are probably logically impossible. But some of the regions in this methodological 8-space contain several methods representing variations on a theme.

Eight Aspects of Clustering Methods

The account of these different aspects of clustering methods will be accompanied by a description of some uses of them. The next section will give details of the most widely used algorithms for defining clusters and of some of the general strategies employed in cluster analysis programs. It goes without saying that although simple cluster analysis is possible by means of desk calculators and even with paper and pencil, almost all work on similarity matrices of any size is nowadays done by computer. It is, therefore, especially important that the user of a clustering program understands what it does to his data.

1. *Agglomerative versus Divisive Methods.*   Starting with a set of t separate entities, *agglomerative techniques* group these into successively fewer than t sets, arriving eventually at a single set containing all t OTU's. Agglomerative techniques are so

frequently employed, especially when limited to sequential, hierarchic, and non-overlapping procedures, that a great variety of methods has been developed. An entire section (5.5) will be devoted to this subject.

By contrast, *divisive techniques* commence with all t OTU's in one set, subdividing this into one or more subsets, and further subdividing each of the subsets. If only because they are simpler to program, agglomerative techniques have been most frequently used in numerical taxonomy and the employment of divisive techniques has been limited. There is a greater danger with divisive techniques of inappropriate allocation of some OTU's that are not later corrected, unless special procedures for reallocation are included (Williams and Dale, 1965). When monothetic classifications are desired, divisive techniques of breaking up the entire group of objects on the basis of successive single criteria are efficient. We cannot agree with Blackith and Reyment (1971) that because of divergences of opinion regarding suitable agglomerative clustering techniques the divisive method should be preferred. Either approach can be tested by a variety of optimality criteria and, for reasons described below, agglomerative techniques are likely to continue to be the preferred clustering techniques in the majority of cases.

The technique of *association analysis* (Williams and Lambert, 1959, 1960, 1961a; Lambert and Williams, 1962, 1966; Lance and Williams, 1965; Williams, Lambert, and Lance, 1966) is the most prominent example of the divisive technique. It has been used mostly in ecological data and employs two-state characters. The total set of OTU's is initially divided into two subgroups based on the two states of a single character i. This character is chosen so that $\sum_{h \neq i} X_{hi}^2$ is a maximum, where h stands for characters other than i, and X_{hi}^2 is computed as $n(ad - bc)^2/[(a + b)(a + c)(b + d)(c + d)]$. It is the well-known X^2-statistic, computed from 2×2 frequency tables (see Sokal and Rohlf, 1969, p. 589). Since the coefficient of contingency, X^2/n, is monotonic with estimates of the correlation between variables h and i, the criterion can be computed as $\sum_{h \neq i} r_{hi}^2$. After the first division on variable i, each subset of OTU's (i.e., that subset whose members have state 0 for i, and that subset with state 1) is now further subdivided in the same way as before. The process ends when a predetermined number of groups has been achieved or when the measure of homogeneity as expressed by $\sum_{h \neq i} X_{hi}^2$ has reached a level below a given criterion. Geometrically, Williams and Lambert's association analysis can be represented in A-space as follows: It is a look at each binary character (normalized by standard error) in turn in order to find the one for which the distance between the centroids (of the two groups defined by it) is maximal. An important theoretical paper on this and related methods is that of McNaughton-Smith (1965). This method has also been adapted to the information statistic $2I$ (Lance and Williams, 1968a).

An approach that seems to be similar to association analysis but may be quicker to compute, especially for many OTU's and on computers with small memories, is the *group analysis* of Crawford and Wishart (1967), also discussed briefly in connection with its graphical representation by Crawford (1969). In this method, a value $W_x' = p_x V_x$ is computed, where p_x = proportion of quadrats containing species x, and V_x = mean number of species in quadrats containing x. W_x' is called the group element potential, GEP, of species x. The group element potentials are summed for all species over a quadrat to form a set element potential. From this value one can test the effect of dividing a set of quadrats based on presence or absence of any one of the n species. By a modified 2×2 table chi-square procedure, those species that give the greatest departure from expectation are chosen to make the first dichotomy, and subsequently other species are chosen by similar criteria to make subsequent dichotomies of the larger subclasses. The details of the procedure can be looked up in Crawford and Wishart (1967). The authors show that their results do not differ greatly from those obtained by association analysis, but the great advantage of their method is that for n species only n tests need to be performed, rather than the n^2 tests required by other methods of dichotomous monothetic classification. The representation of the k resulting groups in a two-dimensional ordination is shown by Crawford (1969).

No really satisfactory algorithms have been produced for polythetic divisive methods. Gower (1967a) points out that the method of Edwards and Cavalli-Sforza (1965), which seeks to maximize the between-sets sum of squares requires the examination of $2^{t-1} - 1$ partitions of t OTU's, an impossible amount of computational labor even when t is a modest number. Approximative solutions are generally found by first employing agglomerative techniques to obtain favorable starting points for reallocations of members among sets. It has also been pointed out by several workers (see Wishart, 1969b) that methods of this kind, including one due to Ward (1963), may divide dense clusters in an unacceptable manner. Gower (1967a) also raises the question of whether the method should maximize intergroup sums of squares of the distances between centroids of groups. The sum of squares method takes into account the sample size of each cluster, and since some samples of equal importance in the overall classification in certain taxonomic configurations will be based on greatly unequal numbers of OTU's, the method based on maximizing distances between centroids may seem to be preferable. A disadvantage of the centroid method as shown by Gower (1967a) is the possibility of the existence of points in one cluster nearer to the centroid of another cluster. With well-separated clusters the maximum sums of squares and maximum distance between centroid methods will yield the same results, but so will most methods of cluster analysis. For techniques to be consistently employed and recommended they must work on the more challenging cases. For a discussion of various sums of squares criteria see Friedman and Rubin (1967).

Orloci (1967b) has devised a criterion for overcoming the heavy computational load of the sums of squares criterion developed for the divisive clustering technique by Edwards and Cavalli-Sforza (1965). He proposes an agglomerative method for joining those OTU's or clusters that maximize a quantity E. This can be defined as $E = (SS_{total} - \Sigma SS_{within})/SS_{total}$ and is analogous, in analysis of variance, to the co-efficient of intraclass correlation (Sokal and Rohlf, 1969). It will be 1 for the disjoint partition and 0 when all OTU's are joined in a single taxon. The quantities SS_{total} and SS_{within} are the total sum of squares and the within-cluster sum of squares, respectively.

Although the optimum junctions at any one level of clustering are computable, there is no evidence that E is maximized for the entire classification over all levels. In fact, as with all agglomerative strategies, once a decision has been taken to join two OTU's into a taxon, they remain associated, and it is unlikely that the classification is optimal except possibly where provision is made for iterative reallocation. Results are represented by phenograms whose measurement axis represents E as a percentage or proportion. It is obvious from the illustration in Orloci (1967b) that E is not a monotonically increasing function and that reversals in its value can occur.

In a divisive method developed by Rubin (1966), the total set of OTU's is sub-divided into a predetermined number of subsets, the membership of which is defined by a criterion that measures the average object stability of the groups. This criterion is given below in Section 5.10. Rubin's method is based on polythetic criteria. For monothetic methods the divisive strategies seem to be preferable, because it would be quite difficult to establish monothetic agglomerative algorithms after the first grouping. Divisive procedures are also of use in establishing keys (see Chapter 8).

Another divisive technique is that of MacNaughton-Smith et al. (1964) called *dissimilarity analysis*. By this technique the t OTU's are examined each in turn to discover which of these will have the greatest dissimilarity with the remaining $t - 1$ OTU's. Having found the OTU most dissimilar to the remaining OTU's, this OTU is placed in a separate group and others are added to it in turn on a trial-and-error basis, the aim being to minimize the distance between the two groups of t' and $t - t'$ OTU's resulting from this procedure. Subsequently the two groups are again further subdivided by an analogous operation. Similar techniques have been developed by Goodall (1966b; the *deviant index*) and by Hall (1965a; the *peculiarity index*).

2. *Hierarchic versus Nonhierarchic Methods.*   We shall define a clustering method as *hierarchic* if it consists of a sequence of $\omega + 1$ clusterings, $C_0, C_1, \ldots, C_\omega$, in which C_0 is the disjoint partition of all t OTU's and C_ω is the conjoint partition. The number of parts k_i in partition C_i must obey the constraint $k_i \geq k_{i+1}$, where k_{i+1} is the number of parts in partition C_{i+1}. Consequently, in such a clustering there are successively fewer taxa. But at any one level of clustering there is no limitation on the degree of overlap of taxa, that is, one OTU may be simultaneously a member of two or more taxa; also, OTU's that are members of a common taxon at

a lower level may again be members of different taxa at a higher level. This definition of hierarchic differs from that of Johnson (1967), whose symbolism we have partially borrowed here, and also from that of Jardine and Sibson (1968a; 1971). However, it follows the *Random House Dictionary of the English Language* definition of hierarchy (any system of persons or things ranked one above another). Thus by the definition adopted here, any member of a lower ranking taxon is also a member of a higher ranked taxon, although not all its associates from the lower ranking taxon will necessarily be included in the higher ranking taxon. Furthermore, the member of the lower ranked taxon could simultaneously be a member of two (or more) equally ranked higher taxa. In this view we are independently borne out by Cole and Wishart (1970). Jardine and Sibson (1968a; 1971) referred to the classifications we have called hierarchic as stratified. And to impose the definition of hierarchies of some other authors, who imply nestedness as well as mutual exclusiveness, we have to adduce nonoverlapping, the next aspect of clustering to be taken up. A nonoverlapping hierarchic classification in our sense is a hierarchic classification sensu Johnson, and Jardine and Sibson. Thus by our definition both classifications illustrated in Figure 5-2 below will be considered hierarchic. However only the first of the trees is also nonoverlapping and creates mutually exclusive taxa. The second diagram permits overlaps of up to two OTU's between different taxa at any one rank.

Classifications that are *nonhierarchic* are those that do not exhibit ranks in which subsidiary taxa become members of larger more inclusive taxa. Set-theoretically, the OTU's do not exhibit partial order. When OTU's are classified to yield only a single partition, the taxonomic structure is nonhierarchical. Also included among the large class of nonhierarchic taxonomic structures are ordination procedures in which OTU's are projected into a two- or three-dimensional space, and graphs among OTU's which, though they may look like trees, are not rooted; i.e., they do not have a beginning from which branches diverge.

The relative merits of hierarchic versus nonhierarchic classifications are difficult to evaluate. For traditional biological taxonomy, hierarchic classifications are required, and even in related fields such as phytosociology it seems desirable to have higher ranking taxa that summarize common information about the majority of the members of the (polythetic) taxon. Nonhierarchic representation is preferred when emphasis is placed on a faithful representation of the relationships among the OTU's rather than on a summarization of these relationships. Permitting overlapping of groups while retaining the hierarchic structure is a compromise between these two positions. First suggested in recent years by Michener (1963), such a device has been implemented by Jardine and Sibson (1968b; see following discussion).

Among the nonhierarchic clustering techniques are those called "clustering systems" by Lance and Williams (1967d). These are equivalent to hierarchic techniques in which one special rank is singled out for the classification and the

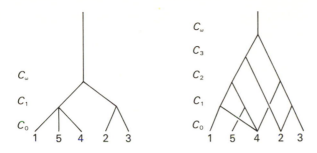

FIGURE 5-2
Dendrograms representing two hierarchic classifications, non-overlapping and overlapping, modified from Jardine and Sibson (1968b). There are five OTU's. On the left $k = 1$, (non-overlapping) on the right $k = 3$ (overlapping). For further details see text. Vertical axis gives clustering levels, $C_0, \ldots, C_\omega$.

particular partition represented by that rank is improved by reallocation of specimens according to certain criteria. Examples of such techniques are the clustering program II by Bonner (1964) or Goodall's probabilistic techniques (1964, 1966c).

A second approach operates on the individual elements or OTU's, as for instance the technique of Rogers and Tanimoto (1960), discussed in detail in Sokal and Sneath (1963, p. 185 ff.).

Lance and Williams (1967d) point out that as studies have become larger, initial sampling of the OTU's for classificatory purposes has become a desirable strategy. A particularly useful system is the k-means system of MacQueen (1967), in which a sample of the OTU's is divided into k groups that are then defined by their mean vectors (centroids). Other OTU's are then added to these groups depending on their distance from the mean vectors. As two groups become too close they are fused and k, the number of groups in the classification, is reduced by one. Similarly as the heterogeneity within one group becomes sufficiently great, the number k is increased through the founding of a new nucleus for classification.

3. *Nonoverlapping versus Overlapping Methods.* In a *nonoverlapping* method, taxa at any one rank are mutually exclusive; that is, OTU's contained within one taxon may not also be members of a second taxon of the same rank. This conventional arrangement has been built into customary taxonomic practice and also into recently formulated taxonomic models based on symbolic logic, such as that by Jardine (1969b). It is simple to conceive of nonoverlapping taxa, and when that concept is combined with a hierarchic classification it gives the familiar nested classifications. However, as we have seen above, these quite frequently distort the phenetic relationships among OTU's. For this reason some workers have preferred to relax the criterion of mutual exclusiveness in taxonomy, and prefer to permit

overlapping in membership at a given rank, rather than to resort to ordination, the antithesis of a nested hierarchic classification.

As an example of *overlapping* clustering techniques we may mention Jardine and Sibson's (1968b) family of overlapping, sequential, agglomerative, hierarchic clustering methods (B_k clustering). These are based on modifications of the single linkage technique, in which overlaps of one or more OTU's per cluster are permitted at any given rank (overlaps of $k - 1$ OTU's to yield k-clusters and k-dendrograms; see Figure 5-2). The more OTU's are permitted to overlap at any given rank, the closer the resulting hierarchy resembles the original dissimilarity matrix. However, it becomes increasingly more difficult to draw and interpret the resulting dendrogram, and graph-theoretical representations become equally complex. These k-clusters (where $k - 1$ is the number of OTU's permitted to overlap in clusters at any given rank) should not be confused with MacQueen's k-means technique discussed above. As Jardine and Sibson (1968b) point out, adequate pictorial presentation of their method requires some ordination procedure, such as nonmetric multidimensional scaling (Section 5.6), to represent the points in the cluster in positions where drawing the edges between vertices of the graph causes the minimum amount of confusion. In view of this it seems to us that representation of taxonomic relationships by ordination is sufficient in most instances to indicate those aspects of the phenetic relationships between organisms that cannot adequately be represented by a nonoverlapping hierarchic system. If a formal overlapping system should be required, this can secondarily be imposed upon the ordination. Both from this consideration as well as from the tediousness and complexity of the overlapping hierarchic clustering method proposed by Jardine and Sibson (1968a,b) we do not feel that it will often be practical in spite of its theoretical elegance. More recently, Cole and Wishart (1970) have developed an algorithm that permits a somewhat more speedy computation of this technique, and Jardine and Sibson (1971) describe several variations of their original methods. Rohlf (1973a) has developed an algorithm which makes this type of cluster analysis computationally practicable for up to a few hundred OTU's.

4. *Sequential versus Simultaneous Methods.* Most clustering methods are *sequential*. By this is meant that a recursive sequence of operations is applied to the set of OTU's that is considered to be either the disjoint partition (in which case the sequential method leads to agglomerative techniques), or the conjoint partition (in which case the sequential techniques are divisive). The possibility of *simultaneous* clustering procedures (of the entire set of OTU's in a single nonrecursive procedure) has been explored by a few workers but has not been generally adopted. Whenever the classification is nonhierarchic in the sense of having only one rank level, an instant division into parts is possible through use of predetermined centers or other logical breaking points. Clearly all ordination techniques are simultaneous. Some trees, such as the procladogram of Camin and Sokal (1965), can be considered

simultaneous techniques in that their construction is based on a single nonrecursive algorithm. Another method of simultaneously clustering all individuals is to have them condense in a series of moves that mimic gravitational attraction. It is therefore analogous to centroid clustering techniques (see Section 5.5) except that it evaluates the gravitational forces of all OTU's simultaneously. Butler (1969) suggests this technique but does not furnish an example of its actual application. Most simultaneous methods are unlikely to lead to optimal clusterings according to some predefined criterion. Hence, iterative techniques of reallocation of OTU's or of rebranching are generally employed to improve such algorithms.

5. *Local versus Global Criteria.*   Rohlf (1970) has pointed out that the clustering solutions usually achieved are not uniformly good at all levels of the taxonomic hierarchy. Thus ordinations by principal component analysis will give reliable representation of intergroup distances over large distances (i.e., between major clusters) but will not be reliable at the tips of the branches (i.e., among closely spaced OTU's within clusters). By contrast, many of the sequential agglomerative clustering techniques do a reliable job of estimating similarities among OTU's within a cluster but become increasingly unreliable as larger and larger taxonomic clusters are considered. Thus the local versus global reliability of various clustering methods may differ considerably; this is related to the degree of tight clustering at different hierarchic levels (Sneath, 1969b). Williams and Dale (1965) have also pointed out that certain clustering methods and combinations of methods may result in locally Euclidean spaces separated by discontinuities or imbedded in non-Euclidean spaces. The attention given by Lance and Williams (1967b) to space-dilating and space-distorting aspects of clustering methods relate to this subject as well. Few definite statements can be made about this problem at the moment but it is certain to become an important aspect of numerical taxonomy in the future. Early attempts in numerical taxonomy assumed an identical similarity metric for an entire study. It now appears possible that unequal metrics will be used in different parts of a taxonomic classification.

Under such a scheme procedures may be developed that could adjust the space as a function of the density with which it is populated or the structure of the cluster inhabiting it. The desirability of such procedures is still in question. However, even if such procedures were adopted in numerical taxonomy, they would be carried out by explicit criteria and not intuitively.

6. *Direct versus Iterative Solutions.*   We shall consider a clustering method *direct* when the algorithm proceeds to the construction of a classification in a straightforward manner, and the solution arrived at is accepted as being optimal in some sense. Most sequential agglomerative or divisive methods are of this type. In sequential methods the solutions seek local optima: that is, a clustering at any one clustering level is generally computed by some criterion of optimality; however,

once structure is established at a certain level, it is not changed during later clustering steps.

By contrast, clustering procedures that aim at optimal solutions over the entire classification are usually impossible to achieve by a direct analytical solution. Such classifications are subjected to self-correcting, *iterative* procedures that aim to improve the measure of local or global optimality or both. These iterative procedures are generally of two kinds. One is the reallocation technique in which objects are successively moved from one part of a partition into another to improve the optimality criterion in a hierarchic classificatory scheme (an example is the object stability criterion of Rubin, 1966). An example of a reallocation technique of this type for a nonhierarchic clustering can be found in Lance and Williams (1967d). Crawford and Wishart (1968) have added a reallocation feature to their earlier method described above.

A second type of iterative clustering rearranges an established dendrogram in order to optimize some criterion such as the cophenetic correlation coefficient or the length of the tree, where length of branches has some quantitative meaning based on character differences. Such techniques have been suggested by Camin and Sokal (1965) for their cladistic work and by Hartigan (1967) in phenetic clustering. Since the number of branching patterns that can be devised from even a very few OTU's is quite large, clearly not all possible trees can be evaluated; instead the initial solutions must be improved by some type of hill-climbing algorithm. In such work there is always the danger of being trapped by local optima, that is, of reaching a structure far from optimal that cannot be changed to a better structure without passing through considerably inferior structures. The problem is like that of a man moving in a random direction trying to climb to the top of the Rocky Mountains through a dense fog: if his strategy is always to climb upward he will most likely be trapped on the peak of one of the lower ranges.

Ball and Hall (1965, 1967) choose arbitrary points as cluster centers and set up equidistant hyperplanes between these. Subsets are divided if they are too diffuse or the OTU's are too numerous, and they are fused if OTU's are too few or too close, which requires new cluster centers be set up as centroids. Ball and Hall's ISODATA program iterates toward finding all discrete clusters and also notes aberrant points. An advantage of ISODATA is that it avoids calculating a complete distance matrix and thus might handle more OTU's than one of the conventional programs.

Another type of iterative procedure is a recursive computation of a similarity matrix from an original similarity matrix. This has been carried out by a number of persons. McQuitty (1967b, 1968a,b) and McQuitty and Clark (1968) show that by repeatedly carrying out this operation, members of a cluster will tend to correlate near $+1$, while members not belonging to the same cluster will show correlation values ranging from less than 1 to -1. Although this method was developed for presence–absence characters it has also been applied to dimensional (measurement) characteristics. It is not clear how it determines the number of clusters that are produced. The mathematical derivation of McQuitty's proof that members of a cluster will yield an intercolumnal correlation of unity is based on the assumption that the correlation of two members of a type (in his sense this means that they are identical in all characteristics) will be equal to that with a third member. In a related technique Bonner (1964) uses an arbitrary threshold value for a similarity matrix, assigning 1's for similarity above the threshold and 0's for similarity below the threshold. He then computes the similarity coefficients of Rogers and Tanimoto (see Section 4.4) between columns of his similarity matrix and, using the same threshold as before, obtains again a similarity matrix composed of 1's and 0's. By iterating this procedure he eventually obtains so-called tight clusters that are maximal complete subgraphs of the similarity matrix graph. These subgraphs or tight clusters form the core clusters around which other OTU's are grouped. Hyvärinen (1962) and Harrison (1968) give similar methods.

7. *Weighted versus Unweighted Clustering.* The various algorithms for clustering procedures provide ample opportunity for weighting certain types of relationships as being more important (closer, tighter) than others. This can be done by transforming the measurement scale, e.g., from correlations to angles.

Weighted clustering is perhaps most frequently used to assign weights to stems joining during sequential agglomerative clustering. In this application the weighting, or lack thereof, is based on the number of OTU's subtended by each stem. Weighted and unweighted clustering of this type are important procedures that are discussed in the next section (5.5) in connection with several clustering methods.

Another type of weighting during clustering is to consider some dimensions of the clusters as more important than others. Thus it seems to many observers intuitively obvious that an elongated dense cloud of points should be recognized as forming a distinct cluster and that distances along the major axis of such an ellipsoid should be scaled down with respect to distances at right angles to this axis. This idea has been voiced in various forms by several authors. Carmichael, George, and Julius (1968) measure relative closeness of new candidates for joining clusters by several criteria—by computing the drop in the average point-to-cluster distance from the last such average (criterion 1); by similarly computing the drop in the distance of the neighbor nearest to any point in the cluster (this is the single linkage distance; see Section 5.5) from the average of the previous successive single linkage distances (criterion 2); and by computing the ratio of the minimum similarity between any pair of points already in the cluster over the minimum similarity between the point being considered for admission and any point in the cluster (criterion 3). Arbitrary values are assigned to these criteria and clusters are terminated when any one (or more) of the three criteria exceed these values. Clusters are also terminated when an acceptable candidate for admission is already a member of another established cluster (criterion 4). During a sequential agglomerative clustering procedure the arbitrary values are successively reduced in a predetermined manner (see Carmichael, George, and Julius, 1968, for details). These authors illustrate their technique with several different shapes of clusters in a two-dimensional space and Carmichael and Sneath (1969) suggest applications to taxonomic problems.

Rohlf (1970) has extended this method by defining a generalized distance function $D_{J \to K} = [(\bar{x}_K - \bar{x}_J)' S_J^{-1} (\bar{x}_K - \bar{x}_J) |S_J|]^{1/2}$ where $\bar{x}_J$ and $\bar{x}_K$ are the column vectors representing the means for clusters J and K, respectively, for n variables, S_J is the variance-covariance matrix of these variables for cluster J, and $|S_J|$ is its determinant, the generalized variance. This function weights distances along axes of a hyperellipsoid inversely to the eigenvalues corresponding to each of these axes, and in this manner would count distances along the principal axis of an ellipsoid as equivalent to the much smaller distances along the minor axes. It is assumed that clusters J and K are each based on several OTU's and are thenceforth treated as the new OTU's. The variance-covariance matrix computed is not pooled as in Mahalanobis' D^2 but is based on one cluster or the other. It is necessary that such variances be known. When there is only a single OTU in a cluster, Rohlf suggests several techniques, among them guesses at an artificial variance-covariance matrix in order to create hyperellipsoids of the desired volume. The details of the method can be consulted in the reference given. Rohlf's linear generalized distance clustering results in elongated clusters being "favored" over hyperspherical ones (if the cluster is elliptical). The method can be extended to the nonlinear case by the use of dummy variables standing for squares and products of the character axes. Examples of parabolic, annular, and ellipsoidal clusters are illustrated in Rohlf (1970). Thus, in a ring shaped cluster, the distances between the OTU's of the cluster and the circumference of the ring are treated as less than the distances from the center (which is the centroid of the cluster but is empty), even if they are greater in the untransformed A-space.

Unweighted clustering is probably best called an equally weighted procedure. It does not prefer any one direction in a swarm of points over another, nor does it weight relationships between taxa on the basis of the size of their membership. Illustrations of its application are given in Section 5.5.

8. *Nonadaptive versus Adaptive Clustering.* Most clustering methods are *nonadaptive*, that is, the algorithm proceeds either directly or iteratively toward a solution in which the clustering method is fixed and interacts with the constellation of points in the A-space to form a cluster. As we have noted, the clustering method imposes a certain structure on the data, and once a given clustering strategy has been initiated it often forces the data into a certain arrangement. Measures of stress or distortion may indicate how satisfactory a given classification is.

An ideal clustering method would be a learning or *adaptive* method, which would initially explore the types of constellations that are probably present in the data, decide on the most likely one, and modify the clustering algorithms in the program to weight some inter-OTU distances differentially. Thus if the program had reason to believe that the cluster was ellipsoidal, Rohlf's linear generalized distance might be employed, which would decrease distances along the principal axis by comparison with those along the minor axes. Several authors have attempted such procedures. Rohlf (1970) has offered a scheme of this sort which, however, is not truly adaptive in the above sense. It is a program that calculates his generalized distance function (defined above), using an initially defined set of functions that determine a series of alternative plausible cluster shapes. A parameter determining the speed with which the program adapts to apparent trends in the data is furnished.

Wishart (1969b,c) has described a method, *hierarchical mode analysis*, that is partly adaptive in that it seeks out clusters that may be elongate, but does not include in them sparse points that may represent "noise" or unwanted variation, and chaining (Section 5.5) is suppressed. An integer k is chosen, as well as a starting radius r. From each OTU one tests whether k or more other OTU's lie within r. If so, the OTU is considered "dense," and "dense" points are clustered (using the single linkage criterion, see Section 5.5). The remaining points are initially unallocated, but may be allocated to nearby clusters under some circumstances. By increasing r, a hierarchic classification is produced; the number of clusters rises at first, and then finally falls to 1 as r is increased. If $k = 1$, the method is the single linkage SAHN method (Section 5.5). Shepherd and Willmott (1968) describe a similar method. A method of P. M. Neely, whose outlines are described in Oxnard and Neely (1969), *neighborhood limited classification*, is also similar to Wishart's techniques, as is one devised by Farchi (1966), used by Hill et al. (1965).

A clustering method with some adaptive features was suggested by Ornstein (1965), in which a continuous updating of the established clusters by new specimens coming into the system (in his case classifications of serum proteins from patients in a hospital) shift the criteria for allocating specimens. We have not seen the method applied subsequent to its publication and from the details provided cannot evaluate its prospects or success.

Another technique with adaptive features is Sneath's (1966e) method for curve seeking from scattered points in many dimensions. The positions are adjusted by a gravitational model so that they fall as nearly as possible into smooth curves, the shapes of which need not be specified prior to the procedure. So far this method has only been able to handle trends in characters as through time or ontogeny, but possibly such a method could be extended to handle clouds of OTU's in multidimensional space. A somewhat similar method has been suggested by Jancey (1966a) and applied to a population of a legume species (Jancey, 1966b).

Clearly much work remains to be done in the field of adaptive clustering techniques. Conceptually it is by far the most attractive method of organizing data without making them conform to one or a limited number of preconceived constellations.

5.5 SEQUENTIAL, AGGLOMERATIVE, HIERARCHIC, NONOVERLAPPING CLUSTERING METHODS

These methods, for which we shall employ the acronym SAHN, are the most frequently employed strategies for finding clusters, especially in biological material. The four separate properties of this class of methods have already been discussed, but now some general properties of SAHN techniques will be taken up.

General Considerations

Although much of the work on clustering will be based on matrices of similarity coefficients, S_{jk}, complements of similarity coefficients, which represent measures of dissimilarity and are analogous to distances, are generally more heuristic. Such coefficients can usually be expressed as $(1 - S_{jk})$, especially when $0 \leq S_{jk} \leq 1$. We shall let U_{jk} stand for a general dissimilarity coefficient of which taxonomic distance, d_{jk}, is a special example. Euclidean distances will be used in the explanation of clustering techniques below because they are easy to visualize geometrically, although not all dissimilarity coefficients are necessarily metric.

In discussing agglomerative clustering procedures Lance and Williams (1967b) make a useful distinction between the following three types of measure (we have altered their symbolism slightly for uniformity with other chapters): (J)-measures are those that define a property of a single group or cluster such as its centroid, its dispersion, its shape, etc; (J, K)-measures estimate similarity or dissimilarity between two groups or between an OTU and a group; finally (JK, L)-measures describe changes in some measure when two groups fuse. An example would be the increase in information resulting from the fusion of two separate clusters.

In all SAHN clustering two considerations govern every clustering step. One is the recomputation of the similarity or dissimilarity coefficient between newly established clusters and potential candidates for future admission and the other is the admission criterion for new members to an established cluster. The first of these is a (J, K)-measure as defined by Lance and Williams (1967b), and it differs for various SAHN techniques; the second may be based on combinations of all three types of measures defined by these authors. However for all pair-group methods treated below the criterion is the same and is based on a (J, K)-measure. For evaluating the first criterion we shall adopt the following uniform symbolism. We shall be concerned with clusters **J**, **K**, and **L** containing t_J, t_K, and t_L OTU's, respectively, where t_J, t_K, and t_L all ≥ 1. OTU's **j** and **k** are contained in clusters **J** and **K**, respectively, and $l \in L$. Given that clusters **J** and **K** join, the problem is to evaluate the dissimilarity between the fused cluster and further candidates **L** for fusion. The fused cluster is designated $(\mathbf{J}, \mathbf{K})$, with $t_{(\mathbf{J},\mathbf{K})} = t_{\mathbf{J}} + t_{\mathbf{K}}$ OTU's. The various clustering methods differ in their computation of the dissimilarity coefficient $U_{(\mathbf{J},\mathbf{K}),\mathbf{L}}$.

Lance and Williams (1967b) have distinguished the following criteria for SAHN clustering procedures. They consider such procedures *combinatorial* if the dissimilarity coefficient $U_{(\mathbf{J},\mathbf{K}),\mathbf{L}}$ can be computed from the previously evaluated dissimilarities $U_{\mathbf{J},\mathbf{L}}$, $U_{\mathbf{K},\mathbf{L}}$, $U_{\mathbf{J},\mathbf{K}}$ and the sample sizes $t_{\mathbf{J}}$ and $t_{\mathbf{K}}$. With combinatorial techniques coarser clusters can always be computed from previous finer clusters and it is not necessary to return to the original dissimilarity matrix while clustering. As these authors point out, such a procedure has obvious advantages over a *noncombinatorial* clustering method in which the original matrix must be retained for the calculation of measures required during succeeding clustering steps. For combinatorial SAHN clustering, Lance and Williams (1966a, 1967b) have developed a formula for linear combinatorial strategy, as follows:

$$U_{(\mathbf{J},\mathbf{K}),\mathbf{L}} = \alpha_{\mathbf{J}} U_{\mathbf{J},\mathbf{L}} + \alpha_{\mathbf{K}} U_{\mathbf{K},\mathbf{L}} + \beta U_{\mathbf{J},\mathbf{K}} + \gamma |U_{\mathbf{J},\mathbf{L}} - U_{\mathbf{K},\mathbf{L}}|$$

This formula is useful for differentiating the various SAHN strategies as long as they are combinatorial. The Greek letters refer to arbitrary constants that determine the nature of the clustering strategy. The formula can also be used as an algorithm for computer programs, although for any given strategy there is usually a more efficient algorithm available.

A second criterion in Lance and Williams (1967b) is *compatibility* of a clustering strategy. In a compatible strategy the metric among coarser clusters is the same as that among finer clusters or even among the original OTU's. Thus the dimensionality of the original space is retained and it is simple to represent clusters in the original A-space. Clustering procedures that do not have this property are called incompatible. For obvious reasons they are generally undesirable.

Even though the measures are compatible, the dimensions of the space may be distorted as new dissimilarities are computed between growing clusters. If this is not so, Lance and Williams (1967b) call the clustering strategy *space-conserving*. In *space-distorting* strategies it appears as though the space in the immediate vicinity of a cluster has been contracted or dilated. We shall encounter such instances in the methods recounted below.

Returning to the criterion of admission for a candidate joining an extant cluster, this is constant in all pair-group methods discussed below. OTU or cluster **J** will join **K** if and only if $U_{\mathbf{J}\mathbf{K}} < U_{\mathbf{J}\mathbf{L}}$, and $U_{\mathbf{J}\mathbf{K}} < U_{\mathbf{K}\mathbf{L}}$, where **L** is any OTU or cluster in the study (at the current level of clustering) other than **J** or **K**. This means that **J** and **K** are the mutually closest pair of OTU's or clusters. Often, as in the examples below, ties will result in situations such that $U_{\mathbf{J}\mathbf{K}} < U_{\mathbf{J}\mathbf{L}}$ but $U_{\mathbf{J}\mathbf{K}} = U_{\mathbf{K}\mathbf{M}} < U_{\mathbf{K}\mathbf{L}}$ (where **L** stands for any taxon other than **J**, **K**, or **M**). In such cases arbitrary decisions must be made, and conventionally for computer programs the choice is the first link, i.e., U_{JK} over U_{KM}. This may result in **M** being delayed in admission to the cluster because the new cluster (**J, K**) may show changes in its relationships to other taxa from the separate relationships of **J** and **K**.

Pair-group versus Variable-group Methods. In the various algorithms used in SAHN techniques, only one OTU or cluster may be admitted for membership at one time (*pair-group method*), or several may join simultaneously (*variable-group method*). The resulting dendrogram will therefore either show bifurcations or multiple furcations, respectively. The justification for permitting a variable number of candidates to join a cluster is that differences in phenetic relationships between the potential candidates are often so slight that if a pair-group technique were used, none of the differences established would be significant (statistically or in magnitude). Also, to restrict dendrograms to bifurcations is a limitation upon the topology of these trees that some workers may not wish to impose. However, the pair-group method has been used more frequently than the variable-group method because the former is easier to program. Whenever the variable-group method is chosen, an arbitrary criterion must be chosen at which the program cuts off further candidates from joining the cluster in any given clustering step.

The practical consequences of choosing one or the other alternative strategy are not very great. Even if the pair-group method is consistently chosen for convenience of programming, small differences in the levels at which successive members join a given cluster may be eliminated and multiple furcations may be reintroduced in the eventual display of the phenogram after the clustering process has been completed. We shall therefore restrict ourselves to pair-group methods in the discussion that follows.

There are numerous algorithms for SAHN clustering. Many of them go back to the earliest applications of cluster analysis to scientific problems. We shall review several of the more commonly employed ones and give passing references to some others.

The best way for the reader to familiarize himself with the clustering procedures is to carry out the computational steps involved in several of these. As an example we shall use a set of artificial data designed to show some of the differences between the various methods. The data are 16 OTU's whose coordinates in a 2-space are given in the top two rows of Table 5-1. Euclidean distances and their squares are represented in the lower and upper triangular matrix in the same table, respectively. The properties of the various clustering techniques are summarized in Table 5-2.

Single and Complete Linkage Clustering

Single Linkage Clustering. This method, introduced to taxonomy by Florek et al. (1951a,b) and Sneath (1957b), has also been called the nearest neighbor technique

TABLE 5-1

The coordinates of 16 OTU's in a two-space defined by axes X_1 and X_2 are given in the first two rows of the table. Euclidean distances between pairs of OTU's are shown in the lower triangular matrix, their squares in the upper triangular matrix

	a	b	c	d	e	f	g	h	i	j	k	l	m	n	o	p
X_1	0	0	1	2	3	2	2	1	5	6	7	5	7	6	6	8
X_2	4	3	5	4	3	2	1	0	5	5	6	3	3	2	1	1
a	×	1	2	4	10	8	13	17	26	37	53	26	50	40	45	73
b	1.0	×	5	5	9	5	8	10	29	40	58	25	49	37	40	68
c	1.41421	2.23607	×	2	8	10	17	25	16	25	37	20	40	34	41	65
d	2.0	2.23607	1.41421	×	2	4	9	17	10	17	29	10	26	20	25	45
e	3.16228	3.0	2.82843	1.41421	×	2	5	13	8	13	25	4	16	10	13	29
f	2.82843	2.23607	3.16228	2.0	1.41421	×	1	5	18	25	41	10	26	16	17	37
g	3.60555	2.82843	4.12311	3.0	2.23607	1.0	×	2	25	32	50	13	29	17	16	36
h	4.12311	3.16228	5.0	4.12311	3.60555	2.23607	1.41421	×	41	50	72	25	45	29	26	50
i	5.09902	5.38516	4.0	3.16228	2.82843	4.24264	5.0	6.40312	×	1	5	4	8	10	17	25
j	6.08276	6.32455	5.0	4.12311	3.60555	5.0	5.65685	7.07107	1.0	×	2	5	5	9	16	20
k	7.28011	7.61577	6.08276	5.38516	5.0	6.40312	7.07107	8.48528	2.23607	1.41421	×	13	9	17	26	26
l	5.09902	5.0	4.47214	3.16228	2.0	3.16228	3.60555	5.0	2.0	2.23607	3.60555	×	4	2	5	13
m	7.07107	7.0	6.32455	5.09902	4.0	5.09902	5.38516	6.70820	2.82843	2.23607	3.0	2.0	×	2	5	5
n	6.32455	6.08276	5.83095	4.47214	3.16228	4.0	4.12311	5.38516	3.16228	3.0	4.12311	1.41421	1.41421	×	1	5
o	6.70820	6.32455	6.40312	5.0	3.60555	4.12311	4.0	5.09902	4.12311	4.0	5.09902	2.23607	2.23607	1.0	×	4
p	8.54400	8.24621	8.06226	6.70820	5.38516	6.08276	6.0	7.07107	5.0	4.47214	5.09902	3.60555	2.23607	2.23607	2.0	×

TABLE 5-2

Formulae and properties of some SAHN clustering methods.

Cluster Method	Synonyms	$U_{(J,K),L}$				$U_{J,K}$
		α_J	α_K	β	γ	
Single linkage	*Nearest neighbor **Minimum method	$\frac{1}{2}$	$\frac{1}{2}$	0	$-\frac{1}{2}$	$\min_{jk} U_{jk}$
Complete linkage	*Furthest neighbor **Maximum method	$\frac{1}{2}$	$\frac{1}{2}$	0	$\frac{1}{2}$	$\max_{jk} U_{jk}$
Average linkage Arithmetic Average						
Unweighted (UPGMA)	*Group average	$t_J/t_{(J,K)}$	$t_K/t_{(J,K)}$	0	0	$\dfrac{1}{t_J t_K}\sum_{jk} U_{jk}$
Weighted (WPGMA)		$\frac{1}{2}$	$\frac{1}{2}$	0	0	$\sum_{jk} w_j w_k U_{jk}$†
Centroid						
Unweighted centroid	*Centroid	$t_J/t_{(J,K)}$	$t_K/t_{(J,K)}$	$-t_J t_K/t_{(J,K)}^2$	0	§
Weighted centroid	*Median	$\frac{1}{2}$	$\frac{1}{2}$	$-\frac{1}{4}$	0	§

*Lance and Williams (1967b).
**Johnson (1967).
†Where $w_j = 1/2^{C_j}$ and C_j is the number of prior clustering steps of OTU j.
§Direct computation of U_{JK} only meaningful for Euclidean distance between weighted or unweighted centroids of **J** and **K**.
‡Monotonic if $\alpha_J + \alpha_K + \beta \geq 1$.

(by Lance and Williams, 1967b, who use a common term from plant ecology) and has been designated the minimum method by Johnson (1967); it is known by yet other names in various other fields. An OTU that is a candidate for an extant cluster has similarity to that cluster equal to its similarity to the closest member within the cluster. Thus, connections between OTU's and clusters and between two clusters are established by single links between pairs of OTU's. This procedure frequently leads to long straggly clusters by comparison with other SAHN cluster methods. A single linkage cluster analysis of the dissimilarity matrix of Table 5-1 reveals the following steps (we shall express our dissimilarities U_{jk} as Euclidean distances Δ_{jk}, but because they are simple integral numbers it will be simpler to search for low dissimilarities by looking at their squares, Δ_{jk}^2, which are mono-

Hierarchical clustering scheme monotonic‡	Combinatorial for	Compatible for	Effects on space
Yes	all (**J, K**)-measures	all (**J, K**)-measures	contracting
Yes	all (**J, K**)-measures	all (**J, K**)-measures	dilating
Yes	all (**J, K**)-measures	all (**J, K**)-measures	conserving
Yes	all (**J, K**)-measures	all (**J, K**)-measures	conserving
No	yes for d_{jk}^2; no for r_{jk}, but yes for variances and covariances; not for any Manhattan metric	all (**J, K**)-measures	conserving
No	yes	yes for d_{jk} and Canberra metric, no for r_{jk}	conserving

tonically related). It may be noted that the single linkage method gives phenograms of identical topology from any monotonic transformation of U, such as rank orders (Jardine, Jardine, and Sibson, 1967).

We first find the mutually most similar pairs (least dissimilar pairs) which turn out to be (**a, b**), (**f, g**), (**i, j**) and (**n, o**), all at a distance of 1.0 from each other. For example, OTU **a** is most similar to OTU **b** ($\Delta_{ab} = 1$), and OTU **b** is equally most similar to OTU **a**. Note that **c** is most similar to **a** ($\Delta_{ac} = 1.41421$), but since we have seen that **a** is most similar to **b**, OTU **c** clearly will not join with **a** in this particular clustering cycle. The geometric result of finding the mutually most similar pairs of OTU's is shown in Figure 5-3, where these pairs have been connected by solid lines. We next examine distances from new candidates for fusion

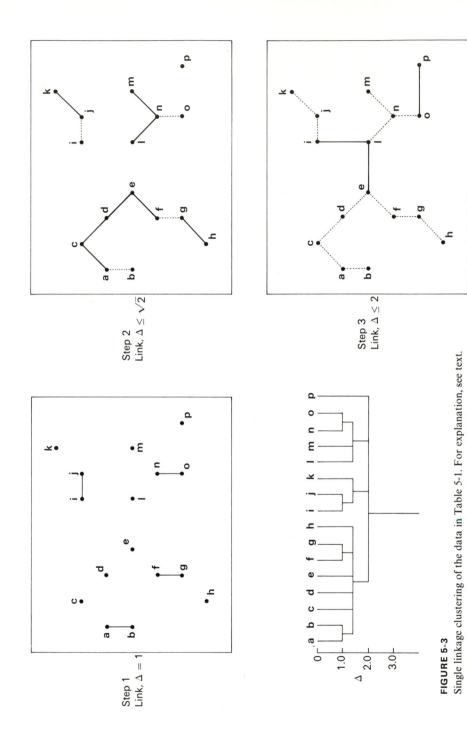

FIGURE 5-3

Single linkage clustering of the data in Table 5-1. For explanation, see text.

with the established clusters. Since in single linkage clustering one can measure the similarity of any OTU to the established cluster (**a, b**) by finding the smaller of the distances between that OTU and either **a** or **b**, it is not necessary to construct a new smaller dissimilarity matrix in which the four established clusters are represented each by a single row and column. Thus the dissimilarity between OTU **c** and cluster (**a, b**) is $1.41421 = \Delta_{ac}$, rather than $2.23607 = \Delta_{bc} > \Delta_{ac}$. However, on searching through the dissimilarity matrix we find that **c** relates to **d** at the same level as it does to **a** and that in turn **d** is related equally to **c** and **e**, **e** relating equally to **d** and **f**. In this simple geometrical example there are far more ties than there would be in a real data set based on many characters where exact ties are rather improbable. The ties are clearly seen as equal distances in Figure 5-3.

The next clustering cycle in the pair-group method would again encounter ties and we would determine arbitrarily that **c** should join (**a, b**), that **e** should join (**f, g**), that **k** should join (**i, j**), and that **l** should join (**n, o**). However, in the very next step and at the same dissimilarity levels, **d** would join (**a, b, c**), **h** would join (**e, f, g**), and **m** would fuse with (**l, n, o**). Finally, still at the same dissimilarity level, (**a, b, c, d**) would join (**e, f, g, h**). Since the dissimilarity level representing the linkage among these OTU's has not been increased it is customary to represent all these fusions as a single clustering step both in the geometric representation in Figure 5-3 as well as in the phenogram also shown in that figure. The new connections have been indicated by solid lines, while those links established in the previous clustering cycle are now shown by dotted lines. Note how the clusters are strung out in characteristic single linkage fashion. The final links that must be made to connect all OTU's into a single cluster are at the Euclidean distance 2.0. The final connection is accomplished by **p** linking into **o**, **e** linking into **l**, and **i** linking into **l**. Formally this would represent two pair-group steps: first the fusion of **e** and **l**, and of **o** and **p**, followed by the fusion of **i** and **l**. Since the links are at the same level they are represented here as a single clustering step, geometrically as well as in the phenogram.

Single linkage clustering has yielded three taxonomic levels for this data set. The most closely related OTU's are (**a, b**), (**f, g**), (**i, j**), and (**n, o**), with OTU's **c, d, e, h, k, l, m,** and **p** remaining single at the same rank. The next higher rank is represented by taxa (**a, b, c, d, e, f, g, h**), (**i, j, k**), (**l, m, n, o**), and **p**, and at the highest rank a single taxon comprises all OTU's in the study. Although all computations were based on the original dissimilarity matrix (this is by far the simplest procedure to use when employing single linkage analysis), we could have computed reduced dissimilarity matrices in which dissimilarity of clusters **J** and **K** could have been represented by $\min_{jk} U_{jk}$, the lowest dissimilarity between any pair of OTU's in these clusters. Thus, for example, $U_{ab,c} = 1.41421$ and $U_{ab,d} = 2.0$. The method is therefore combinatorial in the sense of Lance and Williams (1967b), and one could employ their formula for linear combinatorial strategy as given in Table 5-2.

Single linkage clusters can be obtained agglomeratively as above, or divisively, by finding first the shortest spanning tree (Section 5.7) and then dividing in turn the longest links (Gower and Ross, 1969). Rohlf (1973c) states that the most efficient method for single linkage clustering is based on shortest spanning trees (see Section 5.7).

Complete Linkage Clustering. This method, also known as farthest neighbor clustering (Lance and Williams, 1967b) or the maximum method (Johnson, 1967), is the direct antithesis of the single linkage technique just discussed. An OTU that is a candidate for admission to an extant cluster has similarity to that cluster equal to its similarity to the farthest member within the cluster. When two clusters join, their similarity is that existing between the farthest pair of members, one in each cluster. The method will generally lead to tight, hyperspherical, discrete clusters that join others only with difficulty and at relative low overall similarity values. They are called *homostats* by Cattell, Coulter, and Tsujioka (1966). We illustrate the procedure in Figure 5-4 based on the distance matrix in Table 5-1. The initial linkage is between mutually most similar pairs, which are the same as by the previous method. The distance criterion must be raised to 2.23607 before any other OTU joins the nodal clusters already established. Thus, though Δ_{ac} is equal to 1.41421, Δ_{bc} is equal to 2.23607, and the "level of cohesion" of the cluster is 2.23607. As in clustering by single linkage, some candidates for admission are tied. By a strict pair-group procedure there are more than the six clustering steps shown in Figure 5-4, but in the phenogram all OTU's joining a node at the identical maximal dissimilarity level are shown joining simultaneously. The exact rules for deciding on ties are of consequence in this and most of the other SAHN methods. There can be different rules for breaking ties and these are reflected in the detailed structure of the resulting phenogram. Thus, the frequently used NT-SYS program for complete linkage (Rohlf, Kishpaugh, and Kirk, 1971) produced the phenogram B in Figure 5-4. It differs in some details from phenogram A, which represents the rules for decision that are implied by the graphs connecting the OTU's. In Figure 5-4 all direct links between the OTU's in a cluster are shown including the farthest link, which determined the level of cohesion. Again, the entire clustering procedure can be carried out on the original dissimilarity matrix, but this matrix could have been condensed by representing the similarity between two clusters **J** and **K** as max U_{jk} . Condensation could also be carried out by Lance and Williams' formula for linear combinatorial strategy as given in Table 5-2.

Inspection of the clusters in Figure 5-4 shows their induced compactness by comparison with the loose, strung-out single linkage clusters in Figure 5-3. The data are more structured, showing more taxa and more ranks by the complete linkage method. As Lance and Williams (1967b) point out, the single linkage method is space-contracting, but complete linkage is a space-dilating technique. While the lowest ranked taxa (other than the disjoint partition) are as before (**a, b**), (**f, g**), (i, j), and (n, o), plus the remaining eight single OTU's (step 1 in Figure 5-4), at the next higher rank the following taxa are obtained: (a, b, c, d), (e, f, g), h, (i, j, k), (l, m, n, o), and p (step 2 in Figure 5-4). These six clusters become four in step 3 of Figure 5-4 by the fusion of h with (e, f, g) and p with (l, m, n, o), and this is followed in turn by a reduction to three taxa (a–h), (i, j, k), and (l–p). The two latter taxa fuse next. These steps have been combined in step 4 of Figure 5-4. The final fusion yields a single taxon, the coarsest partition including all OTU's (step 5 in Figure 5-4). Rather than use arbitrary decisions about breaking ties, Sørensen (1948), the originator of complete linkage clustering, prefers fusion to yield the larger group; or if this fails, fusion to lead to as few residual groups as possible; and if this criterion turns out to be indecisive, he recommends choosing the combination with the highest average similarity (that is, the lowest mean U_{jk}).

We would like to relate the methods discussed above to the extensive work on cluster analysis in psychology by L. L. McQuitty, summarized in McQuitty (1967a), where references to his earlier work can be found. His *linkage analysis* is the equivalent of an agglomerative sequential clustering technique employing single linkage criteria. His *rank order typal analysis* is a complete linkage analysis in which the criterion for complete linkage is not a threshold but consistency within the cluster with respect to the order of relationship. That is, when the similarities of any person to all others are ranked, there must be no single similarity rank to a nonmember of the cluster greater than to a member of the cluster. McQuitty expresses this in another way by saying that a cluster must not include a rank higher than the number of OTU's in the cluster. McQuitty recognizes that rank order typal analysis (complete linkage) will yield few types (= clusters). Many persons fail to enter types by his definition, which leaves an unclassified residue. By contrast, linkage analysis (our single linkage) yields "distant cousins": "Some persons are far removed from the reciprocal pairs which initiate the types" (McQuitty, 1967a, p. 29). His assumptions that persons are "imperfect types" as contrasted with pure types are essentialist assumptions that make his psychological theory typological in the philosophical sense. At least for biological material, we doubt whether such assumptions are warranted and suspect that they would distort the true structure of the data.

Modifications of Single and Complete Linkage Clustering.   The elongate growth of single linkage clusters, as shown in Figure 5-3, is known as *chaining*. It is a peculiar feature of a single linkage analysis, especially when there are a number of equidistant points (as in the present artificial example), or near equidistant points (as are frequently found in taxonomy and ecology). The phenograms resulting from such chaining are generally not very informative, because the information on intermediate or connecting OTU's, though of potential interest, is not well shown in phenograms.

To avoid the extremes of single linkage and complete linkage clustering—chaining in the former and small tight compact clusters that leave out many of the

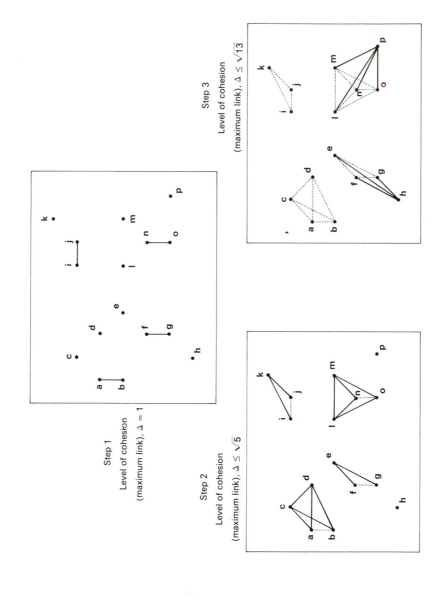

Step 1
Level of cohesion
(maximum link), Δ = 1

Step 2
Level of cohesion
(maximum link), Δ ≤ √5

Step 3
Level of cohesion
(maximum link), Δ ≤ √13

FIGURE 5-4

Complete linkage clustering of the data in Table 5-1. Because of the unusually large number of tied distance values in this simple, artificial example, variation in rules for deciding how to break ties produces substantial differences in the final phenogram. The outcome of the clustering process illustrated by graphs in this figure is shown in Phenogram A; the output produced by the algorithm of the well-known and frequently used NT-SYS program (Rohlf, Kishpaugh, and Kirk, 1971) is shown in Phenogram B. For further details, see text.

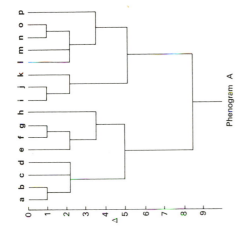

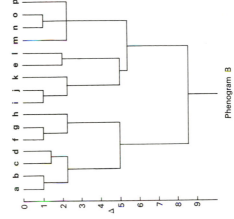

Phenogram A

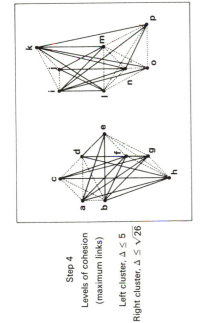

Step 4

Levels of cohesion
(maximum links)

Left cluster, $\Delta \leq 5$

Right cluster, $\Delta \leq \sqrt{26}$

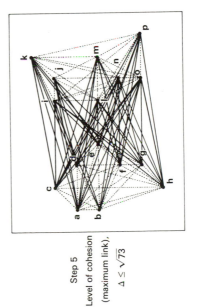

Phenogram B

Step 5

Level of cohesion
(maximum link),

$\Delta \leq \sqrt{73}$

less easily affiliated OTU's in the latter—Sneath (1966b) developed four modifications of the criterion for linking two clusters. *Integer link linkage* specifies the number of links necessary for a new member to join a cluster or for two clusters to join. It can vary between 1 (single linkage) to $t_J t_K$ for complete linkage. *Proportional link linkage* takes into consideration the size of the clusters that are being joined. It specifies the necessary proportion p out of the maximum number of links that can form between any two clusters. Proportional link linkage would include median (Kendrick and Proctor, 1964) and percentile criteria for cluster joining. *Absolute resemblance linkage* allows junction at a level L when the sum of the intercluster similarity coefficients $\sum_{jk} S_{jk}$ reaches a certain fixed level. *Relative resemblance linkage* is a measure in which the sum of the resemblance coefficients is divided by the average total resemblance among all members of the cluster, permitting fusion of clusters when this value is greater than KL. K is an arbitrary constant equalling 1 for average linkage analysis. Thus, this technique includes average linkage analysis. The concepts proposed by Sneath (1966b), especially proportional link linkage, are closely related to measures of connectivity and compactness of clusters as described in graph theory (see Kansky, 1963, and Gabriel and Sokal, 1969; see also Section 5.7).

A modification of single linkage clustering was developed by Shepherd and Willmott (1968) to prevent chaining. They developed a chaining coefficient

$$C = \frac{S_1 - S_3}{S_2 - S_3}$$

where S_1, S_2, and S_3 are levels such that fusion is permitted at S_1 but not at S_2 or S_3. This allows single-linkage to become k-linkage at lower S values.

An interesting approach to overcoming the disadvantages of single and complete linkage analysis has been proposed by Lance and Williams (1967b). They imposed four constraints upon their linear formula for combinatorial strategy, as follow:

$$\alpha_J + \alpha_K + \beta = 1 \qquad \alpha_J = \alpha_K \qquad \beta < 1 \qquad \gamma = 0$$

By adjusting the value of β from 1 to -1 Lance and Williams are able to simulate the results of chaining by single linkage and of extremely compact clustering as observed in complete linkage analysis. It is assumed that some intermediate value of β between 0 and -1 will give a reasonable space-conserving clustering of the data and they suggest a value of -0.25 for general use. See Figure 5-5 for the changes imposed upon the same data set by varying the parameter β under the above constraints. Lance and Williams call this a *flexible clustering strategy*. There

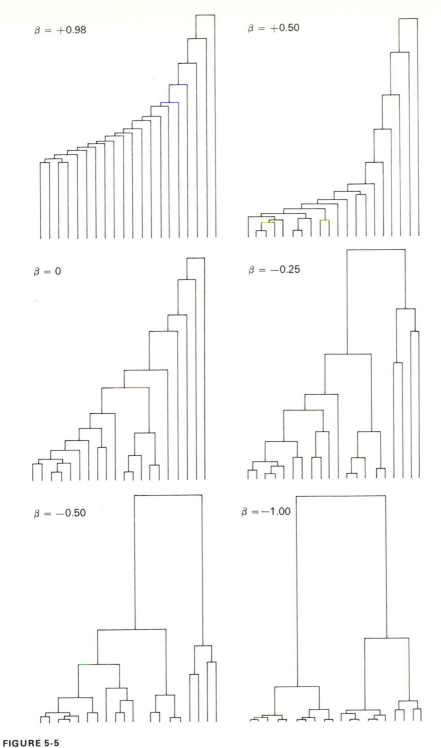

FIGURE 5-5

Flexible clustering. Effect of varying β. Data: 20 OTU's specified by 76 binary attributes. The resemblance measure is Euclidean distance. For explanation, see text. [From Lance and Williams (1967b).]

is, however, some danger in adjusting parameters until one obtains what one likes, rather than choosing some prior criterion and sticking to the results.

Average Linkage Clustering

These techniques (earlier called group methods) were developed by Sokal and Michener (1958) to avoid the extremes introduced by either single linkage or complete linkage clustering. They require computation of some kind of average similarity or dissimilarity between a candidate OTU or cluster and an extant cluster. Since there are various kinds of averages, several average linkage methods have been proposed. We shall place principal emphasis on the four most common methods that result from the four combinations of two criteria each with two alternatives: arithmetic average versus centroid clustering, and weighted versus unweighted clustering. *Arithmetic average clustering* computes the arithmetic average of the similarity (or dissimilarity) coefficients between an OTU candidate for admission and members of an extant cluster, or between the members of two clusters about to fuse. It does not take into account S or U coefficients between members within the cluster; thus, the density of the points constituting the extant cluster (or a candidate cluster about to fuse with it) is not a factor in evaluating the resemblance between the two entities. This is readily seen when inspecting Lance and Williams's (1967b) formula for linear combinatorial strategy for an arithmetic average method. The coefficient β that multiplies $U_{J,K}$ is always 0. By contrast, *centroid clustering* finds the centroid of the OTU's forming an extant cluster in an A-space and measures the dissimilarity (usually Euclidean distance) of any candidate OTU or cluster from this point. Centroid clustering thus has a simple geometric interpretation which, as can be seen from Figure 5-11 later in this section, recognizes the dissimilarities among the points constituting a cluster. For this reason the coefficient β in Lance and Williams's formula for linear combinatorial strategy is always negative in centroid clustering methods. No similar geometric interpretation can be easily found for arithmetic average clustering.

Sokal and Michener (1958) introduced *weighted clustering* in an attempt to give merging branches in a dendrogram equal weight regardless of the number of OTU's carried on each branch. Such a procedure weights the individuals unequally, as can be seen in Figure 5-6. Thus in Figure 5-6,*a*, when **d** joins the cluster (**a, b, c**), we have to decide whether to give each of the three OTU's in the old cluster equal weight or whether to weight **c** equal to the cluster (**a, b**). Sokal and Michener (1958) suggested weighting (**a, b**) and **c** equally in an attempt to give equal importance to phyletic lineages. The subsequent development of a phenetic philosophy made this goal undesirable. It was also noted empirically that *unweighted clustering*, which gives equal weight to each OTU in clusters whose similarity with another cluster (or OTU) is being evaluated, produced less distortion when phenograms were compared with original similarity matrices (for reasons explained by Sneath,

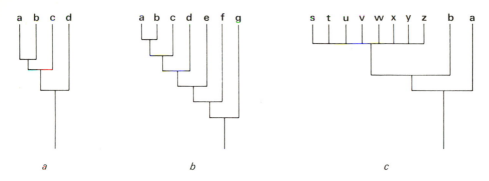

FIGURE 5-6

The weighting of stems in building clusters (for explanation see text).

1969b). Nevertheless, occasions may arise for special purposes where a weighted technique might be preferred. One such occasion may be when a taxonomist is forced to use very disparate sample sizes to represent several taxa. Here the weighted method would lend more importance to the sparsely represented sample and would thus lead to less biased estimates of intercluster distances. It should be pointed out that the weighting discussed above is with reference to individuals composing a cluster and not to the average dissimilarities in Lance and Williams's linear combinatorial formula, in which equal weights apply for weighted clustering and differential weights apply for unweighted clustering. This point has given rise to some confusion (witness several incorrect synonymies in Lance and Williams's 1967b paper), but the terms weighted and unweighted clustering are by now firmly established in numerical taxonomy, and we agree with Gower (1967a) that their past meaning should be retained.

Whenever a weighted pair-group method is to be employed, the weights to be ascribed to any given OTU can be computed simply by the formula $w_j = 1/2^{C_j}$, where C_j is the number of prior clustering steps of OTU j. By the above formula OTU **a** in Figure 5-6,*a* is in two prior clustering (branching) steps before joining **d**, hence it is weighted $1/2^2 = 1/4$; similarly, **c** in the same dendrogram is in only one prior clustering and is weighted $1/2^1 = 1/2$. In unweighted clustering, **a**, **b**, and **c** would each be weighted $1/3$. In a more extreme case shown in Figure 5-6,*b*, **a** in cluster (**a–f**) would differ in weight considerably in comparison with **g**, depending upon whether the clustering process is weighted or not. Thus, in weighted clustering its weight would be $1/2^5 = 1/32$, and for unweighted clustering it would be $1/6$. When a variable group method is employed, weights would also differ. Thus in Figure 5-6,*c*, when **a** joins the cluster (**b, s–z**), the weight of **s** and all other OTU's would be $1/9$ in the unweighted case, but it would be $1/16$ in the weighted case with **b** being weighted $1/2$.

We shall now briefly discuss the four combinations of the above clustering strategies.

UPGMA (*unweighted pair-group method using arithmetic averages*) is probably the most frequently used clustering strategy. Its development was outlined by Sokal and Michener (1958), but it was first employed by Rohlf (1963b) in a classification of mosquitoes. Lance and Williams (1967b) call this the group-average method. The dissimilarity $U_{J, K}$ between any two clusters J and K can be computed as $(1/t_J t_K) \sum_{jk} U_{jk}$; the method works for similarity coefficients as well. It rests on the plausibility of the concept of an average dissimilarity or similarity coefficient. Average correlations between OTU's have usually been computed as averages of their transforms, which are then back-transformed into correlation measure, i.e., $r_{JK} = \cos [(1/t_J t_K) \sum_{jk} \text{arc cos } r_{jk}]$. The method is combinatorial for all (J, K)-measures and, by means of Lance and Williams's (1967b) formula for linear combinatorial strategy, successive similarity or dissimilarity matrices during clustering can be computed from immediately prior ones. The measure is compatible in the sense of Lance and Williams and, since it obeys the ultrametric inequality, is monotonic.

The UPGMA algorithm computes the average similarity or dissimilarity of a candidate OTU to an extant cluster, weighting each OTU in that cluster equally, regardless of its structural subdivision. It is best to illustrate this important method in detail, using the distance data for the 16 OTU's of Table 5-1. The initial clustering step is the same as shown previously for single or for complete linkage analysis. We find reciprocally least dissimilar groups, which turn out to be the clusters (**a, b**), (**f, g**), (**i, j**), and (**n, o**). We shall proceed working with distances only (lower half of dissimilarity matrix in Table 5-1) to enable comparisons to be made among the various methods. However, the computation could have been carried out with squared Euclidean distances as well, yielding somewhat different results. In Table 5-3 we see the first clustered matrix. Dissimilarity $U_{ab,c}$ can be computed simply by averaging $U_{a,c}$ and $U_{b,c}$. Thus $1.82514 = \frac{1}{2}(1.41421 + 2.23607)$. Dissimilarities involving two new clusters, such as $U_{(a,b),(f,g)}$, are computed as $\frac{1}{4}(U_{a,f} + U_{a,g} + U_{b,f} + U_{b,g}) = \frac{1}{4}(2.82843 + 3.60555 + 2.23607 + 2.82843) = 2.87462$. Dissimilarities between OTU's that did not join any cluster are transcribed unchanged from the original dissimilarity matrix; for example, $U_{c,d} = 1.41421$.

The first clustered matrix is examined for mutually most similar (least dissimilar) pairs in the same manner as the original dissimilarity matrix. This results in fusion between **c** and **d**, **k** and (**i, j**), and **l** with (**n, o**). The second clustered dissimilarity matrix in Table 5-3 reflects these fusions. Here the unweighted criterion first comes into play, for in the earlier clustering step the computations were the same by either the weighted or unweighted method. Thus $U_{e,(l,n,o)} = 1/3(U_{e,l} + U_{e,n} + U_{e,o}) = 1/3(2.0 + 3.16228 + 3.60555) = 2.92261$. This is the noncombinatorial approach utilizing the original dissimilarity coefficients. The same estimate could have been obtained by the linear combinatorial strategy of Lance and Williams

(1967b). From Table 5-2 Lance and Williams's formula for UPGMA reduces to

$$U_{\mathbf{L,JK}} = \frac{t_{\mathbf{J}}}{t_{(\mathbf{J,K})}} U_{\mathbf{L,J}} + \frac{t_{\mathbf{K}}}{t_{(\mathbf{J,K})}} U_{\mathbf{L,K}}$$

Applied to $U_{\mathbf{e,(l,no)}}$, this becomes

$$\frac{t_1}{t_1 + t_{(\mathbf{n,o})}} U_{\mathbf{e,l}} + \frac{t_{\mathbf{no}}}{t_1 + t_{(\mathbf{n,o})}} U_{\mathbf{e,no}} = \frac{1}{1+2}(2.0) + \frac{2}{1+2}(3.38392) = 2.92261$$

which is the same value as found before. The combinatorial formula works equally well for computing dissimilarities between two clusters, as for example, the coefficient $U_{(\mathbf{i,j,k}),(\mathbf{l,n,o})}$. However, this coefficient could also be computed noncombinatorially from the original data by summing all coefficients linking one member of $(\mathbf{i, j, k})$ with another of $(\mathbf{l, n, o})$ and dividing by the number of these dissimilarities, $t_{(\mathbf{i,j,k})}t_{(\mathbf{l,n,o})} = 9$.

We proceed in the above manner through the third and fourth clustering matrix. The fifth clustering matrix consists of a single dissimilarity value $U_{(\mathbf{a-h}),(\mathbf{i-p})} = 5.42563$. This is shown as the level of the base junction of the two main clusters in the phenogram in Figure 5-7, which graphically shows the results of the clustering

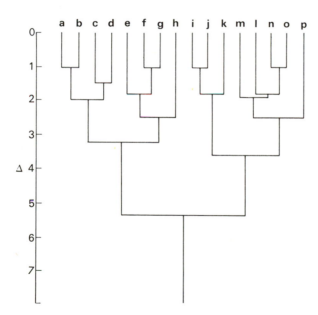

FIGURE 5-7
The phenogram of UPGMA clustering (see Table 5-3) of the data in Table 5-1. The ordinate is in distance scale Δ (see also description in text).

TABLE 5-3

Clustering, by UPGMA, of the Euclidean distance matrix shown in Table 5-1.
For detailed explanation, see text.

	ab	c	d	e	fg	h	ij	k	l	m	no	p
ab	×											
c	1.82514	×										
d	2.11803	1.41421	×									
e	3.08114	2.82843	1.41421	×								
fg	2.87462	3.64269	2.5	1.82514	×							
h	3.64269	5.0	4.12311	3.60555	1.82514	×						
ij	5.72287	4.5	3.64269	3.21699	4.97487	6.73710	×					
k	7.44794	6.08276	5.38516	5.0	6.73710	8.48528	1.82514	×				
l	5.04951	4.47214	3.16228	2.0	3.38391	5.0	2.11803	3.60555	×			
m	7.03553	6.32455	5.09902	4.0	5.24209	6.70820	2.53225	3.0	2.0	×		
no	6.36002	6.11704	4.73607	3.38391	4.06155	5.24209	3.57135	4.61107	1.82514	1.82514	×	
p	8.39510	8.06226	6.70820	5.38516	6.04138	7.07107	4.73607	5.09902	3.60555	2.23607	2.11803	×

	ab	cd	e	fg	h	ijk	lno	m	p
ab	×								
cd	1.97159	×							
e	3.08114	2.12132	×						
fg	2.87462	3.07135	1.82514	×					
h	3.64269	4.56155	3.60555	1.82514	×				
ijk	6.29789	4.62555	3.81133	5.56228	7.31982	×			
lno	5.92318	4.89010	2.92261	3.83567	5.16139	3.48323	×		
m	7.03553	5.71179	4.0	5.24209	6.70820	2.68816	1.88343	×	
p	8.39510	7.38523	5.38516	6.04138	7.07107	4.85705	2.61387	2.23607	×

	abcd	efg	h	ijk	lmno	p
abcd	×					
efg	2.84906	×				
h	4.10212	2.41861	×			
ijk	5.46172	4.97863	7.31982	×		
lmno	5.64839	3.85550	5.54810	3.28447	×	
p	7.89016	5.82264	7.07107	4.85705	2.51942	×

	abcd	efgh	ijk	lmnop
abcd	×			
efgh	3.16233	×		
ijk	5.46172	5.56392	×	
l–p	6.09674	4.64987	3.59898	×

	a–h	i–p
a–h	×	
i–p	5.42562	×

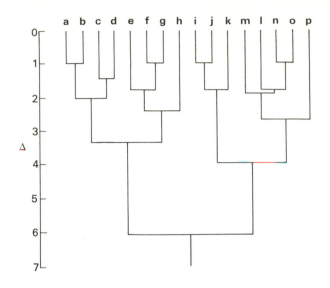

FIGURE 5-8
The phenogram of WPGMA clustering of the data in Table 5-1.
The ordinate is in distance scale Δ (see also description in text).

process. Since the average group methods have no easily discernible geometric interpretation, none can be shown to accompany this phenogram, which should be compared with earlier phenograms of Figures 5-3 and 5-4 and subsequent ones in Figures 5-8 to 5-10. The general taxonomic structure is similar to complete linkage analysis, but there are some fine distinctions that can be noted by the reader.

WPGMA (weighted pair-group method using arithmetic averages) differs from UPGMA by weighting the member most recently admitted to a cluster equal with all previous members. It shares the properties of UPGMA but distorts the overall taxonomic relationships in favor of the most recent arrival within a cluster. The successive clustering matrices can be computed either from the original dissimilarity matrix using the weights $w_j w_k$ defined earlier (see also Table 5-2), or they can be carried out in combinatorial fashion by the formula given in Table 5-2. The results of WPGMA on the data of Table 5-1 are shown in Figure 5-8. The general taxonomic structure is identical to that found in UPGMA, but careful inspection will show that the larger clusters are further apart from each other. This is caused by the greater dissimilarity among the later joiners of the several clusters. Since these are weighted more heavily in WPGMA, they increase the average distances between the clusters. Thus, by WPGMA the final dissimilarity between the two major clusters is $U_{(a-h),(i-p)} = 6.13083$, compared to 5.42562 by UPGMA.

UPGMC (unweighted pair-group centroid method) is attractive conceptually because it computes the centroid of the OTU's that join to form clusters. Distances or their squares (depending on the choice of a dissimilarity metric) are then computed between these centroids. Lance and Williams (1967b) called this the centroid technique. They point out that their combinatorial formula applies only for squared Euclidean distances. The results of UPGMC on Euclidean distances can be inspected in Figure 5-9, which shows the successive clustering stages geometrically, as well as the resulting phenogram (transformed to Δ scale). Because the ultrametric inequality is not met, centroid clustering procedures do not yield monotonic results, as can be seen in Figure 5-9 (and Figure 5-10) by the reversal in the joining of **m** to the cluster (**l, n, o**). A reversal occurs when an OTU (or cluster) joins a cluster after the cluster has formed, but joins at a higher similarity level than that at which the cluster formed. Spearman's sums of variables method (used by Sokal and Michener, 1958, in their first numerical taxonomic study) is a type of centroid method as applied to correlations; however, its frequency of reversals and the relatively high degree of its distortion of the original similarity matrix has led to the abandonment of this technique.

WPGMC (*weighted pair-group centroid method*) has been called the median method by Lance and Williams (1967b) from its linear combinatorial formula, first developed by Gower (1967a). It weights the most recently admitted OTU in a cluster equally to the previous members of the cluster. This procedure shares the properties of UPGMC, including a lack of monotonicity as can be seen in Figure 5-10. There are some differences in the taxonomic structure as implied by the phenogram, which are caused by the heavier weight accorded to late joiners of clusters.

The differences among the single, the complete, and the four average-linkage clustering methods are illustrated graphically in Figure 5-11, which shows a cluster of four OTU's collectively labeled **J** and a cluster labeled **K** containing a single OTU, all about to be joined by another cluster labeled **L** also containing a single OTU. **J** and **K** are assumed to have joined in the last clustering step and it is now desired to compute the dissimilarity of **L** with the newly formed cluster (**J, K**), which we shall call **M** for convenience. The dissimilarities obtained by the various clustering methods expressed as Euclidean distances are laid out along the abscissa. It can be seen that single linkage shows the least value of $U_{L,M}$ since it is the distance between **L** and the closest member of **J**. By contrast, $U_{L,M}$ for complete linkage equals the greatest dissimilarity between **L** and any member of **M**, namely $U_{K,L}$.

The centroid clustering methods measure the length between **L** and **M**, the latter represented by the centroid of clusters **J** and **K**. By the weighted technique (WPGMC) M_1 is the median of line **JK**. By UPGMC the four OTU's of cluster **J** count $4/5 = 0.8$ while the lone OTU of **K** counts only 0.2. Therefore, the centroid for the five unweighted OTU's, M_2, lies closer to **J** (0.8 of the distance from **K** to **J**), and consequently the length $LM_2 < LM_1$. No similar geometric representation is possible for the group methods employing arithmetic averages; however, the dissimilarities obtained by WPGMA and UPGMA are marked off on the abscissa. They

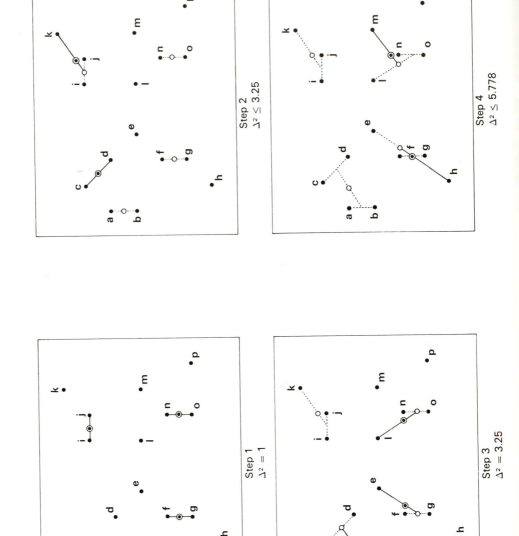

Step 1
$\Delta^2 = 1$

Step 2
$\Delta^2 \leq 3.25$

Step 3
$\Delta^2 = 3.25$

Step 4
$\Delta^2 \leq 5.778$

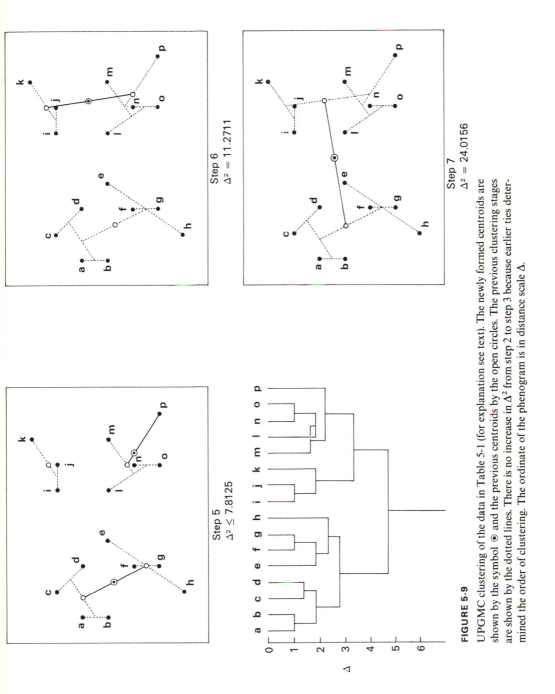

Step 5
Δ² ≤ 7.8125

Step 6
Δ² = 11.2711

Step 7
Δ² = 24.0156

FIGURE 5-9

UPGMC clustering of the data in Table 5-1 (for explanation see text). The newly formed centroids are shown by the symbol ⊙ and the previous centroids by the open circles. The previous clustering stages are shown by the dotted lines. There is no increase in Δ² from step 2 to step 3 because earlier ties determined the order of clustering. The ordinate of the phenogram is in distance scale Δ.

238

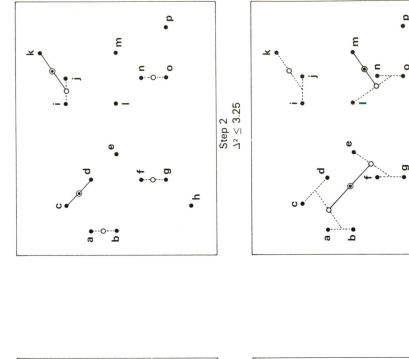

Step 1
$\Delta^2 = 1$

Step 2
$\Delta^2 \leq 3.25$

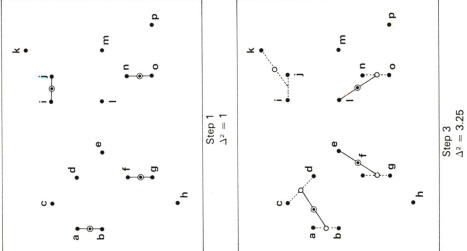

Step 3
$\Delta^2 = 3.25$

Step 4
$\Delta^2 \leq 6.125$

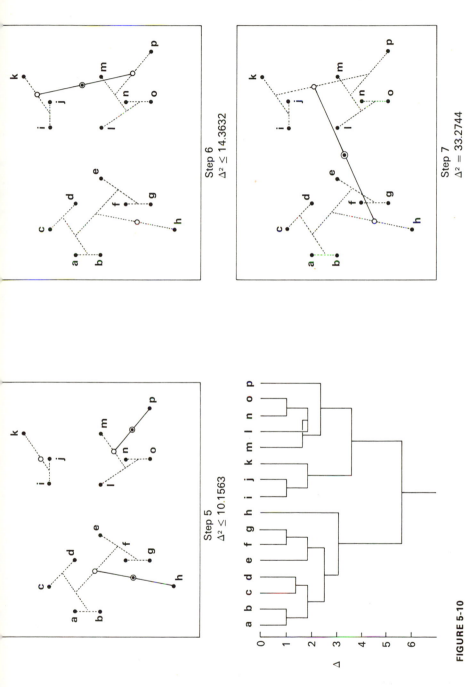

FIGURE 5-10

WPGMC clustering of the data in Table 5-1 (symbolism and other comments as for Figure 5-9). The newly formed centroids are always midway between the previous ones, unlike Figure 5-9.

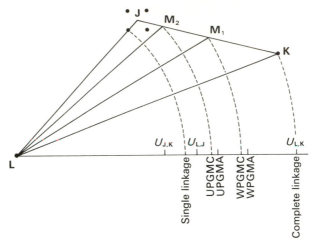

FIGURE 5-11
The effects of several clustering methods (for explanation see text) on the criterion for admitting **L** (containing one OTU) to the cluster formed of four OTU's in **J** plus one in **K**. OTU's are indicated by solid circles. Abscissa is distance Δ.

represent the weighted or unweighted average of the lengths of the vectors from **L** to the five OTU's (four within **J** and one in **K**). It is seen that in each case the dissimilarities are slightly greater than by the corresponding centroid method. This is obvious from an inspection of the respective coefficients for Lance and Williams's linear combinatorial formula in Table 5-2. Both centroid methods have a negative value for the coefficient β while in the arithmetic average methods $\beta = 0$. Since the α coefficients are the same in the two strategies, the centroid methods necessarily result in lower dissimilarities. It may also be noted by comparing the phenograms from the different clustering methods that the levels of linkage between clusters are on the average highest (least dissimilar) for single linkage and lowest for complete linkage, and intermediate for the average linkage methods.

Other SAHN Techniques

Among techniques related to those just discussed is McQuitty's *hierarchic classification* (see McQuitty, 1967a), based on his theory of types where initial reciprocal pairs are isolated and only those characters that are uniform within a reciprocal pair are continued to the next clustering stage. Again, this is based on his reasoning that discordant character states are essentially noise encountered in representing the types. Further clustering is done on those character states common to pairs of reciprocal pairs. In this way the number of characters from which the higher stems are constructed become increasingly less. This seems an undesirable feature. A similar method is that of Lockhart and Hartman (1963).

McQuitty's *multiple rank order typal analysis* is based on clustering types using various subsets of characters that are constant for different clusters or types in the initial clustering procedure. As expected, different typologies emerge depending upon the particular subset of characters chosen.

The *hierarchical grouping method* of Ward (1963) starts out with t separate OTU's, grouping them successively into $t-1, t-2, t-3, \ldots, 1$ taxa and computing at each stage a so-called objective function, which is some measure of the desirability of the particular arrangement of the t OTU's into $k < t$ taxa at any one stage. Such an objective function is the sum of the within group sum of squares (discussed in Section 5.4).

The construction of classifications based on *information statistical similarity coefficients* can be made by applying SAHN techniques to similarity coefficients such as Rajski's metric (Section 4.6), or they can be carried out by a trial and error merging of subsets into larger sets. The basis on which such joining is carried out is to minimize $\Delta I = I(\mathbf{A}, \mathbf{B}) - I(\mathbf{A}) - I(\mathbf{B})$ where ΔI is the increase in the total information of the characters produced when taxa $\mathbf{A}$ and $\mathbf{B}$ are joined. Those taxa are placed into a joint taxon that result in the least increase of information (Section 4.6). Orloci (1969a) gives an outline of such an algorithm. We show an illustration of a simple example of *information analysis* (Williams and Lambert, 1966). The original data matrix is shown below.

Characters		a	b	c	d	e
				OTU's		
1		+	+	−	+	−
2		+	+	−	−	−
3		−	−	+	−	+
4		−	−	+	−	−

The entries are + and − signs for two-state characters but could also be multistate characters. The basic formula for two-state characters is that given in Section 4.6:

$$I_\mathbf{J} = nt_\mathbf{J} \log t_\mathbf{J} - \sum_{i=1}^{n} [a_{i\mathbf{J}} \log a_{i\mathbf{J}} + (t_\mathbf{J} - a_{i\mathbf{J}}) \log (t_\mathbf{J} - a_{i\mathbf{J}})]$$

and

$$\Delta I = I_{\mathbf{J}+\mathbf{K}} - I_\mathbf{J} - I_\mathbf{K}$$

where $a_{i\mathbf{J}}$ is the number of OTU's possessing the + state in group $\mathbf{J}$, and $t_\mathbf{J}$ is the number of OTU's in $\mathbf{J}$. The base of the logarithms is arbitrary; $\log_e = \ln$ is most commonly used and featured here.

The first step is to calculate: (a) the information, I, for each group (initially the five single OTU's); (b) the value of I for every combination of two groups (obtained by combining the groups in all possible pairs); and (c) ΔI, the change in I resulting from combining groups in pairs. A matrix of size $t \times t$ is drawn up. The values for (a) are placed in the principal diagonal. The values for (b) are placed in the lower left triangle, and those for (c) in italics in the upper right triangle. The matrix is the first trial fusion matrix, and is shown below.

OTU's			OTU's		
	a	b	c	d	e
a	0	*0*	*5.545*	*1.386*	*4.158*
b	0	0	*5.545*	*1.386*	*4.158*
c	5.545	5.545	0	*4.158*	*1.386*
d	1.386	1.386	4.158	0	*2.772*
e	4.158	4.158	1.386	2.772	0

The entries (a) are computed as shown in Section 4.6. For a group containing a single member, because one OTU has only one state for each character, the value of I is always zero. Similarly, I is zero for a group of identical OTU's, so that I_{ab} is also zero.

For the group (**a**, **d**), however, I is 1.386, obtained as follows:

$$I_{ad} = 4 \times 2 \ln 2$$

$$- [(2 \ln 2 + 0 \ln 0) + (1 \ln 1 + 1 \ln 1) + (0 \ln 0 + 2 \ln 2) + (0 \ln 0 + 2 \ln 2)]$$

$$= (4 \times 1.386) - [1.386 + 0 + 1.386 + 1.386] = 1.386$$

The terms in the square bracket are obtained by counting, for each of the four characters in turn, the number of OTU's possessing the $+$ and $-$ state, and looking up $f \ln f$ in a table such as Table G of Rohlf and Sokal (1969). The details of this computation were already covered in Section 4.6. It may also be noted that the I values for pairs of single OTU's are equal to $(b + c) 2 \ln 2$ when $b + c$ is the number of mismatches as defined in Section 4.4.

The values of ΔI are found by subtracting the I values for the two separate groups from the I value for the combined groups. In the present matrix (but not in all subsequent ones) I for all separate groups is zero, so that ΔI is the same as I for the combined groups; for example, when joining **a** and **c** $\Delta I = 5.545 - 0 - 0 = 5.545$.

The next step is to fuse the pair of groups that yields the minimum value of ΔI. Here it is minimal for **ab**, being zero. These are therefore fused to give a new group **ab**, and a second trial fusion matrix is computed. This is shown below. Many of the entries are taken from the previous one, but some must be computed afresh : these are shown in bold face.

OTU's·	OTU's			
	ab	c	d	e
ab	0	*7.640*	*1.910*	*5.730*
c	**7.640**	0	4.158	1.386
d	**1.910**	4.158	0	2.772
e	**5.730**	1.386	2.772	0

In understanding the computation it may be helpful to glance back at the original $n \times t$ matrix and imagine that the columns **a** and **b** have been boxed in. It can then be readily realized that the combination **ab** with **c** involves a group of three OTU's with I computed as $4 \times 3 \ln 3 - [(2 \ln 2 + 1 \ln 1) + (2 \ln 2 + 1 \ln 1) + (1 \ln 1 + 2 \ln 2) + (1 \ln 1 + 2 \ln 2)] = (4 \times 3.296) - [(1.386) + (1.386) + (1.386) + (1.386)] = 7.640$. The ΔI values in italics are again equal to the corresponding I values because the principal diagonal still contains only zeros. This matrix shows that **c** and **e** should next be fused, since these yield the smallest ΔI. The third trial fusion matrix is then calculated. The value on the diagonal for **ce** is, of course, just the I_{ce} value from the bottom row of the last matrix (not the ΔI value, though it happens to be the same here).

OTU's	OTU's		
	ab	ce	d
ab	0	*9.182*	*1.910*
ce	**10.568**	1.386	*4.344*
d	1.910	**5.730**	0

It can now be seen that the ΔI values are not always the same as the corresponding I values; ΔI for fusion of **ab** and **ce** is obtained as $10.568 - 0 - 1.386 = 9.182$.

The minimum ΔI is 1.910, yielding the fusion of **ab** and **d**. The final trial fusion matrix is

OTU's	OTU's	
	abd	**ce**
abd	1.910	*9.301*
ce	**12.597**	1.386

Since there is now only one ΔI, fusion will take place in any event, and the analysis is completed. The various I and ΔI values at fusion can be used to construct a phenogram (most commonly by placing the branches at the level of I obtained when a group is formed).

As applications of this method in biological systematics, we may cite Watson, Williams, and Lance (1966) on Ericales, and work on umbellifers by McNeill, Parker, and Heywood (1969a), who also cite several other studies. Information statistic methods are apt to give clusters containing aberrant OTU's that are not very close phenetically, and will give little weight to isolated OTU's (the opposite effect to weighted group methods, Hall, 1967b, 1969a). Many divisive methods behave similarly.

An example of a SAHN technique leaving an unclassified residue is the method of Clifford and Goodall (1967). It employs the probabilistic similarity coefficient of Goodall (1966c), in which character states are weighted in importance based on their rarity. The probabilistic coefficient is expanded to include measures of the probability of obtaining a similarity value at an arbitrary threshold between a candidate to a cluster and the established nodal clusters. Further candidates are admitted to the original clusters by testing whether they fall under the threshold. The similarity between a candidate and a cluster is computed in the manner of Goodall (1968b) or by the deviant index (Goodall, 1966b). In Clifford and Goodall's procedure the largest cluster recognized (or if there are two of the same size, that with the largest probability value) is removed from the set and the residual OTU's are then clustered with probabilities recomputed based on the distribution of character states in the residue. Thus new similarity matrices have to be computed after removal of each cluster. By this method, after a uniform threshold for cluster size has been applied, there will remain residual OTU's that do not cluster with any of the groups. This aspect of the procedure is unsatisfactory from the point of view of traditional classificatory practice and theory, although precedents for similar taxa of dubious validity exist in orthodox taxonomy. The need for ordination of taxonomic data of this sort would seem to be almost self evident. Also, as we have stated elsewhere, we have serious reservations (in agreement with Williams and

Dale, 1965) about the utility of probabilistic similarity indices in biosystematic numerical taxonomy because of the shakiness of the assumptions upon which they are based.

5.6   ORDINATION METHODS

Ordination is the placement of t OTU's in an A-space of dimensionality varying from 1 to n or $t - 1$, whichever is less. Representation of the set of OTU's with respect to more than three characteristics is not possible by conventional means. For this reason means must be found to summarize the information about relation-ships implied by the entire suite of characters. Customarily, two- or three-dimen-sional ordination is employed for ease of inspection and representation.

Principal Component Analysis

A common technique is that of *principal component analysis*. If, as is usual, the basic data matrix is first standardized (unit variance, zero mean) by characters (rows) to give the matrix $\mathbf{X}$, a matrix of $n \times n$ product-moment correlation coeffi-cients r_{hi} between pairs of *characters* is computed as $\mathbf{R} = (1/t - 1)\,\mathbf{XX}'$. Comput-ing the principal components of matrix $\mathbf{R}$ involves the computation of its eigen-values and eigenvectors; that is, we need to solve the equation $(\mathbf{R} - \lambda\mathbf{I})\mathbf{v} = 0$, which will give us the eigenvalues, a set of r nonzero, positive, scalar quantities $\lambda_1, \lambda_2, \ldots, \lambda_k, \ldots, \lambda_r$ where $r \leq \min n,t - 1$. There will be an equal number of associated orthogonal vectors $\mathbf{v}_1, \mathbf{v}_2, \ldots, \mathbf{v}_k, \ldots, \mathbf{v}_r$ called eigenvectors. The num-ber of nonzero eigenvalues that can be extracted from the matrix $\mathbf{R}$ is termed the rank of the matrix, a concept which can be thought of as the number of its independent dimensions (rank, $r \leq \min n,t - 1$). The sum of all the eigenvalues of $\mathbf{R}$ is equal to its rank. The latter is more formally defined as the order (number of rows or columns of square matrix) of its largest nonzero determinant. The impor-tance of these eigenvectors is that they are orthogonal, and they describe the relation-ships between OTU's with economy. In other words, a large proportion of the dispersion engendered by the n characters over the t OTU's may be accounted for by $k < r$ dimensions. Frequently a framework of low dimensionality ($k \ll \min n,t - 1$) may thus account for a large portion of the variation of the original data. When the vectors are normalized (that is, $\sum_n v_j^2 = 1$, which is obtained by transform-ing the original values, v, into $v/\sqrt{\sum v_j^2}$), they can be assembled as principal com-ponent matrices of $n \times k$ dimensions, which we shall designate as $\mathbf{V}$ following the standard notation presented by Anderson (1966). It can be shown that $\mathbf{V}'\mathbf{RV} = \Lambda$ where Λ is the diagonal matrix of eigenvalues. The actual computational methods for obtaining the matrix of vectors $\mathbf{V}$ need not concern us here. There are various numerical methods available at most computation centers for obtaining eigen-vectors and eigenvalues of a symmetric matrix. They are nowadays based on the

Jacobi or Wilkinson-Householder methods. Principal component analysis can also be carried out on a variance-covariance matrix; when this is done the resulting ordinations are based on linear combinations of unstandardized characters.

The k normalized vectors $\mathbf{V}$ give the directions of a set of k orthogonal axes in the A-space and are known as the *principal axes*. The coordinates of these axes are linear combinations of the original variables and summarize the major dimensions of variation. The coordinate points of the OTU's in the new A-space are computed by the equation $\mathbf{P} = \mathbf{V'X}$, where $\mathbf{P}$ is the $k \times t$ matrix of coordinates of the t OTU's on the k principal axes (factors). For further discussion and illustration see Section 5.9. An eigenvalue is equal to the variance along its corresponding axis. Thus the principal axis corresponding to the largest eigenvalue is the dimension that accounts for the greatest amount of variance from the sample (the principal axis of the hyperdimensional "cloud" of OTU's). The second principal axis accounts for the second largest amount of variance from the sample, and so forth. It is customary to extract only enough eigenvectors to remove the majority, say 75 percent, of the total variance of the data matrix. Quite often as few as $k = 3$ principal axes will be responsible for most of the variance, in which case a satisfactory three-dimensional model can be constructed by plotting the OTU's in the 3-space produced by the three axes. Numerous applications of this technique have by now entered the numerical taxonomic literature, and in fact the three-dimensional plots or models of a group of OTU's have become an almost standard procedure that may replace the dendrogram as the most common method of representation of taxonomic results. We need cite only the work of Gyllenberg (1965b, 1967) and Gyllenberg and Rauramaa (1966) on microorganisms, that of Rohlf (1967, 1970), Moss (1967), and Hendrickson and Sokal (1968) on arthropods, and that of Schnell (1970a,b) on birds as examples of this technique. There are numerous other examples from ecology (e.g., Gittins, 1969), zoogeography (e.g., Fisher, 1968), and geographic variation analysis (e.g., Thomas, 1968a).

Harman (1967) makes a distinction between principal component analysis and *principal factor analysis* (abbreviated PCA and PFA, respectively) and this distinction has become generally established. In PCA the diagonals of the correlation matrix are unities: the total variance of the hyperdimensional cluster of points is explained by the vectors obtained. From an $n \times n$ correlation matrix one can extract n principal components (unless there are fewer, $k < n$, independent dimensions; i.e., the rank r of the matrix is less than n). These components will (within rounding error) reproduce the original correlation matrix. In PFA the diagonals are reduced by their *uniqueness* to the so-called *communalities*, the percentage of variation due to the common factors. In numerical taxonomy the differences between PCA and PFA are not very important since only the few largest eigenvalues (usually the largest three eigenvalues for three-dimensional representation) are generally employed.

Principal component analysis is characterized by faithful representation of distances between the major groups or clusters but is notorious for falsifying distances between close neighbors (Rohlf, 1968). In this respect it is quite the opposite of a SAHN clustering technique, which generally will reproduce distances between the adjacent tips faithfully but will show distortion in the distances among members of larger clusters, i.e., at the base level of the stems.

The approach described above results in economy of description, but does not necessarily lead to interpretation of the extracted factors. On occasion, especially in samples of low taxonomic rank, the user may be able to interpret one or the other of the factors as a geographic, climatic, or other ecological variable (see Thomas, 1968a), or a size or typicality factor (Sheals, 1964; Schnell, 1970a,b).

Multiple Factor Analysis

The method of *multiple factor analysis* (Harman, 1967) involves a different philosophy. Once extracted, the factor axes are rotated and also can be permitted to depart from their orthogonal relationship by becoming oblique, i.e., correlated. This makes scientific sense in that the factors underlying the covariation pattern of the characters in nature are themselves undoubtedly correlated. Initiating this technique for numerical taxonomy, Rohlf and Sokal (1962) employed centroid factor extraction with subsequent rotation to simple structure. The criteria for this constellation require that any one factor influence only some of the variables and affect others little or not at all. Such a factor would be called a group factor; it is contrasted with a general factor such as a general size factor, which affects all variables. Simple structure also requires that any one variable should not be affected by all factors. The centroid method employed by Rohlf and Sokal (1962) was at the time the only practical method for extracting factors from a correlation matrix of even moderate dimensions. In recent years the availability of high speed computers and the mathematical neatness of eigenvector extraction has made it the method of choice regardless of whether rotation to simple structure is subsequently desired (Harman, 1967). For matrices that represent an unusually large number of OTU's or variables the centroid method might still be indicated. Multiple factor analysis encounters added problems. The estimation of the communality—the proportion of the variance due to common factors in the study—is a difficult theoretical and practical problem discussed extensively in the factor analytic literature, e.g., Harman (1967), and for which a number of iterative computer programs have been developed. Similarly, the development in recent years of analytical solutions for obtaining simple structure (see Harman, 1967 for a general survey) has made rotation feasible by means of a variety of computer programs. We should point out that rotation to simple structure is not a uniquely defined procedure, and that different criteria and algorithms may lead to differing solutions. Numerous new types of factor analysis have been proposed in recent years. No comprehensive recent review is available. Interested readers are referred to the unpublished report by K. J. Jones, dated 1964 and cited by Harman (1967), and to recent issues of the Research Bulletin of the Educational Testing Service, Princeton, New Jersey. None of these new approaches has yet been applied to taxonomy. Rohlf and Sokal's (1962) study was a Q analysis of the correlation among 40 of the 97 species of the *Hoplitis* complex studied by Michener and Sokal (1957). Rotation to simple structure was carried out by the MTAM method developed by Sokal (1958b). They obtained results that corroborated the cluster analysis carried out by the original authors. From this and other Q studies, factor analysis seems to define the major clusters fairly well but there does not seem to be any special advantage of this method over others. Much of the work in multiple factor analysis in taxonomy has dealt with infraspecific data. Such studies are carried out on suites of characters in populations of organisms and are consequently R studies. For a review of such studies see Sokal (1965). A few examples include Sokal and Thomas (1965), Thomas (1968b), and Fisher (1970).

Principal Coordinate Analysis

An important advance in ordination technique has come about through the *principal coordinate analysis* developed by Gower (1966a). By this technique it is possible to compute principal components of any Euclidean distance matrix without being in possession of either the original data matrix or a variance-covariance matrix of the characters or OTU's. Gower's method can be summarized by means of operations carried out upon a dissimilarity matrix U with individual coefficients U_{jk} for a Q study.

1. Compute $t \times t$ matrix **E** with elements

$$e_{jk} = -\tfrac{1}{2} U_{jk}^2$$

2. Compute $t \times t$ matrix **F** with elements

$$f_{jk} = e_{jk} - \frac{1}{t}\sum_{j} e_{jk} - \frac{1}{t}\sum_{k} e_{jk} + \frac{1}{t^2}\sum_{j}\sum_{k} e_{jk} = e_{jk} - \bar{e}_j - \bar{e}_k + \bar{e}$$

In words, this formula indicates that each element e_{jk} should be corrected as follows. Subtract from it the mean of all elements e_{jk} of its row and the mean of all elements of its column and add the mean of all the elements of matrix **E**.

3. Find eigenvalues and eigenvectors of **F**. Normalize the kth eigenvector so that its sum of squares equals the kth eigenvalue of **F**. The matrix of normalized eigenvectors gives the coordinates of the OTU's on their principal axes.

Gower's method is also applicable to non-Euclidean distance and association coefficients as long as matrix **F** has no large negative eigenvalues. However, in such cases the solution only approximates principal components. The advantages of principal coordinate analysis are several. Not only is it possible to carry out ordination of sets of OTU's found in the published literature for which original data matrices are not available, but also when such matrices cannot be obtained because of the nature of the data, as for example, in matrices of immunological distances. Furthermore, the distances need not be Euclidean. Manhattan distances as well as other Minkowski metric distances can be treated by Gower's method. Principal coordinate analysis also seems to be less disturbed by NC entries than principal components (Rohlf, 1972). The result of principal coordinate analysis on distances obtained from standardized characters is identical with that obtained from principal component analysis of product-moment correlation coefficients (Gower, 1967b). Another form of principal axes analysis that has been used in taxonomy is *canonical vectors* or *canonical variate analysis* (e.g., DuPraw, 1964, 1965b). This is related to methods based on Mahalanobis' generalized distance and hence to discriminant functions (see Section 8.5). Principal coordinate analysis of a D^2-matrix is identical to that of a canonical variate analysis (Gower, 1966b).

Nonmetric Multidimensional Scaling

The most general ordination technique is *nonmetric multidimensional scaling* developed by Shepard (1962, 1966) and Kruskal (1964a,b). In nonmetric multidimensional scaling the elements of a Q-type matrix of observed dissimilarities U_{jk} are ordered by rank from the smallest dissimilarity coefficient to the largest. Thus the basic similarity matrix may be a distance or correlation matrix, or it may only express rank orders of the dissimilarities as evaluated by a subjective observer or an experimental technique. A decision is then taken to represent the t OTU's in a space of $k \ll t$ dimensions in which the distances among the OTU's are computed in a variety of ways, commonly as Euclidean distances, but in general as Minkowski r metrics. If the positions of the t objects in the k-space were such that the distances d_{jk} computed from them were monotonically related to the observed ordinal dissimilarity function U_{jk} (i.e., corresponding elements of matrix U were identically ordered to those in the matrix D of distances d_{jk}), the ordination would be considered perfect. Kruskal (1964a) has developed a measure of stress

$$S = \sqrt{S^*/T^*} = \sqrt{\sum_{j<k} (d_{jk} - \hat{d}_{jk})^2 / \sum_{j<k} d_{jk}^2}$$

where d_{jk} are the distances in the k-space and $\hat{d}_{jk}$ is the value necessary to maintain the monotonicity of the function. The numerator of this expression is the conventional sum of squares of goodness of fit while its denominator is a scale factor to make stress estimates comparable. Coordinates for the k-space are computed by an iterative technique developed by Kruskal (1964b), which has been programmed and is widely employed as the MDSCAL program. The aim of the procedure is to minimize the stress for any given d_r (j, k), the Minkowski distance coefficient (see Section 4.3), and for any dimensionality k. The great advantage of nonmetric multidimensional scaling is that it can consider asymmetrical dissimilarity matrices as well as those with missing or tied dissimilarity values. A first application of nonmetric multidimensional scaling to biological data is by Rohlf (1970) in a numerical taxonomic study of 45 species of mosquito pupae. The stress in a three-dimensional ordination based on 74 characters was 12 percent. In simpler artificial models and in some psychological experiments Kruskal (1964a) was able to obtain lower percent stress. A special advantage of the nonmetric multidimensional scaling method is that it seems better than principal component analysis in giving balance between the large intercluster distances and the fine differences between members of a given cluster (Rohlf, 1970). Related methods are those of Guttman (1968) and Sammon (1969). A recent review by Green and Carmone (1970) is useful.

A method related to principal component analysis is the technique of *pattern analysis* developed by Hayashi (1956), which also computes eigenvalues and eigenvectors of a similarity matrix. The method is especially designed for similarity matrices based on multistate characters that cannot be arranged in a logical order. An application of pattern analysis to biological taxonomy is the two-dimensional ordination of 65 strains of rice (Morishima, 1969b) and of 21 races of the rice blast fungus (Morishima, 1969a). Yet another method related to ordination is *Tryon's key communality cluster analysis* used in the social sciences (reviewed in Tryon and Bailey, 1970) and applied to a botanical example by Crovello (1968b).

Seriation

A simple type of ordination is the process of *seriation* employed in anthropology and archaeology. A recent paper by L. Johnson (1968) reviews earlier work and his own recent studies based on a methodology developed by Craytor and Johnson (1968). Hole and Shaw (1967) have made another extensive study of existing techniques. By seriation is meant the reordering of rows and columns of a similarity matrix in such a way as to place the highest similarities close to the principal diagonal of the matrix with an orderly decrease in similarity values away from the diagonal. In perfect examples of linear ordering, as in certain time sequences in archaeology, such a technique will yield a linear order of development useful for constructing hypotheses about the chronological development of languages, artifacts, or faunal or cultural assemblages. Similar methods have been invented in psychology and sociology, as for example simplex, radex, and circumplex structure of correlation matrices (Guttman, 1954, 1966). Potential applications of the ordering of variables by seriation or simplex structure analysis in biology are the interpretation of growth fields in morphological variation, arrangement of developmental stages of invertebrates, and investigation of evolutionary trends in fossil series.

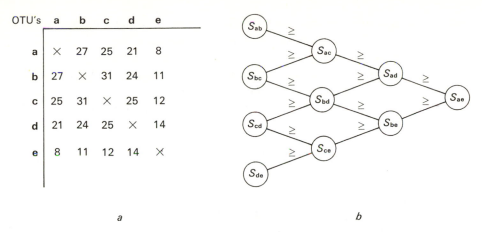

a

b

FIGURE 5-12
Illustration of OTU's arranged in the order giving perfect seriation. *a*, Hypothetical similarity matrix. *b*, Diagram of the inequalities between the similarity values in the matrix. [Modified from Craytor and Johnson (1968).]

Morphometric analyses have been carried out by Alpatov and Boschko-Stepanenko (1928) and Sokal (1952, 1962a) on insects and by Guttman and Guttman (1965) on vertebrate skeletal measurements.

The essence of any seriation method lies in the operational definition of order among the coefficients in a similarity matrix. A perfect ordering of this sort is illustrated in Figure 5-12 taken from Craytor and Johnson (1968). This illustrates a perfect seriation for OTU's **a** to **e**. The optimization of the arrangement among the OTU's has to be performed by an iterative algorithmic procedure since no analytical procedure is known and a trial of all possible arrangements is beyond the capability of present day computers for any reasonable number of OTU's. In addition there are dangers of becoming trapped in local optima (Kendall, 1963). Only in cases of true linear trend development would seriation work as intended. It is thus applicable to very specialized data sets only, and there is a growing view among archaeologists (its principal users) that seriation may be superseded by nonmetric multidimensional scaling or principal axis methods (Cowgill, 1968; Kendall, 1969; Hodson, 1970). Ordination in several dimensions improves interpretation, which is further enhanced by cluster analysis of the archaeological and paleoecological data (Hodson, Sneath and Doran, 1966; Doran and Hodson, 1966; L. Johnson, 1968; Hodson, 1970).

Clustering versus Ordination

Clustering, the results of which are most commonly represented by means of phenograms, and ordination yield summaries of the variation in A-space that differ greatly in appearance and may differ in the taxonomic results to which they lead. The worker may therefore be puzzled about the choice and purpose of each methodology. Williams and Lance (1969) believe that inadvertent chopping of continuous variation into somewhat arbitrary clusters does not usually damage the analysis irretrievably, because the continuity is generally fairly evident. In contrast ordination may not disclose sharp discontinuities if they cannot be displayed in the first few dimensions. For example, Webb et al. (1967a), using ecological data, noted that all the methods of clustering they tried, as well as ordination by Gower's principal coordinates method, revealed the major structure, but the finer divisions, known to be ecologically significant, were less evident in ordinations than in clusterings. This agrees with Rohlf's (1968) observation that distances between close neighbors are not well represented by PCA ordination.

There are as yet no satisfactory methods for telling from the similarity matrix itself whether clustering or ordination is most appropriate, although a high cophenetic correlation (see Section 5.10) may suggest that a phenogram is a reasonable representation of well-clustered phenetic distributions (more often the phenogram is judged by eye, an unsatisfactory procedure though experience often allows the worker to make a fair guess on whether the phenogram is a good one). Cluster methods will yield clusters of some kind, whatever the structure of the data, even if the OTU distributions are random; as noted in Section 5.10 tests of the validity of clusters are under development. The tendency to wish for sharp, tidy classes means that clusters may be accepted, or required, even if the OTU's have been forced together to obtain them. Williams and Lance (1968) note that much material is not too well clustered and express a preference for accentuating clustering by their flexible sorting method. Rohlf (1967, 1970) points out, too, that phenograms give poor representations of relationships between major clusters. Boratyński and Davies (1971) noted in a study on insects that the first few vectors separated higher ranking taxa and the next few vectors separated taxa of lower rank; whether this is generally true does not seem to have been investigated.

Ordination also has disadvantages. With many OTU's and clusters, the ordination may give no simple low-dimensional result, as pointed out by Williams and Lance (1968). Clearly one might have clusters that overlap in 2- or 3-space, though they are quite distinct in hyperspace. Ordination may also be impracticable for very large numbers of characters and OTU's. There is also debate on what axes in an ordination are taxonomically meaningful. Thus, in an early study, Sheals (1964) omitted the first vector of a principal coordinate analysis on the grounds that it represented the departure of the OTU's from the "average" properties of the set—atypicality. Sims (1966) found the first vector to be meaningful taxonomically because he followed a procedure that removed the effect of size. Schnell (1970a) among others noted that the first vector might represent a general size factor, and Seki (1968) found that it was the fourth vector that reflected size. There is no reason why the major axes of taxonomic variation should necessarily lie parallel to the direction that represents size. This will clearly depend on the suite of characters employed, how many of these are size-dependent and by how much they vary in size.

We wish here to draw attention to a danger in working with ordination diagrams: the worker may examine these for clusters and decide on their limits by visual inspection alone. Insofar as it is commonly easy and rapid to do this, it may not greatly matter, because other workers can readily see if they agree. But when the points form clusters of unfamiliar shape the danger arises of interpreting these according to individual whim. Interpretation based on visual inspection alone is counter to our aims of objectivity and explicitness.

Whether first performing an ordination and then clustering on the reduced space is generally desirable, is a debatable matter. Information of largely unknown nature and quality has been lost during the ordination; very low order dimensionality (at the extreme, of one dimension), cannot express much of the multidimensional phenetic relationships; diffuse clusters may overlap. Ivimey-Cook (1969a,b) has clustered on the distances in a 12-dimensional space and found that the resulting phenogram gave more clearcut clusters in his view, though it was very similar to the phenogram from the full 64-dimensional space. Rohlf (1967) obtained identical results in a 9-dimensional space compared to a phenogram in a 74-dimensional space. The results of a microbiological study by Skyring and Quadling (1969b), exploring the use of two-stage ordination, appear less satisfactory, and it may be that they reduced the dimensionality too much before clustering. On the other hand a clustering of bacteria by Hill et al. (1965) on the first five dimensions was satisfactory. There would therefore seem good reasons for basing formal taxonomic groupings upon the phenograms from the full A-space and using ordinations for investigating the general pattern of variation. There are also relations between clusters, ordinations, and graphs, which are explored in the next section.

5.7   GRAPHS AND TREES

Because similarity matrices are composed of generally symmetrical pair functions between all possible pairs of t OTU's, the notion of representing inter-OTU relationships by *graphs* (in the graph theoretical sense), came relatively early in the development of numerical taxonomy. Busacker and Saaty (1965) give a very lucid account of graph theory; a briefer review is given by Ore (1963). Graphs were introduced to numerical taxonomy by Florek et al. (1951a,b) and by Estabrook (1966), and were applied by Wirth, Estabrook, and Rogers (1966) and Moss (1967), among others. Related work in psychometrics is that of Harary (1964). A graph is a set of points (*vertices*) and of relations between pairs of vertices indicated by lines called *edges*. A set of OTU's and their dissimilarities may be represented by a graph, with the OTU's shown as vertices and the dissimilarity relationship between them shown as edges. In graph theory the edges are not directly associated with a real variable such as dissimilarity or distance. Relationship is indicated by the presence, lack of relationship by the absence, of an edge between two vertices. For this reason, in graphs employed in numerical taxonomy the presence of an edge between two vertices (OTU's) has generally indicated a relationship closer (more similar) than a given arbitrary cutoff point. Thus the edges connecting a cluster of OTU's indicate the set of those OTU's that are more similar to each other than an arbitrary similarity criterion S_{crit}. In an informal way, graphs (often incorrectly called network diagrams—the difference between the two is clarified below) have been used for many years in taxonomy as, for example, in simple linkage diagrams with lines connecting OTU's that are phenetically similar, or polygonal diagrams describing the ability of different organisms to hybridize.

The utility of the graph theoretical approach in numerical taxonomy is threefold. First, graphs serve as illustrative devices that enable many investigators to understand a variety of problems connected with cluster analysis. Second, the graph theoretical representation of taxonomic relationships enables us to derive certain properties of clusters from well-established theorems of graph theory and also to employ graph theoretical tools (such as minimum-length tree algorithms) as solutions to specific problems of numerical taxonomy. Third, they provide extra information when superimposed on ordinations.

Graph theory leads immediately to the application of the concepts of connectedness in the study of clusters. A connected graph (Figure 5-13,a) is one in which every pair of vertices is connected by at least one path (although this path need not be direct, for example, **a** and **b** could be connected via **c** as they are in the figure). A *minimally connected graph* is one that contains only one direct or indirect path between every pair of vertices (see Figure 5-13,b). Removal of one edge from such a graph disconnects it into two subgraphs. Such a minimally connected graph, which will have $t - 1$ edges for t vertices, is also known as a *tree*. It should be obvious to the reader that the single link clusters we studied in an earlier section (5.5) are minimally connected graphs. By contrast, clusters formed by complete linkage are *maximally connected graphs* (also known as fully connected graphs), which have direct connections (edges) between every pair of OTU's (Figure 5-13,c). There are $t(t - 1)/2$ edges in a maximally connected graph of t OTU's. The two extremes in connectedness between minimally and maximally connected graphs immediately give rise to the notion of degree of *connectivity*, which is some measure of the amount of connectedness occurring within a graph relative to the maximal possible amount of such connectedness. A variety of such indices of connectivity has been constructed (see Kansky, 1963) for work in regional analysis in statistical geography. Such measures have been applied to various problems of taxonomy (for example, to the measure of tightness developed by Estabrook, 1966, or in the neighborhood-limited classification of P. M. Neely—see Oxnard and Neely, 1969), geographic variation analysis (Gabriel and Sokal, 1969), and to theoretical aspects of intra- and interpopulation relations, as for example, in the analysis of the biological species concept by Sokal and Crovello (1970).

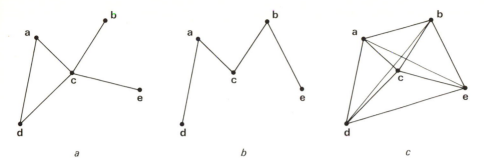

FIGURE 5-13

Three kinds of graph: *a*, a connected graph; *b*, a minimally connected graph; *c*, a maximally connected graph, for the same five OTU's.

Since maximal connectivity is the criterion for complete linkage analysis while minimal connectivity is the criterion for single linkage analysis, the biased results represented by these two extremes may be avoided by specifying intermediate criteria of connectivity that yield clusters neither too compact nor too strung out.

A special family of graphs are *directed graphs*, which are also known as networks (Busacker and Saaty, 1965, p. 237). Other uses of the term network are incorrect, though they are occasionally encountered in the taxonomic literature. Directed graphs imply direction in the edges; that is, relation **a** → **b** is not the same as **b** → **a**. This approach has been especially fruitful in numerical cladistics where the idea of a directed tree ties in naturally with that of a cladogram. A *directed tree* (in graph theory) has edges with direction and a unique path from one vertex V_0, called the *root* of the tree, to all other vertices. A conventional dendrogram is an example of such a graph, if each furcation point, each terminal OTU, and the base of the tree are all considered vertices. Such vertices are often called *nodes* in the cladistic literature, and the edges are called *internodes*.

Although the presence or absence of an edge between any two vertices in a graph indicates only the existence of a connection, it is possible to associate a real number with each edge; this can be called its *length*. These lengths are pair-functions and in numerical taxonomy would signify dissimilarity. One can then ask whether the sum or some other function of these lengths can be maximized or minimized for certain graphs. Two classes of graphs are of special significance here. These are minimum length directed and nondirected trees. In a directed tree one of the OTU's is considered the vertex and is the starting point or root of the tree. *Minimum length directed trees* have been especially applied in numerical phyletics or cladistics (see Section 6.4), where under assumptions of minimal evolution some such tree for a given set of OTU's is considered to be the most likely evolutionary path. Efficient iterative algorithms for finding minimum length (nondirected) trees, also known as *shortest spanning trees*, have been developed by Prim (1957), Kruskal (1956), Whitney (1972), and Rohlf (1973c). Rohlf (1970) and Gower and Ross (1969) found shortest spanning trees to be useful as an additional perspective of taxonomic relationships in an ordination. Since the relationships between close neighbors are frequently distorted in an ordination, especially one based on principal component analysis, a minimum length tree superimposed upon the OTU's (see Figure 5-14) is likely to show up such distortions. In such a figure nearest neighbors will be pointed out by being linked. Several computer programs for producing shortest spanning trees exist and are readily available (see Appendix B).

The results of connecting a set of OTU's with a minimum length tree will depend on the metric among the pairs of OTU's. Euclidean distances and Manhattan distances may give quite different trees and investigators will need to keep their assumptions about the nature of the dissimilarity function clearly in mind before embarking upon such analyses.

Rohlf (1973b) has shown that there is a direct relationship between Jardine and Sibson's B_k cluster analyses (of which single linkage is a special example) and certain types of graphs. If one constructs a graph consisting of just those edges that correspond to cophenetic values identical to the input or original dissimilarity matrix, one obtains a graph with important properties. If $k = 1$ (single linkage), then the graph obtained is the minimum length tree. Gower and Ross (1969) and Rohlf (1973c) have pointed out that the shortest spanning tree contains all the information necessary to compute a single linkage cluster analysis. As already noted, the technique of computing the shortest spanning tree first—and from it the single linkage clusters—is the most efficient procedure computationally. For larger values of k the graphs are more highly connected. They always contain (as a subset) those edges representing distances to the k nearest neighbors of each point. As before, these graphs contain sufficient information to prepare the k-dendrogram. The principal advantage of these graphs is that they are much simpler to represent and publish than the k-dendrograms themselves. Displaying these graphs on the printed page so as to make the length of the edges roughly proportional to their actual length and to minimize the number of crossing lines is in effect a type of two-dimensional ordination. The advantage of this approach is that the distances to the nearest k neighbors are shown quite faithfully, whereas the longer distances are not. This method of ordination unifies the classificatory approaches discussed so far. Given dissimilarities, one can compute a graph. By breaking the graph at various special levels based on the lengths of the edges, one can form clusters according to various rules. If one represents the network geometrically, then one has, in effect, an ordination (Rohlf, 1973b).

5.8   THE RELATION BETWEEN Q AND R TECHNIQUES IN NUMERICAL TAXONOMY

Q technique refers to the study of similarity between pairs of OTU's and R technique refers to the study of similarity between pairs of characters.

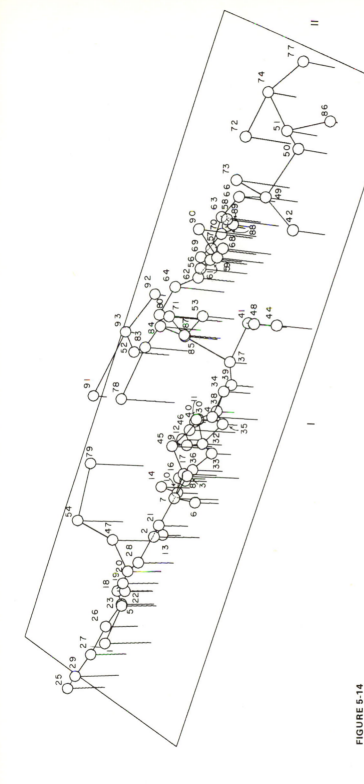

FIGURE 5-14

Three-dimensional projection of 81 OTU's onto the first three principal components, with a shortest spanning tree superimposed upon the OTU's. Distortions can be seen where OTU's are relatively closer together in the three-space than in the full hyperspace. For example, the shorter distance visible between OTU's **1** and **26** than that between OTU's **1** and **5** must represent a distortion of the distances in hyperspace between OTU's **1**, **5**, and **26**; otherwise the tree would have linked OTU **1** to OTU **26** and not to OTU **5**. [From the study of Schnell (1970a) on gulls and allied birds, based on correlations between logarithms of skeletal characters.]

The difference between the two techniques in numerical taxonomy and the representations in A-space and I-space to which they lead were discussed in Section 4.1.

These terms had their origins in factor analysis. Once factor analysts had realized the possibility of analyzing data matrices across rows as well as across columns, obtaining different factor solutions in the process, various investigators attempted to relate the two methods, believing that since both techniques represent different mathematical operations upon the same data matrix they must be transformable one into the other. In fact, Burt (1937) was able to show that under certain restrictive assumptions—unit communalities (ignoring specific factors), data matrix standardized by rows as well as by columns, and factor analysis carried out on the variance-covariance matrix—the principal components factor solutions are directly transformable one into the other. Gower (1966a, 1967b) and Orloci (1967a) have also shown that identical ordinations can be obtained from either Q or R techniques, provided suitable mathematical transformations are applied to principal component and principal coordinate analyses. This permits a worker to choose whichever approach requires least computation. However, Cattell (1966c, p. 228) points out that in principal factor analysis Burt's conditions are not likely to be met and the transformation of R kinds of factor solutions to Q kinds of solutions are likely to be more complex. As a general rule the first (general) factor of one solution is lost in the transposed solution, because of the standardization across rows.

In the clustering methods of numerical taxonomy clearly there must also be relations between clusters of OTU's and clusters of characters, but the mathematical properties of these relations have so far not been studied. They should prove a fertile field of inquiry for the mathematically inclined evolutionist. Some work of this kind in ecology is mentioned in Section 11.1. We know something of the properties of character correlation matrices at higher levels. They are determined by the taxonomic structure of the set of OTU's being investigated, especially by evolutionary patterns affecting the character sets included in the data matrix. Early work in this field at the higher categorical levels included only the study of Stroud (1953). More recently a number of studies at various categorical levels have been made on animals, plants, and bacteria (e.g., Pike, 1965a,b; Gyllenberg, 1965a, 1967; Gould, 1967; Hawksworth, Estabrook, and Rogers, 1968; Clifford, 1969; Johnston, 1969), both by ordination and clustering. It is found that the major R factors or R clusters (i.e., clusters of characters or character suites) correspond to the major Q factors or Q clusters (taxa) as one would expect, but finer details of the taxonomy do not always show in the R studies. This is also true for the studies by Rohlf (1967) on mosquitoes and by Sokal and Rohlf (1970) on bees. The nature of the R clusters seems to be determined almost entirely by the particular configuration of OTU's included in a given study. Adding or excluding OTU's may appreciably change the results of R clusters so obtained. Some R clusters may also recognizably reflect function ; for example, Crovello (1968b) found a cluster of characters concerned in pollination in a study of species of Limnanthaceae.

Factor analyses of R correlation matrices of n characters could be analyzed into k factors, where $k < n$. By isolating independent dimensions of variation among characters, we might remove redundancy from our studies. This has been done in studies carried out by Sokal (1962a), Sokal and Thomas (1965), Rohlf (1967), and Defayolle and Colobert (1962), among others, with interesting implications for numerical taxonomy. If there is redundancy in character information it might be possible to isolate the factors from an R correlation matrix and to calculate factor scores for each OTU on the factors obtained. We could then employ only those characters that provide independent information on the taxa concerned and thereby reduce the number of character scores on which OTU's could be classified. It would, however, not reduce computational effort, since factor analysis must precede the numerical classification. Studies of the elimination of redundancy have not advanced beyond the programmatic stage. We have at the moment little knowledge of how many common factors would be found in an R correlation matrix comprising higher ranking OTU's. Studies of R matrices at the lowest taxonomic levels have generally not produced many factors (for a discussion of this point see Sokal, 1965).

5.9   THE REPRESENTATION OF TAXONOMIC STRUCTURE

The results of the various operations for revealing taxonomic structure, discussed in the past several sections, must be illustrated to enable taxonomists to convey an understanding of the implied relationships to the general scientific public. This step of the numerical classification procedure can itself be subdivided into three distinct stages. First is the graphic representation of the relationships implied by the similarity matrix. Second is an attempt to use this representation to establish taxa at various ranks, i.e., to make a classification in the strict sense from the taxonomic relationships that have been described. Third is to assign formal ranks and names to the classificatory systems developed in this manner. The amount of attention paid by numerical taxonomists to these stages has been inversely proportional to the order of their enumeration here. Much work has been done on methods of representation, less on establishing taxonomic ranks (discussed in Section 5.11), and least attention has been focused on formal recognition and naming of taxa. Such work as has been done in this field will be discussed in Section 5.11 and Chapter 9.

Attempts at representing taxonomic relationships in one or two dimensions generally fail because of the higher dimensionality of these relationships when based on many taxa and characters. Early attempts at creating three-dimensional

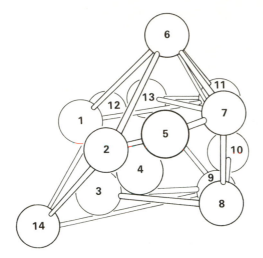

FIGURE 5-15
Taxonomic model of the Enterobacteri-aceae. The OTU's are represented by spheres (balls) connected by rods (sticks) indicating taxonomic distance. [After Lysenko and Sneath (1959).]

models by means of small rubber or styrofoam balls and wires or sticks (*ball-and-stick models*; see Figure 5-15) have generally been only partially successful. To represent fully in a three-dimensional model the taxonomic relationships implied by a dissimilarity matrix, compromises ("bent sticks") have to be introduced.

Representation of taxonomic structure has also commonly been made by shaded similarity matrices as shown in Figure 5-16 (trellis diagrams). We feel that they are rather wasteful of space and would suspect that dendrograms or diagrams of three-dimensional ordination would generally be preferable. If they are employed it is important for the shading to cover the whole range of similarity coefficients.

The representation of taxonomic structure depends on the method for investigat-ing it. The results of cluster analysis have been traditionally represented by *dendro-grams*, which have the advantage that they are readily interpretable as conventional taxonomic hierarchies. Following Sokal and Camin (1965) and Mayr (1965), these have been distinguished into *phenograms* representing phenetic relationships, with which we shall be concerned here, and *cladograms* representing evolutionary branching sequences, to be discussed in the next chapter. Although early practice tended to have the branches of a phenogram pointing upwards, convenience and the ever increasing size of studies have made authors place phenograms almost uniformly on their side with branches running horizontal across the page. The abscissa is graduated in the similarity or dissimilarity measure on which the cluster-ing has been based, and the points of furcation between stems along the scale imply that the resemblance between two stems is at the coefficient value shown on the abscissa. These values are frequently multiplied by a constant to avoid decimal places. It has become customary to place code numbers and frequently names of the OTU's to the right of the tips of the phenograms as shown in Figure 5-17. The

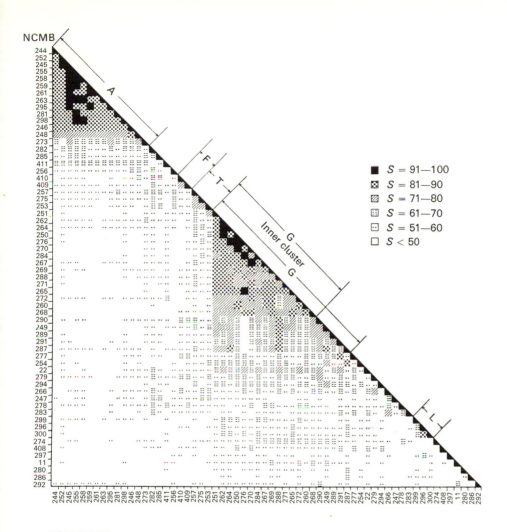

FIGURE 5-16
A shaded representation of relationships between strains of bacteria as OTU's, based on a similarity matrix, after rearrangement of the OTU's by single linkage cluster analysis to bring similar ones adjacent to each other. The similarity coefficient is percent S_J. The letters indicate phenons. [From Floodgate and Hayes (1963).]

ordinate in a phenogram has no special significance and the order in which the branches are presented can be changed within wide limits without changing the taxonomic relationships implied by a given phenogram. To illustrate the multiplicity of ways in which the same phenetic relationships can be represented in a phenogram we show in Figure 5-18,*a* a portion of the phenogram from Figure 5-17, and next to it in Figure 5-18,*b*, an equivalent phenogram in which several of the

262

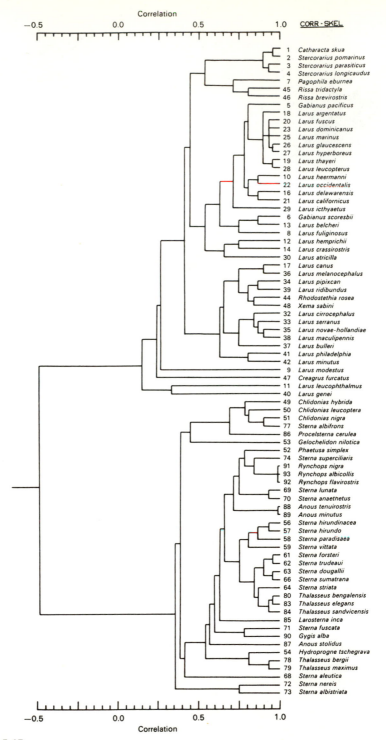

Correlation

| | | | | CORR-SKEL |

1 *Catharacta skua*
2 *Stercorarius pomarinus*
3 *Stercorarius parasiticus*
4 *Stercorarius longicaudus*
7 *Pagophila eburnea*
45 *Rissa tridactyla*
46 *Rissa brevirostris*
5 *Gabianus pacificus*
18 *Larus argentatus*
20 *Larus fuscus*
23 *Larus dominicanus*
25 *Larus marinus*
26 *Larus glaucescens*
27 *Larus hyperboreus*
19 *Larus thayeri*
28 *Larus leucopterus*
10 *Larus heermanni*
22 *Larus occidentalis*
16 *Larus delawarensis*
21 *Larus californicus*
29 *Larus icthyaetus*
6 *Gabianus scoresbii*
13 *Larus belcheri*
8 *Larus fuliginosus*
12 *Larus hemprichii*
14 *Larus crassirostris*
30 *Larus atricilla*
17 *Larus canus*
36 *Larus melanocephalus*
34 *Larus pipixcan*
39 *Larus ridibundus*
44 *Rhodostethia rosea*
48 *Xema sabini*
32 *Larus cirrocephalus*
33 *Larus serranus*
35 *Larus novae-hollandiae*
38 *Larus maculipennis*
37 *Larus bulleri*
41 *Larus philadelphia*
42 *Larus minutus*
9 *Larus modestus*
47 *Creagrus furcatus*
11 *Larus leucophthalmus*
40 *Larus genei*
49 *Chlidonias hybrida*
50 *Chlidonias leucoptera*
51 *Chlidonias nigra*
77 *Sterna albifrons*
86 *Procelsterna cerulea*
53 *Gelochelidon nilotica*
52 *Phaetusa simplex*
74 *Sterna superciliaris*
91 *Rynchops nigra*
93 *Rynchops albicollis*
92 *Rynchops flavirostris*
69 *Sterna lunata*
70 *Sterna anaethetus*
88 *Anous tenuirostris*
89 *Anous minutus*
56 *Sterna hirundinacea*
57 *Sterna hirundo*
58 *Sterna paradisaea*
59 *Sterna vittata*
61 *Sterna forsteri*
62 *Sterna trudeaui*
63 *Sterna dougallii*
66 *Sterna sumatrana*
64 *Sterna striata*
80 *Thalasseus bengalensis*
83 *Thalasseus elegans*
84 *Thalasseus sandvicensis*
85 *Larosterna inca*
71 *Sterna fuscata*
90 *Gygis alba*
87 *Anous stolidus*
54 *Hydroprogne tschegrava*
78 *Thalasseus bergii*
79 *Thalasseus maximus*
68 *Sterna aleutica*
72 *Sterna nereis*
73 *Sterna albistriata*

−0.5   0.0   0.5   1.0

Correlation

FIGURE 5-17

Phenogram of 81 species of gulls and terns based on UPGMA cluster analysis of correlation coefficients on 51 skeletal measurements. The cophenetic correlation coefficient was 0.931. [From Schnell (1970b).]

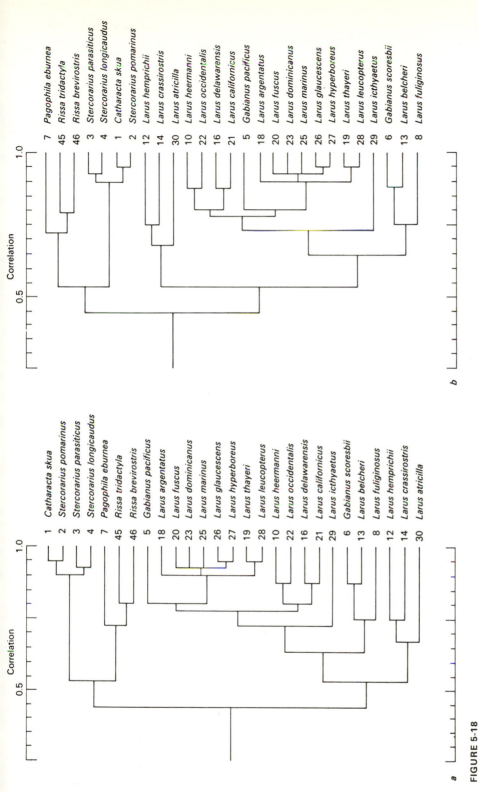

FIGURE 5-18
A part of the phenogram in Figure 5-17, showing *a*, the original arrangement, and *b*, the arrangement after rotation of several stems. The two phenograms are equivalent from topological and classificatory viewpoints.

clusters have been rotated around their axes as though they were mobiles. It is generally convenient to keep code numbers as ordered as possible, especially if they have not been randomly assigned but instead present linear arrangements of generally established taxonomic groups. On the other hand, no special meaning should be read into discontinuities of sequences of OTU numbers because of the very large number of permutations of such rotated positions.

To show the reliability of particular junctions in a phenogram, Rohlf (1970) has produced a program that plots a frequency distribution of similarity coefficients on the basis of which a particular phenogram junction has been made. These are superimposed upon the phenogram and give some indication of the validity of the level of a given furcation (Figure 5-19). Hall (1965a, 1968) and Stearn (1969) have also made suggestions about indicating additional taxonomic information in phenograms. McCammon (1968) has introduced, under the name of the dendrograph, a method of reordering the stems so as to maximize an even triangular shape to the phenogram, and of spacing the stems proportionately to the distance from the tips to the cross bars, thus giving a better visual display of clusters. Sometimes when the tips within a phenon are very numerous they are replaced by an outline.

Another method for representing the results is by the *skyline plot* developed by several authors, among them Wirth, Estabrook, and Rogers (1966) and Ward (1963). An example of such a plot is shown in Figure 5-20. Here the OTU's are listed by code number along the abscissa of the graph and bold horizontal lines give the nested hierarchical clusters for those OTU's whose range they cover. The level of the bold horizontal lines gives the maximum similarity implied by any given cluster along a graduated similarity scale shown along the ordinate. Thus, for example, in Figure 5-20 OTU's **6** and **17** join at a similarity level of approximately 0.95. The numbers immediately below the bold horizontal lines are a special feature of the graph theory model by Wirth et al. (1966) and are not an essential ingredient of a skyline plot. They indicate the numerical value of the moat of a cluster, i.e., a quantification of the gap separating it from adjacent clusters (see Section 5.2).

The results of cluster analysis can also be shown by a series of successive *graphs* (in the graph theoretical sense), as the criterion for OTU's linking up is gradually relaxed. These graphs are often called *linkage diagrams*. Two successive stages from a paper by Wirth et al. (1966) applied to an example from the orchid tribe Oncidiinae, are illustrated in Figure 5-21. These are the same data as shown in the skyline plot of Figure 5-20. Similar graphs were shown earlier in Figures 5-3, 5-4, 5-9, and 5-10, illustrating various clustering procedures. A difficulty of this approach is that in order to get an understanding of the relationships of an entire taxonomic group one needs to have in front of one the various layers or cross sections of the taxonomic hierarchy, which requires a fairly large number of successive graphs showing the increasing interrelationships of the set of OTU's. It is quite difficult to get so many graphs published and even then the comparison between successive stages may be rather tedious, because ideally the clustering process should be

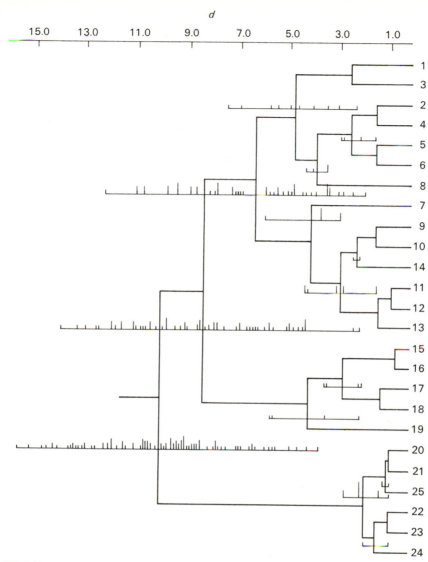

FIGURE 5-19
A phenogram with frequency distributions of similarity coefficients superimposed on the vertical branches. The distributions are of the similarity coefficients between all OTU's in one cluster paired with all OTU's in the other cluster. [From Rohlf (1970).]

shown as a continuous series of images as in a moving picture. However, the author may illustrate only certain of the hierarchic levels that show relationships that especially are meaningful to him.

A somewhat similar method of representation was chosen by Moss (1967), in which an attempt was made to show the various stages of clustering in a single diagram by attaching numerical values to the edges of the graphs and also by using

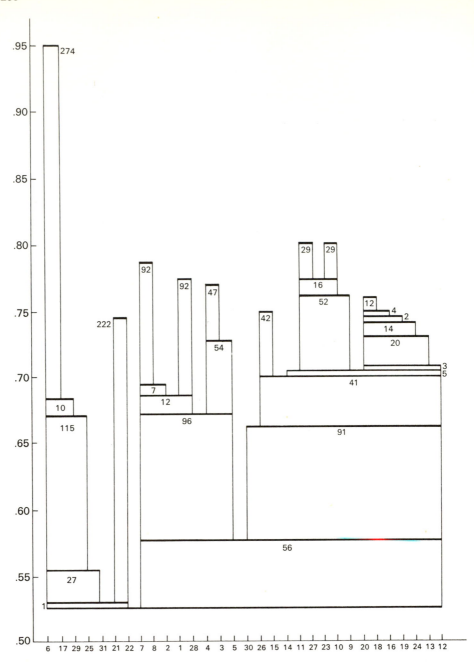

FIGURE 5-20

Skyline plot for 31 members of the Oncidiinae. This summarizes the results of clustering by showing the clusters, and the value of similarity at which they were formed, their hierarchical relationship, and the measure of isolation of each cluster. The OTU numbers are placed along the bottom, and the vertical scale indicates similarity. The bold horizontal lines correspond to the clusters, as follows: the OTU's below the bold line are the members of the cluster. The highest similarity value at which a cluster defined by a bold horizontal line still forms can be read off the scale along the ordinate. The distance from the bold line to the line below is proportional to the "moat" of the cluster, and the value of the moat ($\times 1000$) is entered near the bold line. The lighter vertical lines are given only as guidelines. [From Wirth, Estabrook, and Rogers (1966).]

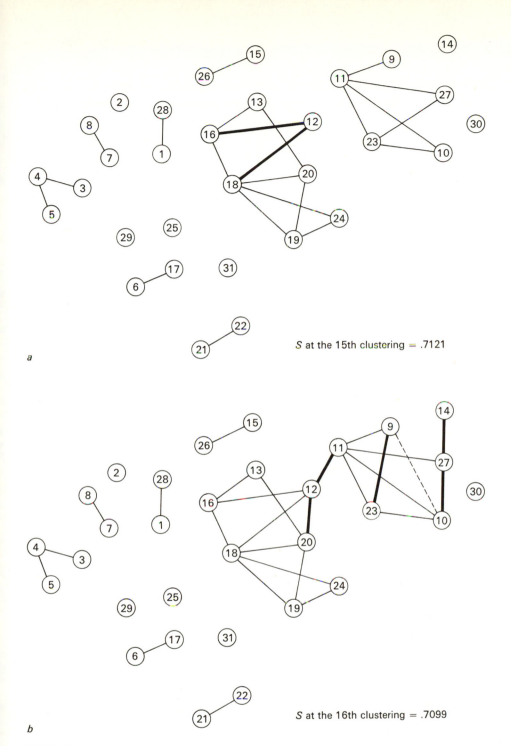

a

S at the 15th clustering = .7121

b

S at the 16th clustering = .7099

FIGURE 5-21

Graph of the relationships of 31 members of the Oncidiinae, expressed as subgraphs, at *a*, the 15th, and at *b*, the 16th clusterings. The value of S_i is the cutoff dissimilarity value that gives rise to the *i*th partition. Heavy lines indicate dissimilarities that changed the previous partition to the present one (i.e., connect previously unconnected groups); the lighter lines represent internal structure that existed prior to the present partition; dashed lines (one is shown in *b*) are connections that will form for dissimilarities greater than those at S_i but less than those at S_{i+1}. [From Wirth, Estabrook, and Rogers (1966).]

different types of lines (solid, broken, and wavy) to indicate strength of relationship (Figure 5-22). The taxometric maps of Carmichael and Sneath (1969) are rather similar.

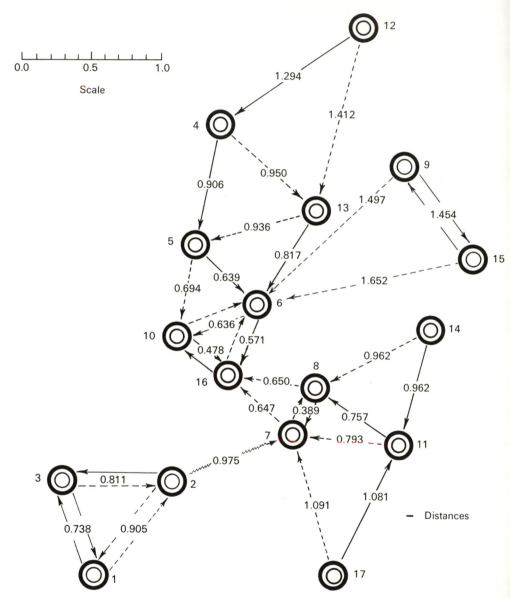

FIGURE 5-22

Graph of relationships of dermanyssid mites based on distances. OTU placement is obtained by triangulation of first-order and second-order distances as shown by solid and broken lines (edges) respectively. Isolated groups, such as the taxon (1, 2, 3), are connected to other groups by means of third-order (or lower-order) distances ("links") as shown by wavy lines (edges). [From Moss (1967).]

Another way of representing the result of clustering is by means of the so called Wroclaw diagrams developed by Polish phytosociologists and anthropologists (for entrance to this literature, see Hubac, 1964) and applied in modern numerical taxonomic work by Hubac (1964) and Moss (1967). Figure 5-23,*a* shows the contour representation of a Wroclaw diagram that may be compared with the graph of the same OTU's in Figure 5-22 and in a phenogram in Figure 5-23,*b*. The clustering contour lines shown in Figure 5-23,*a* correspond to contour lines on a topographic map. In a modified version (Figure 5-24) Moss has also given an indication of the levels at which clustering has taken place. They are shown as straight lines of different types between clusters rather than by the series of concentric contour lines shown in the earlier figure.

One difficulty of representing clustering procedures as graphs, Wroclaw diagrams, or ball-and-stick models, is that the placement of the OTU's becomes quite difficult, and often requires considerable trial and error. The reason for this is that the taxonomic relationships implied by the similarity matrix cannot usually be represented adequately in a two-dimensional space, and without an efficient ordinating procedure even the optimal two-dimensional representation cannot be easily found. For this reason, we would recommend—and Jardine and Sibson (1968b) have explicitly proposed—the prior ordination of such taxonomic units in order to simplify the drawing of edges between OTU's or of contour lines. The ball-and-stick models discussed earlier are also best arranged by three-dimensional ordinations.

When confined to only two dimensions the results of ordination can be represented easily on the printed page. We illustrate an example (Figure 5-25) from plant ecology (Gittins, 1969), where a principal components ordination is shown; the 45 stands are based on 33 characters: 30 represent species composition and 3 represent physical variables. The ordination here is an R analysis ordinated within an A-space and the first two principal components seem to represent soil depth and exchangeable soil potassium. Thus Gittins points out that *Thymus drucei* and *Trifolium repens* differ greatly on principal axis 1. *T. drucei* is known to prefer shallow, well-drained soils, while *T. repens* is found in deeper soils with a more balanced water relationship. Certain special scatter plots are useful in bacteriology (e.g., Quadling and Colwell, 1964; De Ley and Park, 1966).

With ordination on three or more axes, difficulties of representation arise. It is always possible to show aspects of the multidimensional space two dimensions at a time. An example of this is shown in Figure 5-26 (from Hendrickson and Sokal, 1968), which shows three two-dimensional ordinations of 29 species of the mosquito genus *Psorophora* based on principal components computed from 158 adult characters. This is a Q study ordinated in an A-space. However, it is difficult to gain an overall view of relationships from two-dimensional representations that are the projections of the other dimensions on the plane formed by the two axes being considered. A second disadvantage of this approach is that as the number of

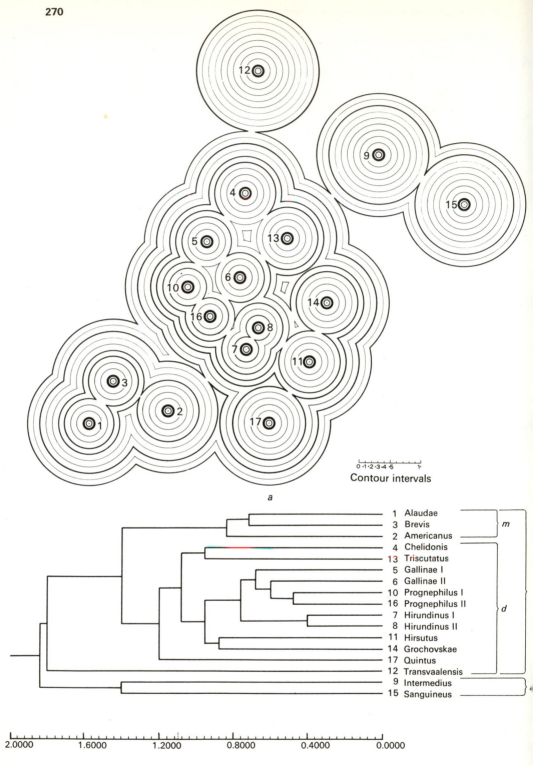

a

0 ·1 ·2 ·3 ·4 ·5   1·
Contour intervals

1	Alaudae
3	Brevis
2	Americanus
4	Chelidonis
13	Triscutatus
5	Gallinae I
6	Gallinae II
10	Prognephilus I
16	Prognephilus II
7	Hirundinus I
8	Hirundinus II
11	Hirsutus
14	Grochovskae
17	Quintus
12	Transvaalensis
9	Intermedius
15	Sanguineus

m

d

2.0000   1.6000   1.2000   0.8000   0.4000   0.0000

b

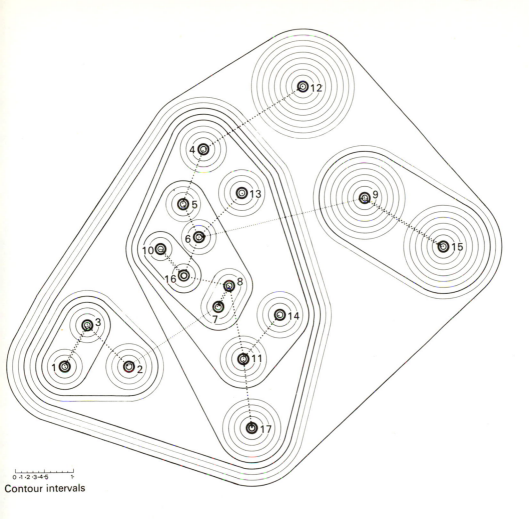

0 ·1 ·2 ·3 ·4 ·5   1·

Contour intervals

FIGURE 5-24

A contour diagram consisting of a modification of Figure 5-23. The levels at which clustering has taken place are shown by dotted lines. Compare this figure with Figures 5-22 and 5-23,*a*. [From Moss (1967).]

FIGURE 5-23 (*opposite page*)

a, Contour representation of the relationships of dermanyssid mites shown in Figure 5-22. Levels at which clustering has taken place are shown as darker lines. *b*, Phenogram of the same relationships shown based on distances and UPGMA clustering. [From Moss (1967).]

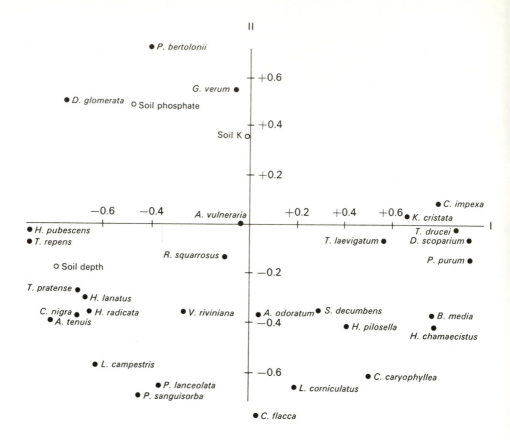

FIGURE 5-25
Ordination plot from principal component analysis. The abundance of 30 plant species and the values of three soil variables (the characters) were recorded in 45 stands (OTU's) of grassland in Anglesey, North Wales. An R analysis is ordinated in an A-space. The diagram shows the positions of the species and soil variables on the first two principal axes. [From Gittins (1969).]

dimensions to be considered increases, the number of combinations taken two at a time becomes quite large and the plots are of little heuristic value. Attempts have therefore been made to construct at least three-dimensional ordination models (when three dimensions are indicated). This can be done by the construction of models using styrofoam or rubber balls attached to wires or hatpins in a styrofoam, wooden, or cardboard base. The dimensions of the base are the first two axes and the height of the balls represents the third. Difficulties arise with the labelling of the balls—often this has to be done numerically—and with the identifying of groups of these balls—frequently this is done by coloring them. The positioning of balls of any size, whose centers are close to each other in 3-space, is often difficult.

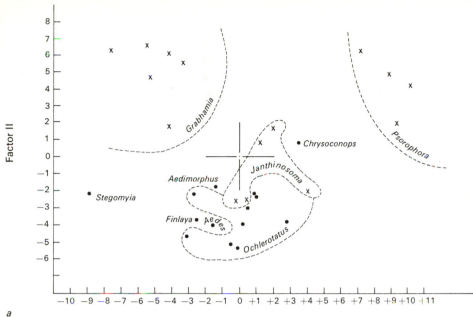

a

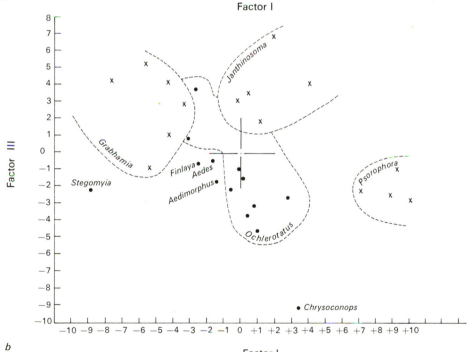

b

FIGURE 5-26

Twenty-nine species of mosquitoes as OTU's, plotted against centroid factors I, II and III, taken two at a time. These factors were extracted from the matrix of correlations between characters and represent linear combinations of the 158 characters employed. The conventional subgeneric names and boundaries are shown. *a*, Factors I and II; *b*, Factors I and III; *c* (overleaf), Factors II and III. [From Hendrickson and Sokal (1968).]

274

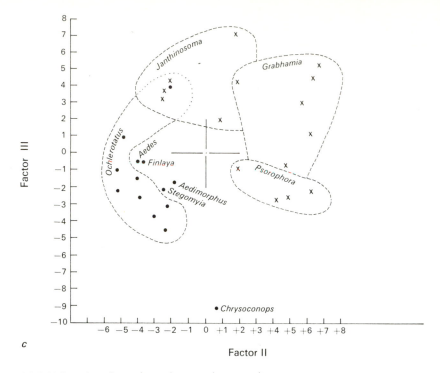

c

FIGURE 5-26 (*continued; see legend on previous page*).

Three-dimensional models are difficult to transport to scientific meetings or to other laboratories, so their utility is limited. Nevertheless they are usually of very great value to the investigator who wishes to examine or explain taxonomic relationships with his colleagues in personal contacts. Photographs of three-dimensional models are usually not successful, unless in color. In order to over-come the limitations of black and white photography, different shaped tips can be added to the wires or hatpins, as was done by Hendrickson and Sokal (1968). However, such devices have so far not been especially esthetic or successful.

Thus, to show a three-dimensional diagram on a two-dimensional page it either has to be shown in projection (see Figure 5-27), as was done by Moss (1967), or as a stereogram. Rohlf (1968) has developed a computer method for plotting three-dimensional stereograms of ordinated points (Figure 5-28). Such figures have to be viewed by means of stereoscopes, but some persons are able to train their eyes to superimpose the images and obtain a three-dimensional stereoscopic image.

Future technology may improve our methods of representing three-dimensional space. Stereoscopic visual displays can now be shown on computer consoles in which program knobs enable the viewer to rotate the image of the constellation, so as to create the impression of a model as it would be seen if the viewer were holding it in his hand and turning it various ways. These computer programs could be transmitted to other interested investigators and could solve the problem of

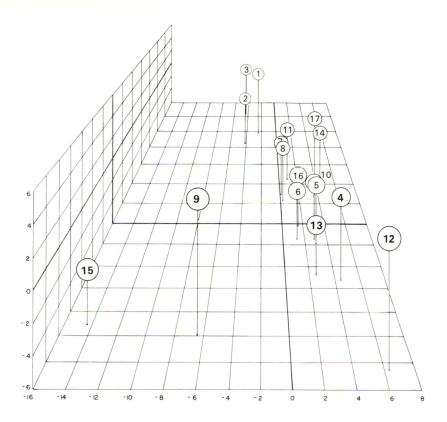

FIGURE 5-27
The relationships among the dermanyssid mites of Figures 5-22, 5-23, and 5-24 shown as a three-dimensional centroid factor ordination, viewed in perspective somewhat obliquely along the first axis. [From Moss (1967).]

transporting models. The difficult problem of plotting points and of choosing three-dimensional subsets of a higher dimensional space would be much simpler also if the processes could be automated with a graphical computer console. Views of the space could also be made from various special perspectives. An early fore-runner of such work is a film prepared by Geoffrey Ball, in which the viewer of the film got the impression of walking through a configuration of OTU's in a three-dimensional space.

5.10 OPTIMALITY CRITERIA AND THE COMPARISON OF CLASSIFICATIONS

We now come to a section where we shall attempt answers to many questions raised earlier in our account of clustering methods. First we shall list these questions and then go into some detail about each of them.

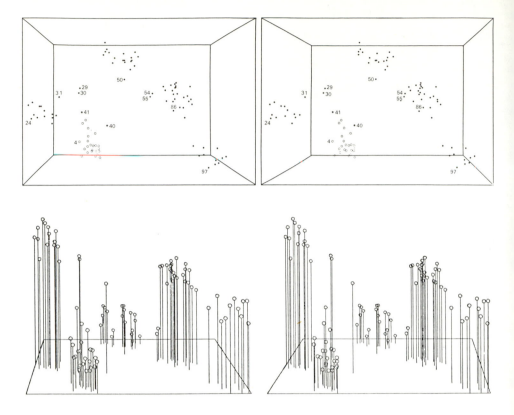

FIGURE 5-28
Stereograms of three-dimensional models of 97 species of bees of the *Hoplitis* complex. A few species of special interest are labelled with their OTU code numbers. The upper stereo pair, *a*, shows the OTU's as free floating points, and the lower, *b*, shows them as balls on pins, and from a different angle. [From Rohlf (1968).]

1. *What is a good classification?* Is there a universal criterion of goodness of a classification or are there separate criteria of optimality for various classificatory techniques? This is perhaps the most fundamental question of taxonomy, yet we shall not be able to present an unequivocal answer to it.

2. *Is classification A better than classification B?* If we can arrive at a satisfactory answer to question 1, the answer to the second question presents no problem. Because no universally accepted criterion of optimality is available, decisions on the relative worth of classifications are more difficult. Yet question 2 is asked more frequently than any other in taxonomy, whether conventional or numerical.

3. *How similar are two classifications?* Attempts to answer question 2 immediately lead to measures of similarity between two classifications, and numerical taxonomists have investigated a number of such measures. These statistics are essential

for taxonomic practice as well as for testing some theoretical questions, such as the nonspecificity hypothesis.

4. *Is there significant structure in a set of OTU's?* To the statistically-minded taxonomist the significance of any given classification is the fundamental problem, rather than question 1, yet this point has been largely ignored in work to date. The reason for this neglect is not so much because numerical taxonomists did not realize the importance of this question but because of the inherent difficulty of answering it. In recent years some mathematical statisticians have begun to interest themselves in this problem but there are many problems to be overcome, as we shall note below.

5. *Do two or more clusters, delineated by a prior procedure, differ in their newly observed properties?* If the previous question of the significance of the overall classification can be bypassed, a test of lesser importance but some intrinsic value may be devised. Do clusters previously delineated by orthodox taxonomic work or by criteria not part of the data matrix differ from each other on the basis of the newly observed characteristics? Except at the population level, this question seems of less interest for biological taxonomists; in other applications of numerical taxonomy, such as soil classification, it may be of great practical importance.

6. *Do two or more clusters, delineated by numerical taxonomy, differ in any of the properties on which they were clustered?* If two or more clusters are delineated by a taxometric method, how can we test that they differ in some specified way, using only the information in the original data matrix?

Optimality Criteria

Returning to the first three enumerated questions we find immediately, as noted earlier (Section 5.2), an intimate relation between optimality criteria and classificatory algorithms. This is because as soon as we define a criterion of optimality for a classification we should be able to develop in turn an algorithm that—even if it could not arrange a set of OTU's optimally (an analytical solution for optimality may be impossible)—would at least develop an iterative procedure for obtaining ever better classifications. In such a case, however, the optimality criterion would guide our selection of the type of classificatory algorithm to be employed. Similarly, if we determine a given method of classification as most suitable for a certain taxonomic task, then the optimality criterion may be hard to specify apart from the procedure used.

As already stated, no general agreement on the optimal classification exists except in cladistics, in which the optimal classification is the one best representing the branching pattern of the organisms through evolutionary history. Yet there may be no unique criteria for the operational steps, such as estimation of minimum evolution. In numerical phenetics such a criterion is not appropriate, and two general types of optimality criteria have been developed. The first considers a classification to be optimal if it represents as closely as possible the original similarity matrix among the OTU's. The other consists of attempts to develop inherent criteria of optimality: certain arrangements of OTU's are good because specified measurable properties of their similarity coefficients are optimized (in this case the similarities implied by the arrangement of OTU's need not be the best fit to the original similarity matrix among the OTU's).

Best fit to the original similarity matrix among OTU's is a natural concept and measures based on this criterion were developed fairly early in the history of numerical taxonomy. The most commonly applied method is that of *cophenetic correlations* developed by Sokal and Rohlf (1962). This method assigns a *cophenetic value* to the similarity between any pair of OTU's implied by a given dendrogram and generates a matrix of such values for any set of OTU's. The cophenetic value C_{jk} between any two OTU's **j** and **k** is the maximal similarity S_{jk} (or minimal dissimilarity U_{jk}) between the two OTU's implied by the dendrogram. Thus in Figure 5-29 we find that OTU's **a** and **c** have a cophenetic value of 4.1 expressed in *d*. When Sokal and Rohlf first suggested their method, computational facilities were relatively primitive and it seemed desirable to use coded class marks for the cophenetic values. However, in more recent work the actual similarity or dissimilarity values implied by a given furcation in the phenogram have been generated within the computer and used as cophenetic values. Similarly, cophenetic values based on ordinations are the actual distances (in any specified metric) in the ordination.

To simplify the subsequent discussion we shall restrict ourselves to discussing similarity matrices and their elements, keeping in mind that the statements made will also apply to their complements, the dissimilarities. A product-moment correlation coefficient is then computed between the elements S_{jk} of the original similarity matrix **S** and cophenetic values C_{jk} of the matrix **C**. This *cophenetic correlation coefficient* is a measure of the agreement between the similarity values implied by the phenogram and those of the original similarity matrix. Cophenetic correlations have usually been abbreviated r_{coph}, although others have symbolized it as r_c and Farris (1969a) has employed CPCC. We shall employ a new symbol here that we hope will help to clarify subsequently the several meanings in which the cophenetic correlation coefficient has been used. We shall use r_{CS} to indicate that it is the correlation between the elements of **C**, the matrix of cophenetic values and the elements of **S**, the similarity matrix. The simplicity of the cophenetic correlation coefficient has led to its extensive application by a variety of authors. It has generally been found to vary from 0.6 to 0.95, depending on the method producing the phenogram or ordination and on the natural structure of the OTU's being classified. For example, a representative set of values is shown in Sokal and

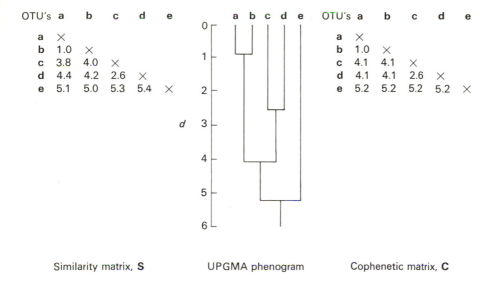

OTU's	a	b	c	d	e
a	×				
b	1.0	×			
c	3.8	4.0	×		
d	4.4	4.2	2.6	×	
e	5.1	5.0	5.3	5.4	×

OTU's	a	b	c	d	e
a	×				
b	1.0	×			
c	4.1	4.1	×		
d	4.1	4.1	2.6	×	
e	5.2	5.2	5.2	5.2	×

Similarity matrix, **S**   UPGMA phenogram   Cophenetic matrix, **C**

FIGURE 5-29
Cophenetic correlations. The similarity matrix S contains the original similarity values between OTU's (in this example it is really a dissimilarity matrix U of taxonomic distances). The UPGMA phenogram derived from it is shown, and from the phenogram the cophenetic distances are obtained to give the matrix **C**. The cophenetic correlation coefficient r_{CS} is the correlation between corresponding pairs from **C** and **S**, and is 0.9911. Other cophenetic measures are derived similarly, e.g., $\hat{\Delta}_{1/2}$ is equal to $\sqrt{0.30}/\sqrt{183.46}$, which equals 0.0404. For explanation, see text.

Rohlf (1970) where UPGMA clustering of different data matrices based on the same set of OTU's resulted in values of r_{CS} ranging from 0.74 to 0.90.

Farris (1969a) has subjected the cophenetic correlation coefficient to an intensive analysis. He finds that pair-group clustering, in particular UPGMA, will always maximize r_{CS} at least as well as variable group analysis (substantiating by mathematical proof the empirical findings of numerous workers over the last few years). He also discovered that for some data sets the cophenetic correlation coefficient will assume its highest value when "reversals" are permitted in a phenogram: $S_{ab,c} > S_{a,b}$; yet $S_{a,b} > S_{a,c}$ or $S_{b,c}$. The clustering procedures advocated by most numerical taxonomists are such that reversals are not permitted and the phenograms consequently imply ultrametrics (see Section 4.2). Farris feels that it is improper to use a coefficient as a measure of optimality that, when employed, leads to an undesirable taxonomic structure. Furthermore, unequal OTU numbers in branches of a phenogram can also lead to reversals if an attempt is made to maximize r_{CS}. We doubt whether these drawbacks are very serious for practical work as long as it is realized that r_{CS} cannot truly be maximized in some circumstances. From our experience it still seems a satisfactory measure of the agreement of a phenogram with a similarity matrix, and gross differences in r_{CS} are generally meaningful. One can also define the problem as one of maximizing r_{CS} with the restriction that no reversals occur.

Related to the cophenetic correlation coefficient are two other statistics : (a), the correlation coefficient $r_{C_1C_2}$ between matrices of cophenetic values representing two phenograms or ordinations, and (b), $r_{S_1S_2}$ between two similarity matrices obtained by different techniques or based on different characters. In a loose sense these, as well as r_{CS}, have in the past all been called cophenetic correlations but probably the collective term *matrix correlations* for all three types would be more suitable.

As an example of the use of these methods we may report the work of Boyce (1969), who found relatively little difference in r_{CS} between UPGMA, WPGMA, and UPGMC (using distances between centroids). In his study of 20 hominoid skulls based on a similarity matrix of correlation coefficients he found that UPGMA gave the best representation when measured by cophenetic correlation coefficients. The weighted method showed the least concordance with the original correlation matrix. Differences between weighted and unweighted clustering methods were greater when applied to a distance matrix between these skulls. Boyce concluded that ordination by principal components analysis reproduced the original similarity values more faithfully and yielded more information about the relationships among the OTU's than any of his phenograms. Also by cophenetic correlations, Sneath (1966b) was able to show that average linkage represents a distance matrix of random points better than either complete or single linkage, the latter being the least representative. Average linkage and complete linkage were closer to each other (by $r_{C_1C_2}$) than either was to single linkage. Schnell (1970b) has developed these methods further by cluster analysis of matrices of matrix correlations, thus obtaining compact summaries about techniques in the form of clusters of methods, or characters sets, that give very similar results.

Associated with matrix correlations are two-way scattergrams of similarity coefficients or of cophenetic values being compared with similarity coefficients. From such scattergrams one can obtain a visual impression not only of the magnitude of the cophenetic or matrix correlation coefficient but also of the distribution of the similarity coefficients or cophenetic values singly and jointly, which often gives insight into the taxonomic structure of the data set (see Figures 5-30 and 5-31).

Another measure of distortion (the complement of agreement as represented by the cophenetic correlation coefficient) is given by a family of measures developed by Jardine and Sibson (1968b). This coefficient is given as

$$\hat{\Delta}_\mu = \frac{\left(\sum_{jk} |U_{jk} - C_{jk}|^{1/\mu} \right)^\mu}{\left(\sum_{jk} U_{jk}^{1/\mu} \right)^\mu}$$

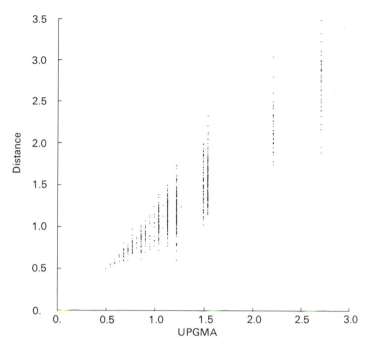

FIGURE 5-30
Two-way scattergram of taxonomic distances in a similarity matrix and of cophenetic values from a matrix of such values based on UPGMA clustering. The cophenetic correlation coefficient r_{cs} is 0.913. The data base is 45 species of mosquito pupae coded for 74 characters. Thus there will be 990 similarity coefficients or cophenetic values. [From Rohlf (1970).]

where U_{jk} is the dissimilarity coefficient between OTU's j and k ($j \neq k$), C_{jk} is their cophenetic value, and μ is an arbitrary coefficient $0 \leq \mu \leq 1$. By varying the coefficient μ the investigator is able to emphasize the smaller or greater differences between the dissimilarities and the cophenetic values, respectively. In their illustrative example, Jardine and Sibson (1968b) employ $\mu = \frac{1}{2}$, which turns $\hat{\Delta}_\mu$ into the normalized root mean square. These measures are Minkowski metrics (Section 4.3). In fact, Kruskal's (1964b) measure of stress in nonmetric multidimensional scaling (see Section 5.6) is a measure of optimality for ordination methods in this same general family. As will be recalled from Section 5.6, improvements for the stress coefficient are computed by an iterative procedure. Another procedure that iterates to a minimum the weighted sum of squares between the original dissimilarity coefficients and the cophenetic values, $\sum_{jk} w_{jk}(U_{jk} - C_{jk})^2$, where $j \neq k$, has

been proposed by Hartigan (1967). Again a weighting coefficient w_{jk} is introduced to afford the opportunity of giving more or less important to some of the differences. Not too much experience has accumulated with either Jardine and Sibson's or Hartigan's measure of distortion and their relative merits by comparison with

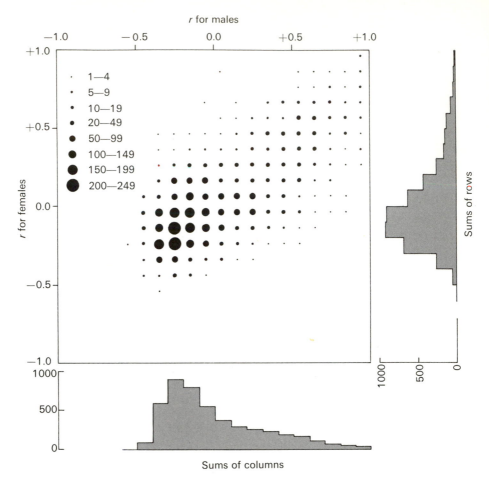

FIGURE 5-31

Two-way scattergram (frequency distribution) of similarity coefficients (in this case r) for the comparison of similarity matrices based, respectively, on male and female characters in the *Hoplitis* complex of bees. [Data of Michener and Sokal (1966).] The abscissa and the ordinate indicate the magnitude of the correlation coefficient from males and females, respectively. The frequencies in each cell of the two-way frequency distribution are represented by solid circles. The size of each circle indicates the magnitude of the frequency according to the key shown in the upper left corner of the figure. The figure is based on 4,656 correlation coefficients. The matrix correlation coefficient $r_{S_1 S_2}$ between the two variables is 0.71.

the cophenetic correlation coefficient. The objections raised by Farris (1969a) would presumably apply to these measures as well. One advantage of Hartigan's formula is that the sum of squared differences may be partitioned into components that express the cophenetic discrepancy due to specified OTU's. From several studies it appears that a few of the OTU's account for most of the discrepancy (e.g., Rohlf and Sokal, 1962; Rohlf, 1964, 1970; Sneath analyzing the data of Hudson et al., 1966; Moss, 1967; Ehrlich and Ehrlich, 1967; Johnson and Holm, 1968). This cophenetic statistic could also be adapted to finding the characters that caused the most discrepancy, following lines suggested by Johnson and Holm (1968). The statistical significance of cophenetic measures is an unsolved problem. It is complicated by the lack of independence of the individual coefficients in a similarity matrix. Recent work by Lerman (1970) has shown that one of his optimality criteria, $m(R)$, which measures the difference between a similarity matrix and a given classification of the data, approaches normality when the number of OTU's is large. There has recently been an interest in topological measures of stress that depend only on the rank order of branching in dendrograms (e.g., Phipps, 1971; Williams and Clifford, 1971).

The second approach to optimality is to attempt to maximize inherent criteria. These are statistics based on the similarity coefficients without comparisons with the original similarity matrix. Yet they are dependent on the metric chosen for this matrix. Of the methods based on some measure of within-group versus among-group differences we mention first of all the hierarchical grouping method of Ward (1963), which iteratively groups t OTU's into $k < t$ taxa at any one stage. At any clustering level Ward takes the optimal solution on the basis of an "objective function." One such function he suggests is the sum of the within-group sums of squares, but we have noted (Section 5.4) that it may correspond to unacceptable partitions. Similar methods have been proposed by Edwards and Cavalli-Sforza (1965) and Friedman and Rubin (1967). A related concept is the *object stability* defined by Rubin (1966) as

$$O_j = \frac{M_j - S^*}{1 - S^*} + \frac{S^* - \overline{M}_j}{S^*}$$

where M_j is the average similarity of OTU j to other members of its group, S^* is the maximum linking level of the classificatory scheme, and $\overline{M}_j$ is the maximum attraction (it is not a mean value by Rubin's notation!) of OTU j to any *other* group. For each classificatory scheme an average object stability $(1/t)\Sigma_{j=1}^{t} O_j$ is computed. The aim is to maximize the average object stability for any specified subdivision of the t OTU's into k groups, which is carried out by a hill-climbing algorithm developed by Rubin (1966) and is described in greater detail in a mimeographed publication referred to in that paper.

Wirth, Estabrook, and Rogers (1966) used two concepts, adumbrated in Section 5.2, that fall within the general class of optimality criteria. They measure the *connectivity* of clusters as

$$\frac{\text{Total number of existing connections} - \text{necessary connections}}{\text{Total possible connections} - \text{necessary connections}}$$

and for each cluster established they calculate its *moat*, which is defined as follows: Moat $= S^* - $ max S_{jk}, where OTU **j** is within and OTU **k** is outside the cluster defined by the linking similarity value S^*.

Significance of Clusters

The methods discussed so far address themselves to the first three questions at the beginning of this section. Answers to the last three questions are considerably more difficult. To know whether there is a significant structure among a set of OTU's requires the postulation of a plausible null hypothesis (as pointed out by Sneath 1967b), a hurdle that has so far not been overcome. That the OTU's studied form a single class seems to Goodall (1966e) to be the most useful null hypothesis in classification. If this hypothesis must be rejected, progressively more complex hypotheses—of two, three, and more classes—are substituted until the simplest acceptable hypothesis is found.

One approach is to test whether the classification arrived at by a clustering method could have been obtained from a random data set. Rohlf and Fisher (1968) examined cophenetic correlations between UPGMA phenograms and similarity matrices based on standardized characters that were random variables from a multivariate normal population. The cophenetic correlation decreased as the number of OTU's increased and also decreased more gradually as the number of characters increased; it settled around a value of $r_{cs} = 0.3$ for a correlation matrix and 0.6 for a distance matrix. For random character values from a uniform distribution (representing a hyperspace randomly filled with OTU's) the cophenetic correlations were even lower (see also Kaesler, 1970a). The phenograms show a rather typical appearance. All the branches are within a narrow range along the similarity scale for the normal distribution and are rather symmetrical for the uniform one, as was also noticed by Bonham-Carter (1965) and Sneath (1967b).

Another way of testing significance is to assume that the cluster has a given shape and to measure the amount of deviation of ordinated points from such a hyper-dimensional surface, as shown by Rohlf (1970).

Switzer (1970) has made some attempts based on simplifying the assumptions about distribution of points in the hyperspace. Bonner (1964) performs an approximate test to determine whether a given cluster could be sampled from a multivariate normal population of OTU's, in which the characters are uncorrelated. Because of the unrealistic assumptions it does not seem to be of special value.

Goodall's (1966c) probabilistic similarity index (Section 4.6) leads directly to tests of significance of subsets of the entire set of OTU's. In an important note Goodall (1966d) points out that a classification is an arrangement of individuals into classes and has no probability value attached to it. Probabilistic statements can be made about specific features of classification: what the probability is that a given procedure will distinguish any classes at all, or that a given subset of individuals will be allocated to a certain class. An important concept developed in this connection by Goodall is that of *utility* of a general-purpose classification, which he defines as the ability to predict the differing states of a character selected at random. Readers will recognize that this relates closely to the concept of the naturalness of a general classification discussed in Section 2.2. As a measure of utility, U, Goodall (1966d) suggests $U = -(1/n)\Sigma_{i=1}^{n} \ln P_i$ where P_i is the probability that a random partition of the states of character i into the same number of classes will result in as much variance as is observed in the classification at hand. This approach, to the extent developed by Goodall, seems to be limited to single level, nonhierarchic classifications only.

When the question is simplified to one of testing the significance of differences between clusters on the basis of their means or some other parameter, we must distinguish between clusters believed to exist on the basis of earlier evidence not part of newly observed characteristics (question 5) and clusters obtained by numerical taxonomy from a data matrix (question 6). This distinction is analogous to a priori and a posteriori multiple comparisons tests in analysis of variance (see Sokal and Rohlf, 1969, p. 226 ff.). Only few tests of an extrinsic, a priori hypothesis have been suggested. Goodall's (1966c) probabilistic approach can also be used for individual characters that are not part of the data matrix and whose concordance with an established classification is to be measured. Another probabilistic approach that seems to have given satisfactory results is that of MacNaughton-Smith (1965). This question is also discussed by Williams and Lance (1968).

Turning to significance tests of clusters established by numerical taxonomy, the simplest methods consider only the distinctness between two clusters. One may look at the depth of the cleft between the lowest side branch of either of two clusters in a phenogram for a rough indication (Hall, 1968); if this is several times the standard error of the similarity coefficient it is fairly safe to consider the clusters are distinct. Hutchinson, Johnstone, and White (1965) suggest plotting the distribution of S values of OTU's of clusters **A** and **B** from the center of one cluster, **A**, and examining the resulting histogram, which should be bimodal, with separated peaks for the members of **A** and **B**. They used as the center the OTU of cluster **A** with the lowest standard deviation of S values when compared with all other OTU's in that cluster, but other central measures would be suitable. Figure 5-32 shows an example generalizing this to Euclidean distances from the centroid to **A**, from an example on bacteria (Sneath, 1972). Tsukamura (1967d) has tested whether the mean similarity of members within clusters was different (greater than) the mean similarity between clusters. His technique used a t-test and pooled variances, and does not take into consideration the (unknown) distribution of the similarity coefficient. Probably a better way to do this would be to study these distributions by Monte

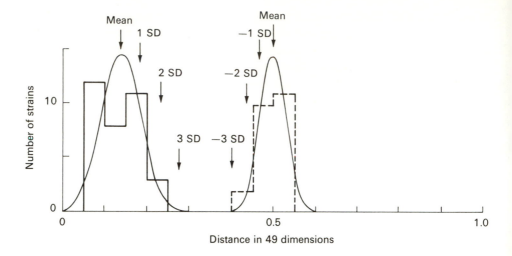

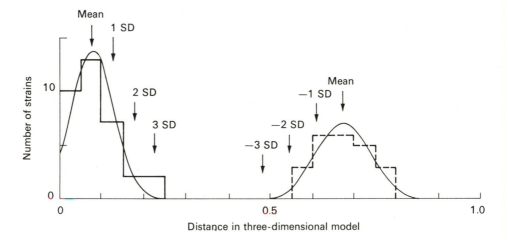

FIGURE 5-32

Histograms of the distances from the centroid of a chosen taxon. The distances are measured from the centroid of a cluster of strains of *Pasteurella pseudotuberculosis*, to the individual strains of *Pasteurella pseudotuberculosis* (solid lines) and to strains of another cluster, *Pasteurella enterocolitica* (broken lines).

Normal curves have been superimposed and above these are marked the means and 1, 2, and 3 standard deviations away from them. It can be seen that the two species do not overlap at the 3 standard deviation level.

The upper figure shows distances in the original 49 dimensions; the bimodality of the histogram on the left is not statistically significant. The lower figure shows distances in a three-dimensional principal coordinate analysis projection: although this differs from the upper figure mostly in a scaling factor, it will be seen that the histogram on the left is skewed to the right, and this is a consequence of the reduction from 49 dimensions to 3 dimensions. [From Sneath (1972).]

Carlo techniques. Sokal and Rohlf (1969, p. 634) illustrate how such a test could be applied in numerical taxonomy, although the criterion discussed there would have to be refined for serious work. Several significance tests for clusters have been developed, based on assumptions of multivariate normal distributions, e.g., Engelman and Hartigan (1969), M. E. Turner (1969), Mountford (1970), Hotellings' T^2 for the significance of a difference between cluster centroids, and tests based on Mahalanobis' D^2 (see Section 8.5, and Anderson, 1958). Although the assumptions on which they are based may not be very safe in taxonomic work, these tests may be useful as rough indications of significance. A posteriori versions of these tests would be needed to establish the significance of previously formed clusters. Bailey (1967) has emphasized the importance of corroboration for numerical taxonomic studies, and Silvestri and Hill (1964) have proposed that phenons should be validated by adding new OTU's and confirming that they fall into the phenons originally found. Although there may be some logical difficulties in this (Sneath, 1967b), one could circumvent these by dividing the material into two random sets of OTU's and comparing the taxonomies (e.g., Rohlf, 1965). We do not at present know of work directed specifically to this end.

The Inherent Error of a Classification

It is obvious that there is an inherent degree of error in any classification, conventional or numerical, which results from several sources: poor sampling of the phenetic hyperspace, error in recording of character states and shortcomings of the cluster-producing method. Sokal and Rohlf (1970) postulate an uncertainty principle for taxonomy that states it is impossible to know the natural (phenetic) classifications of a group of organisms above a certain level of precision, since conventional or numerical classifications by independent taxonomists would always disagree to a certain extent. Are the differences in classification due more to the difficulty of choosing appropriate characters, or to processing extensive and complex information? If the former is true, there is good reason to value deep experience in a particular group of organisms. If the difficulties are mainly those of processing complex information, an inexperienced worker should be able to choose characters that gave a satisfactory classification when processed numerically. For the purposes of the question, Sokal and Sneath (1966) have described the *intelligent ignoramus*: a worker who is ignorant of the particular taxon to be studied but is trained in the methodology of taxonomy and given a limited time to choose and score the characters. Should experiments show this procedure can generally be employed satisfactorily, much of the emphasis on experience would be unjustified. From here it is only a small philosophical step (though a major technical one) to automatic scanning and recording of characters (Section 3.3). These considerations are also of interest to the psychology of perception, though we have found little of direct relevance to numerical taxonomy in the literature of that subject.

A few experiments have been made with artificial organisms. Sneath (1964b) showed that an acceptable phenogram of heraldic beasts could be made numerically from characters chosen and scored by a child of seven, and presented some other artificial organisms with defined numerical relationships. These suggest that one might sometimes be misled by undue emphasis on prominent characters. Another experiment was made to test the ability to detect a time-trend from the pattern of complex objects (Hodson, Sneath, and Doran, 1966 ; Sneath, 1968c), using archaeological artifacts (brooches), which could be dated with fair accuracy on independent criteria. Several subjects found this task difficult (and an anatomist was as good as the best of several archaeologists), which suggests that detecting such trends may be less easy than constructing clusters and phenograms.

Sokal and Rohlf (1970) have made a fuller study of these points. They report on using two technicians untrained in taxonomy who independently described twenty-five species of bees. The results of the analysis of their descriptions were compared with the data of Michener and Sokal (1957), used as a standard, and abbreviated M&S. Three-dimensional ordination revealed that the generic groupings were not nearly as distinct in the data obtained from the technicians as in the M&S data. When the data collected by both technicians were pooled, they yielded much better agreement with the M&S data than did either data set alone. A similar conclusion was derived from examination of the phenograms. Although the phenograms made good sense overall, there were numerous discrepancies between those of the technicians and those of M&S. An analysis of the correlations among the 234 characters from both technicians as well as from M&S revealed no characters that were obvious duplicates (no correlations equal to unity). The same morphological structures were examined, but these were described in different ways. Factor analysis indicated that the character suites described in the three studies shared dimensions of variability. One of the technicians had not described characters loading highly on one of the factors. An important feature of the data is the large number of unique characters (reflecting independent dimensions of variation). These ranged from 18 percent to 40 percent. Sokal and Rohlf suspect that these unique characters indicate the presence of taxonomic "noise." Yet the experienced taxonomists (M&S) had a relatively high noise level. There is, of course, no certainty that the taxonomic structure given by M&S is the "true" standard. The distinct gaps shown by the M&S data may be due to the subconscious selection of characters tending to separate the groups during the previous conventional taxonomic study by Michener.

The results of this study would seem to imply that characters defined by untrained technicians could lead to acceptable classifications if enough characters are considered, because most dimensions of variation would be captured. Consequently Sokal and Rohlf were encouraged about prospects for the eventual automation of the data-gathering process in taxonomy. The work reported by Moss (1971) has similar implications.

The Incorporation of Additional Data into a Classification

After a numerical taxonomic study has been completed, two kinds of additional data are likely to be forthcoming. First, new study of the organisms may reveal characters other than those employed in the earlier work. Second, information may become available on OTU's that are related to those which were previously studied but for one reason or another were not included in the study before. The second situation may arise when persons are studying similarities among a large number of organisms and are unable to process all the data simultaneously because of limitations of the computer. We shall take up these problems in turn.

Adding new characters will be warranted only if the new characters are quite numerous or if the earlier classification has been based on relatively few characters. If the hypothesis of the matches asymptote holds reasonably well and a sufficient number of characters have been employed previously, the new characters should not change the arrangement of the taxa appreciably. The intelligent ignoramus study (Sokal and Rohlf, 1970) has shown that relatively little information was added by increasing an original 119 characters by another 115. We shall have to await more work in this field before coming to definite conclusions on the number of characters that must be added before a revision of a group need be made. In such circumstances records of new characters should probably be deposited in some central agency in electronic files (see Chapter 12) until a sufficient number have accumulated in order to warrant revision of the group. When taxonomic reprocessing of the data is indicated, all characters new and old would be considered. It may be that newly studied characters will have to be correlated with characters that have already been established, and only new information in the sense of character variation not represented by the previously studied characters will be added to the eventual data matrix.

Related to the above problems are situations where two studies are made on the same material but use different samples of characters (even if partially overlapping). The tests of the nonspecificity hypothesis have shown that very close agreement cannot be expected. Yet we may find that the two similarity matrices may at least agree by rank ordering (i.e., changing $r_{S_1 S_2}$ from a product-moment to a rank correlation coefficient; Sokal and Rohlf, 1969, 533 ff.). In such cases the phenograms resulting from cluster analyses of their similarity matrices would resemble each other topologically to a high degree.

Adding new OTU's to a study presents more serious problems. If many OTU's are added, it is obvious that a revision of the entire group is necessary. On the other hand, if only a single one or a few OTU's are added, the computation of the resemblance between these new OTU's and the others is relatively simply carried out. However, since relationships would inevitably be changed to some degree, the advantage of the stability of an analysis is to some degree negated. It is therefore quite important to make efforts to obtain reasonably complete and representative samples of taxonomic groups based on intuitive taxonomy before undertaking a revisional study by numerical taxonomy. Goodall (1968b) has developed a probabilistic criterion for determining whether a given OTU should be added to an established cluster.

Another approach may be to set up a series of standard OTU's, comparing newcomers against them. Although such a method would locate the newly added OTU's in hyperspace by comparison with the standards, it would not necessarily compare them with each other or with other OTU's previously studied. This is therefore only a stopgap measure, which cannot take the place of a complete analysis.

5.11   CRITERIA OF RANK

We believe that rank should be based on phenetic criteria, as indeed we think is the usual practice of taxonomists. These criteria have usually been based on two points: (1) that the internal phenetic diversity of taxa of equal rank should be as nearly equal as possible; (2) that gaps between taxa of equal rank should be as nearly equal as possible. Whenever these two rules are inconsistent, it is necessary to decide which one to employ. Simpson (1961, p. 133) would, for instance, retain as a genus the very diverse taxon *Rattus*, although the divergence among its species is greater than the average distance between *Rattus* and the closely related genus *Thallomys*. Mayr (1969a) has similar views.

At very low ranks another problem occurs: how does one rank taxa that overlap extensively and that consequently cannot be fitted into a nonarbitrary hierarchy? For example, different mutants of one species cannot be fitted into a nonarbitrary hierarchy; one cannot say whether white cats are of higher rank than long-haired cats. Sneath (1961) referred to such groups as "rankless taxa," which are not natural taxa in the usual sense, but in view of the suggestions we shall give below on measuring rank it would be better to refer to them as overlapping taxa. Related to this is the question of deciding on the rank of the species category. The species category is the fundamental unit in nomenclature, but because of the varied and mostly nonoperational definitions of the species it introduces nonphenetic criteria into the phenetic system. This problem is discussed at length in Section 7.1. We may note that one of the few phenetic criteria that could define a special level in a hierarchy is that level below which there are no distinct nonoverlapping clusters of individuals, and above which there are distinct gaps between clusters of individuals. This level would be equivalent to a phenetic species (most often morphospecies). It does, however, give rise to certain difficulties. It could only be established from studies on numerous individuals. The phenetic level would differ from species to species. It would perpetuate as full nomenclatorial species numerous microspecies (taxa that are phenetically distinct, yet extremely similar, but with very little internal variation). In addition, agreed criteria would be needed for deciding borderline cases.

Multivariate statistical methods—particularly, discriminant analysis—could provide these, if some figure were agreed on for use in considering clusters as distinct, for instance, that 95 or 99 percent of individuals must lie within the clusters. Since little work has been done on such approaches to species definition, we shall mainly discuss rank as applied to categories above the species level.

Most taxonomic textbooks—e.g., Simpson (1961), Davis and Heywood (1963), Blackwelder (1967a), Mayr (1969a), Crowson (1970)—have relatively little to say upon the details of how one decides upon taxonomic rank. Among the many principles mentioned are stability, tradition, number of included taxa, size of gaps between taxa, and internal diversity. While the first two are certainly of practical importance in well-worked groups with an extensive literature, they should not, we believe, be criteria for rank in numerical taxonomy; workers may, of course, take them into account in preparing a formal classification based on a numerical study. It would seem undesirable to employ the number of included taxa as a criterion of rank. The device of subcategories and intercalated categories (subtribe, cohort, etc.,) was introduced largely to deal with this problem, and should be used for it. One may also use *ad hoc* categories for dividing up a taxon with very numerous members, especially for identification purposes, but these need not necessarily be embodied in the recognized categories to which the panoply of the codes of nomenclature must be applied; such divisions are often arbitrary, and formal naming may then give the misleading impression that they are natural taxa. This leaves the last two considerations, internal diversity of a taxon and size of gaps between adjacent taxa. These two aspects can be readily represented in A-space. It is seen at once that only the first is meaningful for a single taxon taken in isolation; distances to other taxa are then indeterminate. The distances among taxa are properly a measure of difference in rank, or more precisely, of the rank of the superior taxon that just includes them. Gaps have been stressed as criteria of rank, especially by Michener (1970). However, the size of gaps between taxa is dependent both on the size, i.e., internal diversity of the taxa, and on the distance between their centers. If we consider two well separated taxa and increase the size of one of them the gap decreases. Also, if we decrease the distance between their centers the gap will decrease. The size of the gap is therefore an unsuitable criterion, all the more so for its dependence upon outlying OTU's (which may project erratically from some taxa and whose positions are likely to be a good deal less certain in A-space than are the positions of the taxon centers).

Measures of Rank

We may conclude that the most satisfactory criterion of a rank of a single taxon is some measure of its diversity. One suggestion (Farris, 1968) is to use the maximum dissimilarity between any pair of OTU's within the taxon. Another, and we feel the most natural statistic to use is the standard deviation s of the OTU's about the centroid (a suggestion similar to that of Cousin, 1956a). This is equivalent to the average within-group similarity. We would recommend that numerical taxonomists consider adopting this criterion where the mathematical model is appropriate. If we now consider two adjacent taxa, the rank of the superior taxon that includes both would be the standard deviation of all OTU's of both about the common centroid. The distance between the centroids of the two taxa could also be given a meaning: it is the average difference representing the rank of the members of the taxa, and could be conceptualized as the "rank difference" between the taxa as represented by central exemplars. However, it is less than the diversity representing the rank of the superior taxon itself, as shown in the following pages, for we are using the term "rank difference" to mean the rank that would just include the two centroids; it is not the same as the difference between a given pair of OTU's one from each taxon, for this is simply the inter-OTU distance.

The treatment above assumes that a distance measure is employed, and for angular measures of resemblance some modifications would be required, though in a hyperspace of many dimensions this may not be a significant problem. With angular measures the distances could be treated as the arc subtended by the angle, but if we are to retain the Euclidean properties required for the development given below we need instead the chord on a hypersphere of unit radius. This chord is given by $2\sin(\tfrac{1}{2}\theta)$ for an angle θ. The practical importance of such modifications of the model remains to be explored.

It should also be made plain that this suggested measure of rank does not define an envelope to a cluster, for if the cluster is elongated, the envelope would lie within the extreme OTU's on the longer axes, but outside those on the shorter axes. It does, however, represent a hypersphere of radius s around the centroid, and for a hyperspherical cluster this will indicate an envelope of one standard deviation. An envelope cannot in general be expressed by one parameter, so it cannot be used as a single measure for representing rank. Further discussion on envelopes will be found in Section 8.4.

Despite some drawbacks the model proposed has several advantages. It can be generalized to the case of many taxa. The rank of a taxon is simply the root mean square of the distances from its centroid, $\bar{x}$, to the t OTU's. If we wish to make this independent of the number of characters, n, we can then define rank as

$$s = \left(\frac{1}{nt}\sum_{j=1}^{t} d_{\bar{x}j}^2\right)^{1/2}$$

where

$$\sum_{j=1}^{t} d_{\bar{x}j}^2 = \frac{1}{2t}\sum_{j=1}^{t}\sum_{k=1}^{t} d_{jk}^2$$

It will be noted that this is the sample standard deviation, not the estimated population standard deviation (for which we would use $1/n(t-1)$ instead of $1/nt$).

This is because for a taxon of one OTU the latter would give an indeterminate value of s, with consequences upon the variance equations that follow here. The form adopted gives zero s for a taxon with one exemplar (of one OTU); clearly the *observed* diversity is zero, and this gives no clue to the actual diversity of this taxon in nature. The quantity s can also be calculated from the distances between the OTU's, provided that all t^2 distances between OTU's are included as shown in the formula just above.

One may also note that this expression is a measure of *average atypicality* (Section 4.11) or of its analogues in standardized A-space or in D-space (Sections 4.11 and 8.5), though we need to check whether we have divided by n as well as by t. The value of s above is the root mean square atypicality per character. We may now use the additive properties of sums of squares between and within taxa as follows. We have q taxa $(\mathbf{A}, \mathbf{B}, \ldots, \mathbf{J}, \ldots, \mathbf{Q})$ each with $t_\mathbf{A}, t_\mathbf{B}, \ldots, t_\mathbf{J}, \ldots, t_\mathbf{Q}$ OTU's. For any taxon $\mathbf{J}$ we symbolize its centroid by $\bar{\mathbf{x}}_\mathbf{J}$ and any OTU within it as $\mathbf{j}_\mathbf{J}$. The centroid of all the t OTU's $(t = t_\mathbf{A} + t_\mathbf{B} + \ldots + t_\mathbf{J} + \ldots + t_\mathbf{Q})$ is symbolized by $\bar{\mathbf{x}}$. Then

$$\sum_\mathbf{J} \left(\sum_{\mathbf{j}_\mathbf{J}} d^2_{\bar{\mathbf{x}}_\mathbf{J}, \mathbf{j}_\mathbf{J}} \right) + \sum_\mathbf{J} (t_\mathbf{J} d^2_{\bar{\mathbf{x}}, \bar{\mathbf{x}}_\mathbf{J}}) = \sum_\mathbf{J} \left(\sum_{\mathbf{j}_\mathbf{J}} d^2_{\bar{\mathbf{x}}, \mathbf{j}_\mathbf{J}} \right)$$

where d is the distance between the points denoted by the subscripts. If we have n characters this formula reduces to variances per character by dividing each term by nt to give pooled variances (i.e., weighted according to the number of OTU's).

It may be more useful in some circumstances to treat each taxon as described only by its variance and the distance of its centroid from $\bar{\mathbf{x}}$. This in effect treats them as if we had equal numbers of OTU's in each, so that the clusters are replaced by abstractions representing q new entities, the taxa, as is a common taxonomic practice. If we do this we can simplify the relations between the variances as

$$\frac{1}{q} \sum s^2_{\text{within taxa}} + s^2_{\text{among taxa}} = s^2_{\text{total}}$$

With this formula we estimate the first two terms and sum them to obtain the third. Then s_{total} is the rank of the superior taxon. The quantity $s_{\text{among taxa}}$ can never be negative, and this has consequences of importance for clusters that are not hyperspherical. Suppose we have two elongated clusters, $\mathbf{A}$ and $\mathbf{B}$, lying side by side with the long axes parallel and with equal numbers of OTU's in each. Then the rank of $\mathbf{A}$ is $s_\mathbf{A}$, and of $\mathbf{B}$ is $s_\mathbf{B}$, while their rank difference is $2s_{\mathbf{AB}}$. We have

$$s^2_\mathbf{A} + s^2_\mathbf{B} + 2s^2_{\mathbf{AB}} = 2s^2_{\text{total}}$$

The squared rank of the superior taxon will always be greater than the sum of the squared ranks of $\mathbf{A}$ and $\mathbf{B}$ (unless their centroids coincide when $s^2_{\mathbf{AB}} = 0$, but in this event we would not establish two separate taxa). This also holds for clusters of curious shape that are unlikely to occur in taxonomy, and the important consequence is that at least for realistic clusters, *the rank of the superior taxon will be greater than the ranks of the included taxa.* We can thus avoid the awkward situation where the measured rank of, say, a genus is less than that of one of its highly variable species.

Phenon Nomenclature

Those who have devised techniques for numerical taxonomy have suggested that they can be used to decide the rank of the taxa they yield. These have generally been applied to phenograms, and insofar as the linkage level of a given set of OTU's is an approximation to the average distance between them, it is applicable to the model described above.

The criteria proposed for phenograms are all very similar in essence. The same criterion for a given rank must be applied to all parts of an analysis. That is, where a hierarchical tree has been made, the line delimiting a given rank must be a straight line drawn across it at some one similarity level. The line must not bend up and down according to personal and preconceived whims about the rank of the taxa.

One consequence of the application of numerical taxonomy may be the ease with which we are able to recognize small differences in rank. The traditional categories of rank, such as order, family, or genus, may not be numerous enough even when expanded by intermediate ranks produced by prefixing the terms sub-, super-, and infra-. Words such as "supersubfamily" would be ugly and prone to lead to confusion. They could be avoided by citing the value that characterized the rank of the group in a numerical study, as will be described below.

The groups established by numerical taxonomy may, if desired, be equated with the usual rank categories such as genus, tribe, or family. However, these terms have evolutionary, nomenclatural, and other connotations one may wish to avoid. We therefore prefer new expressions (Sneath and Sokal, 1962). We call the groups simply *phenons* and preface them with a number indicating the level of resemblance at which they are formed. For example, an 80-phenon connotes a group affiliated at no lower than 80 on the similarity scale used in the analysis. Within the context of a given study 80-phenons are a category; *any given* 80-phenon is a group treated as a taxon. Regrettably, Mayr (1969a) has recently redefined the term phenon in a much vaguer sense as "a phenotypically reasonably uniform sample." This differs markedly from established usage in numerical taxonomy.

The term phenon is intended to be general, to cover the groups produced by any form of cluster analysis or from any form of similarity coefficients. Their numerical values will vary with the coefficient, the type of cluster analysis and the sample of characters employed in the study. They are therefore comparable only within the limits of one analysis.

Phenons are groups that approach natural taxa more or less closely, and like the term taxon they can be of any hierarchic rank or of indeterminate rank. Since they are groups formed by numerical taxonomy, they are not fully synonymous with taxa; the term "taxon" is retained for its proper function, to indicate any sort of taxonomic group.

An example of the delimitation of phenons can be seen in Figure 5-33. Drawing a horizontal line across the phenogram at a similarity value of 75 percent creates four 75-phenons: **1, 2, 5, 9**; **3, 6, 7, 10**; **4**; and **8**. We have found it convenient to refer to a given phenon by its first and last member. In the example above, we have 75-phenon (**1 ... 9**). This is an informal system of nomenclature. The dots (ellipses) do not indicate that the OTU's are necessarily in sequential, alphabetic, or numerical order, but merely that one or more are included between the terminal units. It is clear, of course, that such a label must refer to a given phenogram and cannot easily be transferred to another study. If the original taxonomic units had been species, these phenons might be subgenera or genera. A second phenon line at 65 percent forms three 65-phenons. Phenons are more obviously arbitrary and relative groups than the taxa of the Linnean nomenclatural scheme. If some investigator felt that the taxa in Figure 5-33 should be divided into two instead of three groups, the phenon line would have to be drawn at a similarity value between 50 percent and 60 percent; or he might feel that the two phenon lines were too close together

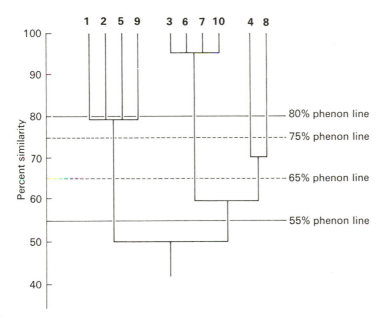

FIGURE 5-33
The formation of phenons from a phenogram.
For explanation, see text.

and did not summarize the main relations very fairly, as a result of which he might draw the first line at the 80 percent level. The designation of the phenons would then change to 55-phenons or to 80-phenons, respectively; however, the relationships among taxa in the phenogram are quite unchanged. Phenons are not only suggested for use with phenograms but with suitable prefixes could be used for phenetic clusters defined on other than dendritic scales.

Adding or removing taxa from a study would not have serious consequences in determining taxonomic rank, we believe, provided that only a small proportion of the OTU's were involved. The methods of cluster analysis employed (see Sections 5.4 and 5.5) will themselves have some influence on the ranks, since different methods summarize the similarity matrix in slightly different ways. In general we expect this effect will be as great as that produced by omitting a small proportion of OTU's. The employment of cladograms for determining the rank of taxa presents few difficulties if purely cladistic criteria are employed. An extensive discussion can be found in Hennig (1966, p. 154 ff.).

Absolute and Relative Criteria of Taxonomic Rank

There appear to be no satisfactory absolute criteria for taxonomic rank at present. It is clear that the difficulties would be formidable in treating widely different organisms. Early hopes that protein sequence studies would provide such criteria have not been fulfilled; different classes of proteins show different degrees of similarity when they are taken from the same pairs of organisms. It is possible that DNA pairing has some promise. Liston, Weibe, and Colwell (1963) have suggested that the phenon level of 75 percent may be a good criterion for the phenetic species level in bacteria, but this must depend greatly on the similarity coefficient and the set of characters; it cannot be taken as an absolute criterion.

The position with regard to relative criteria, that is, uniformity over one large taxon, is somewhat better. Thus it may be possible to compare families of birds with families of other vertebrates. A number of criteria from molecular biology are now available that would assist in such tasks. Thus, although different classes of proteins cannot be compared, one can compare resemblances in amino acid sequence in one functional protein, e.g., cytochrome c from various vertebrates as Fitch and Margoliash (1967) have done. There is some evidence from cytochromes that simple relationships between resemblance, rank, and presumed time to common ancestor may hold for one protein over quite wide ranges (see Section 6.2). Hoyer et al. (1965) have made similar observations on DNA hybridization.

We believe that for the near future each major group will have to be standardized separately. No useful standard can yet be applied to both bees and jellyfish, but within the megachilid bees, or perhaps within all the bees, some worthwhile standardization might be attained. To make this practicable there would have to be agreement on the rank of the whole group under study; we also would have to decide on the rank of the OTU's employed; frequently these will represent species. The other ranks could then be intercalated evenly.

There is some interest in the relationship between conventional taxonomic ranks and numerical measures of resemblance, inasmuch as this bears on two points, the psychological aspects of perception and classification, and relationships found in practice to be useful in systematic work. It has been pointed out to us by Dr. H. K. Clench that in butterflies the conventional categories species, genus, tribe, and family, if given rank orders 1, 2, 3, 4, bear approximately a linear relationship to the square root of taxonomic distance. It will, of course, depend in part on the resemblance measure chosen. We have noted, however, that this tendency holds reasonably well for several studies using distances: those of Ehrlich and Ehrlich (1967), Sokal and Michener (1967), and Fitch and Margoliash (1967). For protein sequences, the number of differences is equivalent to d^2, so it is the fourth root that is roughly linear with the rank category. It may well be that taxonomists intuitively choose a series approximating to the squares of the rank measure s as their basis for successive categories so, for example, the s values of species, genus, tribe, and family form a series such as 1, 4, 9, and 16. Farris (1968) suggests that a series of powers of 2 may be more appropriate.

As work proceeds on the higher ranks there are sure to be changes in the relative ranks of some taxa. For example, it might become apparent when a thorough comparative taxonomy of all insect orders was made that the Blattaria, for example, which had been treated initially as an order, were only of familial rank. For this reason some new ranks might have to be intercalated to express this, or possibly a revision of the system might be forced. This should be a pressing reason for comparative work at various hierarchic levels. Whatever the growth of numerical taxonomy may be, there should be great efforts to build it up by coordinated work between different specialities, calling for much more cooperation than has occurred in the past. To make an imperfect but legitimate analogy with map making, a number of reference points over wide areas will be of as great value in systematics as surveyor's bench marks are in cartography.

5.12   CORROBORATION OF A CLASSIFICATION BY BIOCHEMICAL METHODS

The quantitative relationships obtained from serology and molecular biology and the resemblance measures used in those fields are taken up in this section. Biochemical resemblance can be compared with taxonomic relationships estimated by other means, including numerical methods.

Comparative Serology

There is a general concordance between taxonomic difference, difference in protein structure, and serological cross-reaction as has been noted in Section 3.5. Unexpected serological cross-reactions are usually due to carbohydrates of wide taxonomic distribution. Cross-reactions at and below the generic level may be almost indistinguishable, and between classes and above they may be undetectable.

Pitfalls in the use of serology have been discussed in detail by Boyden (1942, 1958). Despite these the method is valuable as one of the few independent checks that can be made of a proposed taxonomy. It yields phenetic measures of resemblance, though not all its practitioners have acknowledged this in their published work. Hennig (1966, p. 104) has discussed this point thoroughly.

Several measures of serological similarity have been devised and were discussed in *Principles of Numerical Taxonomy*. One of the most useful, *I.D.*, the immunological distance of Mainardi (1958, 1959) is best expressed on a logarithmic scale, so that for the usual doubling dilutions, $\log_2 I.D.$ would be appropriate. This is $\frac{1}{2}$(sum of homologous scores) $- \frac{1}{2}$(sum of heterologous scores) when a score of 1 is given for each tube in the titration showing a maximal reaction and a fractional value is given for a lesser reaction. Moore and Goodman (1968) have proposed a new resemblance measure for gel diffusion studies. For precipitation reactions it appears adequate to plot the curve of the amount of precipitate obtained at various dilutions and to employ the area under the curve (Bolton, Leone, and Boyden, 1948); for complement fixation tests Sarich (1969a) has employed a form of *I.D.*

Since serology summarizes many characteristics of the protein molecules (probably several hundreds) it might be argued that it should be given great weight in a taxonomic study. However, one cannot convert a serological cross-reaction into a single character to be incorporated, with other characters, into a numerical analysis. The serological results are already a matrix of similarity coefficients. This matrix is not symmetrical, because the two heterologous reactions (for example, antihorse serum with cow serum, and anticow with horse serum) may not be the same in degree, though they are usually close. We still have slender evidence for deciding what weight should logically be given to a coefficient of serological cross-reaction compared with the weight of a resemblance value obtained numerically, and it would therefore be difficult to combine the two coefficients. Comparative serology must therefore remain at present a separate technique for assessing phenetic relations (albeit of a specialized kind). Systematic serologists (with a few exceptions, e.g., Butler and Leone, 1967; Kirsch, 1968; Basford et al., 1968; Goodman and Moore, 1971) have been slow to appreciate the need for numerical techniques such as cluster analysis to apply them to the serological affinities in order to yield taxonomic groupings. Also, it is becoming increasingly clear that it is not sufficient to estimate just a few of the relations between the OTU's, because this only allows an incomplete determination of taxonomic structure. The similarity matrix should therefore be as complete as possible. The development of principal coordinate analysis by Gower (1966a) opens the way to ordination of serological data, as does the method of Lee (1968) and Basford et al. (1968), which treats the antisera as characters and the precipitate values as character states. One disadvantage of the latter method is that OTU's that are not similar to each other or to other, clustered OTU's, appear spuriously close together (F. J. Rohlf, personal communication).

Chemosystematics

Alston and Turner (1963) have summarized much chemosystematic work in their book; although they hope for phylogenetic taxonomies, they make some use of resemblance measures. The paired affinity of Ellison, Alston, and Turner (1962) is S_J, and they also suggest other indices that are roughly equivalent to typicality and atypicality. Jaworska and Nybom (1967) suggest a distance using the areas of chromatogram spots as character state values. Hubby and Throckmorton (1965) have shown that chemical data in insects can be usefully represented by S_{SM}, though they did not cluster the matrix as in the botanical studies of Parups et al. (1966) and Simon and Goodall (1968). Ghiselin et al. (1967) also made a study of this kind, but effectively they had very few characters, and we do not believe these can support the wide-reaching deductions they draw. It is very clear that the integration of chemical data into taxonomy requires the use of numerical taxonomy, and only slow progress is being made in this direction.

There is now considerable interest in quantifying the resemblances between electrophoretic and similar patterns, which presents a number of difficulties. Rogoff (1957) made an early attempt to do this in comparing infrared spectra. Some authors have calculated association coefficients on the assumption that bands occurring on two patterns at a given position (or at overlapping positions) constitute a match, though it is known that homologous proteins from different organisms may migrate to different positions. Thus Whitney, Vaughan, and Heale (1968) calculate S_J and describe a relation between it and conventional taxonomic rank. Landau, Schechter, and Newcomer (1968) used this method and also derived forms of d and r. But if no other criteria of homology are available (e.g., enzyme specificities), the position of the bands and their intensities constitute the only information. Shipton and Fleischmann (1969) and Rouatt et al. (1970) were unable to find an entirely satisfactory method, but Johnson and Thien (1970) have used, with good effect, the correlation between pairs of points arrayed along the curves, together with an internal criterion for homologies.

When the presence or quantities of particular chemical substances are used as characters the numerical analyses present no particular difficulties, and a survey of the literature suggests that there is acceptable congruence between chemotaxonomy and morphological data (e.g., Grant and Zandstra, 1968), though no critical review has yet been published. Underwood (1969) considers the appropriate coding of chemical data that add up to 100 percent.

Protein Sequences

Data on protein sequences are now becoming available, and although these can be treated as sets of characters, they raise certain problems of their own. The usual measure of resemblance is Gower's coefficient with qualitative multistate characters (Section 4.4). The resemblance is then expressed as the proportion of positions where the amino acids are the same, dissimilarity as the proportion of positions where they differ. This difference may be considered as a measure proportional to the square of taxonomic distance through the relation between Gower's coefficient and distance, though the unusual nature of A-space raises some unexplored problems (Sneath, 1967b). The proportion of matches is about 1 in 14 in unrelated proteins, and Sackin (1969) found the mean to be 0.0715 with standard deviation 0.0115. However, two modifications of this resemblance measure have been introduced. Because the observed differences may include some that have undergone more than one mutation during the evolutionary pathway leading to the two organisms that are being compared, the observed number of differences is corrected for the estimated number of double or multiple mutations at any site. It is still uncertain what proportion of sites is mutable (Zuckerkandl and Pauling, 1965b; Margoliash and Smith, 1965), but the usual procedure is to estimate the corrected differences as $n \ln [n/(n - D)]$ where there are D observed differences and n sites in the chain. This increases D, but very little for small proportions of difference. Dayhoff and Eck (1968) note that this may still underestimate the evolutionary change, because back mutation may be quite frequent; they propose an empirical correction. The second method has less to commend it; this is to calculate the minimum number of single nucleotide changes in the genetic code that would account for the differences (Fitch, 1966a). This leads to incompatibilities in the nucleotide triplets when many organisms are considered, and introduces additional assumptions on mutations and the like. A theoretical point of some interest is raised by Zuckerkandl and Pauling (1965a): if two different nucleic acid messages code for the same sequence of amino acids, should this indicate dissimilarity? In other words, is the genetic message the nucleic acid or the protein version? This may have relevance in microorganisms, where the nucleotide triplets vary a good deal in frequency (De Ley, 1969a).

There is good congruence between the percent resemblance on a given protein and the taxonomic position of the organisms, noted by Doolittle and Blombäck (1964) and Fitch and Margoliash (1967) and the exceptions, noted by Niall et al (1969), may well be accounted for by sampling error. A single protein cannot be assumed to be representative of the whole genome. Protein sequences are unique in that they allow numerical comparisons between organisms as diverse as mammals and yeast (see Section 3.5).

A special problem in studying protein sequences is to decide on homologous positions on two polypeptide chains when they differ in length. This was discussed in Section 3.4. Needleman and Wunsch (1970) have proposed a new method of calculating resemblances that overcomes several of these problems. There are yet further complications: the work of Ingram (1961) shows that because of gene duplication during evolution it may be uncertain which proteins of a family of related ones are homologous (see Section 6.4). A striking example is the homology of chicken egg lysozyme and cow's milk α-lactalbumen (Brew, Vanaman, and Hill, 1967). The phylogenetic implications of this work are presented in Section 6.4.

Nucleic Acid Pairing

The technique known as nucleic acid hybridization or pairing (Doty et al., 1960, and McCarthy and Bolton, 1963) is capable of yielding resemblances between the DNA of different organisms. This measures the degree to which samples of single-strand DNA from an organism can form a hybrid double-strand nucleic acid with the single-strand DNA (or sometimes RNA) from another organism, which in turn depends on the extent to which the two forms of nucleic acid are similar in their base sequences and homologous in a genetic sense. Much information of this type is now available in microorganisms (for reviews see De Ley, 1969b, and Jones and Sneath, 1970). The study of Reich et al. (1966) is unusual in that a complete resemblance matrix was prepared together with confidence limits on the entries. One surprising finding is the evidence for genetic sequences common to all mammals that also include smaller fractions common to birds and fishes (Hoyer et al., 1965; also see De Ley et al., 1966b, on bacteria). DNA pairing presents certain problems with higher organisms, and Walker (1969) and Bendich and McCarthy (1970) have reviewed this in some detail for animals and plants respectively. DNA resemblances, like serological cross-reactions, should be subjected to cluster analysis or ordination as sufficient data become available.

The percentage of guanine plus cytosine in DNA has found some application in bacterial taxonomy and this is reviewed by Hill (1966) and De Ley (1969a), though it has the peculiar property that it is only marked differences in percent GC that are significant (indicating unrelatedness), because quite diverse taxa can have the same GC ratios.

Congruence between Numerical Phenetics and Biochemical Relationships

There are a few papers that give information on the congruence between numerical taxonomy and serology. The relationships of galliform birds inferred from myology (Hudson et al., 1966) can be compared to the serological data of Mainardi (1963). We have calculated the matrix correlation between $\log_2 I.D.$ from Mainardi and the sum of scores for wings and legs; $r = 0.493$ for the nine OTU's included in both studies, and the agreement is slightly less for correlations ($r = -0.382$). Arnheim, Prager, and Wilson (1969) found that lysozyme sequences and serology agreed well with the numerical results of Hudson and his colleagues. Downe (1963) reported on the serology of a few of the species of *Aedes* studied by Rohlf (1962, 1963a,b). The serology of the adults is quite congruent with the numerical taxonomy of the adults, though not wholly so with that of the larvae. There is good concordance of phenetics and serology when different species of microorganisms are compared (Sneath and Buckland, 1959; Jarvis, 1967; Jarvis and Annison, 1967; Campbell, 1969).

Work with nucleic acids has been mainly on bacteria. Here there is excellent congruence between phenetic similarity and DNA pairing (Heberlein, De Ley, and Tijtgat, 1967; Jones and Sneath, 1970) which is little less than that expected from the experimental and sampling errors of the two methods. The relationship between the two resemblance measures is not linear (because the DNA pairing falls to a low level around 40 percent S_{SM}), though suitable transformation may make it roughly linear (Sneath, 1971a, 1972). This work is important in demonstrating close correspondence between two methods that claim general validity on theoretical grounds. Another line of supporting evidence is the excellent concordance between bacterial phenetics and GC ratios (within the limitations of the latter) and many papers have noted this (for reviews see De Ley, 1969a; Jones and Sneath, 1970).

5.13 SUMMARY OF RECOMMENDATIONS FOR CARRYING OUT AND PUBLISHING A NUMERICAL TAXONOMIC STUDY

The general advice on how to undertake a numerical taxonomic study contained in this section will, we hope, draw together the major conclusions of discussions of topics in other parts of the book. Much of this advice is in the nature of comments on a desirable format for a taxometric publication. Since taxonomic work is not carried out in an ideal world—one must contend with scarcity of specimens, difficulty in obtaining characters and shortage of funds for preparatory work or for computing, not to mention potential lack of adequate computing facilities—an optimal strategy for obtaining the best possible classification under a given set of circumstances, or at a minimal cost, is essential. In this connection the nature of the classificatory task to be performed must be clearly specified. Exploratory data analysis, such as the classification of units not previously ordered (e.g., stands from a new association or bacterial strains not already extensively investigated) and the revision of an established group using new or more extensive material, will be considered among the points that follow. Exploratory work of this kind should be distinguished from other types of classificatory endeavor that require different strategies. For example, trying to fit data to a preexisting classification or to a preconceived taxonomic structure, or exploring nature by an adaptive clustering strategy that seeks to reveal phenetic configurations in A-space with minimal bias, are tasks that need to be approached in different ways.

The following points summarize the main requirements for a published numerical taxonomic study.

1. The OTU's should be chosen to represent appropriately the variation to be studied (Section 3.1).

2. Enough characters should be employed, properly chosen (Sections 3.5–3.8).

3. A resemblance measure appropriate to the aims of the task should be employed (Chapter 4) and clearly defined or described, and prior transformations of the crude data (e.g., standardization and scaling, Section 4.8) should be specified. Advice on handling very large sets of data is contained in Section 3.1 and Appendix B.

4. An explicit method should be used to represent taxonomic structure: normally clustering or ordination—not simply visual evaluation of the similarity coefficients.

The merits of many clustering methods have been discussed at length in Section 5.4 and 5.5, and we simply note a few major points here. If the OTU distribution in A-space exhibits tight clusters with large gaps between them (note that this cannot be known without prior study) then most of the common methods will give very similar results. In such cases the choice of a resemblance coefficient is more critical than the choice of clustering method. In general we advocate space conserving cluster methods, in agreement with Lance and Williams (1967b) and Hall (1969c). An average linkage method is probably the most useful, especially if a formal, non-overlapping hierarchy is desired. Lance and Williams's flexible linkage method is another suitable alternative, as it embraces a wide range of linkage strategies. Single linkage is apt to give the chaining effect, but may be useful for exploring elongated clusters.

5. For publication we recommend both a phenogram and an ordination. Williams and Lance (1968) prefer to cluster first, and make an ordination if the phenogram appears unsatisfactory. Phenograms are simple to prepare and give useful summarizations of the taxonomic relationships, particularly when the OTU's are well clustered and fit readily into a hierarchy. Ordinations, preferably in three dimensions (either as stereograms or as several two-dimensional diagrams) are especially valuable for understanding the taxonomic structure in more detail. The scale of all axes should be clearly indicated.

The preferred ordination methods have been discussed in Section 5.6. We generally recommend principal component or principal coordinate analyses, but the worker should keep in mind the effect of transformations (e.g., character standardization) on general factors such as size (see Section 4.11). Nonmetric multidimensional scaling may be the method of preference when problems of scale are encountered, or even for all instances, since it frequently gives results very comparable to principal component analysis. However, we have not as yet enough experience with this method in numerical taxonomy and its application may also be limited by constraints of the considerable amounts of computer storage space required by this technique. If an ordination is to be a satisfactory representation of taxonomic structure a high value (≥ 0.8) of the cophenetic correlation coefficient r_{CS} (or other suitable criterion) between the ordination and the original similarity matrix is desirable. This may or may not be accompanied by an appreciable (≥ 40 percent) proportion of the variation accounted for by the few dimensions that are displayed. We would feel that an ordination is unsatisfactory if the variation accounted for is less than 40 percent unless there is evidence to the contrary from cophenetic correlations as shown by Rohlf (1967, 1968). Sometimes it is not clear whether the axes of ordination diagrams have been scaled so that the variance on each axis is the same (which usually distorts distances), or whether the scales of the axes are proportional to the square roots of the eigenvalues, so this point should be clarified in the accompanying account. The dangers of establishing formal taxa simply by inspection of ordinations have been noted in Section 5.6.

6. Some indication of distortion should be given. Such measures must always be given in any numerical taxonomic study to enable the reader to get some impression of how well the classification represents the similarity matrix. The adequacy of a phenogram in representing taxonomic structure is indicated by a high cophenetic correlation, and if r_{CS} is over 0.8 the phenogram is likely to be fairly satisfactory in this respect.

Furcations on a phenogram at differences of less than about one standard error of the resemblance coefficient (where this can be estimated) are of little significance, and to avoid an impression of spurious accuracy, the branches may be grouped into an appropriate number of different levels. Another approach might be to indicate variation around stems in the manner of Rohlf (see Figure 5-19).

Measures of distortion between the resemblance matrix and the structural representation (phenogram or ordination) have been discussed in Section 5.10. The cophenetic correlation coefficient r_{CS} and its more general analogue, the matrix correlation coefficient $r_{S_1 S_2}$ can be generally recommended for phenograms or for ordinations. A stress coefficient can also be computed for all ordinations including nonmetric multidimensional scaling. The proportion of the total variance accounted for by each of the represented principal axes should be given.

7. Taxonomic conclusions should be based as far as possible on the representation of structure. If there are aspects of the results that are unacceptable to the taxonomist, he should try to explain his reasons for rejecting them.

8. Phenetic and cladistic conclusions (or diagrams) should be carefully distinguished.

9. Identification schemes (Chapter 8) should preferably contain an indication of success; i.e., a rating of how well they work in practice on *new* material.

10. Either the primary data should be given (for small studies) or locations where data are deposited should be stated. It is the scientist's responsibility to deposit

such data matrices in as permanent a form as is possible (punched cards and print-ups of data matrices including OTU names and character descriptions) and to make such information freely available to other investigators. These considerations are as important in numerical taxonomy as the secure deposition and accessibility of type specimens in orthodox taxonomy. Where character variation must be summarized for publication, Jardine and Sibson (1970) have some useful advice. Taxometric methods should be described in sufficient detail for other workers to be able to repeat the analyses. With the commoner methods, references given to standard methodology may be sufficient. The names and sources of any computer programs employed should be stated, and the computer center at which the study was carried out should also be noted.

5.14   THE DISTRIBUTION OF OTU's AND TAXA IN
PHENETIC SPACE

Although it has long been known that distribution of OTU's and clusters of OTU's (taxa) in phenetic space would reveal much about the evolutionary mechanisms that have given rise to the observed diversity and also about the nature of the ecological forces that govern the evolutionary patterns, very little work has so far been done in this field. This is so largely because we have not yet reached unanimity on methods of ordinating OTU's in phenetic space. We must therefore restrict ourselves to speculation in this field and to inspection of frequency distributions of the similarity coefficients. Michener and Sokal (1957) found bimodality in the distribution of correlation coefficients in their study of bees of the *Hoplitis* complex. The primary mode indicated intergeneric relationship, the secondary (higher values of r) was due to closely related species. Similar findings were made by Rohlf and Sokal (1965) on mosquitoes and by Sneath (1957b) on bacteria.

Of the few studies that actually investigate the distribution of organisms in hyperspace the earliest dates back to the end of the last century. Heincke (1898) investigated the distribution of individuals of races of herrings in what was in effect a phenetic character space, and obtained results giving the following generalization for any one homogeneous race: if the sum of squares of deviations of all characters from the centroid is calculated (equivalent to a squared taxonomic distance) these sums are, according to Heincke, approximately the same for every individual of the race. He formulated this as a law, sometimes called "Heincke's Law." Zarapkin (1934) and Smirnov (1925) had reservations on the validity of this concept, but it can be shown that it will usually be approximately true for clusters in phenetic spaces of high dimensionality, and is also approximately true for the unsquared distances. If one has a cluster of points in one part of a space of high dimensionality, with the points fairly evenly scattered in the n dimensions, and sweeps out hyperspheres of increasing radius r from the centroid, it can be realized that the hypervolumes increase very steeply, in proportion to r^n. At first very few points will be within the distance r, but as the radius increases, the ever larger hypervolumes will contain more and more new points, until the edge of the cluster is reached, when the number of new points will rapidly decrease. A histogram of the number of points within specified distance intervals from the centroid will therefore show a steep peak, or in other words, many points will lie at approximately the same distance from the centroid. Furthermore, for a wide class of distributions likely to occur in nature (including the multivariate normal distribution discussed below) the resulting histogram will approximate a normal curve: only when the points are clustered extremely densely and fall off very swiftly with distance, or when they are distributed in only a few dimensions (i.e., when the character values are extremely highly correlated) will this generalization break down. Heincke's law holds in fact even better for unsquared distances than for squared ones.

We may therefore expect that Heincke's law will hold for most clusters in practical applications. It will also generally hold where the distances are measured from outside the cluster (e.g., from the centroid of another cluster), though in such cases the histogram may not yield a normal curve. It must however be noted that it will hold less well with distances measured in a space of few dimensions such as two- or three-dimensional ordination (e.g., Sneath, 1972).

Little additional data are yet available in numerical taxonomy to test these points, but the similarity values found by Sneath (1957b) for strains of two species of bacteria, considered in turn, showed approximately a normal distribution. Similar conclusions were reached in work on bacterial species by Liston, Weibe, and Colwell (1963) and Hutchinson, Johnstone, and White (1965), and an example using Euclidean distances is shown in Figure 5-32. We do not yet have experience with correlation coefficients and other angular resemblance measures, but we expect on theoretical grounds that they would show similar behavior. In many applications it is likely that clusters will represent samples from homogeneous populations that will be distributed approximately in a multivariate normal manner, but this will be less often true at generic or higher levels.

Willis (1922) noted that in many families of animals and plants a histogram recording the number of genera containing respectively one, two, three species (and so on) gave a "hollow curve"; that is, there was a marked excess over expectation of monotypic genera and also an excess of genera containing a great many species. Willis and Yule (1922) found that plotting the logarithm of frequency, against the logarithm of the number of species these genera contained, yielded straight lines with slopes close to 1.5. At high species numbers the number of genera fell off (possibly because taxonomists tend to subdivide large, unwieldy genera into several genera). A similar pattern was found in higher ranks when the number of genera per family was studied. It is independent of "lumping" and "splitting" if done consistently (C. B. Williams, 1964). The interpretation of these curves has been much debated (see Yule, 1924; Wright, 1941; Stebbins, 1950, p. 531), but in the absence of objective criteria for what should be a family or a genus, and so on, the problem has been difficult to study, and the regularities have never been satisfactorily explained.

If the branching in phenograms reflects phylogenies, and if it occurs at random (so that there is an equal chance of every stem giving off one or more side branches in a given interval of the phenon scale), the number of side branches per stem in a given interval will obey a Poisson distribution, the number of OTU's being the total number of side branches plus one.

The data of Michener and Sokal (1957) on bees show that there is, at most levels, an excess over expectation (on the basis of the Poisson distribution) of stems that do not branch and also of stems that branch many times. This gives an excess of phenons with only one species and with many species. For example, at the 80-phenon level (800 level in the original paper) the distribution of the numbers of branches, always one less than the number of OTU's in the phenon, are as follows: phenons with zero branches, 16; with one branch, 4; with two branches, 1; with three branches, 2; and 6 phenons with, respectively, 4, 5, 7, 8, 12, and 20 branches each. The variance, 19.06, is much greater than the mean number of branches per phenon, 2.31, and the expected number with no branch (monotypic phenons) is 3.8 instead of the observed 16. Other studies that include a good cross section of the species of one taxon show a similar pattern, seen in the dendrograms of species of *Oryza* (Morishima and Oka, 1960), *Bacillus* (Sneath, 1962), *Ononis* (Ivimey-Cook, 1969a), and *Ditrigona* (Wilkinson, 1968). Thus "hollow curves" seem a general rule with the usual clustering methods.

The general form of the "hollow curves" is a clustered distribution, in which the variance is higher than that expected for a random distribution. This has a bearing on the construction of hierarchies, as has been discussed in Section 5.3 and it offers an explanation for the validity and usefulness of hierarchic classifications in systematics. It also relates to the common tendency toward a large proportion of "aberrant" forms (e.g., monotypic genera). However, it should be noted that information analysis may not give "hollow curves," though it is not yet clear whether this is due to the special properties of information analysis (see discussion following the paper of McNeill, Parker, and Heywood, 1969a and compare the dendrograms in that paper). The cumulative curve of the pattern of branching could be studied using Kolmogorov-Smirnov statistics (Sneath, 1967b).

Turning to results from ordinations above the population level, for example, among the species of one genus, we may cite the work of Rohlf (1967), Hendrickson and Sokal (1968), and Ivimey-Cook (1969a). It would seem that species within generic clusters are somewhat more evenly spaced than points in a multivariate normal distribution. Also there are some aberrant forms. There is little information on bacteria, but no strong evidence that they are different from the pattern described above (e.g., Rovira and Brisbane, 1967; Skyring and Quadling, 1969c). Sometimes, as in mosquito genera, clusters are very elongate, but we do not know whether this is in fact the way the species are distributed phenetically or whether this depends on our acceptance of traditional genera. Gilmartin (1969a) has listed what are in effect the cluster diameters of numerous taxa of Bromeliaceae.

Above the population level at all ranks the Willis curve is quite consistent. Phenetic distributions seem to have the following properties: (a) There are some dense clusters of lower taxa (containing, say, 20–50 lower taxa each). (b) There are a great number of isolated monotypic taxa. The evolutionary implications of this, although suggested by Yule (1924), remain to be worked out. (c) The consistency of the slope of the logarithm of the number of included taxa may reflect either some constancy of organized nature or some constancy in the way in which taxonomists have constructed rank levels. At all levels the "diameter" of clusters without clear internal subdivision seem to vary a great deal. This is probably to be expected, but the point to be noted is that there seems no tendency in nature for clusters to be of equal diameter.

Natural taxonomic groups are formed by the restriction of evolution to certain regions of the phenetic hyperspace and the accumulation of genetic differences through time. The small stepwise nature of genetic changes suggests that most characters of descendants are the same as those of their immediate ancestors. The taxa may be polythetic, for there is no assurance that any given character of an ancestor will persist in all its descendants. They are in any event operationally polythetic, since constant characters may not be among those available for our study. We do not yet know fine genetic structure sufficiently to say how frequently and to what degree the OTU's in natural taxa are genetically alike, but it seems probable that appreciable parts of the genotype are constant in all members of many natural taxa, at least in taxa of lower rank, so that these taxa are probably not fully polythetic.

Most conceivable intermediates between actual organisms are likely to be non-viable. The successful organisms will therefore possess complexes of correlated characters, and these correlations allow the recognition of distinct taxa. In addition, the constant evolutionary divergence leads to overall phenetic divergence, though at different rates in different lineages.

The question of how full is phenetic space is difficult to answer because we have to define the space and also state what are to be considered as OTU's: individuals, or species. If we were able to agree on these terms we could in theory work out the density in an n-dimensional hyperspace, which would have to be limited by scaling each dimension. Raup (1966) has shown that only a small part of the phenetic hyperspace defined by parameters of shell form is occupied by animals.

6

The Study of Phylogeny

As soon as phylogeny becomes a consideration to be dealt with during the classification of a group of organisms, we must be concerned not only with the phenetic relationship among the end points of the branching sequence, but also with phenetic relationships among any points that have at one time or another been occupied by organisms belonging to the phyletic branch under consideration. We have furthermore to concern ourselves with the sequence of branching as well as the time dimension. The nature of these relationships and the problems inherent in their analysis have been discussed in detail in Chapter 2. In Section 6.1 we shall be primarily concerned with the time dimension. This leads to a discussion of rates of evolution in Section 6.2, followed by the theory and practice of cladistic analysis in Sections 6.3 and 6.4. Finally in Section 6.5 we cover the applications of numerical taxonomy to paleontology.

6.1 PHENETICS AND THE TIME DIMENSION

Although we have already defined our terms carefully in earlier sections it will be useful to review them briefly for a discussion of phenetics and time dimensions. Phenetic taxonomy of recent organisms estimates their resemblances on the present time plane. Note that phenetic taxonomy is not restricted to comparisons at a single time plane; as we shall see in Section 6.5, it is possible and of some interest to estimate phenetic relationships between OTU's from different time planes. Cladistic taxonomy in the generally accepted sense reconstructs the branching pattern leading to the present phenetic positions but does not deal with the phenetic relationships among the ancestral forms, except to estimate the cladograms. Phylogenetic taxonomy would aim to reconstruct not only the branching pattern itself, but also the detailed phenetic resemblances between the stems in the past. In theory, therefore, phylogenetic taxonomy aims to reconstruct the characters of the ancestral organisms and thus to obtain such properties (e.g., evolutionary rates, convergence, parallelism) as would be obtained from a complete fossil record. Developments on these lines are now being attempted based on conventional morphological data and studies of protein sequences.

Can Ancestral Forms Be Included in Phenetic Classifications? What Would Be Their Rank? Plainly any division of continuous lineages is to some extent arbitrary. When two taxa ranked as classes are separated at a given line, the species on either side of this line, though both genetically and phenetically closely related, will be grouped in different classes. In discussing this problem, Remane (1956) points out that in lineages such as the one in Figure 6-1 (representing the phylogeny of a family) the species **H** and **I** have a dual but partly contradictory relationship. They are very similar both in properties and closeness of ancestry and should therefore be placed in one genus. However, they are also ancestors respectively of the genus (**A, B, C, D**) and the genus (**E, F, G**), and the closeness of their phenetic and phyletic relationship indicates they could legitimately be included in these two genera to give the genus (**A, B, C, D, H**) and the genus (**E, F, G, I**).

Establishing taxa (**A, B, C, D**), (**E, F, G**), and (**H, I**) is horizontal classification. The second method—establishing taxa (**A, B, C, D, H**) and (**E, F, G, I**)—is classifi-

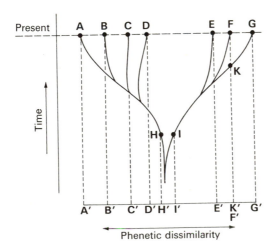

FIGURE 6-1
A phylogenetic tree as commonly represented, with a time dimension and one phenetic dimension. For explanation, see text.

cation by *clades* (Huxley, 1958). In such complex relationships it is a serious problem to know where to draw the dividing lines (made even more difficult if lineages
fuse by hybridization, which is often true of plants). Those taxonomists who prefer
monophyletic lineages would divide the group into clades wherever possible,
though by phenetic criteria the OTU's would cluster to give the most cohesive
taxa. Mainly such phenetic clusters would be monophyletic, but need not be so.

One misleading point about a figure such as Figure 6-1 is that it does not represent the phenetic relations at all accurately. In Figure 6-1, for example, the form
(species?) **K** is directly below **F**; it is, however, exceedingly unlikely that it would
be phenetically identical with **F**. The phenetic relations of the forms in this diagram
are their relations in the horizontal plane, as would be shown by throwing a
shadow of them onto the base line. This is equivalent to making projections of all
the points on the dendrogram onto the abscissa, as is shown for a few points in
Figure 6-1, for in general a single dimension is insufficient for representing phenetic
relationships. These are represented somewhat more clearly in Figure 6-2, where a
three-dimensional model of a substantially similar diagram is shown. Here the
horizontal plane shows phenetic dissimilarity in two dimensions; the vertical
dimension represents time, as before. Now the phenetic relations can be shown by
the shadow of the phylogenetic "tree" projected onto the base plate. This shadow
is shown again in Figure 6-3,*a*. It is a fronded figure in which the organisms of all
time periods are shown without overlapping one another (on this scale of taxonomic
discrimination). If we wished to divide it phenetically, we would divide it into two
main groups, genera perhaps, roughly as shown. The exact place of the division
line could be determined mathematically if we wished, though for most purposes a
division at the point of branching would suffice. We would rarely have enough
fossil data to have a complete shadow, however; we would be more likely to have
an incomplete set of organisms giving a shadow such as in Figure 6-3,*b*, and most

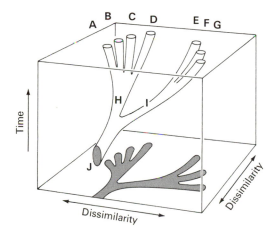

FIGURE 6-2
A phylogenetic tree in three dimensions, one of time and two of phenetic
dissimilarity. The "shadow" of the tree
on the base indicates the purely
phenetic relationships. For explanation, see text.

312

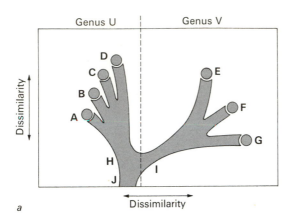

a

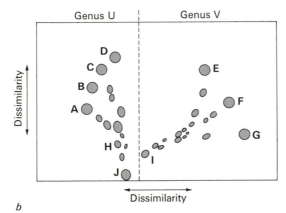

b

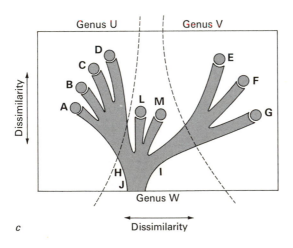

c

FIGURE 6-3
The "shadow" from Figure 6-2. The phenetic relationships are represented as shadows on the horizontal plane. *a*, The shadow from Figure 6-2 with a dashed line dividing it into two phenetic taxa, such as two genera. *b*, A patchy shadow. This is a more realistic representation because of the usual scarcity of fossils. *c*, Division into three phenetic taxa (such as genera) when branches **L** and **M** have been added.

often we would be lucky to get this amount of information. The exact position of the dividing line would then not be worth much argument.

We might have had some subsidiary branches (say, **L** and **M**) near the common ancestor, as shown in Figure 6-3,*c*. If so, one would, on phenetic grounds, divide it more or less as shown into the three genera **U**, **V**, and **W**. Note that in all cases the phenetic divisions—divisions made on the basis of the shadows—are, as we would expect, fairly close approximations of monophyletic groups or single lineages, though divergent clusters of branches may be excluded from the basal taxon. This seems to the authors the only honest thing to do; we believe that phylogenies are deduced necessarily from the phenetic relations.

6.2   RATES OF EVOLUTION

When it is possible to study fossil material, thus obtaining data from several known points of time, the resemblance coefficients will allow estimates of overall evolution rates. The dissimilarity between ancestral and descendant forms will be the measure of the overall evolution that has occurred in the intervening period. Simpson (1944) has discussed the great advantages of measuring the overall rate of evolution (what he calls the "organism rate"), as well as the rate of evolution in one or a few characters ("character" and "character complex" rates such as those studied by Haldane, 1949; Kurtén, 1958, 1959; and Buzzati-Traverso, 1959). Numerical taxonomy therefore offers a solution to many of the problems propounded by Simpson and by Huxley (1957). Simpson (1944) uses the terms tachytelic, horotelic, and brady-telic to describe rates that are respectively rapid, moderate, and slow. Little is known about tachytelic evolution, since the changes are so rapid that there is small chance of finding fossils of the relevant period. Bradytelic evolution is the kind shown by "living fossils" such as *Lingula*, *Ginkgo*, *Metasequoia*, and the coelacanth *Latimeria*.

Character Rates

Haldane (1949) has suggested a measure of the rate of evolution of a single character (for example, the length of an organ), so that the unit rate, the *darwin*, corresponds to a change by a factor of *e* in one million years—that is,

$$\frac{\ln X_T - \ln X_0}{T} = 1, \qquad \text{or} \qquad \frac{X_T}{X_0} = e^T$$

when the character has the value X_0 at time 0 and X_T at time T, with T measured in *crons* (the units of X are immaterial). (The word *cron* (Huxley, 1957) is a convenient term for one million years.) A darwin is approximately equivalent to a change by a factor of 1/1,000 in 1,000 years. If the allometry equation log y = log

$a + b \log x$ is used, the constant b should be used without transformation into logarithms since it is itself effectively a logarithm. In the examples studied by Haldane, the rate of change in horotelic evolution was around 0.04 darwins (40 millidarwins), but he noted that domestic animals have changed at rates of kilo-darwins, so that increased selection can evidently greatly increase the usual rate of evolution.

In a study of fossil horses, Downs (1961) found rates of 12.3 to 124.3 millidarwins for tooth characters, their mean being about 57 millidarwins. These rates seem fairly typical of horotelic evolution, though Simpson (1953) notes that there is considerable variation between horotelic lines. The rate may, of course, vary within any one line as well. It may be that in the work to date there has been a bias to select for measurement the more rapidly changing characters.

The darwin cannot be thought of as an absolute measure of evolution, since its value depends on the manner of scaling of the character (Sokal and Sneath, 1963, p. 239). Haldane (1949) also suggested measuring evolution rates as the time for the mean of a given character to change by one standard deviation, but this seems unduly dependent on the variability of the character, which may perhaps change erratically with time or, as has recently been suggested, may in fact be determining the rate of evolution (see Section 6.3).

Recent work on protein sequences has thrown some light on the rates of evolution as inferred from the time of separation of phyletic lines from geological evidence. Margoliash and Smith (1965), Zuckerkandl and Pauling (1965), and Kimura (1969) showed that the estimated rates were fairly constant over long periods of geological time, and suggested corrections to account for repeated mutations at the same site (see Section 5.12). There are some uncertainties with regard to very long periods, such as time back to the common ancestors of animals and plants. The rates vary widely with different types of protein: the average number of mutations becoming established in a lineage for a sequence of 100 amino acids in a period of 100 crons varies from about 90 for fibrinopeptides to 0.06 for histones (McLaughlin and Dayhoff, 1969), with more typical values such as 12 for hemoglobin and 3 for cytochromes.

Little is yet known of the relation of these rates of protein evolution to those of DNA. Hoyer et al. (1965) have shown that there is a roughly logarithmic relation between the degree of nucleic acid pairing and the time to a common ancestor, with a drop in pairing by one power of 10 in every 300 crons. Studies on the relations between phenetic change and change in DNA include those of Sneath (1964d) and Laird, McConaughy, and McCarthy (1969). The existence of numerous single mutation differences between almost all proteins of closely related mammals poses questions for population biology, because there are several thousand different proteins in mammals. The replacement of such large numbers of alleles by selection implies very heavy evolutionary cost (see Kimura, 1968).

Character Complex Rates and Mosaic Evolution

Incongruence between numerical taxonomies based on different stages of the life cycle or different organs implies that there have been different rates of evolution in various character complexes. It is easy to imagine, for example, that the larva of an insect species might become adapted to a special habitat, giving rise to marked phenetic changes, while the adults of this and closely related species might remain in much the same habitat as before and retain high taxonomic similarity. When applied to different rates of evolution in different organ systems this phenomenon has been called mosaic evolution (de Beer, 1954). A good discussion of this problem in human evolution is given by Campbell (1964). If there is great difference in the evolutionary rates of different character complexes, this clearly raises problems in measuring overall evolution (organism rates, discussed next) analogous to the problems in making a taxonomy when incongruence is marked. Marcus (1969) has used Mahalanobis' D^2 to measure character complex evolution rates in teeth of the orangutan.

Organism Rates

When we come to a discussion of evolutionary rates in organisms, we enter a subject fraught with many pitfalls. In a simple, superficial sense it seems quite clear that different groups have evolved at different rates. For example, the statement that coelacanths have evolved more slowly than horses since the Eocene appears self-evident and is not likely to be challenged. However, when we attempt to analyze in detail the meaning of statements such as these, we run into considerable difficulty because we do not have any absolute scale of phenetic similarity. We think that we can make two definite statements. First, the absolute evolutionary rates of no group should be investigated until its characters and phenetic resemblances have first been investigated and evaluated with relation to the higher ranking groups to which they belong, and also to neighboring taxa. Second, a paleontologist, in comparing the rates of evolution of coelacanths and horses, has in the back of his mind an idea of the range of characters within vertebrates that have to his knowledge changed, and the kinds and degrees of change that have happened in all the vertebrate classes. He is evaluating the changes in the coelacanths and in horses against this unexpressed standard. A possible mode of evaluation may be the following: once a large group, such as the mammals, has been sufficiently studied by means of numerical taxonomy, a series of marker taxa may be chosen, with which evolutionary change can be compared. Thus a single representative of each order of mammals might be appropriately included in the matrix of resemblances and serve to furnish the proper scale for a group, such as the horses, for example. A fairly comprehensive standard set of taxa of this sort would comprise a reasonably stable standard of comparison.

An advantage of organism rates is that they are likely to be more steady than character rates, since bursts of rapid change in individual characters will tend to be smoothed out. Simpson (1944) has estimated organism rates (the "taxonomic rates" of Kurtén, 1959) by measuring the time for a phyletic lineage to change morphologically (phenetically) from one genus to another. From the data of Simpson (1944, 1961) and Kurtén (1959) and a consideration of the time of appearance of different taxa in vertebrate evolution, we may estimate that the time corresponding to change in rank in horotelic evolution in vertebrates is approximately as follows: morphospecies, 0.5 crons; genus, 7 crons; family, 20 crons; order, 45 crons; class, 80 crons. Myers (1960) discusses the rate of evolution of fishes after their introduction into lakes. With the exception of one lake in the Philippines (where very rapid evolution of several genera may have occurred in as little as 10,000 years), the usual pattern is the evolution of a few new species and subspecies after about 1/10 cron, many new species and some new genera after 1/2 cron, and many new genera and some new families after one to two crons. These rates are somewhat faster than those given above.

The rates appear to be much slower in some other phyla, such as insects (Crowson, 1958) and many lines of mollusks. In flowering plants rates have also been slower (Stebbins, 1950, pp. 529, 547 ff.) though there has been rapid evolution of some groups during the Pleistocene, with new species arising within one cron, including some arising in historic times. It should be made clear that by the phenetic change corresponding to a genus (for example) we mean the minimum phenetic difference between two forms that would just necessitate the placing of the two forms in different but closely similar genera instead of placing both in one genus (according to the criteria established by the investigator).

Early work on organism rates includes that of Westoll (1949) on lungfishes. Kurtén (1958) suggested that the percentage of significantly differing allometric growth gradients between two populations could be used as a measure of the organism rate. He calls this the "differentiation index." The index increases as a geometric series, with the limit value of 100; for example, the steps from 0 to 50, from 50 to 75, and from 75 to 87.5 are all equivalent. It runs parallel with taxonomic (phenetic) change but has the disadvantage of not taking into account the magnitude of the differences in the gradients (except so far as the magnitudes make the differences statistically significant). In most instances in mammals the rate of change was about 0.2 percent per thousand years, but periods of more rapid evolution also occurred. He found that the morphological difference between two subspecies was equivalent to an index of about 50 percent and that between two species was about 75 percent (Kurtén, 1959). It is clear from the context that Kurtén here uses subspecies to indicate a major morphological subdivision of a species rather than a trivial variant, and that by species he means a category approximating a morphospecies.

We should emphasize that all the above considerations are based on conventional judgments of taxonomic rank and are only as precise as these evaluations. Lacking better ones, we cite them to give a general indication of the nature of the problem. Numerical taxonomy will yield resemblance values between chronologically successive organisms, which can be used as measures of evolutionary rate. Over small ranges the change in the resemblance values compared with time will be satisfactory expressions of the rate. Over larger ranges this may be unsatisfactory inasmuch as the similarity values may be in a scale (such as the index of Kurtén, just described) in which dissimilarities are not additive, so that for three OTU's, **a**, **b**, and **c**, with **b** midway between **a** and **c** in the agreed phenetic space, $U_{ab} + U_{bc} \neq U_{ac}$. If dissimilarities are not additive (this includes cosines of angular coefficients) it will be necessary to make appropriate transformations. Alternatively one can divide the resemblance scale to define different ranks and express the evolutionary rate as the time taken for a phyletic line to pass through the degrees of similarity values applicable to these ranks, as Simpson suggested.

Several numerical taxonomic studies on organism rates have recently been published. Grewal (1962) examined the rate of phenetic divergence in inbred lines of mice based on numerous skeletal characters, using a resemblance measure that corrects for sampling error. Lerman (1965a,b) reports on several paleontological examples, using as a distance measure the square root of Mahalanobis' D^2. His evolution rate measure is D/T, where T is the time between taxa **J** and **K**.

Lerman used rather few characters, but there is some evidence that the evolution rates were different in different parts of the lineages that he studied (horses, oreodonts, and molluscs). There would seem no strong reason to prefer what is in effect discriminant analysis units over other phenetic measures, such as taxonomic distance, because the estimation of evolution rates is not primarily a problem of discrimination. It may be mentioned that the most appropriate measures require some further investigation. If evolution is predominantly by replacement of successive single mutations the total evolutionary pathway in A-space is represented in Manhattan metric rather than taxonomic distance, since the pathway consists of successive displacements on each character axis in turn and not on the diagonal joining the starting and ending positions. There is some support for this from protein evolution data, because the difference in two sequences is roughly proportional to time, and this difference is equivalent to the distance along successive sides of a hypercube, whereas the taxonomic distance is the square root of this (see Sections 4.4, 5.12).

Whether evolution rates are constant and divergent throughout all parts of a phylogeny has important consequences for cladistic analysis, which is taken up in the next two sections. The evidence on whether rates are constant and divergent may be briefly discussed here. It is clear that for morphological characters evolution rates have been far from uniform, but in the absence of numerical phenetic data little more can be said about this. There is however a good deal of numerical evidence from protein sequences that bears on the question. Various authors, as mentioned above, have concluded that for any given protein, evolution has usually been divergent and fairly constant in rate, although except for Kimura (1969) little attention has been paid to sampling error, which must be considerable, or to uncertainty in the paleontological dating of divergence between lineages, probably equally considerable. Kirsch (1969) points out that strictly constant evolution would rule out the possibility of biochemical "living fossils." One would be reluctant to deny their existence, although at present we have no way of demonstrating it. Exceptions to constant evolution rates are in fact now becoming evident in the insulins (Dayhoff and McLaughlin, 1969), for example, and mutations are far from random (Clarke, 1970). Similar considerations apply to serological evolution, which presumably reflects fairly closely the evolution of the proteins concerned; Sarich (1969a,b) and others (e.g., Read and Lestrel, 1970) consider that serological rates have been fairly constant, but Farris (1972) has shown very large heterogeneities in rates of albumin evolution in carnivores, on the basis of immunological evidence.

However, Kirsch (1969), Jardine, van Rijsbergen, and Jardine (1969), and Moore (1971), have made an important observation: if rates are constant and divergent, then the resemblances between extant organisms will be an ultrametric; by the usual methods of clustering the cophenetic correlation will be 1, and single linkage and complete linkage (as well as most forms of average linkage) will yield identical phenograms. Kirsch points out this is demonstrably not so for most matrices of serological resemblance although the evidence from serology may be questioned on the basis of inappropriate resemblance measures or techniques, or lack of correspondence between serological reactivity and protein differences.

Nor has evidence from protein sequences about constancy and divergency of evolution been carefully assessed either. The matrices of differences between organisms in Dayhoff (1969a) show, for many proteins, the long rows and columns of almost constant values that would be expected from an ultrametric with superimposed sampling error. J. S. Farris (personal communication) has recently shown that such seemingly ultrametric structure is to be expected in protein sequence data, whether the rates of evolution in the phyletic lines are really homogeneous or not. This is so because sites differing between modern taxa **A** and **B** are likely also to become modified in the evolutionary path connecting **X**, the common ancestor of **A** and **B**, to a third taxon, **C**. If **C** is quite distant from **X**, then the number of amino acid sequence differences between taxa **A** and **C** and taxa **B** and **C** tend to approximate equality—even though these differences are substantial for taxa **A** and **X** and **B** and **X**. It is of critical importance to investigate this carefully, including the consideration of alternatives like local constancy of rates (Jardine et al., 1969). The squared discrepancies between the resemblance matrix and the UPGMA cophenetic matrix could be compared with the sampling variances, because UPGMA clustering might be expected on theoretical grounds to give the closest approximation to the correct cladogeny (see Moore, 1971). Kirsch (1969) has made some simulation studies along these lines. If a hypothesis of constant divergence can be strongly supported it would open the way for attack on many intractable cladistic problems.

Colless (1970) has pointed out that less stringent conditions than constant and divergent evolution may still permit reconstruction of cladogenies; phenograms may be quite good estimators of cladistics, provided that (1) later stocks do not evolve very much faster than earlier ones for appreciable periods of time, (2) sister species do not divide again before they have diverged appreciably, and (3) later lines do not converge upon each other less than upon earlier lines. Colless gives some reasons for preferring average linkage cluster methods for constructing the phenograms, while Moore (1971) has constructed a set of assumptions under which UPGMA clustering yields optimal estimates of cladograms.

6.3 CLADISTIC ANALYSIS

The problems of making inferences about cladistic relationships from data on recent organisms (even when fossil evidence is available) have been detailed in Sections 2.5 and 2.6. In spite of the manifold difficulties discussed there, considerable progress has been made in recent years in developing techniques of numerical cladistics. In this section we shall discuss in some detail the types of evidence on which phylogenetic work has largely rested as well as the philosophical bases of cladistic inferences.

The sources of data for cladistic inferences are invariably phenetic. This is obvious when cladistic analysis on morphological data follows the methods of Hennig (1966) or Maslin (1952). Even phylogenetic inferences from fossil material are based on phenetics; Colless (1967b) and Rowell (1970) have pointed this out. In certain special instances the data are not immediately recognized as being phenetic. These include cytogenetic data, especially those on chromosome inversion sequences in some species of diptera, which are believed by some (Stalker, 1966) to yield incontrovertible evidence of branching patterns. Yet when the reasoning about inversion sequences is examined carefully it is found that a linear order of inversions must be postulated before any cladistic inferences can be made. This linear order is based on the (phenetic) homology of bands in the chromosomes, on postulates about the likelihood of inversions (simultaneous breaks in the chromosomes with concomitant inversion of a middle segment), and on the relative improbability of such an event happening in the identical chromosome segments two or more times in any one lineage. The phenetic evidence must be employed together with the postulations of linear order and the likelihood of inversions before cladistic inferences can be made. When such information is subjected to a numerical cladistic procedure (that of Camin and Sokal, 1965 in this instance) in which the cytogenetic characteristics have been recoded in a linear ordering following the assumptions about their order of origin, the resulting cladogenies match those established by cytogenetic methods (unpublished work by John Hendrickson, Jr. on data by

Stalker, 1966) or are more parsimonious than previously published evolutionary trees (analyses of blackflies; unpublished information by R. Hansell about studies by Rothfels, 1956, and Basrur, 1959, 1962). The trees postulated by Rothfels and Basrur differ from the numerically obtained cladograms by allowing hypothetical intermediate forms in which alternative banding patterns were permitted to co-exist. This is a model similar to one developed in a cladistic computer program by J. H. Felsenstein.

Genetic information is often thought to be a sounder way of establishing cladistic relationships than phenetic characteristics. However, we are rarely in a position to establish genetic relationships among representatives of fairly divergent taxa (for exceptions see the work on protein structure and on DNA pairing discussed in the next paragraph). In genetic relationships of low ranked taxa the problems become more those of population structure and species definition than of cladogeny in the strict sense, although the general line of argument would still hold. However, even evidence on crossability is a phenetic character as will be obvious on some reflection. Morishima (1969b) has employed interstrain cross-ability of rice in just such a manner.

Biochemical evidence is often thought to be a "true indicator" of cladistic relationship. Yet the two major criteria, namely mutation distances found in comparing two sets of proteins and the degree of DNA pairing are only different types of estimates of phenetic resemblance between the genomes or their products. We have already discussed DNA pairing in Section 5.12 but shall deal in some more detail with matching and analysis of protein sequences in this and the next section. A difference in an amino acid at a given site in a protein is clearly a phenetic difference whether expressed in number of mismatches of the amino acids or in the inferred number of mutational steps necessary to bring about a change in the nucleotides coding for the given amino acid. Fitch and Margoliash (1968) recognize that the trees they reconstruct are phenograms because differences between amino acid sequences in proteins are no more or no less phenetic than conventional morphological characters.

We have already seen in Section 2.5 that an ordering of character states into an evolutionary sequence is at best a tricky procedure and at worst can be grossly misleading. Any attempt to reach decisions about the order of the evolutionary sequences in character states leads inevitably to a statistico-phenetic taxonomy (Colless, 1968a) and some of the protein chemists have more frankly relied upon what they have descriptively termed decision by majority vote (Dayhoff and Eck, 1969).

Direct phenetic evidence is not the only kind used for making decisions on cladistic relationships. Traditionally other evidence such as the recapitulation hypothesis is employed to provide ancillary support. Biogeographic evidence is frequently used at the species level and for higher ranked taxa. There is little doubt that such evidence can lend support to a given cladistic hypothesis, or weaken it, as the case may be. However, it is unlikely to serve as primary evidence. Some function needs to be constructed for incorporating ancillary evidence of various types into the decision algorithm for arriving at cladistic sequences. No attempts at such an incorporation of ancillary evidence has yet been made. It would involve a procedure for weighting these ancillary characters in relation to orthodox characters and would yield modified cladograms.

Several principles are either explicit or implicit in the operations and reasoning of cladistic systematists. Those who would largely use phenetic similarity as evidence for recency of cladistic ancestry must assume at least some uniformity of evolutionary rates in the several clades, although Colless (1970) has shown that the requirements need not be as stringent as is commonly thought (see Section 6.2). We stress here uniform rather than constant rates of evolution. As long as rates of evolution change equally in parallel lines it is unimportant whether these rates are constant through an evolutionary epoch. We may use the analogy of multiple clusters of fireworks, smaller clusters bursting from inside large clusters, a familiar sight to most readers. While the small rays of the rocket have not "evolved" at all until the small rocket exploded, their rates of divergence from the center of their rocket are identical but not constant, since they were zero during the period of the early ascent of the rocket.

Clearly all the evidence at hand indicates evolutionary rates in different clades are not uniform. Different lines do evolve at different rates. Taking this fact into account Farris (1966) has argued that present intrataxon variation is inversely related to conservatism of the character in the taxon. Kluge and Farris (1969) have used this assumption for weighting characters in their method of constructing Wagner trees (see Section 6.4). A. G. Kluge (personal communication) has recently obtained evidence from a variety of animal groups that tends to bear out the correlation between past evolutionary rate and present character variation within a taxon. This hypothesis is a modern version of the doctrine of uniformitarianism that, if it can be verified, would be of great importance to evolutionary theory. Some evidence tends to contradict it. Thus Selander et al. (1970) find that *Limulus*, a "living fossil," is as variable as other organisms. In any case, for such a hypothesis to be operationally useful one must be able to define taxa of the desired rank in order to measure their intrataxon variability.

A principle that has almost uniformly guided evolutionists in devising the most likely evolutionary trees has been that of minimum evolution. This can be construed as the minimum number of evolutionary steps (Camin and Sokal, 1965), the minimum number of mutational steps (Fitch and Margoliash, 1967), or a minimum length tree (Edwards and Cavalli-Sforza, 1964; Kluge and Farris, 1969). It is not easy to justify minimum length evolutionary trees in other ways than through the generally accepted principle of parsimony. However, Cavalli-Sforza and Edwards (1967) feel that such a principle cannot be accepted as an article of faith but must be justified on independent evidence. It is this type of evidence that by the nature of the problem is so hard to come by.

One must also decide whether minimum length trees should be rooted and directed. If so, one must postulate the nature of the ancestor and also whether the characters should be ordered from evolutionarily primitive to advanced and whether reversal of states should be permitted. Another implicit assumption that is often made, resting in part on the principle of parsimony, is that the common ancestor most probably possessed character states close to the central values of the states in present day organisms (i.e., it was similar to the centroid of extant OTU's). Certain other related concepts were discussed in our earlier volume (Sokal and Sneath, 1963, p. 230 ff.). On none of these issues have we found solid evidence either at the morphological or the molecular level, and some of the difficulties in the operational procedures for numerical cladistics described in the next section stem from this fundamental difficulty.

One aspect of cladistic analysis that is well known to evolutionary taxonomists, but whose implications are often neglected, is that ancestors are unknown: almost all of the work deals with recent organisms. Even if some recent organisms are shown to be close to the common ancestors of a set of OTU's, it is unlikely that the organisms living at present actually are unmodified descendants of these common ancestors. Some uncritical statements are frequently made for this reason. If cladogenies are deduced from present day organisms only, so that these are the only OTU's, then all earlier forms must be to some extent hypothetical and none of the OTU's should be shown as an ancestor of any other OTU. The caution must extend to fossil material. It is rare that a known fossil can be assumed to be a direct ancestor of a later fossil or extant form, though it may be phenetically close to the ancestor. More likely it is a form more similar to the common ancestor of both forms than is the extant form. Yet the construction of cladogenies of necessity involves branching points that are assumed to be the last common ancestors of the diverging lines. Whenever such cladogenies are based on suites of characters there is a temptation to reconstruct a description of the common ancestor from the postulated states for the entire suite of characters. Farris (1970) has called such postulated ancestors HTU's, for *hypothetical taxonomic units.* We must realize that the observed OTU's are points in a hyperspace that we are trying to connect in some defined optimal manner. The lines connecting these points are hypothesized as are most of the branching points (HTU's). But once a cladogram is constructed, the lines are very apt to assume some sort of reality in the minds of the beholder and so do the HTU's. One should always be on guard against this attitude.

A second problem is what lines or points should be studied. Should these represent individuals? The ancestry of a furcating line comprising many individuals gives a very complex network of reticulations and furcations, as Hennig (1966) has correctly pointed out. Do we therefore wish to make statements about some average of the mass of individuals when this mass has undergone a splitting process? Although it may sometimes be naively assumed that changes in sets of characters take place simultaneously in the internodes between the OTU's in a cladogram, this is in fact unlikely. Any given mutation may arise several times, dying out each time but eventually becoming established, or different mutations may occur at different times in any one line and thus heterochronous divergence or heterochronous parallelism may result. We have reasonably good evidence on these phenomena from studies of amino acid sequences in proteins. It is also possible that the cladistic separation at the species level occurred prior to the phenetic separation, as may well be the case in two sibling species that could be isolated by a sterility barrier before substantial phenetic differentiation.

6.4 NUMERICAL APPROACHES TO CLADISTIC ANALYSIS

General Considerations

In the published work on various numerical approaches to cladistic analysis there seem to be four main schools. Edwards and Cavalli-Sforza (1964) measure differences among OTU's in terms of gene frequencies at one or more loci expressed as angles. Distances between OTU's thus are continuous variables. These authors construct what seem to them plausible models of the evolutionary process and then suggest a variety of statistical procedures for fitting evolutionary trees to the data, given the models. Frequently such solutions are approximated by the construction of shortest length trees from these dissimilarity matrices. Camin and Sokal (1965) worked with discrete characters and attempted to find cladograms requiring the minimum number of evolutionary steps to represent the character state vectors of each OTU. Farris and coworkers in an extensive series of papers discussed at length below have used Manhattan distances between pairs of OTU's based on both continuous and discrete characters to form minimum length graphs and directed trees that estimate evolutionary sequences. Finally, investigators estimating protein phylogenies, for example, Fitch and Margoliash (1967) or Dayhoff (1969a,b), have computed directed trees in which the dissimilarity between amino acid sequences of proteins is expressed in nucleotide minimal mutation distances. Although the approaches of these schools differ, their results are frequently similar. They yield or approximate minimum length nondirected or directed trees based on a variety of similarity coefficients between OTU's. Some methods use similarity coefficients based on all characters, others weight characters by lowering the weight of those presumed to be convergent to obtain a better estimate of patristic similarity, because the latter should be more likely to reflect cladistic relationships than would overall phenetic similarity. We shall discuss such weighting methods later in this section.

Most similarity coefficients and clustering algorithms employed in numerical cladistics are also employed in numerical phenetics. The important distinction between phenetic and cladistic analysis lies not in the similarity coefficients or clustering algorithms, therefore, but in the assumptions underlying their use in numerical cladistics and in the conclusions drawn from the results of the study.

In cladistics we encounter *minimally connected graphs* or *trees*. Three examples are shown in Figure 6-4. A new feature is that not all of the nodes (vertices) of the graphs are OTU's. Some will be postulated common ancestors (hypothetical taxonomic units, HTU's). Thus in Figure 6-4,*a*, we would interpret the nodes **a**, **c**, **f**, and **g** as OTU's (presumably extant organisms), while the nodes **b**, **e**, and **d** are HTU's (postulated ancestors; see Section 6.3). This situation differs from the phenetic graphs considered in Section 5.7 (Figure 5-14) where all nodes are (Recent) OTU's. As pointed out by Farris (1970) these graphs can be defined by a *connection function* $g(j)$, where $g(j)$ defines the node (OTU or HTU) connected to node **j** in some consistent manner. Thus in each of the three graphs in Figure 6-4, $g(a) = b$, $g(b) = d$, $g(c) = b$, $g(d) = e$, $g(e) = g$, $g(f) = e$. The trees all contain a unique base element **g**, reached by repeatedly applying connection function $g(j)$ from any node. Function $g(g)$ is undefined. Connection functions do not imply directionality. This is shown by the nondirected graph in Figure 6-4,*c*, Although the three graphs in Figure 6-4 give the appearance of being different, they represent topologically equivalent relationships. The trees can therefore be defined as ordered pairs $W = (N, g)$ where N is a collection of nodes (OTU's or HTU's) in the graph and g is a function as defined above.

In a cladogram, directionality is added to yield a *minimally connected directed graph* or *directed tree*, and one of its nodes must be postulated to be the root. In the first two graphs in Figure 6-4 nodes **d** and **e**, respectively, are assumed to be the roots. A rooted connection function $f(j)$ is called an *ancestor function* by Farris (1970), where $f(j)$ defines the immediate ancestor of node **j**. In Figures 6-4,*a* and 6-4,*b* the directionality has been indicated by arrows. Repeated application of the ancestor function starting from any point will terminate at the root of the tree, which in a cladogram is the common ancestor. Thus the cladogram of Figure 6-4,*b* is rooted in the common ancestor **e**. Such a directed tree is called an *evolutionary tree* by Farris (1970) and defined as an ordered pair $T = (N, f)$ where N is a collection of nodes (OTU's or HTU's) on the tree, and f is the ancestor function defined above. Estabrook (1968) has called $T = (N, f)$ an *evolutionary hypothesis*. Readers wishing to consult Farris (1970) will be helped by knowing that he uses "simply connected network" for tree, and "tree" for directed tree. We have retained below his terms "Wagner network" and "Wagner tree" with the meanings coined by him to avoid terminological confusion.

For any sizeable number of Recent OTU's a large number of combinations for any one topology and a large number of different topologies can be formed. Cavalli-Sforza and Edwards (1967) report that $(2t - 3)!/[2^{t-2}(t - 2)!]$ different rooted trees may be recognized for t OTU's. When $t = 10$, this equals 34,459,425 trees. They also state that $(2t - 5)!/[2^{t-3}(t - 3)!]$ different nondirected trees can be found; this equals 2,027,025 when $t = 10$. Another way of looking at these

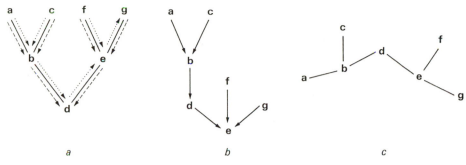

FIGURE 6-4

Three topologically equivalent trees (minimally connected graphs). Graphs in *a* and *b* are also directed trees (evolutionary trees) with roots at **d** and **e**, respectively. The direction of these trees is given by the ancestor function and is indicated by dashed arrows in *a*, and by the arrows in *b*. The connection function is shown in *a* by dotted arrows. For further explanation, see text. [Modified from Farris (1970).]

formulae is the observation by Fitch and Margoliash (1968) that the number of rooted trees for t OTU's equals the number of unrooted trees for $t + 1$ OTU's. The number of topologies (tree forms irrespective of which OTU's are placed on the terminal branches) is considerably less than the above. Cavalli-Sforza and Edwards computed 98 ways in which ten OTU's can be connected into a tree. The above computations limit the connections of one node to maximally three others. If more than three nodes can be connected to another one, the number of possible trees becomes even larger. Yet another complicating factor is the possibility of reticulate evolution, discussed at the end of this section. This greatly increases the number of possible configurations. Reversibility of evolution, discussed in the previous section, and the large number of possible tree forms are two of the fundamental problems of numerical cladistics. How can one give direction to evolutionary change as reflected in character state differences so that rooted trees rather than unrooted trees (simply connected graphs) are produced? And among the very large number of possible trees how is one to choose the most plausible one? The various techniques of numerical cladistics described below have made different approaches to a solution.

Trees

These nondirected graphs seem the most conservative representation of cladistic relationships. By setting up a minimum length graph among OTU's we are stating that the transition from one to the other most likely occurred along evolutionary pathways indicated by the edges or internodes of the tree. By leaving the relationship as a tree rather than as a cladogram (directed tree) the investigator is able to assign experimentally one node or the other as an ancestor and to make judgments about the reasonableness of the resulting cladogenies. The network among OTU's will include additional nodes that help to minimize the overall length of the graph. The additional nodes will rarely if ever be extant organisms, and will normally be either fossils or HTU's. Such graphs are known in graph theory as *Steiner minimal trees* (Gilbert and Pollak, 1968). They are minimal length trees for t vertices (OTU's, terminal nodes). In order to achieve minimum length they may contain other vertices (HTU's, nodes, or in graph theoretical terminology —Steiner points) in addition to the t vertices.

The earliest numerical method for designing an evolutionary nondirected tree is that of Edwards and Cavalli-Sforza (1964), which was elaborated in several subsequent papers of which Cavalli-Sforza and Edwards (1967) is representative. The methods developed by these authors are applicable to continuous characters, but are primarily applied to gene frequencies at several loci. Edwards and Cavalli-Sforza prefer gene frequencies to other continuous characters such as morphometric ones, because they feel that they understand the underlying genetical mechanisms of these characters and therefore can make statements about the likelihood of selection, mutation, and random drift, which they could not do for morphometric characters. The gene frequencies are transformed by the cosine transformation (see Section 4.13).

Cavalli-Sforza and Edwards (1967) consider evolution as a branching process with recent OTU's in a "now" plane or hyperplane determined by the gene frequencies, with a normal to this hyperplane representing time. Branching points represent ancestral OTU's that are rarely known and are in fact HTU's. The projection of the tree on the hyperplane (the graph) is considered an estimate of the evolutionary topology. Although these authors considered various evolutionary models, they have mainly dealt with evolution as a branching Brownian-motion process at a constant rate for all characters at all times; populations do not become extinct but bifurcate at random time intervals.

Cavalli-Sforza and Edwards (1967) have worked on three basic approaches to a solution of the estimation problem. For uniform Brownian motion as applied to a Yule process they suggest the method of maximum likelihood. They treat this method in considerable mathematical detail but in fact are able to solve it only for the simplest cases, likely to be trivial in any actual evolutionary study. The reason for the difficulty is that the log-likelihood surfaces contain numerous singularities that lead to difficulties of estimation. In a recent paper Edwards (1970) showed that a maximum-likelihood solution can be found in principle but that considerable computational difficulties still stand in the way of an actual numerical solution. Because of these difficulties Cavalli-Sforza and Edwards have resorted to "intuitive approaches" that effectively are "minimum evolution" techniques. They wish to obtain a tree with minimal length on the phenetic hyperplane by introducing suitable branching points, thus giving a Steiner minimal tree in Euclidean space. To this end they suggest obtaining a shortest spanning tree based on the distances between the OTU's and HTU's, but this requires finding the optimal positions for nodes (HTU's). Cavalli-Sforza and Edwards (1967) stress that the minimum-evolution approach is successful not because evolution actually proceeds parsimoniously but that the minimum-length tree solution is probably close to that of the projection of the maximum-likelihood tree on the "now" plane. Since the technique of computing a shortest spanning tree among a set of OTU's has already been discussed in Section 5.7, there is no need to deal further with it here. A third method by Cavalli-Sforza and Edwards is called the additive tree model, which is a least squares fit to a plausible tree structure. They believe that its justification is similar to that of the minimum evolution method. They also point out that hybridization, convergence, or parallelism cannot be handled by their model.

Applications of these methods have been made to human blood groups in 15 populations (Edwards and Cavalli-Sforza, 1964) and in four populations (Cavalli-Sforza and Edwards, 1967) from different racial stocks. Similarly, Goodman et al. (1971) describe a method that minimizes the difference between patristic differences and mutational differences in protein sequences.

Although various authors have used minimum length trees to generate evolutionary hypotheses, we owe a formal treatment of the subject to Farris (1970). In addition to the t terminal nodes (OTU's) the Steiner minimal tree will have a variable number of $t^* - t$ additional nodes (HTU's or Steiner points), where $t^* \geq t$, yielding a total of t^* nodes for the entire tree. He defines the length of the internode between two nodes $\mathbf{j}$ and $\mathbf{k}$ as the Manhattan distance

$$d_1(\mathbf{j}, \mathbf{k}) = \sum_{i=1}^{n} |X_{ij} - X_{ik}|$$

between the two nodes (see Section 4.3). The character states may be any real numbers and thus include continuous characters as well as cases where character states can assume only integral numbers, the "evolutionary steps" of Camin and Sokal (1965). The length of a tree θ with t^* nodes is defined as

$$L(\theta) = \sum_{\mathbf{j} \neq \mathbf{q}} d_1(\mathbf{j}, g(\mathbf{j})),$$

which is the summation of the $t^* - 1$ internode lengths $d_1(\mathbf{j}, g(\mathbf{j}))$ within the tree. There is one less internode than the number of nodes t^*, since $d_1(\mathbf{q}, g(\mathbf{q}))$ is undefined for the unique base element $\mathbf{q}$. Shortest spanning trees can be computed, minimizing the length of the network defined as above. An efficient algorithm and a FORTRAN IV program for their computation is furnished by Farris (1970).

A *Wagner network* as defined by Farris (1970) is a Steiner minimal tree, measured in (multidimensional) Manhattan distance. Such trees in two dimensions were considered by Hanan (1966). Wagner networks also permit reversibility of character states. Thus if the character states for a given character i for two nodes are

$X_{ij} = 2$ and $X_{ik} = 3$, respectively, the distance $d_1(\mathbf{j, k}) = |X_{ij} - X_{ik}| = 1$ may imply the change $2 \rightarrow 3$, as well as $3 \rightarrow 2$ within the same network. Wagner trees are based on concepts developed by W. H. Wagner (e.g., Wagner 1963, 1969) and illustrated by work such as that of Mickel (1962), Lellinger (1964), and Kesling and Sigler (1969). Farris shows that the character state $X_{i\mu}$ for any character i of HTU $\boldsymbol{\mu}$ connecting a triad of OTU's $\mathbf{a}$, $\mathbf{b}$, and $\mathbf{j}$ (see Figure 6-5) is the median value of the three character states X_{ia}, X_{ib}, and X_{ij}, if the sum of the Manhattan distance lengths of the internodes $d_1(\mathbf{a}, \boldsymbol{\mu})$, $d_1(\mathbf{b}, \boldsymbol{\mu})$, and $d_1(\mathbf{j}, \boldsymbol{\mu})$ is to be minimized. In this manner the HTU for these OTU's can be constructed as the vector of median states for all characters of the three connected OTU's.

An algorithm for constructing Wagner networks is given by Farris (1970) as follows. We find the pair of OTU's $\mathbf{a}$ and $\mathbf{b}$ whose distance $d_1(\mathbf{a, b})$ is greater than that between any other pair of OTU's in the study. We compute the distance of all other OTU's $\mathbf{j}$ with the internode $(\mathbf{a, b})$. This is done by computing the Manhattan distance $d_1(\mathbf{j}, \boldsymbol{\mu})$ between each OTU $\mathbf{j} \neq \mathbf{a}$ or $\mathbf{b}$ and the median character state vector $\boldsymbol{\mu}$ for each triad $\mathbf{a, b, j}$. To simplify this process, Farris suggests computing the distance between $\mathbf{j}$ and a point representing the internode $(\mathbf{a, b})$. This point is called an *interval* (symbolized by INT). This is given as

$$d_1(\mathbf{j}, \text{INT}(\mathbf{a, b})) = \tfrac{1}{2}[d_1(\mathbf{a, j}) + d_1(\mathbf{b, j}) - d_1(\mathbf{a, b})]$$

Farris has shown that $d_1(\mathbf{j}, \text{INT}(\mathbf{a, b})) = d_1(\mathbf{j}, \boldsymbol{\mu})$. The OTU, say $\mathbf{c}$, with the greatest distance $d_1(\mathbf{c}, \boldsymbol{\mu})$ is chosen and an HTU $\boldsymbol{\mu}_1$ is constructed as the median of OTU's $\mathbf{a}$, $\mathbf{b}$, and $\mathbf{c}$ (see Figure 6-6,a). The connection function for the network is defined as $g(\mathbf{a}) = \boldsymbol{\mu}_1$, $g(\mathbf{b}) = \boldsymbol{\mu}_1$, $g(\mathbf{c}) = \boldsymbol{\mu}_1$ and one proceeds to look for the unplaced OTU, say $\mathbf{e}$, with the next greatest distance $d_1(\mathbf{e}, \boldsymbol{\mu})$ to internode $(\mathbf{a, b})$. We now find the "interval" on the network constructed so far that is closest to the new OTU $\mathbf{e}$. As before, we construct an HTU $(\boldsymbol{\mu}_2)$ to represent this "interval" and connect the triad through $\boldsymbol{\mu}_2$. For example, if $d_1(\mathbf{e}, \text{INT}(\mathbf{a}, \boldsymbol{\mu}_1))$ had been less than either $d_1(\mathbf{e}, \text{INT}(\mathbf{b}, \boldsymbol{\mu}_1))$ or $d_1(\mathbf{e}, \text{INT}(\mathbf{c}, \boldsymbol{\mu}_1))$, we would have constructed the internodes $(\mathbf{a}, \boldsymbol{\mu}_2)$,

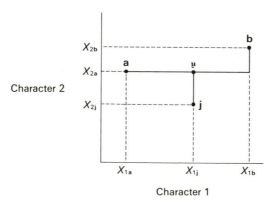

FIGURE 6-5

The calculation of character states of an HTU. The three OTU's are $\mathbf{a}$, $\mathbf{b}$, and $\mathbf{j}$, with character state values on characters 1 and 2. The HTU $\boldsymbol{\mu}$ is given the median values of the states of $\mathbf{a}$, $\mathbf{b}$, and $\mathbf{j}$ for each character, and the total length of the three internodes (solid lines), is then a minimum. For further details, see text.

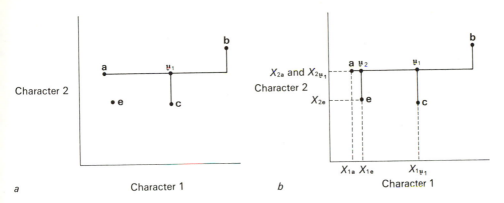

FIGURE 6-6
Construction of HTU's in making a Wagner network, as explained in the text. *a*, Construction of an HTU, μ_1 from OTU's **a**, **b**, and **c**. A fourth OTU, **e**, remains unconnected at this stage. *b*, Addition of OTU **e** to the network, with the construction of a new HTU, μ_2. This HTU is constructed as the median of the states for **e**, **a**, and μ_1 since μ_2 is closer to **e** (using the Manhattan metric) than the alternative new nodes for the triads **e**, **b**, μ_1, or **e**, **c**, μ_1. These alternative new nodes would be situated at the same point as μ_1 and just above **c** respectively (if the first had been required it would not have received a new symbol).

(μ_1, μ_2), and $(\mathbf{e}, \mu_2)$, with new connection functions defined as $g(\mathbf{a}) = \mu_2$, $g(\mu_1) = \mu_2$, and $g(\mathbf{e}) = \mu_2$ (see Figure 6-6,*b*). Functions $g(\mathbf{b}) = \mu_1$ and $g(\mathbf{c}) = \mu_1$, as before. The algorithm continues until all OTU's have been connected to the network. Another version of the algorithm searches for the OTU with the greatest distance to any of the available internodes. After the first cycle these will be internodes $(\mathbf{a}, \mu_1)$, $(\mathbf{b}, \mu_1)$, and $(\mathbf{c}, \mu_1)$. The new OTU is then linked to that internode with which it has the smallest Manhattan distance, again using the combinatorial formula given above. A new hypothetical taxonomic unit μ_2 is constructed intermediate between the two OTU's bounding the chosen internode and the most recent candidate for joining the network. The process continues until all candidate OTU's have been attached to the network by means of HTU's.

To illustrate the procedure for finding a Wagner network, we apply the first of the two algorithms proposed by Farris (1970) to the data matrix illustrated in Table 6-1. These are a subset of the Caminalcules (Camin and Sokal, 1965) mentioned in several places in this book. For reasons that will become apparent later we shall delete character (row) *4* from the data matrix for the computations to follow and base the Wagner network to be constructed on characters *1* through *3*, and *5* through *7*. Table 6-2 gives the successive steps necessary to obtain the Wagner network and in Figure 6-7 a graphic representation can be found. The length of the resulting network, $\Sigma d_1(\mathbf{j}, \mathbf{k}) = 15$, is minimal, since the sum of the absolute values of the ranges of the characters (the last column of Table 6-1) is indeed 15.

Necessarily, however, distances along the network between pairs of OTU's may be longer than in the initial distance matrix based on their character states. Thus, although $d_1(7,25) = 4$, the length of the path along the network between these OTU's adds up to 6.

TABLE 6-1

Data matrix for seven selected Caminalcules (Group A).
[From Camin and Sokal, 1965.]

Characters			t OTU's					Number of character states, m_i	Range or minimum number of steps $m_i - 1$
	7	8	13	14	15	25	28		
1	1	0	1	0	1	1	1	2	1
2	1	1	0	0	0	2	0	3	2
3	1	1	1	2	0	2	2	3	2
4	3	2	0	0	3	1	0	4	3
5	1	1	2	1	0	1	3	4	3
6	1	0	0	0	0	-1	0	3	2
7	0	0	0	0	-1	0	1	3	2

$$\sum_{i=1}^{n} (m_i - 1) = 15$$

TABLE 6-2

Finding a Wagner network

A. Manhattan distance matrix based on data matrix for Group A of the Caminalcules (Table 6-1) with character 4 omitted. The HTU's, numbered $\mu_1, \ldots, \mu_5$, are constructed as explained below.

OTU's and HTU's				OTU's and HTU's									
	7	8	13	14	15	25	28	μ_1	μ_2	μ_3	μ_4	μ_5	
7	×												
8	2	×											
13	3	3	×										
14	4	2	3	×									
15	5	5	4	5	×								
25	4	4	5	4	7	×							
28	6	6	3	4	7	6	×						
μ_1	3	3	2	1	4	3	3	×					
μ_2	2	2	1	2	3	4	4	1	×				
μ_3	1	1	2	3	4	3	5	2	1	×			
μ_4	2	2	1	2	3	4	4	1	0	1	×		
μ_5	3	3	2	1	4	3	3	0	1	2	1	×	

TABLE 6-2—(*continued*)

B. Matrix of character states for HTU's.

Characters	μ_1	μ_2	μ_3	μ_4	μ_5
1	1	1	1	1	1
2	0	0	1	0	0
3	2	1	1	1	2
4	×	×	×	×	×
5	1	1	1	1	1
6	0	0	0	0	0
7	0	0	0	0	0

(Header: Characters | HTU's)

Step 1.
Take $d_1(\mathbf{15,25}) = 7$. This is greater than other distances except for $d_1(\mathbf{15,28})$, which equals it. In this instance and all others when decisions among criteria of equal magnitude must be made, the algorithm arbitrarily chooses the distance involving an OTU or HTU lowest in numerical order. The first internode is drawn. See Figure 6-7,*a*.

Step 2.
Compute the distance $d_1(\mathbf{j},\boldsymbol{\mu})$ for all OTU's **j** other than **15** and **25** to the HTU μ_j representing the interval INT(**15,25**) for the given **j**. We employ the formula given in the text. For example, $d_1(\mathbf{7},\text{INT}(\mathbf{15,25})) = \frac{1}{2}[d_1(\mathbf{7,15}) + d_1(\mathbf{7,25}) - d_1(\mathbf{15,25})] = \frac{1}{2}[5 + 4 - 7] = 1$. Choose the highest $d_1(\mathbf{j},\boldsymbol{\mu})$, which is $d_1(\mathbf{28},\text{INT}(\mathbf{15,25})) = 3$.

Step 3.
HTU μ_1 is constructed between OTU's **15**, **25**, and **28**. The character states of this vector are the median states for the OTU's. Thus for character *1* the respective states are 1, 1, 1; the median is clearly 1. For character 2 the states are 0, 2, 0; the median state is 0. For character 5 the states are 0, 1, 3; the median state is 1. In this manner we obtain the character state vector of μ_1 as $\{1, 0, 2, 1, 0, 0\}$. The HTU μ_1 is drawn in to connect the three OTU's **15**, **25**, and **28** in Figure 6-7,*b*.

Step 4.
Choose the next highest $d_1(\mathbf{j},\boldsymbol{\mu})$ to the internode (**15,25**). Since all remaining OTU's, **7**, **8**, **13**, and **14**, have the same $d_1(\mathbf{j},\boldsymbol{\mu})$ we employ the arbitrary rule stated above and take the distance in order of OTU number. Therefore **7** is chosen and we compute $d_1(\mathbf{7},\text{INT}(\mathbf{j},\mu_1))$ for all existing internodes (that is, **j** = **15**, **25**, or **28**). We choose the internode for which $d_1(\mathbf{7},\text{INT}(\mathbf{j},\mu_1))$ is least. This is $d_1(\mathbf{7},\text{INT}(\mathbf{15},\mu_1)) = 2$.

Step 5.
HTU μ_2 is constructed between OTU's **7** and **15** and HTU μ_1, in the same manner as in Step 3 above. The new character state vector for μ_2 is $\{1, 0, 1, 1, 0, 0\}$. The new HTU is drawn in to connect **7** to the network. See Figure 6-7*c*.

Step 6.
These procedures are continued, with OTU **8** joining internode (**7**,μ_2) via HTU μ_3, OTU **13** joining internode (**15**,μ_2) via HTU μ_4 and OTU **14** joining internode (**25**,μ_1) via HTU μ_5. When all OTU's are connected the network looks like Figure 6-7,*d*.

Step 7.
Since two pairs of HTU's, μ_1 and μ_5, and μ_2 and μ_4, have zero distance between them they are deleted from the network that can then be represented in its simplest form as in Figure 6-7,*e*. The sum of the lengths of the internodes is 15.

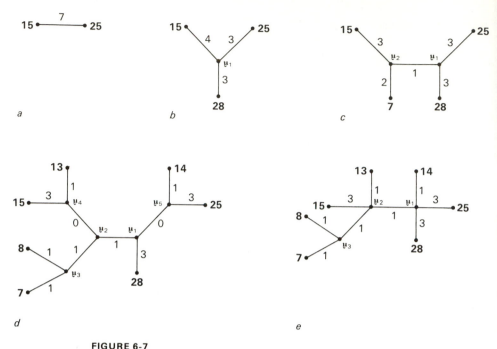

FIGURE 6-7

Construction of a Wagner network for Group A of the Caminalcules. OTU's are numbered, HTU's are subscripted μ. The various steps are explained in detail in Table 6-2.

In other data sets the HTU's produced during the successive clustering of OTU's to form the network are not likely to be optimal ones in the sense of minimizing the length of the entire network. There will be local minima but the tree may not be of minimum overall length. An iterative procedure for optimization can then be applied (Farris, 1970). When the internodes are arbitrarily assigned directions, the method is equivalent to the optimization procedure described below for Wagner trees. A worked application of Wagner networks in paleontology is given by Kesling and Sigler (1969), who also used character weighting similar to p_i of Farris discussed later in this section. It is important to repeat that this method permits the reversal of character states.

Cladograms

Cladograms are evolutionary reconstructions in the form of rooted directed trees. Although Camin and Sokal (1965), in their numerical cladistic method, dealt only with discrete characters, Farris (1970) generalized this to characters described by any real number. Camin and Sokal made the following four assumptions about their data sets. (1) Characters can be expressed in discrete states differing among at least some of the OTU's of the study. (2) The character states for any one character can be arrayed in evolutionary order; thus any one character yields a character state tree (see Figure 6-8). (3) The most ancestral state in any study arose only once within the collection of OTU's under study. (These primitive character states, conventionally coded zero, are the unique character states of Wilson, 1965). (4) Evolution is irreversible. Thus a descendent character state cannot revert to an ancestral character state.

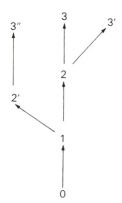

FIGURE 6-8

A character state tree. In this character the primitive state is 0. State 1 has given rise to two different states, labelled 2 and 2'. The latter has given rise to 3", and the former has changed to both 3 and 3'.

Camin and Sokal (1965) phrased their methodology in terms of "evolutionary steps," the number of changes in discrete character state values needed to proceed from a prior (ancestral) OTU in a cladogram to a subsequent (descendent) OTU. This number of evolutionary steps was evaluated over all characters examined in a given study and was shown graphically by cross marks on the cladogram (see Figure 6-9). However, it is readily apparent that the number of evolutionary steps between an ancestral HTU **j** and a descendent OTU **k** on a cladogram is nothing but the Manhattan distance $d_1(\mathbf{j}, \mathbf{k})$ between these nodes over all characters. As an example let us examine the number of evolutionary steps or the Manhattan distance between OTU **25** and its immediate ancestor **a**, based on the data matrix in Table 6-1 and illustrated in Figure 6-9 (from which the states of **a** can be obtained). The character state vectors for these two OTU's are the following:

Character	Character state vectors for OTU's	
	25	**a**
1	1	1
2	2	1
3	2	1
4	1	1
5	1	1
6	−1	0
7	0	0

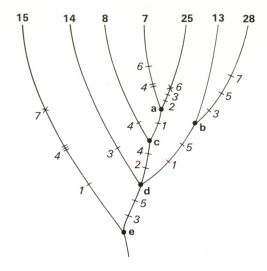

FIGURE 6-9

Reconstruction of a cladogram. The OTU's are at the tips, and are the Group A Caminalcules in Camin and Sokal (1965). They are identified by the numerals used in that paper.

   The HTU's that are ancestors of the OTU's at the tips are indicated by black circles and identified by letters **a** to **e**. Their probable character states can be easily obtained from the cladogram by going from the base (whose states are all zero) to the HTU. A cross-bar indicates a change in character state for the character whose number is placed beside it, producing an increase by one unit, e.g., from state 0 to state 1. A cross-sign indicates a decrease by one unit, e.g., from state 0 to state −1. Thus for **c**, characters *1* through *7* will be 0, 1, 1, 1, 1, 0, 0 respectively. [Modified from Camin and Sokal (1965).]

By inspection, their Manhattan distance $d_1(\mathbf{25}, \mathbf{a})$ can be computed as 3, since the two OTU's differ by one evolutionary step (absolute differences of one) for characters *2, 3*, and *6* only.

   The character state vectors for any OTU can be thought of as points in a hyperdimensional lattice in which, because of the irreversibility assumption, the OTU's follow a monotonically increasing ($\geq$) or decreasing ($\leq$) trajectory for any one dimension. This concept will become clearer when looked at in a two-dimensional diagram (see Figure 6-10). OTU's **a** and **b** represent the following character state vectors.

Character	Character state vectors for OTU's	
	a	**b**
h	1	2
i	2	1

Panel *a* shows independent evolutionary trajectories for these two OTU's. Parallel-ism is present in characters *h* and *i*. Panels *b* and *c* show equivalent minimum length cladistic paths. Both OTU's share the evolutionary path part of the way. The cladistic representation of panel *a* is shown in cladogram *d* of Figure 6-10, that of panels *b* and *c* in cladogram *e*. Obviously, the maximum length of a cladogram will be the sum of the independent paths of each OTU, emerging as a separate line from a common ancestor—V-shaped for two OTU's in Figure 6-10,*d*, brushlike when there are many OTU's. Thus, using the convention that all character states in the common ancestor are equal to 0, the maximum distance from any OTU **j** to the common ancestor **o** must be $d_1(\mathbf{j}, \mathbf{o}) = \sum_{i=1}^{n} |X_{ij}|$. Consequently the maximum length of the entire cladogram comprising *t* OTU's will be the sum of the separate evolutionary lengths for each OTU, $L_{max} = \sum_{j=1}^{t} \sum_{i=1}^{n} |X_{ij}|$.

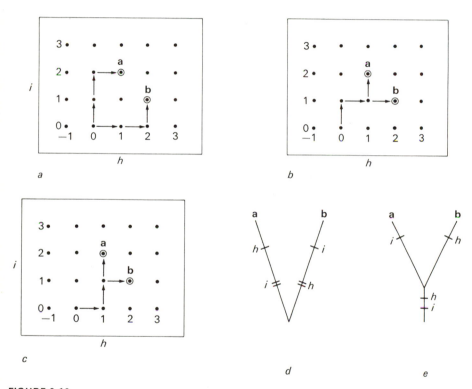

FIGURE 6-10

Three representations of an equivalent cladistic sequence. Two characters, *h* and *i*, are shown, and − 1, 0, 1, 2 and 3 are character states. The two OTU's **a** and **b** have states (1,2) and (2,1) respectively. *a*, A maximum-length cladistic path (reversals not permitted). *b*, A minimum-length cladistic path, using the point (0,1) as an intermediate vertex to represent an HTU. *c*, The other minimum-length path, using (1,0) as an intermediate vertex. *d*, The cladogram representation equivalent to (*a*). *e*, The cladogram representation equivalent to *b*; that for panel *c* differs only in the order of characters *h*, *i* at the lowest internode, and since this order is arbitrary, *e* can represent both *b* and *c*. [Modified from Hendrickson (1968).]

Certain evolutionary steps can be *shared*, that is, changes in some OTU's occurred before their last common ancestor and can be placed on a single internode. Then the total length of the tree is diminished by the sum of the lengths of the shared internodes, each multiplied by $(t_J - 1)$, where t_J is the number of terminal OTU's subtended by the shared internode. Thus, to compute the length of the cladogram in Figure 6-10,*e* we compute $(1 + 2) + (2 + 1) - (1 \times 2) = 4$. At the other extreme, the minimum length for a cladogram assuming the maximum possible number of shared steps for all OTU's, no parallelisms, and complete consistency among the characters, must be $L_{min} = \Sigma_{i=1}^n (m_i - 1)$. In other words the absolute theoretical minimum length cannot be less than the number of character states less 1, summed over all the characters. In virtually all cladograms the actual number of evolutionary steps will be somewhere between L_{max} and L_{min}, and it is the aim of numerical cladistic methods to minimize the overall length of the cladogram (or the number of evolutionary steps).

In a search for most parsimonious (shortest length) cladograms, Camin and Sokal (1965) developed a series of initial approaches that constructed cladograms close to a most parsimonious solution. A variety of clustering algorithms can be used to obtain first approximate cladograms (called procladograms by Camin and Sokal). One such procedure is a single linkage cluster analysis (see Section 5.5) of the pair function describing the number of common evolutionary steps (Camin and Sokal, 1965). This quantity has been called the total number of derived steps shared by OTU's **a** and **b**, and it has also been called the advancement index $h(\mathbf{j_{a,b}})$ of the common joint ancestor $\mathbf{j_{a,b}}$, which is the most recent common ancestor of OTU's **a** and **b** (Farris, Kluge, and Eckardt, 1970). We have adopted here the simpler symbolism consistent with the rest of our notation rather than the $h(\mathbf{J}(\{\mathbf{a}, \mathbf{b}\}))$ and $\mathbf{J}(\{\mathbf{a}, \mathbf{b}\})$ employed by Farris et al. (1970).

Camin and Sokal (1965) proposed the monothetic method for obtaining procladograms. Its algorithm runs as follows.

1. Count the number of zero character states for each OTU. (This provides some measures of "primitiveness," as OTU's with more characters of state zero should branch off the main trunk of the cladogram near its base). Go to 2.

2. Remove the OTU with the greatest number of zeros from the data matrix. Test the remaining data matrix for rows (characters) without zeros. If no such rows are found, replace the removed OTU and remove another one tied with it by number of zeros or by possession of the next largest number of zeros. If removal of any one OTU with a high number of zeros does not produce nonzero character rows, combinations of two or even three OTU's with large numbers of zero character states should be removed from the matrix. Once a nonzero row (or rows) appears, go to 3.

3. Draw a branch subtending the removed OTU or OTU's from the base of the cladogram. Go to 4.

4. Subtract unity from each character state code in each nonzero row (repeatedly

if necessary) until each row contains at least one zero. Rows that have become all zero are dropped from the matrix and from further consideration. Go to 5.

5. Are any OTU's unplaced on the branches? If yes, go to 1; if no, the procladogram is complete.

An example of such an algorithm is illustrated in Table 6-3. In this table characters having both negative and positive character states are, for convenience, recoded into all positive character states. Character *4* has been omitted as unreliable for reasons discussed later in this section. The algorithmic steps in Table 6-3 are self-explanatory. At the end of this procedure a procladogram is obtained, shown in Figure 6-11,*a*.

Camin and Sokal (1965) reduced the length of the procladogram by an iterative procedure. They first removed all internodes that do not bear any evolutionary steps (or whose lengths are zero on the Manhattan metric). This is done because there is no reason to assume separate branching points in the absence of intervening evolutionary steps. Next, all common evolutionary steps found on adjacent branches are moved to the base stem of these branches. This is followed by a trial and error moving of branches illustrated in Figure 6-11. In the procladogram resulting from the monothetic method (Figure 6-11,*a*), we move one of the terminal branches (that bearing OTU's **25** and **28**) down one branching level to level 3. This now becomes the terminal level (level *ω*) and OTU's **25**, **28**, **7** and **13** all emerge from this point (Figure 6-11,*b*). The length of the original procladogram was 17 (17 evolutionary steps), but the new length is 18 and is thus moving away from the intended direction. However, members of the cluster (**25**, **28**, **7**, and **13**) can now be examined for common steps that can be removed to the internode at the base. In Figure 6-11,*c* we note that OTU's **25** and **7** can be placed together with a step for character *2* in common and OTU's **28** and **13** can be joined with a step for character *5* in common. This necessitates parallel steps in OTU's **25** and **28** for character *3*, which previously was a common step, but we have now reduced the number of steps for the cladogram to 16, i.e., its length is 16. This is the same number required by the true cladogram, but Figure 6-11,*c* is not the correct solution. Moving the branch that bears OTU's **7** and **25** in Figure 6-11,*c* from branching level 3 to level 2 (Figure 6-11,*d*) and shifting OTU **8** to share a step of character *2* with (**7**, **25**) yields the correct cladogram of Figure 6-9. This illustrates that different but equally parsimonious solutions (trees of equal length) may occur and it is impossible to decide among them unless evidence from other characters can be relied upon.

The entire cladistic method of Camin and Sokal—compatibility matrix computation (see below), procladogram construction, and the iterative search for the most parsimonious tree—has been programmed (Bartcher, 1966) and has been found to yield generally satisfactory results. Its application to a cladogeny of the horses, the Caminalcules, and several other groups has been promising (Camin and Sokal, 1965).

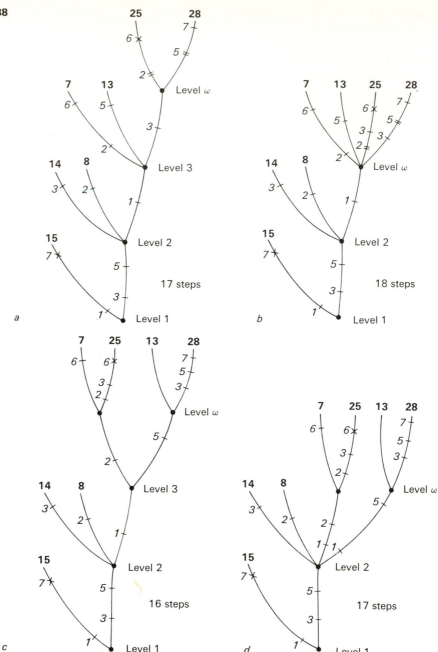

FIGURE 6-11

Steps in the reconstruction of the cladogram of Group A (Figure 6-9) by the monothetic method. Symbolism as in Figure 6-9. Level numbers and ω refer to levels of furcation from 1 to terminal level. *a*, Procladogram resulting from monothetic technique illustrated in Table 6-3. Total number of evolutionary steps is 17. *b*, OTU **25** moved down one branching point; 18 evolutionary steps result. *c*, OTU's **25** and **7** grouped, as are OTU's **13** and **28**. Achieved parsimony of 16 steps. This is equally parsimonious but not identical to the cladogram in Figure 6-9. Character *4* has been omitted from these cladograms. Further adjustments for parsimony result in a cladogram identical to Figure 6-9. *d*, Branch bearing OTU's **7** and **25** moves down one branching point; 17 evolutionary steps result. If OTU's **8, 7**, and **25** are now rearranged so that they share their common step for character *2*, the cladogram of Figure 6-9 is obtained with 16 evolutionary steps. [Modified from Camin and Sokal (1965).]

TABLE 6-3

Camin and Sokal's monothetic method for reconstructing cladograms.

Data matrix for Caminalcule Group A (Table 6-1) with characters 6 and 7 recoded and character 4 omitted.

Characters	OTU's							Cycle 1, Step 2
	7	8	13	14	15	25	28	
1	1	0	1	0	1	1	1	OTU's **14** and **15** have 6 zeros each. Removal of OTU
2	1	1	0	0	0	2	0	**14** leaves unchanged the number of "nonzero" rows.
3	1	1	1	2	0	2	2	Removal of OTU **15** leaves rows 3 and 5 nonzero and
5	1	1	2	1	0	1	3	7– all zero. Therefore, remove OTU **15**.
6+	1	0	0	0	0	0	0	
6–	0	0	0	0	0	1	0	
7+	0	0	0	0	0	0	1	
7–	0	0	0	0	1	0	0	
Number of p 1 zeros	3	5	5	6	6	3	4	

Data matrix A with OTU **15** removed.

Characters	OTU's						Cycle 1, Step 4
	7	8	13	14	25	28	
1	1	0	1	0	1	1	Subtract unity from rows 3 and 5; delete row 7–.
2	1	1	0	0	2	0	
3	1	1	1	2	2	2	
5	1	1	2	1	1	3	
6+	1	0	0	0	0	0	
6–	0	0	0	0	1	0	
7+	0	0	0	0	0	1	
7–	0	0	0	0	0	0	

Data matrix B with unity subtracted from rows 3 and 5 and row 7– deleted.

Characters	OTU's						Cycle 2, Step 2
	7	8	13	14	25	28	
1	1	0	1	0	1	1	Recompute number of zeros for remaining OTU's.
2	1	1	0	0	2	0	OTU's **8** and **14** have 6 zeros each. Removal of either
3	0	0	0	1	1	1	**8** or **14** leaves unchanged the number of nonzero rows.
5	0	0	1	0	0	2	Removal of both **8** and **14** leaves row 1 nonzero.
6+	1	0	0	0	0	0	Therefore, remove OTU's **8** and **14** together.
6–	0	0	0	0	1	0	
7+	0	0	0	0	0	1	
Number of p 1 zeros	4	6	5	6	3	3	

TABLE 6-3 (*continued*)

D. Data matrix C with OTU's **8** and **14** removed.

Characters	OTU's				Cycle 2, Step 4
	7	**13**	**25**	**28**	
1	1	1	1	1	Subtract unity from row *1*. Row *1* becomes all zero, s
2	1	0	2	0	delete.
3	0	0	1	1	
5	0	1	0	2	
6+	1	0	0	0	
6−	0	0	1	0	
7+	0	0	0	1	

E. Data matrix D with row *1* deleted.

Characters	OTU's				Cycle 3, Step 2
	7	**13**	**25**	**28**	
2	1	0	2	0	Recompute number of zeros for remaining OTU
3	0	0	1	1	OTU **13** has 5 zeros. Removal of OTU **13** makes
5	0	1	0	2	nonzero rows. Similarly for OTU **7** with 4 zer
6+	1	0	0	0	Removal of OTU's **7** and **13** leaves row *3* nonzer
6−	0	0	1	0	Therefore, remove OTU's **7** and **13**.
7+	0	0	0	1	
Number of Step 1 zeros	4	5	3	3	

F. Data matrix E with OTU's **7** and **13** removed.

Characters	OTU's		Cycle 3, Step 4
	25	**28**	
2	2	0	Subtract unity from row *3*. Row *3* becomes all ze
3	1	1	so delete. OTU's **25** and **28** are a terminal bifurcatio
5	0	2	
6+	0	0	
6−	1	0	
7+	0	1	

[Modified from Camin and Sokal, 1965.]

However, the method of Camin and Sokal, while providing the user with numerous equally parsimonious solutions, had no criterion that indicated to the user when a minimum length tree was found or how many such minimum length trees there might be. This problem was solved by Estabrook (1968); using the topological technique of partial orders, he was able to find a general solution to obtaining the entire set of minimum length trees for a given collection of OTU's and characters. Since this set can sometimes be very large, one may need a substantial number of characters in order to reduce its size. We do not as yet have enough experience with the Estabrook procedure (programmed for computer processing by its author) to enable us to judge its usefulness, and to know whether increasing the number of characters substantially would reduce the number of equally parsimonious solutions for any given collection of OTU's.

The *weighted invariant step strategy* (WISS) of Farris, Kluge, and Eckardt (1970) consists of finding pairs of mutually highest advancement indexes (those pairs that have the highest number of shared derived steps) and replacing each pair by their most recent joint common ancestor. This approach has also been suggested by J. Hendrickson, Jr. (personal communication). The clustering cycle is repeated until all OTU's have been clustered. This method is thus a modification of a weighted pair group analysis using the total number of derived steps shared by OTU's j and k as the similarity coefficient. Since the similarity coefficient based on highest number of shared derived steps is a Manhattan metric if the primitive character is coded zero, we cannot recompute the dissimilarity coefficient between a recent joint common ancestor and a new branch by the combinatorial formula of Lance and Williams (1967b; see Section 5.5). However, Farris, Kluge, and Eckardt (1970) provide a computational formula for this operation,

$$h(\mathbf{j_{a,b}}) = \tfrac{1}{2}[h(\mathbf{a}) + h(\mathbf{b}) - d_1(\mathbf{a}, \mathbf{b})]$$

similar to Farris' earlier mentioned formula for $d_1(\mathbf{j}, \boldsymbol{\mu})$. The clustering technique of Farris, Kluge, and Eckardt (1970) is very similar to that of Camin and Sokal (1965), except that the candidates for joining clusters are defined by their advancement index, $h(\mathbf{j}) = \Sigma_{i=1}^{n} |X_{ij}|$, which is the sum of the column vector of their absolute character states, whereas Camin and Sokal choose their candidates by the number of zero character states in the column vector. Thus it is possible that an OTU that was coded 1 for each character state might be preferred by the WISS method, although one that has many zeros but a few large character states might be preferred by Camin and Sokal's monothetic method.

Methods for finding *Wagner trees* are also given by Farris (1970). Since a Wagner tree is simply a directed Steiner minimal tree in Manhattan distance for a given set of OTU's (a Wagner network being a nondirected minimal Steiner tree for the same set of OTU's), any tree generated from a Wagner network by assuming one of the OTU's or HTU's to be the ancestor is a legitimate candidate for a Wagner tree. Another algorithm suggested by Farris (1970) would assume an ancestor e for a given study. The ancestor could be an OTU (considered very primitive) or it could be a vector representing a hypothetical taxonomic unit. We compute all distances $d_1(\mathbf{j}, \mathbf{e})$ from the t OTU's to the ancestor. The OTU, say **b**, with the smallest $d_1(\mathbf{j}, \mathbf{e})$ is chosen and linked to **e**, forming internode (**e**, **b**). At this point the algorithm becomes very similar to that outlined in detail in Table 6-2, except that new candidate OTU's are chosen in the order of increasing $d_1(\mathbf{j}, \mathbf{e})$. As before, they are linked to that interval to which they are closest.

Using this algorithm and setting the assumed ancestor **e** as the null vector $\{0, 0, 0, 0, 0, 0, 0\}$ and following Camin and Sokal's coding scheme, we obtain a tree that is similar, but not quite topologically equivalent, to that shown in Figure 6-9. It is equally parsimonious in terms of number of evolutionary steps. The HTU's generated during such a stepwise algorithm may not be globally optimal. Farris (personal communication) has described the following minor modification of his method for optimizing the HTU's for a tree with a fixed branching form (Farris, 1970). For each HTU that subtends two OTU's write down a vector whose elements are "character state sets" where each bracketed set is the range of character states for one character over the two subtended OTU's. Thus, referring to Table 6-1, notice that the HTU ancestral to OTU's **13** and **28** would possess, for the six characters, the following vector of character state sets: $\{[1,1], [0,0], [1,2], [2,3], [0,0], [0,1]\}$. HTU's that subtend either one or two other HTU's are characterized by a state set vector where each element or set is the intersection or overlap of the character state sets of the two subtended taxonomic units (thus, [0,1] and [1,3] yield [1,1]; or [0,2] and [1,3] yield [2,2]), or if the intersection is empty, i.e., if no state is held in common, the HTU's are characterized by the least range enclosing elements from both sets (thus [0,1] and [3,4] yield [1,3]). These rules are applied to successive HTU's descending the tree toward the root. A second procedure is to replace the character state sets by their intersections with the corresponding sets in the immediately ancestral HTU. Thus, for example, a state set of [0,1] with an ancestral state set of [1,2] is replaced by [1,1]. This second procedure is applied at two stages in the optimizing process. During the first, descending pass down the tree, after the state sets of each HTU have been computed, all immediately descendent HTU's are examined for state sets whose range is not zero (i.e., whose pair of states is not identical). Procedure 2 is then applied to these state sets. The same procedure is also applied in a second, ascending pass through the tree, from the root to the distal HTU's. Thus the HTU ancestral to OTU's **13** and **28** in group A of the Caminalcules is optimized to the following vector of character state sets: $\{[1,1], [0,0], [1,1], [2,2], [0,0], [0,0]\}$. Since the ranges of the state sets are 0, the character states of the HTU can be recorded as $\{1,0, 1,2, 0,0\}$.

When reversal of character states is permitted, as by Kluge and Farris (1969), the techniques of finding minimum length trees become somewhat more complicated. Ideally, character state trees should be worked out for the entire cladogram, which

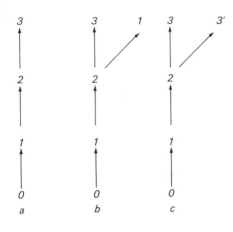

FIGURE 6-12
Recoding of character states. *a,* Original operational coding. *b,* Coded to represent presumed evolutionary sequence including reversed steps. *c,* Recoded to permit retention of the hypothesis of irreversibility. [Based on character *3* in the horse data of Camin and Sokal (1965).]

in effect means recoding characters so that reversed steps at a subsequent time are given a new character state (as suggested by Camin and Sokal, 1965). This is illustrated in Figure 6-12. If such character state trees cannot be constructed, then characters can be designated as primitive only in small local regions of the clado-gram, as pointed out by Farris, Kluge, and Eckardt (1970). To carry out clustering of OTU's into cladograms under these circumstances they suggest a dissimilarity coefficient $U_{(a,b),e} = \frac{1}{2}(U_{a,e} + U_{b,e} - U_{a,b})$, where values of U are in Manhattan distance, d_1. Note that the form of this coefficient is quite similar to the computation of the advancement index of the most recent joint ancestor of OTU's **a** and **b** except that here a local immediate ancestor **e** is assumed. The Manhattan distance should not be found for OTU's that are quite far apart in the network, for this would hide difference due to evolutionary reversals. Jardine, van Rijsbergen, and Jardine (1969) and van Rijsbergen (1970) have described clustering methods that will yield cladogenies if evolution rates are locally constant, as is assumed in much cladistic work.

In recent years there has been much progress in the construction of cladogenies from amino acid sequences in proteins. Early analyses of this type were carried out by Doolittle and Blombäck (1964) on fibrinopeptides and by Horne (1967) on erythrocyte catalase tryptic peptides. The classic paper in the field is that of Fitch and Margoliash (1967), about their work on amino acid sequences of cytochrome *c* for 20 OTU's ranging from yeast to man. Their phylogenetic (cladistic) recon-struction is based on the *minimal mutation distance* between two cytochromes. This distance is the sum of the *mutation values* for each pair of amino acids sequenced in the protein. A mutation value at a given site in the amino acid sequence is the *minimum number* of mutational changes in the nucleotides of a given codon neces-sary to produce the change from one amino acid to the other (the actual number may well be greater). Fitch and Margoliash furnish a table of mutation value for pairs of amino acids based on the known properties of the genetic code. Thus, for example, a mutation from Cysteine to Lysine would require three mutational steps (mutation value = 3) since a UGU or UGC triplet would have to be replaced by an AAA or AAG triplet. By contrast, the mutation value from Aspartic acid to Glycine is 1, since a single nucleotide base substitution (for example, GAU to GGU or GAC to GGC) would suffice to bring about the change in the codon. Mutation distances by their nature are absolute values and can be quantified as $d_m(\mathbf{j}, \mathbf{k}) = n/(n - g)\sum_{i=1}^{n} m_{i\mathbf{jk}}$ where $m_{i\mathbf{jk}}$ are the mutation distances at site i in a given protein of OTU's $\mathbf{j}$ and $\mathbf{k}$. The mutation distances are adjusted proportionately to compensate for variable number of comparisons that involve gaps or deletions in the amino acid sequences. There are n sites, and comparisons are impossible (equivalent to NC in conventional numerical taxonomy) at g sites. Thus while the comparison in the study by Fitch and Margoliash was made over 110 amino acids, the mutation distances were multiplied by a proportionality factor of $110/x$ where $x = n - g$ is a number of positions ≤ 110 over which the comparison between any pair of OTU's was made.

Although Fitch and Margoliash (1967) provided a clustering algorithm for finding the evolutionary tree based on the mutation distances (basically single linkage clustering with the average of the as yet unclustered OTU's), these same authors in a more recent treatment (Fitch and Margoliash, 1970) deemphasize that algorithm and recommend any of several taxometric procedures for obtaining suitable dendrograms. Given several alternative trees, Fitch and Margoliash (1967) chose among them by an optimality criterion based on absolute differences of the mutation distances obtained in the original dissimilarity matrix from those implied by the cladogram. Each difference is divided by the observed mutation distance and multiplied by 100, i.e., it is expressed as a percentage. These percentages are treated as variates and the sample standard deviation of the $t(t - 1)/2$ variates is computed assuming a mean of zero. This standard deviation is employed as a measure of goodness of fit. There is no analytical or combinatorial solution to yield a minimum length tree, but the methods of Camin and Sokal (1965) as modified by Estabrook (1968) would clearly be applicable, the mutation values for any one site being equivalent to the number of evolutionary steps. There is close analytic similarity between the Fitch and Margoliash optimality criterion and the least squares fit measure of the additive tree model of Cavalli-Sforza and Edwards (1967). Recent phylogenetic work by Goodman and Moore (1971) and Goodman et al. (1971) has employed this criterion.

The purported cladogeny for cytochrome c of the 20 OTU's is shown in Figure 6-13. Note that it contains negative distances. This is because the optimizing procedure between the phenetic (mutation) distance matrix and the distances implied by the dendrogram (patristic differences, Farris, 1967a) forces the distances implied by the dendrogram into small negative distances that do not make biological sense but result from optimal fits. This is the phenomenon analogous to the reversals noted by Sokal and Michener (1958; and see Section 5.5).

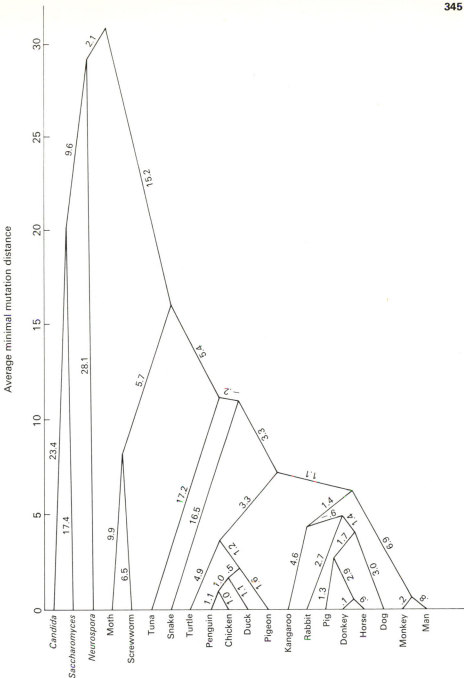

FIGURE 6-13

Phylogeny as reconstructed from observable mutations in the cytochrome *c* gene. Each number on the figure is the corrected mutation distance along the line of descent as determined from the best computer fit so far found. Each apex is placed at an ordinate value representing the average of the sums of all mutations in the lines of descent from the apex. [Modified from Fitch and Margoliash (1967). Copyright © 1967 by the American Association for the Advancement of Science.]

The method can be used not only for showing presumed cladogenies of proteins of different species, but also of genes. For example, Fitch and Margoliash (1967) publish a cladogeny of the ancestral gene for hemoglobin and its evolution into myoglobin and the α, β, γ, and δ chains of hemoglobin. In a later paper Fitch and Margoliash (1968) point out that the information they possess is basically phenetic information and that from that point of view their dendrograms are really phenograms. Only if certain assumptions are given about the evolutionary probability of given changes (or of the order and likelihood of certain nucleotide substitutions) can these dendrograms be turned into cladograms. The validity of their model was demonstrated on a simulated model in which the probability of mutation from one nucleotide to another, as observed in the first two positions of the codon, was employed. The authors caution about the difficulties inherent in double mutations (fixation of a mutation in two of its three nucleotides in the interval between two successive nodes).

The method of Dayhoff and Eck (1968) is similar in general approach, searching for a minimum length unrooted tree in terms of mutation distances between two proteins. A distinctive feature of their method is that when reconstructing ancestral sequences they mark as unknown any portion for which the amino acid is very doubtful, rather than forcing a decision on slender grounds. However, the validity of the reconstructed sequences is undermined by the dependence of the method on the exact set of OTU's, because of the majority voting principle used to decide on ancestral sites. They too analyze cytochrome c. Evolutionary sequences of other proteins are found in papers by Dayhoff and Eck (1969) for the globins; Dayhoff, Sochard, and McLaughlin (1969) for immunoglobulins; and Dayhoff and McLaughlin (1969) for other proteins, all in the volume edited by Dayhoff (1969a). The techniques for analyzing amino acid sequences and their nucleotide encodings are still in their infancy. Among the important aspects to be considered are the *frame shift mutations*, which involve deletion of a single nucleotide and insertion of a single nucleotide elsewhere in a sequence, making the entire message code for a new sequence of amino acids. Such a change would greatly alter the gene products over the region considered. The cladistic problem here is that the change in amino acids, when considered as normal mutational changes, would require a very large mutation distance, though when considered as a frame shift mutation it could be accomplished in only two mutational steps. Fitch (1970c) has estimated the probability that a specific frame shift mutation might be selected in the course of evolution. A very informative review of the entire subject of reconstructing evolutionary sequences on the basis of amino acid and nucleotide sequences is furnished by Fitch and Margoliash (1970). It is evident, however, that some aspects require critical examination; for example, the homologizing of bacterial and mammalian cytochrome c sequences cited in that article would lead to the conclusion that the horse is closer to one bacterium than that bacterium is to a second bacterium.

Character Weighting

In cladistic analysis the problem of character weighting assumes new importance. Since there is a "true" dendrogram of relationships—i.e., the true cladogram—any weighting scheme is desirable that improves our chances of obtaining a closer estimate to the true cladistic relationships. This has been recognized by orthodox phylogenetic taxonomists who have traditionally weighted some characters more heavily than others in arriving at putative phylogenetic classifications (and this practice has been widely abused by taxonomists who weighted some characters with very little justification by claiming that their classifications were phylogenetic). Weighting has also been employed by the more critical phylogenetic systematists such as Hennig (1966), by considering corroborating evidence from large numbers of compatible apomorphous character states and neglecting evidence from few discordant ones. We have already discussed this issue in Section 2.5. This point is also discussed in some detail as axiom A IV by Farris, Kluge, and Eckardt (1970).

The problem of weighting characters also brings up that of the numbers of characters to be used in a cladistic study. In this connection an important distinction between phenetic and cladistic work must be stressed. It is almost intuitively obvious that an adequate measure of overall phenetic similarity cannot be determined with the use of only a few characters. This is so, not only because by definition overall similarity cannot be represented by few characters, but also because there is no parametric measure of overall similarity and, as we have seen in Section 3.8, relative stability in character hyperspace is obtained only when the space is of high dimensions. By contrast it may be believed that if there is a single true cladogeny for any given set of OTU's, then in principle it would be possible to find just a single character that uniquely defines this cladogeny. This would be true only if all branches of the tree were defined by unique character states. It is unlikely that for any large number of OTU's there would be an equal number of unique character states that could be logically ordered. If there are fewer character states than OTU's because of parallelism in several branches, then it would be impossible to arrive at a cladogram on the basis of this character alone. However, it might be feasible to define a cladogram uniquely on the basis of a few divergent characters. It is for this reason that weighting of characters (in favor of those showing derived similarity) is suggested for cladistic analysis by many authors.

Several approaches to character weighting have been suggested in numerical cladistics. Camin and Sokal (1965) developed the method of *compatibility matrices*. For any given set of OTU's and character states they draw for each character the most parsimonious character state tree covering all OTU's. They then fit all other characters in turn to the *pattern* defined by this character state tree and compute the number of extra steps needed to fit each character to the pattern. By extra steps is meant any number greater than $m_i - 1$, where m_i is the number of states for character i. This is the minimum number of evolutionary steps necessary to attain the terminal state for any given character. For each of the patterns we find out whether the character is *compatible*, i.e., needs no more than its minimum number of evolutionary steps, or requires extra steps (parallelisms) to fit to the pattern. If $C(h,i)$ is the minimum number of extra steps for character h on a pattern (character state tree) that fits character i exactly, a nonzero $C(h,i)$ indicates some inconsistency between h and i. From these computations emerges a compatibility matrix as shown in Table 6-4. The numbers of *compatibilities* (no extra steps required) for each pattern and for each character are given in the margins of the table, as are the numbers of extra steps. Patterns that have many compatibilities and few extra steps seem promising starting points for the reconstruction of a cladistic sequence, although as we have seen there are better methods for forming an initial topology.

TABLE 6-4
Compatibility matrix for seven selected Caminalcules (Group A)

Characters (h)		Patterns (i)							Compatibilities	Extra steps
	1	*2*	*3*	*4*	*5*	*6*	*7*			
1	×	2	2	2	2	1	1		0	10
2	1	×	1	2	0	1	0		2	5
3	2	2	×	4	1	2	1		0	12
4	2	3	4	×	3	3	3		0	18
5	1	1	1	3	×	1	1		0	8
6	0	0	0	0	0	×	0		6	0
7	0	0	0	0	0	0	×		6	0
Compatibilities:	2	2	2	2	3	1	2		14	—
Extra steps:	6	8	8	11	6	8	6		—	53

Values of $C(h, i)$ are given in the body of the matrix. Compatibilities in the margins are the frequencies of $C(h, i) = 0$ in a row or column.
Extra steps are $\sum_h C(h, i)$ for either rows or columns.

[From Camin and Sokal, 1965.]

However, those characters whose patterns require a large number of extra steps when the other characters are fitted to them, and which have few compatibilities with patterns made by other characters, are poor choices for the construction of minimum length trees. Characters may be incompatible because they have been incorrectly coded, and in fact the computation of a compatibility matrix may point out incorrect coding to the investigator. Such a case actually occurred in the coding of character 3 (average anteroposterior crown lengths of the fourth upper premolar and the second upper molar) in the fossil horses analyzed by Camin and Sokal (1965). Characters may also emerge as undesirable on the basis of compatibility matrices if they have an unusually large number of parallelisms. The practice in the numerical cladistic method of Camin and Sokal has been to examine the compatibility matrix and to remove from it those characters that seem to be discordant with the majority of the others. This is why character *4*—clearly the most discordant in Table 6-4—was removed from consideration in the discussion of cladistic techniques (Tables 6-2 and 6-3). There is, of course, a danger in such a procedure. If most of the characters that are analyzed, which seem to be mutually compatible, are misleading, and only one or two characters incompatible with the others are the true indicators of the cladistic sequence, erroneous results are likely to follow. However, we know of no method that would reliably guard against this type of error. The philosophy of preponderance of evidence guides us here as in the more traditional approaches. Farris (1969b) has also pointed out that the relationship between the consistency of character h with the tree based on all characters and the number of extra steps for any given character h over all patterns as computed by $\sum_i C(h,i)$ is not a simple one, and one may not safely assume the cladistic reliability of character h from this value.

Le Quesne (1969) has shown that if the characters of a study are two-state, useful information on compatibility can be obtained by tallying the frequency of OTU's possessing character state combinations (0,0), (0,1), (1,0), and (1,1) for characters h and i in a customary 2×2 contingency table.

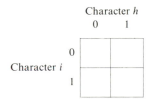

If all four cells of the contingency table are filled, the two characters cannot be compatible. If only three or fewer cells are filled, the two characters are compatible for this set of OTU's, and *might* be "uniquely derived characters," that is, both might have arisen only once in a phylogeny. He suggests removing all characters that are incompatible with others and using the remainder to construct a cladogram by assuming that the rarer state is derived. Le Quesne (1969) developed a *coefficient of character state randomness*, which can be defined as $\sum_i H(h,i)$, divided by the sum of the probabilities that all four states will occur, where $H(h,i) = 0$ when $C(h,i) = 0$, and $H(h,i) = 1$ when $C(h,i) > 0$. The lower this ratio the greater the proportion of presumed uniquely derived characters, and by implication the more reliable the cladogeny. He also extends this to characters of three ordered states. J. S. Farris (see Crovello, 1971) has shown that for binary characters Le Quesne's character-pair matrix is a compatibility matrix. M. J. Sackin (personal communication) points out to us that if there are n two-state characters with all pairs presumptively uniquely derived, then there can be up to only $n + 1$ different OTU's. Furthermore, a rootless tree can be constructed in which every character is uniquely derived (i.e., each character mutates only once throughout the tree). If there are $n + 1$ OTU's then the tree is unique; if there are less, $n - x$ OTU's, then the tree is still unique, but $x + 1$ characters are nonindependent in the OTU's. A similar observation was made by J. S. Farris (see Crovello, 1971).

Farris (1969b) has shown that Le Quesne's method can be applied to multistate characters by recoding them by means of additive binary coding (see Section 4.8), except that there are m binary states for an m-valued character, not $m - 1$, and that the coding follows the steps in the character state tree, not necessarily the order of magnitude of the states (as when the tree branches, see Farris, Kluge, and Eckardt, 1970). One method of weighting proposed by Farris using Le Quesne's methods (and applied to two-state characters) is to use for each character h the weight $w_h = (n/N_h)^3 - 1$, where n is the total number of characters in the study, and N_h is the sum of $H(h,i)$ over the other $n - 1$ characters.

Another weighting criterion is the ratio s_W^2/s_B^2, where the variances describe variation of characters within and between OTU's. Other weighting coefficients proposed by Farris (1969b) relate to a statistic called by him *unit character consistency*, which is defined as $C_i = r_i/l_i$ for character i where r_i is the range of the character and l_i is the patristic unit character length. This is simply the sum of the lengths of the internodes for that particular character state tree. If the length of the internodes over all the OTU's of the study is no longer than the range of the character, then the character will be fully compatible and the unit character consistency $C_i = 1$. To obtain a weighting criterion Farris suggests $p_i = l_i/(t^* - 1)$ where t^* is the number of nodes (OTU's and HTU's) on the tree. Thus p_i is the average length of the character state tree for a given character i. Since in a two-state character $r_i = 1$ we can rewrite $p_i = 1/(t^* - 1)C_i$.

Farris (1969b) experimented with various weighting procedures based on the concepts defined above, using an artificial tree with 31 nodes and 30 consistent characters. From 5 to 150 poorly consistent characters had been assigned at random to the nodes. The data matrix (including the inconsistent characters) was at first weighted by a weight of $w_i = 1$. A shortest spanning tree was then constructed based on Manhattan distances and the number of evolutionary steps (patristic unit character length, network length) for each character was recomputed. This led to a computation of a weighting function $w_i = f(p_i)$, which was applied to the data matrix. A new shortest spanning tree was produced from the weighted data matrix, and this in turn yielded new parameters for the weighting function. The iterative procedure was terminated when the ancestor function of the tree remained unchanged on successive runs. Of various weighting functions employed the concave and unbounded function $w_i = p_i^{-3} - 1$ appeared most successful. It reconstructed the correct cladogram even when as many as 150 inconsistent characters had been added to the data matrix.

Finally, we need to consider a type of weighting introduced into the computations of cladogenies from protein sequences. Relying on knowledge of chemical structure and function, but mainly on the frequency of presumed mutations in the large number of proteins that have been sequenced, one can calculate the probability of a given point mutation from one amino acid to another. When such a study is made (see Dayhoff, Eck, and Park, 1969) great discrepancies in the likelihood of transition from one amino acid to another can be shown. In the reconstruction of ancestral sequences of proteins it seems reasonable therefore to take these probabilities into consideration as weights, and this has been done in some recent work as detailed by these authors. However, there is the slight danger of a logical trap. If we are to investigate the evolution of an as yet unstudied protein or taxon, the use of a matrix of probabilities or weights that effectively summarizes past history based on other proteins and organisms is likely to prejudice our conclusions in the direction of prior knowledge.

Numbers of Characters

We should briefly return to the question of the numbers of characters required for cladistic analyses, which were mentioned earlier in this section in connection with character weighting. This is an area where our knowledge is as yet rather scanty. It is clear that even if characters can be appropriately weighted there is a limit to the number of OTU's whose cladogeny can be established from n characters. As noted earlier, with two-state characters, at the most $n + 1$ OTU's can be arranged in a unique (but unrooted) tree, even if all of them are presumed to be uniquely derived using the methods of Le Quesne (1969), and Le Quesne observed that substantial fractions of characters had to be discarded because they were not uniquely derived. Hendrickson (1967) found empirically that it was necessary to use $2t$ to $3t$ characters to give most parsimonious cladograms of t OTU's if these cladograms were to be regarded with much confidence. It is not very satisfactory to have to choose between cladograms whose parsimony differs in only one or two evolutionary steps, and work is needed on the statistical significance of such differences if appropriate statistical models are to be developed.

Kidd and Cavalli-Sforza (1971) have made a first attempt in this direction by determining the probability of recovering the cladogram of correct topology for a set of four OTU's, using cladograms that were generated by a Monte Carlo technique. The OTU's branched from a common ancestor under conditions of random mutation at a constant rate. The authors showed the difficulty of obtaining the correct cladogram if the branch points were close together, and also demonstrated some differences in the efficiency of their different methods described earlier in this section. What is not so obvious from their presentation, but can be deduced with reasonable certainty by graphical integration of their findings, is that if all configurations are equally probable, then all the methods have much the same performance. Furthermore, even when the number of characters is many times larger than t the chance of recovering the correct cladogram is not very high; e.g., with $t = 4$ and even 100 characters the chance of recovering the correct tree would be only about 70 percent for the rooted tree and a little over 90 percent for the unrooted tree. There are evidently severe limitations on the certainty of reconstructing cladogenies. Haigh (1970, 1971) has made a statistical study of the probability of determining the true root of a branching tree, which is a step in the desired direction (though perhaps his model is not entirely appropriate).

The methods described so far have dealt mostly with the problem of constructing estimated cladograms on the basis of character state information about recent OTU's. Estimating phylogenetic trees when character state information is not available, and when information about the OTU's is represented only in the form of a distance matrix between the OTU's, as for example, in immunological and DNA pairing studies, presents rather different technical problems, and has been little studied. The maximum likelihood inference model of Cavalli-Sforza and Edwards, mentioned above, formally depends only on a distance matrix between recent OTU's. This method, however, is not appropriate in general for phylogenetic studies, since it can be justified only by recourse to Cavalli-Sforza and Edwards's (1967) model of evolution by genetic drift. A promising new approach is suggested by Farris (1972).

Reticulate Evolution

The cladistic methods discussed so far were based upon the fundamental assumption that lineages may branch but never fuse. As we shall demonstrate, this assumption is quite critical. Fusion of lineages leads to what has been termed reticulate evolution, and introduces difficulties of an entirely different order of magnitude.

Is reticulation an important problem in most biological work? We may ignore sexual reproduction, which involves reticulation on a fine scale, and consider only reproductively isolated lineages with occasional hybridization. Reticulate evolution is usually considered rare in animals, at least in the better known groups (e.g., Mayr, 1963) although this may not be equally well founded in lesser known groups of invertebrates and in some fishes and amphibia (see Dobzhansky, 1970, p. 409 for some instances). In plants, however, hybridization is common between quite distinct species and even between genera, leading to persisting and evolving lineages. Hybridization usually occurs through allopolyploidy in sexually reproducing plants, and in some genera there are a large proportion of species that are presumptive allopolyploids (Davis and Heywood, 1963; Dobzhansky, 1970). There is also increasing evidence for reticulate evolution in microorganisms (e.g., Jones and Sneath, 1970). To the botanist and microbiologist, therefore, the question is not simply academic. The problems introduced by permitting fusion of lineages lie at two levels: first, the number of alternative cladogenies that must be considered and the quantity of information needed to distinguish the various hypothetical cladograms; second, the type of information that is required for this purpose.

If fusion of lineages is permitted the number of possible cladogenies rises extremely steeply with the number of fusions. This is illustrated in Figure 6-14 where fusions are imposed on very simple branched cladogenies. Fusions can, in theory, occur between any segments of a branched cladogeny, and even with a single permitted fusion there are numerous possibilities. Furthermore, the direction of gene transfer is pertinent, at least if there is transfer of only part of the genome: a fusion from segment i to segment j generally yields a cladogeny different from the fusion from segment j to segment i. Even if one excludes a small number of possibilities on the grounds that they are trivial, that they yield cladograms identical with those from other fusions, or that they are forbidden by the time scale of the cladogeny, there remains nevertheless a great number of possible cladograms. For example, with a single fusion, there are almost n^2 combinations to add to a branched cladogram containing n segments. When there are two fusions we must consider a number of alternatives that is of the order of the square of this, i.e., n^4 (more accurately $n^2(n + 3)^2$, as new segments are erected by the first fusion). The number of possible alternatives thus becomes impracticably large. The number of characters required to distinguish between the cladograms must also be very large.

Information indicative of reticulate evolution is of the following type. A taxon **C** is discovered to have proteins some of which are identical (or almost so) to proteins of another taxon **A**; others are identical to proteins of a third taxon **B**. Evidence of this quality would provide extremely strong support for the hybrid origin of **C**, since in no other way could the properties of **C** be accounted for plausibly. Relative uniformity of individuals of **A** and **B** would suggest that both **A** and **B** are monophyletic. Evolutionary parsimony suggests the hybrid origin of **C**, since convergence in fine details of several proteins would seem very improbable. But the principle of parsimony is used in a new way: one searches for local similarity among the diversity *within* genomes. One of us (Sneath, 1971a) has suggested that this principle of "similarity among diversity" would have to be applied so that some numerical function of the separate evolutionary changes of the parts of the genomes was minimized. The appropriate methodology remains to be developed, but the need for it should be clear. The ultrametric property of proteins that evolve at sufficiently constant rates could be used to detect reticulate evolution, for if some proteins had a longer evolutionary pathway than others this could be detected (Figure 6-15).

One further point merits attention. It is possible that this principle could be applied to cases where only isolated morphological characters are available, instead of fine structure within sections of the genome or its direct products. In fact P. A. Wells (personal communication) in a cladistic study of hybrid species of *Arctostaphylos* has shown that the introgressed characters show up as parallelisms in a monophyletic cladogram. Yet it would be much more difficult to apply in such

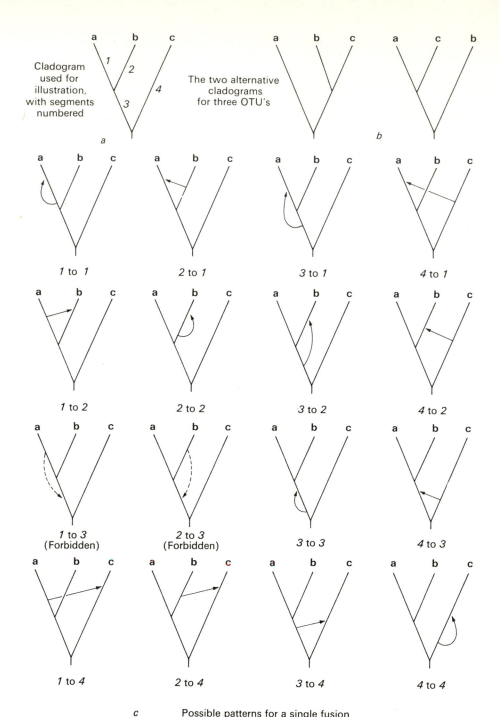

FIGURE 6-14

Alternative cladistic patterns for a single fusion during reticulate evolution in a cladogram of three OTU's. The cladogram at *a* is one of the three possible for OTU's *a*, *b*, and *c*, the other two being shown at *b*. The sixteen theoretical patterns of a single fusion for cladogram *a* are shown at *c*, the arrows showing the direction of transmission of genetic material. Two are forbidden, because they entail time reversal. It is assumed that the pathways contributing to fusion can exist separately for sufficient time for significant evolutionary change to take place (for example, in an isolated geographic area).

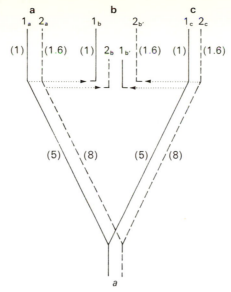

Dissimilarities between proteins

b Initial constitution of hybrid **b** = $1_b 1_{b'} 2_b 2_{b'}$

	1_a	1_b	$1_{b'}$	1_c		2_a	2_b	$2_{b'}$	2_c
1_a	X				2_a	X			
1_b	0	X			2_b	0	X		
$1_{b'}$	10	10	X		$2_{b'}$	16	16	X	
1_c	10	10	0	X	2_c	16	16	0	X

c Later constitution of hybrid **b** = $1_b 2_{b'}$, by loss of $1_{b'}$ and 2_b

	1_a	1_b	1_c		2_a	$2_{b'}$	2_c
1_a	X			2_a	X		
1_b	2	X		$2_{b'}$	19.2		
1_c	12	12	X	2_c	19.2	3.2	X

20	10	0		20	10	0

Phenogram from protein 1   Phenogram from protein 2

FIGURE 6-15

Incongruence between different proteins due to hybridization assuming constancy of evolution rates. *a*, Two organisms **a** and **c** have hybridized to form **b**, and two proteins, 1 and 2, are studied. Evolution is steadily divergent and constant in rate for each protein, although protein 2 is evolving 1.6 times as fast as protein 1. The rate of change, in dissimilarity units, is shown as figures in parentheses on the pathways. *b*, Initially the hybrid contains both genomes, and would thus contain two forms of each protein, 1_b and $1_{b'}$ and 2_b and $2_{b'}$, identical with the proteins in the parental organisms. The dissimilarities are shown for each protein. *c*, The hybrid loses one protein from each parent, and now has the constitution $1_b 2_{b'}$; after a short period of further evolution it is examined and the dissimilarities found are those of the remaining ones, below which are shown the phenograms for each protein. These illustrate the incongruence.

cases, both because the number of available characters would often be too few to demonstrate statistical significance, and because much of the force of the argument rests on the fact that the similarities are between spatial and functional blocks of the genomes.

We have dwelt at some length upon the problem of reticulation for three reasons. First, as cladistic methods are developed and refined, workers will start to employ them in groups other than those where there are strong grounds for believing that reticulation is unimportant. Indeed, this is happening already in the field of human biology; it seems most unsafe to assume that racial and tribal lineages are without significant fusions (e.g., Fitch and Neel, 1969). Workers should therefore be aware of the new problems they may face. Second, it would be a mistake to commence building an extensive corpus of numerical cladistic work in biology upon unsound foundations. The whole subject of phylogeny has greatly suffered in the past for this reason. Third, the statement of the problems may lead to attempts to devise more appropriate numerical strategies to solve them, as has been briefly indicated above.

6.5   NUMERICAL TAXONOMY IN PALEONTOLOGY

In paleontological studies, the importance of exact, numerical methods is even greater than it is with living material. In extinct forms there can be no appeal to genetic analysis, the material may be scanty and incomplete, and unsuspected heterogeneities may complicate what at first sight seem to be single phyletic lines. Reviews of numerical taxonomy in paleontology are given by Rowell (1970) and Kaesler (1970b). Reyment (1963) has discussed the application of multivariate analysis to fossil material, and Van Valen (1969) has reviewed intraspecific variation in animal fossils.

The most obvious application of numerical taxonomy in paleontology is to fairly complete, well-preserved fossil material, in which the hypothesis of non-specificity is likely to hold well enough to obtain reasonably good estimates of overall phenetic similarity. Examples of such studies are those of Rowell (1967) on brachiopods, Kaesler (1969b) on ostracods, Cheetham (1968) on bryozoa, and Lange, Stenhouse, and Offler (1965) on conifers. Ordination has also been used (e.g., Reyment, 1965; Pitcher, 1966; Gould, 1967). Some studies have been primarily directed toward discriminant analysis of previously defined groups (e.g., Reyment and Naidin, 1962). Kesling and Sigler (1969) used Wagner networks (Section 6.4) to reconstruct the cladogeny of certain crinoid genera, and found good agreement with independent geological data.

Fossil material poses several special problems, many of them stemming from inadequacies in the primary data. Specimens may be rare, fragmentary, or not found in critical parts of the geologic record. They may have been sorted by environmental effects, or distorted (though transformation grids might prove useful then—see Sneath, 1967a). In fossils of simple shape it may be difficult to obtain more than a few characters. It is especially unfortunate that many well documented lineages are based on material of this kind (e.g., ammonites, oysters, sea urchins). The application of the allometric equation where possible (Section 4.9), is important in paleontological work because one may have no way of knowing (or estimating) the age of the specimens at death, and the crude character sizes or ratios will sometimes be partly dependent on age and other factors. This has been well discussed by Joysey (1956) and in relation to numerical taxonomy by Fry (1964) and Anstey and Perry (1970). In many instances a fossil form is known from only a single specimen, which may represent a young or an old individual, and it may then be very difficult to know how to code its characters. The same problem, of course, occurs in orthodox taxonomic studies.

Problems of homology (Section 3.4) may be greater with fossils than with living forms because of the lack of supporting evidence from embryology or soft parts. Jardine (1969c) has given an illuminating account of this subject as illustrated by skulls of fossil fish. Benson (1967) has suggested a novel technique for studying shape that may be pertinent here. Nevertheless, most of the studies reported above appeared to be generally satisfactory. One may note, however (as Kaesler, 1967, has pointed out), that cluster analysis may not always be well adapted to analyzing the variation pattern of fossils if evolving sequences are present. Ordination methods (Section 5.6) may often be more appropriate and particularly graphs and trees (Section 5.7), because these are well fitted to finding branched sequences.

A special application of numerical taxonomy that, though it resembles ecological studies, is directed toward a taxonomic problem, is well illustrated by the paper of Kohut (1969). This is the use of clustering methods to group together isolated specimens (in this case conodonts) or fossil fragments that are thought to come from a single organism, but which have been scattered in the sediments. One hopes in this way to distinguish the assemblages derived from different species.

Numerical cladistics and phyletics have been reviewed in earlier sections of this chapter. It remains here to take up some points particularly pertinent in paleontology. The reconstruction of phyletic lineages is largely based on connecting the forms that are most similar phenetically into sequences showing an even progression in characters. Certain restrictions are imposed by the stratigraphy of the fossils, but this is often of little practical value; gaps in the fossil record may make it difficult to know whether a given form had already evolved at a given time, so it may be uncertain whether it might have been an ancestor or a descendant of some other form. The criteria for determining whether a lineage has branched require, in addition, methods of deciding upon the distinctness of the resulting lineages and, usually, some consideration of the principle of evolutionary parsimony (see Sections 6.3 and 6.4). Methods of ordination should be particularly useful in determining sequences in lineages. The techniques employed by archaeologists, including multidimensional scaling (see Cowgill, 1968 and Section 5.6) would also repay closer study. Single-linkage clustering may occasionally be more useful than average-linkage methods because of its tendency to find chains of closely related OTU's.

Besides its importance for studying rates of evolution (Section 6.2), fossil material can contribute to the study of patterns of evolution. Speciation and branching is the most obvious, but there are other patterns that merit investigation. Some points may be seen in Figure 6-16, which is taken from a pilot study (Sneath, 1961) on the fossil fish *Knightia* based on data of Olson and Miller (1958). The four samples are thought to represent one phyletic line, and the individuals have been displayed in two major phenetic dimensions and in time. It may be noted that the comparison of individuals considered as geometric shapes is fairly straightforward. However, when one wishes to draw conclusions about the populations to which the fishes belonged, much more data are required, including allometric transformations to account for possible age differences in the individuals, and also evidence that only one population is represented in each sample. The study showed that the phenetic means did not show a regular displacement with time. This may represent some degree of reversal of evolution, or what Henningsmoen (1964) has described as zig-

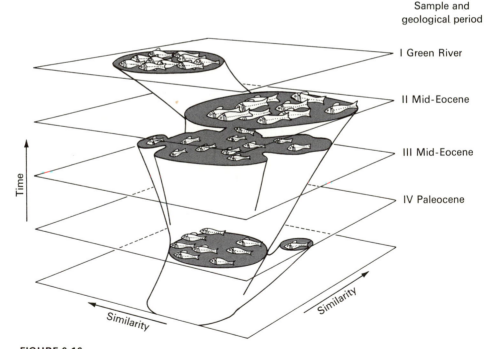

FIGURE 6-16

Schematic and speculative diagram of phylogeny in *Knightia*. The position in the horizontal plane indicates phenetic relationships among individual specimens. The time cuts are equally spaced although the actual time intervals are not equal. The axes labelled "similarity" approximate principal component factor axes. They should properly be labelled "dissimilarity". [From Sneath (1961).]

zag evolution. Alternatively, the ancestors of each group might have been an atypical section of their contemporaries, so that perhaps repeated burgeonings of forms well adapted to the prevailing conditions had occurred, rapidly dying out and contributing little to the succeeding part of the phyletic line.

A recent, especially thorough study of phenetic relationships in a time dimension has been carried out by Rowell (1970) on ten species of Upper Cambrian pterocephaliid trilobites conventionally assigned to three genera, whose stratigraphic position had been fairly well established by previous work. In addition to examining the problem of applying numerical taxonomy to paleontological work in great detail (this paper should be recommended reading to all paleontologists who wish to become acquainted with numerical taxonomy), Rowell elaborates a technique for ordinating the species in a three-dimensional space in which the two horizontal dimensions are the best phenetic representation in a 2-space and the vertical axis represents time (Figure 6-17). The OTU's are also connected by a shortest spanning

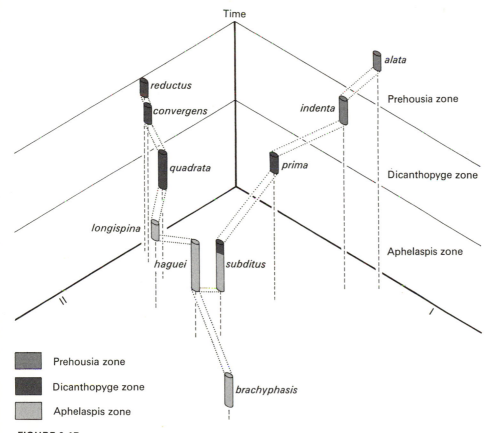

FIGURE 6-17
Inferred phylogenetic relationships between ten species of trilobites. The three axes represent two phenetic dimensions (principal axes) and approximate age. Dotted lines are the shortest spanning tree in the full *n*-space modified by some paleontological considerations. [From Rowell (1970).]

tree that shows the considerable congruence between the presumed phyletic relationships and phenetically closest OTU's. Rowell raises questions about the phyletic relationship of the earliest forms and reviews the evidence in favor of alternative hypotheses. From the figure shown in 6-17 Rowell develops a generalized diagram (Figure 6-18) that is in agreement with known phenetic and chronistic facts. He finds that such a model has high heuristic value, depicting the main features of the phylogenetic history of the group, readily displaying the regions where overall convergence has occurred, and providing information on relative rates of evolution. Although the model does not define the limits of generic taxa it will assist systematists in making such decisions.

Other evolutionary patterns that can only be properly studied by numerical methods are convergence and parallelism, as noted in Section 2.4. Eldredge (1968) has reported a case of convergence in fossil snails by using D^2, though rather few characters were employed. The investigation of mosaic evolution requires techniques for measuring congruence, as noted in Section 6.2. It is possible, too, that numerical methods can be applied to hypotheses of neoteny and macroevolution (de Beer, 1951).

Many of these points are illustrated in a paper by Cheetham (1968) on fossil bryozoa. The cladograms constructed from phenetic and stratigraphic data showed some evidence of zig-zag evolution, and there were also signs that a small degree

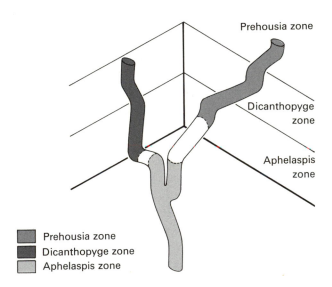

FIGURE 6-18
Inferred lineage of some Upper Cambrian trilobites based on ten species. Phenetic relationships are shown by projection on first two principal components, relative age is proportional to distance along the third, vertical axis. Limits of conventionally recognized taxa of genetic rank shown by shading. [From Rowell (1970).]

of overall convergence, as well as marked parallel evolution of certain characters, had occurred in parts of the phylogeny. Also notable were changes in overall evolution rates; periods of slow evolution without branching (stasigenesis) alternated with bursts of speciation when lineages diverged rapidly (cladigenesis). Cheetham's conclusions are much firmer than those of most studies of this kind, because of their strong quantitative basis, although in some places he departs somewhat from the phenetic findings when they do not entirely fit the presumed cladistic relationships.

Stratigraphy is discussed in Section 11.4. However, the comparison of fossils in different strata would be based on numerical taxonomy of the fossils, even though the results of the taxonomy may be intended to enable geologists to identify the strata in different localities, across geological faults, and so on. Such work is akin to ecological studies in that many different kinds of organisms may be included as components of one OTU (in stratigraphy the OTU's would be strata instead of organisms). Such studies therefore incorporate a new element: strata are similar or dissimilar not only in the number of species of fossils that they share but also in the degree to which the fossils of one higher taxon are similar in the two strata. For example, pertinent evidence on the degree of similarity between the strata **a**, **b**, and **c** might be obtained from the degree of similarity between three species of a given genus, species **x**, **y**, and **z**, each characteristic, respectively, of strata **a**, **b**, and **c**. There is an increasing awareness (e.g., Shaw, 1969; Hazel, 1970; Hughes, 1971) that finer details of stratigraphy can be obtained by numerical taxonomic means such as this than by the traditional methods.

7

Population Phenetics

Except for work in microbiology, where the application of population and species concepts is difficult and the OTU's have usually been individual strains or clones, the earlier development of numerical taxonomy dealt mainly with OTU's at the intermediate and higher categorical levels. Species were grouped into genera and genera into higher groups. The grouping of individuals into populations and species was not attempted until several years had passed, primarily for two reasons. First, the finest levels (the tips of the taxonomic hierarchy) comprise the greatest number of OTU's. In the early days of numerical taxonomy, computational difficulties deterred workers from analyzing such data. Second, while the arbitrariness of delimiting the higher taxa is easily recognized, it was believed that, at least in higher plants and animals, the definition and delimitation of species was based on objective grounds (that it was nonarbitrary).

It has taken some years for the lessons of phenetic taxonomy to penetrate to the species level. This required the confluence of three separate developments, in addition to the development of more powerful computers and computer programs. First, it became increasingly obvious, with the new emphasis on empiricism and operationism in taxonomy, that the biological species definition was difficult to apply and in practice was nonoperational. Section 7.1 deals with this aspect and also discusses the general problem of phenetic patterns in natural populations.

The second development stems from the consistent application of phenetic principles to taxonomy at all levels. Special problems of the methodology of applying numerical taxonomy to species and populations are discussed in Section 7.2. The conclusions about genetic and evolutionary structure of populations that can be obtained from phenetic evidence are discussed in 7.3. The third development was the realization that phenetics was not only a necessary approach to taxonomy at the finer hierarchic levels because of the difficulties of ascertaining breeding structure of populations, but also that the phenetic approach in population biology has considerable inherent interest for the evolutionist. This quite new approach is discussed in Section 7.4. The final section, 7.5, leads to a related topic, the biometric analysis of geographic variation of local populations or species.

7.1   PROBLEMS OF DEFINITION

When considering taxonomic studies at the species level we run head on into one of the most perplexing problems of taxonomy, the problem of what to designate as species. The origin of this problem stems from the recognition that there are relative discontinuities in nature among organisms and individuals, be these phenetic, geographic, or reproductive, or some combination. While the species concept has been a central tenet of biological belief since the early origins of biology as a science, the implications of this term have changed over the years. Different authors at different times have used the term species to denote different concepts and yet, because Linnean nomenclature applies a binomen to these different concepts, the reader is often uncertain about the nature of the entity to which a species name has been given.

At the base of the taxonomist's problems is his uncertainty of what kind of species concept he should employ and whether this should be one that is applicable through all biology. His views often depend on the kind of organisms he studies, but the use of widely differing concepts in different taxa is not conducive to clarity in biology as a whole. Many authors have pointed out that there are three major species concepts that depend on criteria differing in kind rather than in degree. These criteria are genetic, phenetic, and nomenclatural, and various separate terms have from time to time been suggested to distinguish species based upon them. Thus Ravin (1963) refers respectively to genospecies (a group of organisms exchanging genes), taxospecies (a phenetic cluster), and nomenspecies (organisms bearing the same binomen), while Blackwelder (1967a, p. 170) lists many similar terms. No consistent terminology has yet become widely accepted. The paleontologist views the species as an entity extended in time, yielding yet another kind of species concept, the "paleospecies," most fully considered in a symposium edited by Sylvester-Bradley (1956). As a segment of an evolving lineage the paleospecies obviously has no sharp borders in the time dimension, and though intended to be in principle a genetic entity, it is in practice phenetic.

Of these species concepts only the phenetic and nomenclatural are universally applicable, and the latter is readily seen to be secondary to the former. We may note, however, that the phenetic species concept itself is open to different definitions. This topic is taken up below.

A taxonomic species based on morphologically similar populations located in a definite geographic area and morphologically distinct from other populations assigned to different species is perhaps the most commonly employed concept by taxonomists in the plant and animal kingdoms (Davis and Heywood, 1963; Black-welder, 1967a; Michener, 1970). Such a species concept is essentially a phenetic one. As employed in traditional taxonomy the recognition and delimitation of species have frequently been subjective, arbitrary procedures. But the introduction of numerical methods has enabled one to quantify these operations and to make them explicit.

Among species definitions the so-called biological species concept holds a special place. It is intimately tied up with the development of the New Systematics starting in the 1930's. The foremost advocate of the biological species concept, Ernst Mayr, has called it multidimensional (Mayr, 1963) because it deals with populations distributed through space and time, interrelated through mutual interbreeding, and distinguished from others by reproductive barriers. As defined by Mayr a biological species comprises "groups of actually or potentially inter-breeding populations which are reproductively isolated from other such groups." In a more recent statement (Mayr, 1969a,b) he has dropped the much criticized phrase "potentially interbreeding" from the definition. Other biological species definitions are that of Emerson (1945): "... evolved (and probably evolving) genetically distinctive, reproductively isolated, natural population," or that of Grant (1957): "a community of cross-fertilizing individuals linked together by bonds of mating and isolated reproductively from other species by barriers to mating." Objections to these definitions have been made on several grounds. Blackwelder (1962) and Sokal (1962b) maintain that the employment of the bio-logical species concept misleads us into viewing species described by conventional phenetic criteria as having the genetic properties ascribed to the biological species. This in turn leads to unwarranted implications about reproductive relationships from data about which little or no evidence on population structure or breeding structure is available. Sokal and Crovello (1970), in a detailed analysis of the operations required to determine whether a population is a biological species, show that the biological species concept as defined by Mayr or others is in practice nonoperational. Phenetic procedures must be resorted to in order to arrive at decisions on the status of the populations. These procedures in turn do not meet the definition as stated. In the conclusions of their paper they state that just as phenetics is necessary even when delimiting biological species, a biological species concept is not necessary even in evolutionary theory, since much of the theory is based on local interbreeding populations rather than on biological species. The desirable fundamental taxonomic unit would seem to be the phenetic species of conventional taxonomy, the delimitation of which may be improved by methods of numerical taxonomy. Some considerations of a methodology for so doing are discussed by Rogers and Appan (1969).

A phenetic definition of species does nevertheless raise certain other problems. Some of these are discussed in other sections (e.g., 5.11, 7.3, 9.2) but we may note that a phenetic definition can result from consideration of two principal alternatives. We may regard as a species (a) the smallest (most homogeneous) cluster that can be recognized upon some given criterion as being distinct from other clusters, or (b) a phenetic group of a given diversity somewhat below the subgenus category, whether or not it contains distinct subclusters. Whether genetic, phenetic, or nomenclatural concepts are employed, meanings of the term "species" have always been closer to the first alternative; it indicates a *distinct kind*, and by implication the smallest distinct kind. Yet in practice it is the second that is usually chosen if the two alternatives clash seriously, as they do in apomictic taxa containing great numbers of microspecies. Not only might a researcher be faced with naming innumerable microspecies, but one species might show more internal diversity than an adjacent polytypic genus. It is the traditional requirement that a species must bear a binomen that causes taxonomic difficulties at this level, for otherwise subspecific or *ad hoc* infrasubspecific designations could be used for the forms thought worthy of recognition. On the other hand, all categories above species are based on the degree of internal phenetic diversity, and consistency would require one to treat the species category in the same way. But then the species could not be equated with the smallest clusters, because these can be compact or diffuse, and moreover there might be no clear boundaries between them. The solution to this difficulty evidently depends on reforms in our systems of nomenclature (see Section 9.1). The worker who wishes to employ numerical taxonomic methods at the population level can postpone most of these problems until the final stage of a formal taxonomy, and will therefore be inclined to use the first alternative (smallest distinct homogeneous cluster) when sampling OTU's to be employed in the study.

The problems of definition multiply when applied to taxa in categories lower than species. The most common of these is subspecies, the existence, definition, and naming of which has been the source of considerable controversy in recent years. A definition such as that by Mayr (1969a) ("a subspecies is an aggregate of phenotypically similar populations of a species, inhabiting a geographic subdivision of the range of a species, and differing taxonomically from other populations of the species") is not operational in practice—it cannot be used to define critically any such unit in nature. Belief that subspecies exist widely as distinct and definable units and that their putative role is as incipient species in evolutionary processes has led to a description of subspecies that is widespread, especially in vertebrate zoology and the Lepidoptera. Such subdivisions are often based on few characteristics that overlap considerably in their distributions between supposed subspecifically distinct populations. In order to overcome problems of overlapping unit characteristics, "overlap rules" for recognizing subspecies were established (Amadon, 1949); these have been criticized by a number of authors such as Pimentel (1959), Sokal and Rinkel (1963), and Sokal (1965). The fact that an assemblage of populations subdivided on the basis of one character need not always correspond to a similar subdivision of the same assemblage on the basis of a second character is by now evident (see Wilson and Brown, 1953; Gillham, 1956; Sokal and Thomas, 1965; Thomas, 1968a).

Blackwelder (1967a, p. 174) has usefully subdivided the problems encountered in defining the subspecies concept as "(1) whether there is in nature enough diversity to be usefully studied; (2) if so whether this diversity can be treated in the taxonomic system; and (3) if so whether the segregates should be named in the formal system of nomenclature..." He replies affirmatively to the first question and points out that the second question has scarcely been adequately investigated. But in saying that the diversity *might* be treated in the taxonomic system he implies that such a system permits arrangement of the individuals within populations and of populations within the species in hierarchic and mutually exclusive groups, for if one names the groups within the Linnean system of nomenclature one would in effect be implying an affirmative answer for the second question; Blackwelder has recognized this and correctly notes that his third question has "clouded the second."

This again demonstrates our tenet that naming organisms in a system with a predetermined structure may often preclude the correct description of the phenetic variation in nature. Certainly at the infraspecific level it is even less likely that tight hierarchically ordered phenetic clusters, separated by wide gaps, are the correct model for representing populations.

An analysis of the problems of definition of species and subspecies leads inevitably to an inquiry into the nature of populations. The existence of biological populations discrete in some sense from other such populations has been tacitly assumed in theory, and in fact whole branches of science, such as population ecology, population genetics, and now population biology imply that populations of this sort exist. However, the definition of populations and the demonstration of their existence is not easy. There are clearly nonuniform distributions of phenetically similar organisms over any sufficiently large environmental area or space. Note that we have had to include the qualification "phenetically similar" because without this the problem of definition of a population becomes almost insoluble. Sokal and Crovello (1970) distinguish between (1) "localized population samples," essentially a concept from statistical geography, which depends on the place where organisms are found and the ranges they are likely to attain during their life span, and (2) "interbreeding local populations," which are those generally of interest in population biology. A third criterion for delimiting populations is a phenetic one, aiming to establish clusters of phenetically similar individuals distinguished from others by phenetic gaps. The three concepts of distribution, interbreeding, and phenetics are clearly interrelated, yet clusters established on the basis of each will not necessarily be coincident; differences between groupings based on all three should lead to important insights into evolutionary phenomena. Transformations of one set of relationships into the other may offer means of defining evolutionary forces leading to nonequivalence relationships. To assert that one or the other of these viewpoints regarding populations is paramount and that others should be subordinated to it is to exhibit an unwarranted prejudice in favor of one philosophical point of view in taxonomy and is not likely to lead to a deeper understanding of evolutionary mechanisms.

These considerations lead to the conclusions reached by Ehrlich and Holm (1962) in their important paper that advocated study of patterns by which organisms are related in space and time rather than investigation of the concordance in nature between facts and preconceived concepts such as species, subspecies, communities, and even Mendelian populations.

7.2   NUMERICAL TAXONOMY AT THE POPULATION LEVEL

If the problems outlined in the last section are to be investigated fruitfully, comparisons between the phenetic, distributional, and reproductive properties of populations must be made. This will require phenetic analyses of populations at infraspecific levels. Numerical taxonomic work at this level, however, brings with it an entirely new series of problems.

As discussed in Sections 4.3, 5.6, and 8.5, a considerable literature in multivariate analysis of populations dating back into the 1930's is available for such work. Many of these are methods for establishing distances between populations and for ordinating these populations in a parsimonious manner based on as few dimensions as possible. Others attempt to find the best discriminant axes between populations. Relatively few methods concern clustering and grouping of similar populations. The early work in numerical taxonomy was largely at the supraspecific level, and hence did not need to rely on the assumptions of the multivariate methods (multivariate normality, homoscedasticity of variance-covariance matrices), but resorted to the by now familiar techniques of establishing similarity matrices and clustering these to form hierarchic groups. The increasing interest in the application of numerical taxonomic methods to problems in systematics led to the extension of these techniques at the infraspecific level without considering whether clustering methods are necessarily most suited for these problems. No clear answer can be given, even at this date. We are left with the problem of whether taxonomic methodology should primarily use ordination or clustering techniques. In the former case we might represent the phenetic relationships among populations and individuals within populations with a minimum of distortion, but if a large number of populations is sampled and if many individuals are employed for each population sample the resulting dispersion of points in a hyperspace or even in a three-space would probably be unintelligible. Yet any attempt to make order out of such patterns by clustering them according to a given algorithm imposes a structure on these data that may not be justified. Fundamental problems in this regard have already been discussed (see Section 5.3) and need not be restated here.

Various methods of condensation would be appropriate for introducing structure in large ordinated samples, following lines suggested by Sneath (1966e). Not too much has been done about condensing points ordinated in hyperspace and we must therefore consider mostly instances of population phenetic analysis at the infraspecific level based on clustering approaches. If clustering and ordination approaches are to be integrated, problems of scale, character coding, and of dispersions of the characters at the various levels must be investigated. The development of scale free character coding methods (see Section 4.8) should be encouraged. Important also will be the problem of congruence (see Section 3.6) that, as we have seen, may present the investigator with different results at the infraspecific level than at higher taxonomic levels. Problems exist also in estimating the phenetic resemblance of very similar OTU's that may be identical in most of a large suite of character states. Thus the measure of their similarity may depend on differences in one or two character states, an unsatisfactory condition from the point of view of statistical reliability. The problem of resolving dimorphic or polymorphic populations into their components (discussed at the end of Section 5.2) is of great importance at low taxonomic levels.

There is some doubt even about whether meaningful hierarchic structures can be obtained below a given categorical level. Whether this level is below that of the species (whatever its exact definition may be), or whether nonhierarchic phenetic relations begin at a higher level needs further investigation. In fact, if it can be clearly shown over a wide range of organisms that phenetic structure changes markedly at a level commonly designated that of the species, this very fact would aid in the definition of the species category and might further the causal analysis of the phenomenon of speciation. We know that, at conventional specific levels, replicate individuals almost always cluster before they group with nonconspecific individuals (unless the problem is confounded by the introduction of different morphs or developmental stages as distinct OTU's in the same study, as for example, by Boyce, 1964). Thus, in a study of 55 conventionally accepted species of meliponine bees (da Cunha, 1969), each species based on three representatives serving as OTU's, only in 12 species did the conspecific triads not join before clustering with other species. A similar finding was reported by Funk (1964) with 25 apparent species of euzerconid mites and by Moss (1968b) on 15 species in the mite genus *Dermanyssus*.

Turning to results at lower categorical levels, an account of early investigations is furnished by Sokal and Sneath (1963, p. 243). Among the earliest applications of cluster methods by numerical taxonomy at the population level was the analysis by Ehrlich (1961b) of the butterfly species *Euphydryas editha* and *E. chalcedonea*. This study showed that though supposedly conspecific populations clustered together, individuals proximate geographically did not emerge from a cluster analysis of individuals of *E. editha* based on 75 characters. This study was employed by Ehrlich to substantiate his views that patterns of relationship in nature may be inconsistent with their formal taxonomic placement and has been criticized by a number of systematists for being based on only 13 individuals. Although a greater number of OTU's would have been desirable, later studies with larger samples in other groups (reported below) seem to bear out Ehrlich's contentions that phenetic variation may not always be concordant with geography. Furthermore, there may perhaps be significant variation between samples from different years (Mason, Ehrlich, and Emmel, 1968). Among other early studies at the population level could be included a number of the microbiological studies in which the bacterial strains may correspond to individuals (see, for instance, the work of Liston, Weibe, and Colwell, 1963).

Comparisons of clusters based on phenetic analysis of geographic samples yield varying results. There is some correspondence between geographic proximity and phenetic similarity, but the correspondence is not complete. Such findings could be shown by Sneath (unpublished) in data on 20 populations of the house mouse published by Berry (1963) based on the incidence of 35 skeletal variants. Similar results were found by Fujii (1969) in eight populations of the Azuki bean weevil based on 13 ecological characteristics, by Thomas (1968a) for 16 characters of the rabbit-tick *Haemaphysalis leporispalustris*, and by Johnston (1969) in house sparrows based on morphological characteristics of the skins and skeletons.

A study by Soulé (1967a) on lizards showed that island populations were relatively homogeneous phenetically, and that they were most similar to the mainland populations that were closest to each island. A further study (Soulé, 1972) illustrates that general variability of local populations increases with the logarithm of the area of the island. The author believes this to be due to selection for stable genetic polymorphisms in the complex diverse communities of large islands as contrasted with the impoverished ecosystems of small islands. A similar study is that of Berry (1969) on field mice. A dimorphism observed subjectively by the collector (Sokal) in a field sample of 118 galls of the aphid *Pemphigus populi-transversus* could be shown to be supported but not entirely congruent with clusters based on external shape of galls and on the morphology and biology of the aphids (Takade, 1971). In plants, a broad correspondence between phenetics based on chemical characters and large scale geography has been noted by La Roi and Dugle (1968) and Thielges (1969). Similar findings are reported by Hickman and Johnson (1969), though they found that the effects of plant size introduced complications. It may be necessary to grow plants under identical conditions (e.g., Hubac, 1964) to avoid environmental effects.

In studies of population phenetics at the infraspecific level ample thought must be given to the allocation of samples at various hierarchic levels. It is almost always impossible to process as much material as is available, because it is generally easy to obtain population samples based on rather large numbers of individuals. This is complicated by doubt concerning the definition of a local population sample and how frequently such population samples should be taken for a given species. In clustering studies, computing time increases generally as the square of the number of OTU's, so one must do separate analyses for various subunits or take samples limited to the capacity of a program.

7.3   PHENETIC PATTERNS AND EVOLUTIONARY STRUCTURE

The comparison of phenetic relationship and interfertility is of considerable interest. Close correspondence between cross-fertility and phenetic resemblance may give one confidence in the phenetic method. We believe that these considerations should not be stressed too heavily. It is already clear that complete concordance can never be expected and discrepancies are of even greater interest. Some mechanisms that can explain discordance are already well known, such as the pollen incompatibility genes that prevent self-fertilization of many flowers. We may hope that others can be found through comparison of numerical taxonomic with biosystematic studies.

Genetic relationship, like taxonomic relationship, may refer to a number of different concepts. We only take up those that have been studied by numerical taxonomic methods. A few such studies have been made in anthropology, particularly in comparing identical twins with fraternal twins. Thus Vandenberg and Strandskov (1964) found that the within-pair variance of measurements on fraternal twins was several times greater than that of identical twins (and was somewhat greater for boys than for girls). The clear distinction between genetic identity and sibship shows the potential sensitivity of phenetic methods. Huizinga (1965), using head measurements, reports briefly on fathers and sons compared to unrelated persons; the phenetic differences here were less marked.

Other studies examining the phenetics of close blood relations include those of Grewal (1962), Berry (1963), and Berry and Searle (1963) on mice. One point that emerges from their work is that though highly inbred lines are less variable than wild populations, they nevertheless show much phenetic variability. Rhodes and Carmer (1966) in similar work on maize noted that those cultivars (inbred lines) that were closely related by pedigree were clustered together in phenograms. Four large phenons represented the major sweet corn stocks named "flint," "dent-flint," "dent," and "bantam," with some suggestion (as one might expect) that the first three consist of a partial continuum of phenetic forms. Goodman (1967a, 1968b) has made similar studies on maize. Oka (1964) reports some preliminary work of this kind on rice.

Several authors have investigated the effect of polyploidy on phenetics. The members of autoploid series are usually close phenetically (Hubac, 1964; Parups et al., 1966; Bidault and Hubac, 1967), while Heiser, Soria, and Burton (1965) noted that in such a series the hexaploid was closer to the tetraploid than to the diploid parent. Bidault (1968) studied the grass *Festuca ovina* and found the major clusters corresponded to ploidy; there was a cluster of diploid forms and another of tetraploid forms, and within each of these were smaller clusters corresponding to diploid and tetraploid parts of subspecies. For example, the first major cluster contained a subcluster of diploid forms of the subspecies *glauca*, and the second a subcluster of tetraploid forms of the *glauca* subspecies. These findings are probably not due to general size effects, inasmuch as Bidault used correlation coefficients. Such observations raise problems of formal taxonomic treatment similar to those raised by apomictic forms (see Section 9.2). The position with alloploids is less clear. Heiser et al. (1965) noted that experimental *Solanum* alloploids were often very different from both parents, and Katz and Torres (1965) reached similar conclusions on presumed alloploid species of *Zinnia*. This phenomenon is, of course, well known in a nonquantitative way, and relates to the phenetics of hybrids, further discussed later. The protein electrophoretic patterns of allopolyploids may be very close to the appropriately weighted sum of those of their parents (Johnson and Hall, 1965). Ising and Fröst (1969) found that clones of the same cytotype of *Cyrtanthus* species usually clustered together on numerical chemotaxonomy.

As one would expect from the existence in plants of mechanisms to prevent close inbreeding, there is often little congruence between ease of hybridization and phenetic resemblance, and it seems unwise to make hybridization the main test of the validity of a numerical taxonomy, as some authors have done. There is nevertheless some congruence as shown by the studies of Morishima and Oka (1960), Soria and Heiser (1961), Heiser et al. (1965), and Rhodes et al. (1968). For example, Morishima and Oka found that the species of rice in a cluster designated "sativa" all crossed easily, and Soria and Heiser had similar findings on *Solanum*. Rhodes and his coworkers found correlation coefficients of over 0.8 between phenetic similarities and cross-fertility scores among species of *Cucurbita*. Some information is available for bacteria. The relations between genetic behavior of bacterial genera and their phenetic resemblances are reasonably congruent (see reviews by Sneath, 1964a, and Jones and Sneath, 1970) including bacteriophage host range. Dr. H. Morishima (personal communication) tells us that she is at present comparing phenetic resemblance with the pattern of cross-fertility in rice, as judged by behavior of OTU's on crossing with a standard series of tester stocks. This pattern of cross-fertility is in some respects like a phenetic pattern.

Much work is now being done on the application of phenetics to plant breeding. Some of this is related also to the study of character variability and the effects of the environment discussed in the next section (e.g., Goodman, 1968a, 1969; Goodman and Paterniani, 1969; Bhatt, 1970). Other work is concerned with studying the phenetic positions of hybrids in relation to their parents. Hybrids are usually phenetically close to one or both parents, as one would expect, although there are occasional exceptions (e.g., Ramon, 1968). It is often possible to guess the identity of one or other parent from the phenetic position of the hybrid (Bemis et al., 1970; Kaltsikes and Dedio, 1970b). Numerical methods are also powerful in detecting hybrids (e.g., Smith, 1969; see also Rising, 1968, on birds) and this continues the tradition associated with the use of the Hybrid Index (Anderson, 1936), which is still often useful for simple hybrid identification problems (e.g., Goodman, 1967b).

Hybrids are commonly thought to be morphologically intermediate between their parents (a belief based largely on early work with the Hybrid Index), but numerical taxonomic studies on this show that they are not usually intermediate in the sense of lying on the line in phenetic hyperspace that joins the two parents (Cousin, 1956b; Heiser et al., 1965; Katz and Torres, 1965; Wirth, Estabrook, and Rogers, 1966; Whitehouse, 1969). They appear more often to be well to the side of this line. The same was true of a graft chimera studied by Sneath (1968a) that, though not a hybrid, is analogous in combining two genotypes. Alloploids appear to behave in much the same way. This displacement from the line indicates that some genes from one parent are dominant and others are recessive, or it may be the result of overdominance. Using canonical analysis on barley and beans, Whitehouse (1969) found that F_1 hybrids lay on the average midway between their parents but displaced laterally about 35 percent of the interparental distance; the offset was rather less for F_2 hybrids. The scatter in hyperspace of independently formed hybrids of the same parents has not yet been much investigated, but they are apparently usually fairly close phenetically (Heiser et al., 1965; Ramon, 1968; Casas, Hanson, and Wellhausen, 1968). Hubac (1969) observed that hybrids of *Campanula* were closer to the seed parent than the pollen parent, but no sex influence has been noted by other workers. These findings on the position of hybrids have been generally supported by parallel work in chemotaxonomy (e.g., Olsson, 1967; Ising and Fröst, 1969; Dedio, Kaltsikes, and Larter, 1969b). Certain problems in such work when two-state characters are used have been discussed by Sneath (1968a). In particular, considered in Euclidean space, the hybrid cannot lie on the line joining the parents, because it must occupy a corner of a hypercube (this has not been studied for angular measures of resemblance). This may distort the relationships that might have been obtained if the characters could be measured quantitatively and is particularly important with chemical characters that are commonly codominant (see B. L. Turner, 1969) and usually scored only qualitatively. This distorting effect rapidly diminishes as the average number of states per character is increased.

The application of this new knowledge of phenetics of hybrids is being used in plant breeding in several ways. Whitehouse (1969, 1971) suggests it can help in the choice of appropriate parents to produce hybrids with desired properties (the scaling of character axes to reflect economic importance might be also considered). Bhatt (1970) suggests that phenetically dissimilar parents are most likely to yield useful novelties. Errors in pedigree records may often be located by phenetic studies (Rhodes and Carmer, 1966; Rhodes et al., 1970). Such work may also assist in the choice of stocks for preservation in banks of germ plasm (Rhodes, Carmer, and Courter, 1969; Whitehouse, 1969), since in general plant breeders require the greatest variety, that is, stocks with the greatest atypicality values (see Section 4.11). Finally this work may help the understanding of selection in crop plants (Eshbaugh, 1970; Vaughan, Denford, and Gordon, 1970).

The resemblance detected by nucleic acid pairing is in a sense a genetic relationship because it is close to being an expression of the genetic message, though in a sense it is also phenetic; the very good congruence in bacteria between the degree of nucleic acid pairing and the resemblance from numerical taxonomic studies has been discussed in Sections 3.5 and 5.12.

The patterns of distribution of organisms in phenetic hyperspace is no doubt related to the patterns of speciation in different groups. It is possible that the phenetic distribution of individuals is different in sexually reproducing and in apomictic groups of organisms (see Sokal and Sneath, 1963, p. 244). At present there is little pertinent information available about sexually reproducing forms, but some is at hand for apomictic organisms, chiefly bacteria (see Sneath, 1968c; Jones and Sneath, 1970). In bacteria it appears likely that the phenetic pattern consists of compact clusters of individuals (representing a clone or a number of extremely similar clones), with isolated individuals or small clusters (representing uncommon clones) scattered between them. These isolated forms may perhaps originate from occasional episodes of hybridization, or from unusual occurrences of mutation and selection. It is beginning to be recognized that some means of gene recombination is almost universal in living creatures, and in this sense the analogue of the sexual species may be seen in most groups of organisms. However, the mechanisms are often so unlike the well-known sexual mechanisms that it is no easy matter to define and delimit the populations that are undergoing gene exchange. Heslop-Harrison (1962) has discussed this at some length and has suggested some reasons for these taxonomic patterns. This is an area of phenetics in which we expect to see rapid progress very soon.

7.4 PHENETICS AND ENVIRONMENT

In this section we discuss the relation between phenetics and environmental factors. The way in which the environment affects the phenotype is considered first, followed by a discussion of classification of environments on the basis of the responses of phenetic characters to them. The effect of environment on choice of characters has already been discussed in Section 3.6. A good general discussion of this field in plants is given by Davis and Heywood (1963, pp. 335–416).

We should make clear a basic difference between phenetic studies and the more usual studies of environmental effects. Usually one studies the effect of one variable of the environment upon one character of the organism. But in phenetics we study the overall response of the phenotype to different environments. It is thus necessary to calculate the resemblances between OTU's under different environmental conditions. As in more usual taxonomic applications we will wish on occasion to separate environmental effects on the size of the organisms from the effects on shape and to distinguish other appropriate components of phenetic resemblance.

These applications of numerical taxonomic methods can form an experimental science, for which various experimental designs can be adopted. Thus we might change one variable of an environment at a time to find one with the greatest effect on phenetics (or more explicitly, to find the magnitude of the environmental changes that give unit phenetic change). The phenotypic responses form a multidimensional response system; this is an area that awaits exploration by mathematical methods. Factor analysis is an obvious technique for discovering the most important environmental factors and the major components of the phenetic response; analysis of variance and covariance is another important method. In other experimental designs one might not be able to control the environment directly but would be able to observe and analyze "experiments" made by nature. This field seems particularly important for future work on the plasticity of the phenotype. It may also be noted that the phenotype can affect the environment— most obviously perhaps in the case of vegetation, where the macroenvironment determines phenetics, which in turn determines microenvironments.

There has been little work so far in these areas. Phenotypic plasticity is very marked in plants, and studies of clones and purebred seed lines are obvious ones. It would be of special interest to quantify the phenetic responses to environment and ecotypes by following up, for example, the classic work of Clausen and his colleagues (Clausen, Keck, and Heisey, 1940; Clausen and Heisey, 1958; and see discussion by Davis and Heywood, 1963, pp. 390–398). Goodman and Paterniani (1969) have investigated the plasticity of different characters in maize and adduce evidence that some (but not all) characters associated with reproduction are less affected by the environment than are morphological characters. Abou-El-Fittouh, Rawlings, and Miller (1969) used cluster analysis to classify cotton-growing areas into regions where environmental effects on cotton are similar. In using D^2 to study locusts Gillett (1968) found some marked phenetic changes due to changing environmental conditions. A little work has been done in microbiology. Melville (1965) investigated the same bacteria under aerobic and anaerobic growth conditions. Although the direct effects of environment on phenotype were not recorded, the two different environments gave the same phenons. Small changes in phenetic relationships were found by Davis et al. (1969) on altering the growth temperature of certain bacteria. It would be of interest to know whether the relative similarities between OTU's were more stable in the changed environment than the actual similarity values between given OTU's under the different conditions. One can envisage classificatory investigations where the OTU's would be very different phenetically in two environments, yet classifications made from the different environments would be highly congruent, and there may be general biological grounds for thinking this likely. Some notes on the effect of growth temperature and age of culture of bacteria will be found in Sneath (1968d); these two variables have effects analogous to a general size factor.

The use of phenetics to discover the optimal environment for crop plants could be of economic potential. Such an application is not the same as controlling the environment to maximize crop yield, because one might not be able to achieve the ideal state of every attribute of economic importance—higher yields might be of poorer quality, for example. Where numerous attributes of economic significance were invoved one might be able to make a numerical taxonomic approach in the following way. An ideal but hypothetical OTU is constructed and represented as a point in attribute space, whose dimensions are the relevant attributes suitably scaled (not necessarily linearly) to reflect their economic importance. The search then would be for that environment giving OTU's as close as possible to the ideal one, with the hope of finding the best compromise. Such an approach is very close to the rationale of operations research (see Section 11.5) and is similar to the way in which the best genotype (in breeding programs) can be found. Such approaches differ from the more familiar multiple regression and variance-covariance analyses in that the phenetic response is not assumed to be linear. Also, if we consider hill-climbing techniques we find they suggest the best experiments to perform next. For instance, we may consider the change in environment to be the difference between two environmental vectors v_1 and v_2, which correspond to two phenetic vectors φ_1 and φ_2. The search procedure would then be to discover the relations between these vectors, in order to discover the required environment giving the optimal phenetic vector. The technique of response surface analysis (e.g., see Peng, 1967) seems appropriate to obtain the desired results for such problems.

Hutchinson (1957) has suggested that ecological niches should be formulated within the framework of a multidimensional system of environmental requirements. The concept of patterns in phenetic space associated with configurations of ecological space suggests itself very readily (see, for example, Maguire, 1967 and Wuenscher, 1969). Hutchinson (1968) himself has pursued this line of reasoning by illustrating patterns of ecological space occupied by strains or species of rotifers. Thus one might measure the diversification of ecological niches by the phenetics of the organisms that inhibit them. Notable in this connection is a study by Fujii (1969) defining niches from biotic parameters of strains of the bean weevil *Callosobruchus*. Other studies are underway.

The concept of ecological niches is vague. A given habitat may be subdivided into a great many smaller habitats using rather uncertain criteria, which depend in part on the phenotypes of the organisms themselves, since the organisms can change the habitats. Indeed one might say that this method is the only way we recognize one niche as being distinct from any other; if there were no difference in phenetics we would consider them the same, despite differences in physical variables. For example, one school of ecological thought would not consider designating the north and south sides of sand dunes in a desert as separate niches if no differences could be found in the organisms, even though the temperature might be noticeably different. There have been a number of publications noting the difficulty of measuring such ecological entities (e.g., Ehrlich and Holm, 1962; Bock, 1963).

Studies of different environments can be of two kinds. In some, the different environments will affect the phenetics of the same species of organisms (for example, by affecting size, color, frequency, etc.). In other studies the environments will contain different taxa of organisms and the phenetics of these taxa will be an object of study. Studies of this second kind border closely on ecological classification (Section 11.1).

7.5   ANALYSIS OF GEOGRAPHIC VARIATION

Readers of the previous sections in this chapter will have realized that many of the techniques for numerical taxonomy at the population level grade directly into those phenetic methods commonly referred to as geographic variation analysis. These methods have in the past been almost exclusively univariate; that is, investigators would describe the variation over space of a single characteristic at one time, although more than one characteristic might have been investigated. The study of numerous characteristics provided new problems inasmuch as the large amount of data collected needed to be simplified and made interpretable. Although the methods of geographic variation analysis sensu strictu do not fall within the province of numerical taxonomy (since they relate primarily to the variation of characters in a geographic 2-space, or 3-space if topography is included), the points of contact and transition between the two methodologies are so numerous that a brief discussion seems appropriate here. The purposes, problems, and methodologies of testing and of representation will briefly be outlined and suitable references for further study and detailed exposition of the methods will be provided.

The purposes of geographic variation analysis are first of all the description and summarization of patterns of variation and covariation of characteristics of organisms distributed over an area. Any phenetic characteristic regardless of its nature may be so studied. Geographic variation analysis may be carried out on continuous as well as discontinuous variables and also on categorical ones (attributes). Although in general variation patterns have been stressed in relation to spatial distribution whenever the material justifies it, variation in time, as in fossil deposits in any one locality, may be similarly analyzed. A step beyond mere description of variation patterns is categorization. This refers to the grouping together of localities (perhaps by cluster analysis) that are geographically adjacent and whose populations are similar in their characteristics. This may be desired merely for purposes of simplification and summarization, or for the formal or semiformal recognition of a population or a series of populations in terms of the Linnean system. Both aims are consonant with the general aims of taxonomy as we have outlined them in this book. Other purposes of geographic variation analysis include causal analysis of the geographic variation patterns in order to interpret these as adaptations to variation in known environmental factors, such as climatic, topographic, or edaphic variables, or differences in the distributional, reproductive, or ecological patterns within the populations. Finally, geographic analysis may lead to the allocation of unknown specimens to a given population or a geographic locality with a stated probability of success.

In studying geographic variation, problems arise due to errors of three kinds: (1) sampling error due to natural variability of organisms at a given locality; (2) measurement error for the sampled organisms; and (3) errors in representation of the entire area of study by means of a particular set of localities. These errors are discussed in somewhat greater detail by Gabriel and Sokal (1969). The third source of error has been least investigated, but it is of the most profound importance, since very often trend lines bounding contours of equal phenetic values are established without any evidence that the contour lines they imply have any statistical validity. The geographic variation literature abounds with so-called *isophenes* and *isarithmic lines* whose validity is only as good as the validity of the individual points on which they are based and the degree to which the sampling of points over a geographic area represents the trends actually occurring in the area. In recent years statistical geographers have begun to concern themselves with this question (e.g. Haggett, 1965, p. 214; Stearns, 1968), but the field remains essentially open.

The approaches to testing the significance of the results obtained relate also to the method of representation. If the data are plotted as continuous trends with emphasis on the construction of isophenes then the method of *trend surface analysis* will be preferred. By this method a polynomial or Fourier surface is fitted to the local observations for each character by least squares, and contour lines are plotted on a map for easy visualization of the trends. These contour lines estimate the loci of all points with a given value of the character. A first study of geographic variation by such means is due to Marcus and Vandermeer (1966) and an application to zoogeographic data has been made by Fisher (1968). A general discussion of trend surface analysis can be found in Krumbein and Graybill (1965), and in Harbaugh and Merriam (1968). Local errors of estimating parameters of characteristics for any locality can have serious effects on the confidence that may be placed in the fitted trend surfaces. Errors due to nonregular distribution of localities over the geographical area studied can also be serious; this is discussed by Mandelbaum (1963), who also suggests criteria for deciding what is the highest order of the polynomial in the trend surface that should be accepted as "significant."

If one cannot assume that the characters are continuously distributed one may consider that the area consists of separate homogeneous regions differing from each other. This *categorization* approach leads to significance testing by multiple comparisons techniques (see Sokal and Rohlf, 1969, for a discussion of these approaches). Gabriel and Sokal (1969) have explored this approach in depth using Gabriel's simultaneous test procedures, which can be applied to continuous uni-variate data, nonparametric tests (ranked variables), or categorical data, as well as multivariate continuous data. The method consists of finding subsets homogeneous for the characters in question that belong as well to regions that are contiguous in a geographic sense. Those sets of localities that are statistically homogeneous and geographically contiguous (as defined in graph theoretical terms by Gabriel and Sokal) are categorized as being biologically homogeneous. The difficulty with Gabriel and Sokal's approach, as with other multiple comparisons testing, is that often a multiplicity of results emerges, yielding overlapping homogeneous sets of localities rather than a simple categorization of mutually exclusive groups. How-ever, while this may offend one's customary sense of order and neatness and run counter to the common trend of establishing allopatric, named populations, it is undoubtedly more representative of the true relations in nature, where complex breeding patterns exist among various populations distributed in an area. Sokal and Rinkel (1963), Sokal and Thomas (1965), Rinkel (1965), and Sokal, Heryford, and Kishpaugh (1971) applied multiple comparison techniques to geographic variation in the aphid *Pemphigus populi-transversus*, and Thomas (1968a) applied them to the rabbit tick *Haemaphysalis leporispalustris*. Johnston (1969) and Flake, von Rudloff, and Turner (1969) have employed simultaneous test procedures in a geographic variation study of the European sparrow, and a study of clinal variation in junipers, respectively.

Difficulties will arise when clines or time trends occur in the data. If pure clines for any one character are found, the method of trend surface analysis is clearly superior to any categorical method. However, usually either because of inadequacy of sampling or because of the known discontinuity of environmental or distribu-tional factors, breaks in such trends occur, making the categorization approach advisable. Difficulties also arise when the phenetic distribution pattern is mosaic or crazy quilt, in which case again the categorization approach is likely to be more representative of the true pattern of variation. Biogeographical data will rarely be of the quantity and quality needed for simple contouring of the kind commonly employed in geography and related disciplines. Indeed this is a major reason for using the techniques discussed above. Computer methods for contouring are reviewed by Crain (1970).

Representation of the results such as the geographic variation analysis depends on the method employed. When multiple comparisons approaches are used, various shading techniques to indicate the magnitude of a given variable have been employed by Sokal and Rinkel (1963) and Sokal and Thomas (1965), as well as by Thomas (1968a). Gabriel and Sokal (1969) superimposed upon these the require-ment of geographic contiguity, and readers are referred to their paper for methods of representation. Examples of trend surface analysis can be found in the publications referred to earlier. Frequently, three-dimensional models represented in A-space (the ordination techniques discussed in Section 7.2 and described in detail in Section 5.6) are constructed. These models are an attempt to summarize geographic variation patterns described for numerous characters n in a smaller dimensional system ($k < n$ dimensions). Such work becomes necessary when dealing with a large number of characters. Sokal and Rinkel (1963), Sokal and Thomas (1965), Rinkel (1965, in work on the aphid *Pemphigus populi-transversus*), and Thomas (1968a, in work on the rabbit tick *Haemaphysalis leporispalustris*) all plotted factor scores of the character correlation matrices over the geographic area of study and performed various multiple comparison tests on the data. In a similar manner Fisher (1968) plotted trend surfaces of zoogeographic factors. Thomas (1968a) and Gould (1969b, on the pulmonate snails *Cerion* and *Tudora*) ordinated the localities in a factor space summarizing the attributes. In Gould's case, factors were based on only four characters and it is dubious whether the extraction of three factors from such a correlation matrix is worthwhile. When localities are ordinated in A-space the comparison of the two-dimensional ordination graphs or the three-dimensional models with geographic distribution patterns of the localities may frequently prove of interest. Quite commonly at least one of the axes of variation in the model gives a geographic direction (either longitude or latitude, or possibly topography). One would like to be able to transform the phenetic ordination into a geographic ordination and to impart some biological and evolutionary significance to the entries in a transformation matrix between these two models. Holloway and Jardine (1968) have measured the goodness of fit between a two-dimensional ordination and geographic position in an ecological study, and their methods could be readily applied to phenetic data. Johnston (1969), in a study of the geographic distribution of the European sparrow, has attempted to correlate phenetic dispersion with geographic distribution pattern and has shown that while no 1:1 correspondence exists, there is sufficient correspondence to engender useful hypotheses about the evolution of these sparrows.

Only a few general conclusions have yet emerged from this work. Over large geographical areas there is commonly good correspondence between phenetics and geography or climate (e.g., Thomas, 1968a; Power, 1969; Morishima, 1969b; Hickman and Johnson, 1969; Banks and Hillis, 1969) and in some cases the variation is sufficiently discontinuous to give reasonably distinct phenetic clusters. If variation consists of even clines, however, such clusters may not be found, and on a small geographic scale the pattern of variation appears to be extremely diverse (e.g., Berry, 1963; Petras, 1967). Size and shape coefficients may give notably different results in work of this kind (Rees, 1969a,b). In a microgeographic study meaningful clusters did not emerge (Sokal, Heryford, and Kishpaugh, 1971), and it is possible that such absence of structure may be used to indicate an important taxonomic rank level. It may well indicate panmictic populations or a homogeneous environment. For a comprehensive review see Gould and Johnston (1972).

   As computer technology increases, automated mapping of operational bio-geographical units (Soper, 1964) will become commonplace and the development of optimal representation techniques for summarizing geographic variation is therefore an important goal of systematic research. In a few years we also should have a sufficient study of various models that will permit us to generate evolutionary hypotheses about the origins of those variation patterns. Undoubtedly this will require substantial advances in statistical methodology as well as considerable theoretical work in the form of model building and computer simulation.

8

Identification and Discrimination

As we began Chapters 4 and 5, so we shall also begin this chapter by presenting, in Section 8.1, the form in which the data are given—in this chapter, for purposes of identification. We must, of course, already have groups of individuals or OTU's against which to identify an unknown, and these groups will normally be taxa. General considerations for identification and discrimination follow in Section 8.2, and sequential and simultaneous keys are considered in Sections 8.3 and 8.4, respectively. We conclude the chapter with a discussion of discriminant functions in Section 8.5.

Work on identification has not been as intensive in recent years as has been the work on classification. Thus much of the discussion that follows must be tentative and programmatic rather than definite. However, we hope that just as our earlier outline of procedures for classification (Sokal and Sneath, 1963) led to an increased development and improvement of such methods, so the sections that follow will stimulate biologists, mathematicians, and computer scientists to produce a theory and technology of taxonomic keys compatible with our present knowledge and capabilities. There are already signs that the field will advance swiftly.

8.1   THE IDENTIFICATION MATRIX

A data matrix arranged for purposes of identification may be called an identification matrix $\mathcal{I}$. It is shown in Table 8-1. It consists of a number of submatrices, that is,

TABLE 8.1.

The identification matrix $\mathcal{I}$ and vector **u**.

Characters	Taxa			
	OTU's in taxon **A**	OTU's in taxon **J**	OTU's in taxon **Q**	Unknown OTU
	$a_A, \ldots, j_A, \ldots, t_A$	$a_J, \ldots, j_J, \ldots, t_J$	$a_Q, \ldots, j_Q, \ldots, t_Q$	u
1	$X_{1aA}, \ldots, X_{1jA}, \ldots, X_{1tA}$	$X_{1aJ}, \ldots, X_{1jJ}, \ldots$	$X_{1aQ}, \ldots$	X_{1u}
2	$X_{2aA}, \ldots, X_{2jA}, \ldots, X_{2tA}$	$X_{2aJ}, \ldots, X_{2jJ}, \ldots$	$X_{2aQ}, \ldots$	X_{2u}
$\vdots$				
n	$X_{naA}, \ldots, X_{njA}, \ldots, X_{ntA}$	$X_{naJ}, \ldots$	$X_{naQ}, \ldots$	X_{nu}

it is partitioned vertically into q blocks, each block representing a taxon $J = (A, B, \ldots, J, \ldots, Q)$. Within any block **J** are the individuals or OTU's that provide the information on the taxon, i.e., they are the sample of organisms that represent the taxon. These OTU's are numbered $a_J, \ldots, j_J, \ldots, t_J$ within each block **J**. The rows of the matrix represent the n characters $(1, 2, \ldots, i, \ldots, n)$ for the OTU's. A character state value in this matrix has three subscripts; thus X_{ijK} is the value for the ith character of the jth OTU of the Kth taxon. One or more subscripts will be omitted when the meaning is clear. The matrix may, of course, be partitioned differently if the taxonomic rank of the taxa to be considered is changed; thus in studying a family, for example, one partition might be into tribes, another into genera. The $\mathcal{I}$ matrix may often be the same as the original data matrix (Section 4.1) except that the OTU's are reordered and grouped into taxa.

On the right of Table 8-1 we have a column vector for the unknown OTU (an individual) to be identified, symbolized by **u**. Its elements are character state values symbolized as X_{iu}, where $i = 1, 2, \ldots, i, \ldots, n$, as before. Capital letters symbolize taxa to emphasize their resemblance to matrices rather than to vectors. Thus, though we may replace the taxon by character averages (for example), these are obtained by operating on an $n \times t_J$ block of the $\mathcal{I}$ matrix; the vectors representing the averages are collected into a new matrix, in which column vectors represent taxa as averages of character values. Clearly, too, we may have many unknowns to identify, but at any one instant we normally have only one. Successive identifications thus formally entail replacing **u** by other unknowns in turn.

In most applications the matrix will not be partitioned horizontally; the same n characters will normally be recorded for all taxa. However, though n may be the number of characters studied in the whole numerical taxonomic study, we may discard some of them as being of little value in identification and to reduce unnecessary computation. Where it is pertinent we will use $m < n$ to show that n has been reduced to a smaller character set (this is a use of m different from that in Section 4.4, where it means the number of matches in an association coefficient, and is different from m used for the number of states of a character in various sections).

Since characters are quite properly weighted for identification, we require a symbol for this, and use w_{iJ} for the weight of the ith character when testing **u** as a member of taxon **J**, or w_{iJK} when deciding between taxa **J** and **K**. This is principally used in discriminant analysis (Section 8.5). Over all characters the weights constitute a vector **w**. Moreover, certain characters should be preferred over others because they are readily and constantly observable, so that an additional weight, e_{iJ} (expressed as a vector $\mathbf{e_J}$), may be given to symbolize ease of observation of character i in taxon **J**.

Missing values in the identification matrix may be coded NC, but they are of two types that may sometimes require separate symbols. These may simply be unrecorded values (e.g., petal color blue, but unrecorded) or they may be inapplicable (e.g., petal color when there are no petals). They need distinguishing, because a specimen with blue petals can be excluded as a member of a species without petals, but it could belong to a species whose petal color had not been recorded.

The identification matrix is often transformed into some other matrix before constructing a scheme for identification. There are two main types of transformed matrix. The first replaces the t_J columns of a taxon **J** by one or two columns representing some simplified summary of the character values, such as their means, ranges, standard deviations, and for 0,1 characters in particular, the proportion of OTU's with a given state. The second main form is a variance-covariance matrix (or a correlation matrix) between characters together with vectors of means, the starting point for discriminant analyses.

8.2   GENERAL CONSIDERATIONS

The objects of any identification scheme are ease and certainty of identification (Davis and Heywood, 1963). All other considerations are secondary. If one identifies an unknown specimen, this presupposes that one already has taxa with which to identify it. The form of the identification matrix shows this clearly. We distinguish therefore between classification in the sense of making classes, clusters, or taxa, and identification. The use of the word classification by many statisticians to mean identification is particularly confusing, and this is why we emphasize the point. There are some strategies that combine the two procedures, usually by successively "identifying" new individuals, but these also require criteria for deciding when identification with an existing class is unacceptable, so that new classes may be started (e.g., Ornstein, 1965; Rosen, 1967). These methods then become effectively cluster analyses, and we believe the distinction is a useful one.

The main methods used in identification are keys and discriminant functions. By far the commonest and most versatile are the former. Two differences between identification keys and classifications may be noted. Keys are not necessarily natural classifications in any of the usual senses. The divisions of the key may be quite arbitrary, as long as they are convenient for identifying specimens. Also, the same taxon can key out many times in different parts of the key; it need not have a unique position. Discriminant functions are more restricted in scope and much less often used. The various subdivisions of these methods are described later in this section, so we digress now to some general points about discriminatory characters.

We noted in Section 8.1 that characters are quite properly weighted for purposes of identification. There are two main approaches to calculating the appropriate weights, w_{iJ}. The most usual is based on the frequencies of various character states in different taxa but ignores correlations between characters. A detailed discussion is given by Ledley and Lusted (1959a). Since highly correlated characters tend to behave as a single character, this approach is likely to give overestimates of the probability that a given identification is correct (a good study of this is that of Mosteller and Wallace, 1964). The other approach considers the correlations between characters and is employed particularly in discriminant analysis. It is theoretically more powerful and precise.

Another, different form of weighting noted in the previous section is weighting according to ease of observation of different characters, e_{iJ}. Characters that are prominent, unlikely to be confused, and found in all specimens and during much of the life cycle (or in plants, throughout the year) are to be preferred. These weights, though unavoidably subjective in part, should also take account of the chance of loss of organs through damage or the cost of obtaining a given measurement.

In practice it is usual to reduce the original list of n characters to a smaller list of m characters, (the smallest effective number). The choice of characters for discrimination may be carried out in many ways. Inspection of the original tables of data after rearranging the columns to give the $\mathscr{I}$ matrix is the most usual (e.g., Steel, 1965; Moss, 1968a). A character that is invariant throughout is clearly useless, but unless the data being used are part of a larger study, such characters will have already been deleted. For two-state characters one can use the algebraic difference between the frequencies in two taxa of the 1 state symbolized as G by Sneath (1962); the most discriminatory characters have the highest values of G (positive or negative). This is a simple method that Hall (1965b) found useful in a botanical study, but nonadditive scoring (Section 4.8) causes difficulty.

Gyllenberg (1963) obtains the 0,1 characters most useful as discriminators (on the average) as follows. The proportion of the 0 or 1 values for character i, whichever is the greater, is noted for each taxon, and Gyllenberg calls this C. The sum of C over all q taxa, ΣC_i for character i, is then a measure of the value of i for separating groups. Characters that are least variable within taxa score highest, and $\frac{1}{2}q \leq \Sigma C \leq q$.

Next a separation figure, S_i, is calculated for the character, which is the product of the number of taxa, q_1, in which the character is predominantly 1 and of the number of taxa, q_0, in which it is predominantly 0 (using chosen cutoff levels such as 0.9 and 0.1). The value of S_i is greatest for characters that divide the taxa as nearly as possible into equal halves. The general usefulness of a character as a discriminator is indicated by the rank figure, R_i, which is $\Sigma C_i \times S_i$. Characters with the highest R_i are preferred for constructing identification schemes. It is often sufficient to calculate the values of S_i and an example of its use to select new tests in bacteriology is given by Lapage and Bascomb (1968).

Maccacaro (1958), Möller (1962a,b,c) and Jičín, Pilous, and Vašíček (1969) use rather similar methods but employ information statistics. This can be generalized to multistate characters, and the expression $\Sigma^q_{J=1}[-(\Sigma^m_{g=1}p_{gJ}\log_2 p_{gJ})]$, where p_{gJ} is the proportion of the gth of the m states of the character in the Jth taxon, may be useful. Niemalä, Hopkins, and Quadling (1968) give two methods for 0,1 characters. The first is to compute for each character the quantity $\log(q_1 + q_0)!$ $-(\log q_1! + \log q_0!)$. The highest values are given by the characters that are the best separators. By an extension of this last formula they also obtain the m characters that are jointly the best (which are not necessarily those with the highest values when considered singly). The second method is to operate on the $\mathscr{I}$ matrix and to delete characters in turn, providing the deletion does not make any pair of taxa indistinguishable. The character states for the taxa are recoded as 1 and -1 (it is implied that the commoner state is used), and $A_i = |\Sigma_q X_{iJ}|$ is calculated for each character. The characters are then discarded in diminishing order of A_i (this deletes the least useful characters first) until the chosen number, m, remain.

Another way to rank characters is the method developed by Bonham-Carter (1967a), who does so by the magnitude of chi-square values. His null hypothesis is the independence of the taxa from marginal totals of the characters summed over all taxa. Several of the information-theoretic methods can also be adapted for this purpose (e.g., Estabrook, 1967; Bisby, 1970b).

The classificatory method of Lockhart and Hartman (1963) and association analysis (Williams and Lambert, 1959; Section 5.4) extract discriminatory characters in the course of constructing monothetic groups. Although most taxa are at least partly polythetic, one may find some character states that sharply distinguish any two taxa; that is, they are present in all members of one taxon and absent in all members of the other. There may be no single states of this kind, but it may be possible to distinguish the taxa by using several character states that occur with different frequencies in the two taxa. This latter situation, phenetic overlapping, is found in taxa that are fully polythetic (see Section 2.2). It is here that discriminant analysis (Section 8.5) is particularly valuable.

The minimum number of characters for discrimination is easy to calculate. No more groups can be distinguished than the product of the number of character states. Thus three characters, two of three states and one of four states, allow at the most the distinction of $3 \times 3 \times 4 = 36$ groups. In general: log (number of distinguishable groups) $\leq \Sigma$(log number of character states). Actually, because of character correlations only rarely will the number of distinguishable groups be as large as the product of the number of character states. In practice many more characters are required than the theoretical minimum: examination of various dichotomous keys in the literature shows that a given character seldom serves as a convenient separator in more than one part of the key, so that one needs about as many characters as there are branches, even if one character (and not more) is used at each branch point. Since for a dichotomous key $q - 1$ branches are needed to separate q taxa, the ratio of characters to taxa, m/q, is usually over 1 and may be as high as 2 or 3 if the taxa are difficult to separate, or if the author wishes to make a very reliable key. The contrast between theory and practice is seen by the fact that the theoretically minimum number of characters required to separate an estimated ten million species of living organisms is only 24 two-state characters, whereas Munroe (1964) believes that about 500 characters would be needed in practice. Munroe's figure is probably an underestimate, but since it is not likely that ten million characters would be needed, this suggests that the relation of m to q is still poorly understood (see Ledley and Lusted, 1959a,b; Osborne, 1963a,b) and needs further study.

It should be noted that there are two possible errors in identification. First, an unknown may be identified as a member of taxon **J** when it should be identified with another taxon in the scheme, **K**. Second, an unknown is identified with **J** but belongs to a taxon outside the study entirely. Some schemes use a criterion for successful identification and include a provision for recording "no identification made" to guard against the second danger. But this possibility is still a serious danger, because such tests will not always work. If, for example, some very similar taxa were inadvertently omitted it is likely the characters that discriminate between them and the included taxa might not have been chosen. Identification schemes should therefore be comprehensive with regard to taxa, and limitations of age, sex, life stage, etc., must be clear. A third type of error is of course the exclusion of an unknown from any of the known taxa when in fact it is a member of one of these.

Identification methods overlap a good deal, but we divide them into two main types, the sequential and the simultaneous. The *sequential methods* are the usual diagnostic keys and certain related schemes like multiple entry keys. Sequential methods can be divided into monothetic and polythetic ones. The *simultaneous methods* include discriminant functions and also others where some measure of agreement over all characters is employed, so that the identification can be made at one step. The synoptic table is an informal device of this kind. In many of these methods the unknown is in effect placed in a phenetic space and its closeness to (or inclusion within) known clusters is determined; they can therefore be considered as phenetic distance models. But not all involve distances, so the broader term of simultaneous methods is preferred here. Simultaneous methods are almost always polythetic.

Discriminant functions are probabilistic by design, and any of the others can be made so. By probabilistic we mean that some measure is given of the likelihood that the identification of a given specimen is correct. The need for this is least at high ranks, where taxa are sharply distinct.

This aspect should be given more attention and identification schemes should wherever possible be thoroughly tested with specimens that were not used in their construction. Probabilistic considerations also enter in another way. Schemes can be devised that identify members of some taxa with higher probability than others. In some work the Bayesian approach, in which the probabilities of correct identification are highest for the commonest taxa, might be desirable on the grounds that occasional misidentification of a rarity was less serious than misidentification of common forms; but in other work a converse approach might be desirable. It may be noted that these probabilities are not necessarily related to the weights given to the characters, for in monothetic keys the effective weight at a given couplet is infinite, because all other characters are ignored at this division.

Yet another consideration of probability that will affect the identification procedures, especially in large scale screening, is the Bayesian consideration of the likelihood of a given taxon being found in nature. We are less likely to identify an unknown OTU u as belonging to taxon J if we know that only very few individuals belonging to J have ever been collected. These considerations, which have not so far been extensively applied to identification schemes, are relevant to both sequential and simultaneous procedures.

Identification schemes, like other algorithms, can handle only a certain limited number of taxa conveniently. If there are too many taxa they must be divided into several schemes, and a sequential strategy is then superimposed on that of the schemes themselves. In this the schemes are much affected by practical considerations—length or complexity, the number of characters demanded, etc.—all of which must be balanced against their success rate.

Computers will increasingly be used in this field as electronic data processing comes into use in systematics. The only practicable way of originally calculating discriminant functions is by computer. Some taxometric programs now provide lists of characters of high discriminatory value. We need, however, to distinguish two different uses of computers in this connection. First, the computer may make a key or discriminant function that can be printed and used independently. This may be their major use for some time to come. Second, the identification scheme may be stored in the computer so that it is used "on line". This is most promising for simultaneous methods with large collections of data (e.g., Goodall, 1968a; Lapage et al., 1970). We have pointed out elsewhere (Sokal and Sneath, 1966) that this will make acute the problem of standardizing descriptive terms throughout large taxa.

8.3   SEQUENTIAL KEYS

Sequential keys can be constructed according to any desired sequence of divisions of the set of taxa into successively smaller subsets. The first decision is whether the sequence is to follow the established taxonomic hierarchy, or whether the most efficient but probably quite artificial system is to be used. Since the objectives of identification are different from those of classification, there is no strong reason why the taxonomic hierarchy should be embodied in the key, although in large studies it may be convenient to set up successive keys based on selected rank categories. Thus in a family one may have a key to genera, and for each genus a separate key to species, but the key to genera need not show the subfamilies and tribes.

As noted earlier, a given taxon may occur many times at the tips of the key (though this may lead to the suspicion that it is not a very natural taxon). This is better than to construct the key so that the taxon occurs only once with many cross-references from other parts of the key (Davis and Heywood, 1963). Osborne (1963a,b) believes that such repetition (he calls keys of this type reticulated) is likely to be inefficient on mathematical grounds.

Although keys can have more than two alternatives at each step, the clarity of dichotomous keys is a considerable advantage, and we therefore restrict our discussion to them. Osborne discusses several aspects of the branching structure and how one may make the key as short and efficient as possible. A dichotomous key for q taxa has $q - 1$ branch points (unless taxa occur more than once at the tips). The number of furcations is thus the same however it is arranged (Figure 8-1), but if the OTU's are split off one at a time, using distinctive characters, then the key requires $q - 1$ different characters (Figure 8-1,b). Furthermore the average number of characters that must be examined to identify an unknown specimen is higher (and very much higher for large q), than if the paths bifurcate repeatedly (Figure 8-1,a). Osborne notes that the latter type of key is generally easiest to use, most rapid and most reliable. There are occasional exceptions (because it may sometimes be convenient first to dispose of a few highly distinctive taxa using characteristic features), but in general each division should be made on a character that as nearly as possible divides the taxa under consideration at that point into equal halves. This conclusion is also reached on grounds of information theory (Maccacaro, 1958; Rescigno and Maccacaro, 1961; Möller, 1962a,b,c) and underlies several of the methods for choosing diagnostic characters mentioned in the last section. With repeated branching one can key out 2^m taxa in m levels of the key, theoretically using only m characters. Osborne points out that if the chance of making a mistake in answering a question is the same for each character, this key will give fewest errors. These considerations may not be important in practice, however. The procedure of finding characters that give division into equal numbers of taxa could lead to unreliability at later branches. The distinctive characters may be less

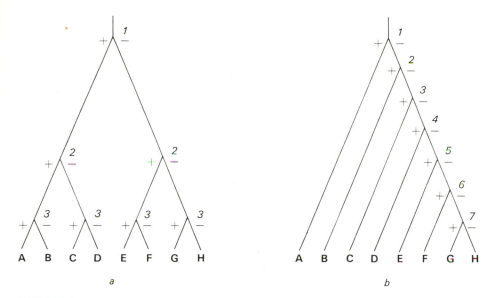

FIGURE 8-1

Two arrangements of a dichotomous key for eight taxa, **A** to **H**. The characters used are symbolized as *1* to *7*, each with two states, present (+) or absent (−). *a*, The paths branch repeatedly, and only the theoretical minimum of three characters is required. The number of characters to be examined to identify an unknown specimen is three. *b*, One taxon is split off at a time on the basis of a distinctive character. Now seven characters are required and the average number that must be examined is $28/8 = 3.5$.

liable to errors than the others. Also, as noted in the last section, it is rare that anything like the theoretical minimum will be sufficient.

The "bracket" and the "indented" keys are the most common forms, though the terminology is confusing since either can be indented; they differ mainly in typographic layout and are illustrated by Mayr (1969a, p. 278). The bracket key is the most generally useful, because it can be worked in reverse so that one can retrace a false lead. We do not describe the details of making keys and refer the reader to the articles of Ainsworth (1941), Voss (1952), Metcalf (1954), Stearn (1956), and Mayr (1969a). It may be quite difficult to make a good key that is simple, short, and efficient. Blackwelder (1967a) notes that at high taxonomic ranks the choice of characters may be very difficult, because some exceptions are likely to occur with most characters (for example, there are arthropods without legs).

Monothetic Sequential Keys

Monothetic sequential keys have a single contrasting statement in each couplet, referring to only one character, to be answered (in principle) by a single yes or no. They are, of course, vulnerable to exceptions, as are all monothetic schemes.

Osborne (1963a) suggests that the characters should be scaled on an integer scale of 1 to 4 so chosen that the compiler can be fairly sure the user will recognize the correct integer from his examination of the specimens to be identified. With this scheme the key will be reliable if every taxon differs from every other by at least a score of 2 upon one or more of the characters.

Several on-line computer schemes are now being developed. Rypka et al. (1967) and Rypka and Babb (1970) have incorporated Gyllenberg's scheme (Section 8.2) into a computer program with some additional modifications. They first compute Gyllenberg's S for each character and select the character with the highest value. This is the best initial separator; they then choose as the next character the one giving the highest joint S with the first. The third character chosen as divisor is that with highest joint S with the previous two, and so on. The joint S is calculated as follows

$$S_{\text{joint}} = \tfrac{1}{2}[q^2 - (q_a^2 + q_b^2 + \ldots + q_z^2)]$$

where q is the total number of taxa, and $q_a + \ldots + q_z$ are the numbers of taxa possessing the z various unique combinations of 0,1 states in the n characters considered. With binary characters as here, $z = 2^n$, but many of the combinations will probably not occur. Rypka and his coworkers note that one can compute directly the best pairs, triples, quadruples, etc., of characters, using all possible combinations of the characters, but this makes heavy demands on computing time. Although intended to identify bacteria, this method is likely to make insufficient provision for exceptional isolates, and may be more suited to higher organisms.

Multiple entry keys are another group of keying schemes that are generally monothetic and sequential. A good description of one form is given by Leenhouts (1966). Each taxon is listed against each character arranged under the two leads, for example:

Leaflets	Taxa
(a) Entire	A B D (F) (G*)
(b) Not entire	C E (F) (G*)

Taxa that can possess either state are in parentheses, and those whose state is unknown are given also asterisks. To use the key one chooses any character and excludes taxa that do not agree, then chooses another character, and continues until only one taxon is left. The principle is readily applicable to superimposed punched cards ("peek-a-boo" systems), of which a good example is a key to the families of flowering plants produced by Hansen and Rahn (1969). Each card represents a character with a fixed position on it for each taxon; selections of cards are superimposed until only one perforation remains, indicating the required taxon.

For example, overlapping cards No. 8 (tendrils present), No. 53 (flowers zygo-morphic), and No. 133 (carpel 1) leaves only family—No. 29 (Papilionaceae). The principle is also readily applied to on-line computing, and descriptions of systems that follow a similar strategy have been given by Boughey, Bridges, and Ikeda (1968), Goodall (1968a), and Morse (1971). An advantage of multiple entry keys is that the user may employ any of the key characters that are available on the specimen; with the usual keys he must have the very characters required for each couplet in turn.

Although monothetic groups are seldom required in taxonomy, monothetic cluster methods (e.g., Maccacaro, 1958; Williams and Lambert, 1959; Lockhart and Hartman, 1963) may be useful for constructing keys because they yield charac-ters that are likely to be near optimal for key making. For this purpose Gower (1967a) has suggested subdividing on the character that, at each dichotomy, maximizes the multiple correlation between it and all previously unused characters.

Polythetic Sequential Keys

Polythetic sequential keys are keys in which at least some couplets consist of several statements about different characters. These are the commonest form of taxonomic key. The reasons for using several characters are threefold: (1) one or more characters may be unobservable on some specimens (for example, damaged specimens or plants not in flower); (2) there are a few taxa (or individuals) excep-tional in the most readily observed characters; and (3) the user may make a mistake in deciding about a character. In each case the other characters help the user to decide which branch to take. The basic idea is that of the majority vote; unless the key says otherwise, the user is best advised to follow the majority verdict (we note, however, that this is rarely stated explicitly, and some workers intend the first character to be more important than the others). In other words, the user gives preference to the alternative most similar to his specimen, but no one character is essential. The strategy is thus basically polythetic, consisting of a comparison of the specimen with the statements in the couplet, followed by choice of the best match.

This procedure affords many advantages, not the least of which is a better pros-pect of accurate probabilistic estimates. It does have the disadvantage of being somewhat less clear-cut. Also, the procedural rules we have just mentioned are not self-evident. Polythetic keys, therefore, require some formalizing. With monothetic keys, at any branch point the single character has decisive value (all others have zero weight). In polythetic keys the characters require weights, either differential weights or else a specific statement that they are equal. Hall (1965b) gives an illustra-tion of a key in which such weights are attached to each character in the couplets. These weights are not only the statistical discriminatory weights, w_i, but also the ease of observation values, e_i. The latter cannot be so readily estimated as the former, for they depend a good deal on the experience of the user, so the practical problems of using keys are clearly relevant here. For example, are the difficulties due more to poor character descriptions (perhaps they are described in highly technical language without a diagram, one of the biggest stumbling blocks for the in-experienced) or to difficulty of observation, as when special microscopic prepara-tions are essential?

Several methods have been developed for producing keys automatically by computer (Morse, 1968, 1971; Pankhurst, 1970a,b; Hall, 1970). Pankhurst gives extensive details of the technique he uses. The key is constructed from a table of character state values for the taxa. These values can be qualitative or quantitative, and provision is made for missing data, although if there are many missing entries this greatly increases the difficulty of making a key. The number of characters at each branch point can be controlled. Either an "indented" or "bracketed" key can be produced (either of them with or without typographic indentation) in a form ready for use (Figure 8-2). When a taxon keys out, all remaining distinctive characters for that taxon are furnished by the computer program; they are useful as a check on the identification. The user can allocate weights, w_i or e_i, to each character, or he can weight the taxa to obtain short identification routes to taxa of his choice (such as commoner ones). Taxa are allowed to key out several times if this is the only way to get a key, and highly distinctive taxa key out early (but this is allowed to happen only rarely because it interferes with the attempt to optimize the key). The basic procedure is to find characters that divide the taxa into equal halves, with preference given to dichotomies over polychotomies. Pankhurst uses a separation function $F = F_1 + F_2$, where

$$F_1 = (k - 2)^2 \text{ and } F_2 = \sum_{b=1}^{k} |1 - (q_b\, k/q_a)|$$

and where, at a given branch point, a, with q_a taxa under consideration, there are k subgroups each containing q_b taxa. The divisions on different possible characters are tested, and that with minimum F is preferred subject to certain accessory con-ditions, such as that characters with high w_i are considered first. The methods of Morse and Hall use rather similar principles. Morse (1971) makes special provision for characters that are variable within taxa. The character for the initial couplet is the one giving the highest value of $DV \times \exp\{CV\}$ provided the character is not unknown or inapplicable for any of the taxa under consideration. DV is calculated as $2q_T q_F + \frac{1}{2}q_V(q_T + q_F)$ where q_T, q_F, and q_V are the numbers of taxa for which the answer to the first lead is true, false, or variable, respectively. The value CV is a "convenience value" given to the character by the user. This procedure is repeated for subsequent branches of the key.

Computer-made keys are generally as short or shorter than manual ones, and if appropriate values of w_i are chosen they appear comparable to manual keys in quality. It is likely that in the near future they will become superior to those

1	Stem 0–10 cm	2
2	Sterile rosettes absent, capitula more than 3 cm	17 J. fontqueri
2	Sterile rosettes present, capitula up to 3 cm	3
3	Capitula obconical, involucral bracts lax, patent or recurved	15 J. humilis
3	Capitula subglobose, involucral bracts appressed	16 J. taygetea
1	Stem more than 10 cm	4
4	Pappus shorter than achene	11 J. polyclonos
4	Pappus longer than achene	5
5	Involucral bracts lax, patent or recurved	6
6	Capitula more than 3 cm	7
7	Stem leafy throughout	10b J. mollis. ssp. moschata
7	Stem leafy at base	8
8	Involucral bracts lanceolate	10 J. mollis
8	Involucral bracts linear	14 J. glycacantha
6	Capitula up to 3 cm	9
9	Stem woody at base	6 J. albicaulis
9	Stem herbaceous	10
10	Basal leaves subglabrous above, tomentose beneath, achene more than 5 mm	9 J. eversmanii
10	Basal leaves puberulent above, tomentose beneath, achene 2–5 mm	12 J. ledebouri
5	Involucral bracts appressed	11
11	Basal leaves subglabrous above, tomentose beneath	12
12	Distal crown of achene inconspicuous	13
13	Capitula subglobose	8 J. cyanoides
13	Capitula hemispherical	13 J. consanguinea
12	Distal crown of achene conspicuous	14
14	Rhizome absent	2 J. stoechadifolia
14	Rhizome present	3 J. tzar-ferdinandi
11	Basal leaves arachnoid tomentose	15
15	Sterile rosettes present	16
16	Basal leaves pinnatifid, capitula obconical	4 J. pinnata
16	Basal leaves entire, capitula hemispherical	7 J. kirghisorum
15	Sterile rosettes absent	17
17	Stem woody at base, basal leaves entire	1 J. linearifolia
17	Stem herbaceous, basal leaves pinnatifid	5 J. tanaitica

FIGURE 8-2

A computer generated key of an "indented" type for European species of *Jurinea* (Compositae). The figure has been arranged to reflect, for the most part, the format a computer line-printer would adhere to, though in some details (typeface and line width), line-printer output would differ. [From Pankhurst (1970b).]

generally made by hand, and can be made even when the number of taxa is otherwise discouragingly large. It is, however, necessary to provide considerable amounts of accurate data in the form of the matrix of taxa and characters, though this would generally be easy after a numerical taxonomic study. It is especially important for the range of within-taxon variation to be known.

Estimates of the probability of correct identification are likely to be more accurate with several characters to a couplet than one. The methods of estimating these are discussed in the next section, for they are basically the same as for simultaneous keys (but on a restricted character set). If the characters are quite few, it may be feasible to do direct counts of OTU's with different character combinations and use these as rough estimates of the underlying natural phenetic distributions. Because the manual testing of keys is a laborious business, it would be useful to have a computer program that would generate hypothetical specimens by a Monte Carlo process from plausible frequency distributions of the character states. It could then test computer-made keys for their success rating and also pick out taxa that are not readily separable, which need further attention. Polythetic sequential keys should not list more characters than are necessary in practice, or much of their convenience will be lost.

Any sequential key can be made probabilistic, though this is rarely done. The main attempts have been made by Möller (1962a,b,c) and Hill and Silvestri (1962). Each tip of the key has an associated probability that a specimen that keys out to this position will be identified correctly. These probabilities should be as close to 1.0 as possible. A major problem is the accurate estimation of the probabilities; with monothetic keys one would expect that errors of estimate would accumulate rather readily. It is likely that Monte Carlo methods would have to be used in the way mentioned above.

8.4   SIMULTANEOUS KEYS

Simultaneous keys are those in which the unknown is compared in turn with all the taxa in the hope of obtaining an unambiguous identification with one of them in a single step. Their form is most often a table of m characters against the q taxa, in which entries are the typical or commonest values for the taxa. The vector $\mathbf{u}$ of the unknown is compared in turn with each column, and the taxon with which it shows closest agreement is taken as the correct identification. The underlying concept is thus polythetic, and a simultaneous key is formally the same as a multiple branch point in a sequential polythetic key. Indeed, in any large study it becomes necessary to adopt a sequential strategy by breaking the full table into sections and to make the identification first to the major groups of taxa, and then to individual taxa. This is because too large a table is inefficient, as many of the characters are redundant for any one attempted identification; it is therefore best to use successive small tables. A good example is the work of Cowan (1965) and Cowan and Steel (1965), where a table is used to identify two major groups of genera, and for each such group a separate table is provided to carry identification to the genus. Such a scheme is, of course, virtually a multiple choice sequential polythetic key, but with numerous characters.

Any resemblance measure can be used to assess the best match. The usual one (as in Cowan and Steel's work) is the number of agreements for 0,1 characters. Cowan and Steel note, however, that there are difficulties with this simple method (which is analogous to using S_{SM} with equal weighted characters). Among these is the problem that for some characters the 0 states may have dubious significance; they may indicate a clear-cut negative or that chemical tests may not have been performed properly. And of course the characters are not explicitly weighted. Corlett, Lee, and Sinnhuber (1965) have used this method in a computer-based scheme where a punched card with 19 test results is fed in to afford identification of a bacterium. A similar computer method is described by Walker et al. (1968) for pollen grains. Elimination of taxa as possible answers must be made on some chosen low value of the resemblance measure. Increased power would come from giving each character value X_{iu} a weight w_{iJ} and an ease of observation value e_{iJ}, so that during computation the contribution to the resemblance due to X_{iu} was multiplied by $w_{iJ} \times e_{iJ}$; what is effectively weighting of this kind is used in some of the methods described below and in the next section.

A related concept is that of giving each taxon a limiting envelope and treating it as if it occupied a definite volume in A-space. An unknown is then identified with the taxon closest to it. Furthermore, one can tell if the unknown is outside any taxon or is intermediate between two taxa. This concept is mainly associated with discriminant analysis but need not be restricted to it. Any phenetic space can be used, and angular measures of resemblance can be treated as great circle distances on a unit hypersphere (see Firschein and Fischler, 1963). Ordinary Euclidean distances are easier to handle, however. This model therefore extends the idea of a central position of a taxon to include also a measure of its size, most readily as the measure of the radius of a hypersphere. This is satisfactory if the taxa are roughly hyperspherical, but if they are markedly elongated because of pronounced correlations between characters, then discriminant analysis is better (discriminant analysis effectively makes the taxa as nearly hyperspherical as possible in a transformed phenetic space). The model will break down if intermediate forms are very numerous, so it is assumed that they are relatively uncommon.

Problems for which this model is suited occur in bacterial taxonomy, and Gyllenberg (1964, 1965b) has proposed a detailed scheme. Examples of its use are given by Gyllenberg and Rauramaa (1966). Gyllenberg actually used correlation coefficients, and also reduced the A-space to three dimensions by principal component analysis, but here we describe a more general form.

A taxon is defined by the coordinates of its centroid ($\bar{\mathbf{x}}_J$) and by a radius r_J. Different measures for r_J have been mentioned in Section 5.2, including that suggested by Gyllenberg, which is twice the root mean square of the distances of the OTU's of the cluster from the centroid, and which is likely to overestimate the effective radius. It is preferable to determine a radius empirically as described in Section 5.2, such that it encloses a chosen percentage of OTU's. The identification matrix is thus converted into a new matrix, $\mathcal{L}$. This has m rows and q columns, recording the centroids as the mean value of the characters within each taxon, together with an additional row vector giving the radii of the taxa.

The distance between the unknown $\mathbf{u}$ and the centroid of each taxon is calculated. If the distance of $\mathbf{u}$ from the centroid of a taxon J, $d_{\bar{\mathbf{x}}_J, \mathbf{u}}$, is less than r_J, the unknown lies within that taxon. If the unknown lies outside any taxon it is recorded as unidentified (or possibly as an intermediate if it lies between two taxa). If $\mathbf{u}$ lies within only one taxon it is identified as belonging to it. Some taxa may overlap, and $\mathbf{u}$ may then lie within the hyperspheres of two or more taxa. Gyllenberg suggests that the unknown is then best identified with the taxon J for which the ratio $r_J/d_{\mathbf{x}, \mathbf{u}}$ is greatest. This is not necessarily the same as the taxon whose centroid is nearest to

u, as can be seen in Figure 8-3 for the unknowns marked $\mathbf{u}_2$ and $\mathbf{u}_3$. This figure also illustrates the other points of the scheme.

Clearly the system will break down if there is considerable overlap between hyperspheres, perhaps necessitating reclassification. Overlap is readily found by testing if any distance between centroids is less than the sum of the appropriate radii. The system is likely to be satisfactory if the taxa are approximately hyperspherical, that is, if character correlations are not, on the average, great.

The center of a taxon is best represented by the centroid, although other central measures can be used (Section 5.2). The hypothetical median organism of Liston, Weibe, and Colwell (1963) has been used as the center of taxa by Bogdanescu and Racotta (1967), and identifications were made by calculating distances from these. Hutchinson, Johnstone, and White (1965) used as the cluster center the OTU with minimal variance of distances to other members of the cluster. We expect that in spaces of high dimensionality other central measures like the centrotype would also be suitable.

In models not explicitly conceived as distance models, but closely analogous, there have been several attempts at calculating probabilities of correct identification.

Macnaughton-Smith (1965) suggests for 0,1 data a criterion for identification that appears to have several advantages. For each taxon **J**, the constant $C_\mathbf{J}$ is calculated:

$$C_\mathbf{J} = (n - 1)\log t_\mathbf{J} - \sum_{i=1}^{n} \log(t_\mathbf{J} - t_{\mathbf{J},1i})$$

where

$$t_{\mathbf{J},1i} = \sum_{j_\mathbf{J}=1}^{t_\mathbf{J}} X_{ij\mathbf{J}}$$

Also in each taxon one calculates for each character the quantity A_{ij}

$$A_{ij} = \log(t_\mathbf{J} - t_{\mathbf{J},1i}) - \log t_{\mathbf{J},1i}$$

For an unknown, **u**, one calculates for each taxon in turn the sum of C and all the A_i's that refer to those characters scored 1 in the unknown. Identification is with the taxon for which this sum is least. The logarithm of the probability of misclassification is proportional to this sum. The arbitrary choice of zero for log 0 may be needed to avoid indeterminacy.

Goodall's deviant index (Goodall, 1966b) can also be used to estimate the probability of correct identification, as can his probabilistic similarity index (Goodall, 1964, 1966c).

Considerable success has been achieved in the difficult field of bacterial identification by using a method based on conditional probabilities of 0,1 characters (Dybowski, Franklin, and Payne, 1963; Dybowski and Franklin, 1968; Lapage

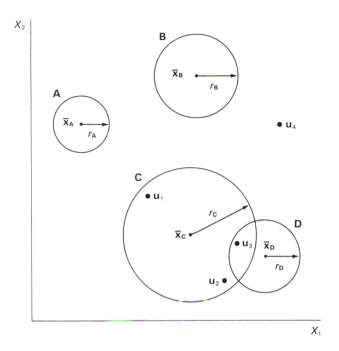

FIGURE 8-3
Identification as a process in A-space. The four taxa, **A**, **B**, **C**, and **D**, are represented by circles. They have centroids on the two character axes X_1 and X_2 shown by the central dots, $\bar{x}_A$, $\bar{x}_B$, $\bar{x}_C$, and $\bar{x}_D$ and dimensions shown by the radii of the circles r_A, r_B, r_C, and r_D.

An unknown u_1 lies within circle **C** and no other, and is identified with **C**. The unknown u_2 would also be allocated to **C**, although it is closer to the center of **D** than the center of **C**. The unknown u_3, which is within both circles **C** and **D** would be regarded as an intermediate form, or else allocated to **C** by the ratio rule given in the text. This is because the ratio of r_C to the distance u_3 to $\bar{x}_C$ is about 1.4, greater than the ratio of r_D to the distance of u_3 to $\bar{x}_D$ (about 1.1). The unknown u_4 is outside any circle and remains unidentified.

et al., 1970). A matrix $\mathscr{P}$ is stored in the computer that contains the proportion of state 1 for each of the m characters (mostly biochemical tests) in the q taxa. The entries p_{iJ} lie between 0 and 1, but they are never set exactly to 0 or 1 for two reasons: (a) some exceptional bacterial strains must always be expected, as well as occasional mistakes in performing tests; and (b) values of 0 or 1 will rule out a possible identification completely if an atypical result occurs because this leads to multiplication by zero in the process described below. In practice limiting values of 0.01 and 0.99 are suitable. The basic logic is as follows: if in taxon **J** the proportion of state 1 of a given character h is, say, 0.2, then for that character the probability that an unknown that scores 1 belongs to taxon **J** is taken as p_{hJ}, which here equals 0.2. If the unknown scores 0, it is taken as $1 - p_{hJ}$, here 0.8. Similarly if a second character, i, is considered, with p_{iJ} of 0.7, then the probabilities associated with state 1 and 0 are taken as 0.7 and 0.3 respectively. On considering both characters the probabilities are multiplied, so that in this example the probability for an unknown with the character states 1 and 1 is $0.2 \times 0.7 = 0.14$. The highest joint probability is given by an unknown possessing the majority states (in this example 0 and 1 respectively for characters h and i, giving $0.8 \times 0.7 = 0.56$).

The unknown is therefore compared with each taxon in turn and the individual probabilities are multiplied together for as many characters as are available, to obtain L, the joint likelihood values:

$$L_J = \prod_{i=1}^{m} |X_{iu} + p_{iJ} - 1|$$

The above formula assumes independence of characters.

As the number of characters is increased the joint likelihood becomes vanishingly small for a misidentification. For a correct identification it also falls, but more slowly. To compensate for this Lapage et al. (1970) calculate $L_J/\Sigma_{J=1}^{q} L_J$, and call this the *identification score*, which seems superior to earlier proposals by Dybowski and Franklin (1968). If the score reaches a sufficiently high level, such as 0.999, for the comparison of **u** with one taxon, this is accepted as a successful identification. If this level is not reached the program prints out the most likely candidates, and also valuable information on what tests should be made next in order to clinch the identification if possible. Figures 8-4 and 8-5 show examples of computer output.

Your ref. no. 85   Patient's name or source   Computerlab. no. W 96/97 Run 2
                                                                       Control 201

The Director
The Public Health Laboratory

Growth at 37	+99	Growth on MacConkey	+99	Oxidase	− 1	Gelatin 1–5 days	− 1	
Gelatin after 5 days	− 1	Simmons' citrate	+99	KCN	+95	Gluconate	− 1	
Malonate	−40	Urease	− 1	Indole	− 5	H₂S Iron media e.g. TSI	−99	
H₂S paper	+99	Arginine decarboxylase	+99	Lysine decarboxylase	− 1	Ornithine decarboxylase	−15	
Methyl red 30/RT	+99	Voges-Proskauer 30/RT	− 1	Gas from glucose	+99	Glucose	+99	
Cellobiose	+99	Dulcitol	−55	Lactose	+85	Maltose	+99	
Mannitol	+99	Salicin	−25	Sorbitol	+99	Sucrose	−15	

Group	Identification Score	CIB only	28 Tests done
1 Citrobacter freundii	0·999955		
2 Klebsiella ozaenae	0·000045		

Identification level reached
Citrobacter freundii

Differs from expected results for this organism

H₂S Iron media e.g. TSI

FIGURE 8-4
Computer identification: complete identification. The unknown has been identified as *Citrobacter freundii* with probability of over 99.99 percent. The next alternative, *Klebsiella ozaenae*, has a probability of less than 0.01 percent. The percent values of p_{iJ} for *Citrobacter freundii* on the 28 tests done, the results (+ or −) found with this unknown, and an aberrant test result, have also been shown. The figure has been arranged to reflect the format of a report such as a computer line-printer generates. Actual line-printer output would differ in some details. [From Lapage et al. (1970).]

Your ref. no. 85　　　　　Patient's name or source　　　　　　　　　　Computerlab. no. W 96/67.Run 1
　　　　　　　　　　　　　　　　　　　　　　　　　　　　　　　　　　　　　　　Control 201

The Director,
The Public Health Laboratory

Growth at 37	+	Growth on MacConkey	+	Oxidase	−	Gelatin 1–5 days	−
Simmons citrate	+	KCN	+	Malonate	−	Urease	−
Indole	−	H₂S paper	+	Gas from glucose	+	Glucose	+
Dulcitol	−	Lactose	+	Maltose	+	Mannitol	+
Salicin	−	Sucrose	−				

Group	Identification Score	CIB only	18 Tests done
1 Citrobacter freundii	0·983566		
2 Hafnia alvei	0·013304		
3 Arizona	0·002403		

Test suggested	Value in set	Value alone
Gluconate	2	2
Arginine decarboxylase	2	2
Lysine decarboxylase	1	2
Ornithine decarboxylase	1	2
CIB only (set value = 6 Key = 6)		
Cellobiose	2	2
Sorbitol	2	2
Gelatin after 5 days	2	2
CIB only (set value = 6 Key = 6)		
Methyl red 30/RT	2	2
Voges-Proskauer 30/RT	2	2
CIB only (set value = 4 Key = 6)		
H₂S Iron media e.g. TSI	2	2
CIB only (set value = 2 Key = 6)		

Remaining tests have zero value

FIGURE 8-5

Computer identification: the unknown has not given an identification score that is high enough, although the most probable answer is *Citrobacter freundii*. Four sets of new tests are suggested, and the user may then perform any or all of the four. Commonly the first set is sufficient. The relative value of the new tests is also indicated (for details see the original article). The figure has been arranged to reflect the format of a report such as is generated by a computer line-printer. Actual line-printer details would differ. [From Lapage et al. (1970).]

The scheme implemented by Lapage and his colleagues is now receiving extensive testing and has shown itself to be extremely powerful for identifying bacteria of medical importance. This is despite the fact it does not take character correlations into account, does not use a criterion to exclude misidentification of strains of taxa not represented in the matrix (i.e., there is in effect no critical radius of the taxa if the system is viewed as analogous to a distance model), and may be sensitive to vigor and pattern differences. The number of characters required for a high percentage of identifications is about 30, but this is quite economical since it represents a ratio of m/q of about 0.5, whereas with conventional methods the ratio is about 1. Very few misidentifications occur, and the identification rate appears satisfactory in view of imperfections in the present classification of bacteria (and consequently in the $\mathscr{P}$ matrix) and the frequency of aberrant bacterial strains in nature (for further discussion on these points see Sneath, 1969a, 1972).

Simultaneous keys are well suited for use on-line with a computer, as the data tables can be easily stored and the computations swiftly made. Hall (1969a) describes a modification that makes provision for excluding numerous very unlikely possibilities, thus increasing the speed of identification. Simultaneous keys are less useful in printed form, because matching of the unknown on the columns is troublesome. Several mechanical devices have been suggested for use with them (Cowan and Steel, 1960, 1965; Olds, 1970), and "peek-a-boo" punched cards can also be adapted to them (Yourassowsky et al., 1965); such modifications overlap with sequential techniques described in the previous section.

Other possibilities in simultaneous identification methods are the use of automatic scanning devices, the output of which is discussed in Sections 3.3 and 3.4. These may one day allow identification directly from the specimen. There is also rapid advance in automated methods of biochemical analysis, which could be coupled to an on-line identification scheme.

Related to simultaneous identification techniques are programs developed by Lance, Milne, and Williams (1968), which take hierarchical classifications or ordinations (and the data matrices underlying these) and output mean differences in desired characters for any specified groups. Although much information could be obtained as a by-product of the classificatory procedures, Lance, Milne, and Williams recommend that it be done as a separate run inasmuch as obtaining all the possible comparisons would be far too time-consuming and produce excessive printed output. Also it is impossible to know which particular comparisons will be of interest until the classifications have initially been obtained and examined by the investigator.

8.5   DISCRIMINANT ANALYSIS

In previous sections of this chapter we have mentioned the idea of weighting characters for identification. When the weighting is done in a manner that maximizes the probability of correctly identifying unknown specimens from a few close or overlapping taxa it leads to the branch of multivariate statistics that we discuss in this section. Most of these methods can be viewed as extensions of taxon distance models, such as the model illustrated in the last section, where the character axes have been transformed.

Discriminant Functions

A linear discriminant function is a linear function z of characters describing OTU's that weight the characters in such a way that as many as possible of the OTU's in one taxon have high values for z and as many as possible of another have low values, so that z serves as a much better discriminant of the two taxa than does any one character taken singly. The n characters are almost always reduced to a smaller set,

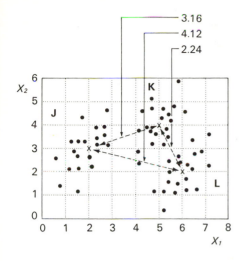

FIGURE 8-6

Three clusters of OTU's representing three taxa, **J**, **K**, and **L**, for two character axes X_1 and X_2. The crosses represent the centroids, which are $\bar{x}_J = 2.0,\ 3.0$; $\bar{x}_K = 5.0,\ 4.0$; and $\bar{x}_L = 6.0,\ 2.0$. The pooled variance-covariance matrix **W** is:

	Characters	
	1	*2*
Characters *1*	0.5	0.4
2	0.4	1.0

and the inverse matrix W^{-1} is:

$$\begin{array}{rr} 2.941 & -1.176 \\ -1.176 & 1.471 \end{array}$$

The Euclidean distances between the centroids are shown.

m. The function is also such that it has maximal variance between groups relative to the pooled variance within groups.

As originally described (Fisher, 1936), the discriminant function was applied to two taxa, and was later generalized to many taxa. It is calculated from the pooled variances and covariances between the *m* characters within each taxon. In its original form this is a weighted average of the variances and covariances of the characters in taxa **J** and **K**. When generalized, the variances and covariances for all the taxa being considered are pooled. In addition, the means of each character for each taxon are required, representing the centroids of the taxa.

The method is illustrated by Figures 8-6 to 8-8. Figure 8-6 shows three taxa, **J**, **K**, and **L** for two character axes, X_1 and X_2. The taxa are shown as clusters of individuals (OTU's), but we are not now concerned with the values for the individuals, but only with descriptive parameters of the three clusters. These are their centroids and their dispersions.

The variances and covariances between the characters are first calculated, yielding the three $m \times m$ matrices (here $m = 2$); these are then averaged to give a pooled within-groups variance-covariance matrix **W**. This is shown in the legend to Figure 8-6 together with the values of the centroids of the taxa.

To calculate the discriminant function between **J** and **K** the inverted **W** matrix is multiplied by the vector $\delta_{JK} = [(\bar{X}_{1J} - \bar{X}_{1K}),\ (\bar{X}_{2J} - \bar{X}_{2K}),\ \dots,\ (\bar{X}_{mJ} - \bar{X}_{mK})]$. This gives the discriminant function as a vector **z**.

$$z_{JK} = W^{-1}\delta_{JK}$$

This vector consists of a series of weights, w_i, which we here symbolize as $z_1, z_2, \dots, z_m$ for characters $1, 2, \dots, m$. In the example $z_{JK} = -7.647, 2.057$. These weights are then multiplied by the observed character values of the unknown individual, and summed to give a discriminant score, DS.

$$DS_u = z_1 X_{1u} + z_2 X_{2u} + \ldots + z_m X_{mu}$$

This score is used for discriminating between members of **J** and **K** by calculating three reference scores for the centroid of **J**, for the centroid of **K**, and for the point midway between them—DS_J, DS_K, and $DS_{0.5}$ respectively. These are obtained from the equations

$$DS_J = \bar{X}_J z'$$

$$DS_K = \bar{X}_K z'$$

$$DS_{0.5} = \tfrac{1}{2}(\bar{X}_J + \bar{X}_K)z'$$

$$= \tfrac{1}{2}(DS_J + DS_K)$$

The midway point assumes that the frequency of members of **J** and **K** in the population are equal. The values for the example are shown in Figure 8-7, which also shows the geometric effect the transformation has on the original character

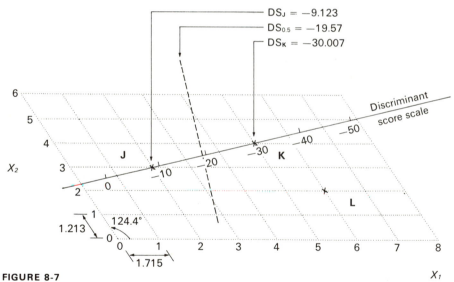

FIGURE 8-7
The effect of the discriminant analysis transformation upon the geometry of Figure 8-6, and the discriminant score scale for taxa **J** and **K**.

The transformation has two effects: the mean intrataxon variance is equalized for each character axis, and the axes are skewed (according to functions of the covariances), so as to make the clusters as nearly hyperspherical as possible. The angle of skewing is shown together with the new scales of the transformed axes. The discriminant score scale for separating **J** and **K** is also shown. The scores are calculated as $(-7.647 \times X_1) + (2.057 \times X_2)$ as explained in the text. For example, an unknown at $X_1 = 3$, $X_2 = 4$ has a score of -14.713. The points representing individual OTU's have been omitted for clarity. The discriminant function vector $z = -7.647, 2.057$ is obtained by multiplying $\mathbf{W}^{-1}$ by δ_{JK} which is $(2.0 - 5.0), (3.0 - 4.0) = -3, -1$. See Figure 8-6.

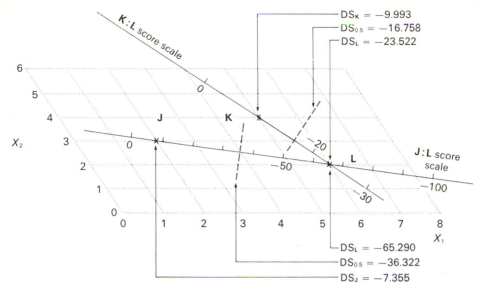

FIGURE 8-8
The discriminant score scales for separating taxa **J** and **L**, and **K** and **L**.

scales. The value of $DS_{0.5}$ is halfway between the other two scores and defines a plane midway between the centroids and perpendicular to the line joining them, shown by the dashed line in Figure 8-7. If the observed score for an unknown, DS_u, lies on the DS_J side, the unknown is allocated to taxon **J**, and if on the DS_K side, to taxon **K**. The length of the line between the centroids of **J** and **K** measured in discriminant function units is the square root of Mahalanobis' D^2 and the absolute difference between the scores DS_J and DS_K is also equal to D^2.

It should be noted that a different discriminant function, and discriminant scores, are calculated for each pair of taxa. Figure 8-8 shows the other two discriminant lines; it will be seen that the scales are different in size and orientation. It will also be evident from Figures 8-6 to 8-8 how discriminant functions are valuable when character values overlap, particularly when many characters are involved and one cannot draw scattergrams of the clusters.

Blackith (1965) has pointed out that the vector angle between two discriminant function vectors measures contrasts of form of the taxa. If the angle is small the functions measure similar contrasts of form, and large angles represent distinct contrasts.

There are two important uses of the distance D^2_{JK} between the scores DS_J and DS_K. First, one can test whether this indicates that the centroids are significantly different. For this, one uses an F test with m and $(t_J + t_K - m - 1)$ degrees of freedom and tests the ratio

$$\frac{D^2_{JK}(t_J t_K)(t_J + t_K - m - 1)}{(t_J + t_K)(t_J + t_K - 2)m}$$

This ratio is related to Hotellings' T^2, as follows (Rao, 1952, p. 74):

$$T^2 = \frac{t_J t_K}{t_J + t_K} D_{JK}^2$$

Second, one can determine the contribution that each character makes to D^2, and hence, see if any of them have such little discriminatory power that they are unlikely to be worth using. Alternatively, one can choose the best few characters from the set, and ascertain by the F test whether enough have been selected. The percent contribution of character i is $100 \times (z_i \delta_i / D^2)$ where z_i and δ_i are the ith elements of vectors $\mathbf{z_{JK}}$ and $\boldsymbol{\delta_{JK}}$. This criterion does not consider correlations between characters; if two or more characters are correlated they contribute to D^2 to a greater extent than this test suggests.

Although it is usual to take the midpoint between centroids as the criterion for identifying an unknown, there is no reason why one need do this. If it were very important to be sure of identifying all members of taxon $\mathbf{J}$ even at the price of misidentifying some members of $\mathbf{K}$ by allocating them to $\mathbf{J}$ in error, one can choose a criterion lying closer to the center of $\mathbf{K}$ than of $\mathbf{J}$. $DS_{0.5}$ gives equal probability of misclassification of unknowns from either taxon.

The probability of misclassification can be calculated on the assumption of a multivariate normal distribution and also that the unknown does belong to $\mathbf{J}$ or $\mathbf{K}$ (and not to some distant cluster). A primary purpose of a discriminant function is to minimize the probability of wrong assignment of unknown individuals. If an unknown lies upon the DS_J side of $DS_{0.5}$, we can ask how many standard errors it is from DS_K and consult tables of the normal distribution. If it is many standard errors, then it is very unlikely to be a member of $\mathbf{K}$ misclassified as a member of $\mathbf{J}$. The standard deviation of a taxon in D-space is taken as 1.0 in every direction. The square root of the difference between DS_J and DS_K, i.e., D itself, gives the number of standard deviations the centroids are apart, and half of this corresponds to the $DS_{0.5}$ plane. An unknown lying on the DS_J side therefore is over $\frac{1}{2}D_{JK}$ standard deviations from $\mathbf{K}$.

Figures 8-6 to 8-8 illustrate several important points. The use of a given discriminant function implies that the unknown does belong to one or the other of the two taxa being considered. If instead it belongs to a quite different cluster, located far off in the space, it may have almost any discriminant score and may thus appear to belong to one or the other of the two taxa under consideration, when it really belongs to a third. Also, when there are many taxa one has to test against a large number of discriminant functions. These two problems are largely overcome by the use of D^2 as described below. Harder to overcome is the fact that all the usual methods of discriminant analysis assume that the dispersion matrices of the taxa are homogeneous (that is, the clusters all have much the same size, shape, and orientation in phenetic space) and that the clusters have multivariate normal distributions.

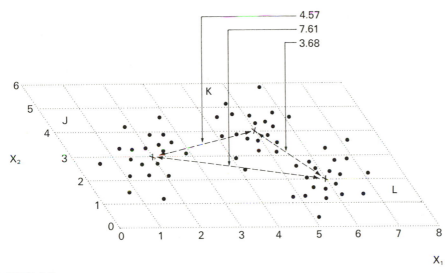

FIGURE 8-9
The transformation into discriminant analysis space from Figure 8-6 shown in more detail. The points representing the OTU's are indicated, together with the Euclidean distances in the new space between the centroids. These distances are the values of D (i.e., the square roots of D^2).

Mahalanobis' D-space and Multiple Discriminant Analysis

Figure 8-9 shows the original clusters in the transformed D-space, but without the discriminant score scales. It was noted earlier that D^2 could be obtained between the centroids from the discriminant scores. However following Mahalanobis (1936), one can calculate it between any pair of points f and g by the equation

$$D_{fg}^2 = \delta_{fg}' \mathbf{W}^{-1} \delta_{fg}$$

Then the square root of D^2 is simply the Euclidean distance in the D-space. This can be seen in Figure 8-9, where the distances between the centroids are marked. The method therefore transforms the original space into a new space, in which the original axes are stretched and also skewed so that they are no longer at right angles. The length of a unit in dimension i is p times the original units, where p is the square root of the a_{ii} element of the matrix $\mathbf{W}^{-1}$. The direction cosine between the dimensions h and i is equal to $a_{hi}/\sqrt{a_{hh}a_{ii}}$. Gower (1967b) gives a representation with correlation coefficients.

In the special case where none of the characters are correlated (so that all covariances are zero) the transformation simply stretches the axes but leaves them at right angles; the length of a unit in dimension i is then $1/s_i$ times the original units, where s_i is the mean within-cluster standard deviation of character i. That is, the axes are stretched in inverse proportion to the standard deviation. If in addition all variances are equal to s^2, then D is $1/s$ times the ordinary Euclidean distances. The D units are of a kind that can be called "ease of discrimination units." Confidence limits in D-space can be found from the sampling variance of D^2, which is $D^2/(t_1 + t_2 + \ldots + t_q - q)$.

Canonical Variates

Because D can be represented in an orthogonal system of axes (though the orientation is arbitrary), one can perform analyses on distances between taxa or individuals, in particular by principal coordinate analysis (Gower, 1966a, 1967b). Canonical variates and multiple discriminant analysis are equivalent except in minor particulars. The coordinates of the points in an orthogonal system can be obtained, and the orthogonal axes are the canonical variates, which can be used also for discrimination. In Figure 8-10 the orthogonal D axes are shown, obtained by principal coordinate analysis of D distances between the centroids of J, K, and L. It is clear that one can then readily identify unknowns by seeing whether they fall within critical distances of taxon centroids. Gower (1966a) points out that though there is rarely need to do so, one can transform into D-space even if one has only a single taxon: one considers the whole set of OTU's as one large group. The positions of OTU's will then be made such that the entire set forms a hyperspherical cluster, within which, of course, there may be subclusters. Note that the relations between vector lengths are preserved in D-space. An OTU twice as big as another but of the same shape will lie on the same line from the origin in Figure 8-9, but twice as far away. Discriminant functions and D^2 are less sensitive to general size factors than taxonomic distance; but an unknown of the same shape as a member of a taxon may appear outside that taxon if it differs much in size.

There are certain difficulties with discriminant analysis. If any character is invariant in each of the taxa, the matrix **W** cannot be inverted, unless "generalized inverses" are used. Yet the character might have a unique state for each OTU, and by itself be a perfect discriminator. There are also difficulties in choosing the limited set of best characters from the large number that should be employed in numerical taxonomy. Characters with means that are well separated in relation to the variances and that are not highly correlated with other characters are in general the best, but optimal methods of choosing them pose statistical and computational problems (see Feldman, Klein, and Honingfeld, 1969). We believe that for most taxonomic work it will be possible to choose a nearly optimal set by inspection. It has been shown by Dunn and Varady (1966) that rather large numbers of individuals are required in each taxon for reliable discriminant functions. The gain in discriminatory power over simpler methods may not be very great (Sokal, 1965) particularly with 0,1 characters (Gilbert, 1968; Kurczynski, 1970) and simple discriminants based on equal weight for each character can be quite effective (e.g., Kim, Brown, and Cook, 1963). Discriminant functions have most value for very close clusters

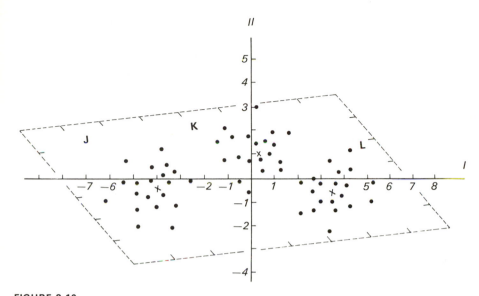

FIGURE 8-10
Canonical variates I and II superimposed on the positions of OTU's in Figure 8-9. The axes are principal axes, and the new origin is at the centroid of the three taxon centroids, 4.33, 3.0. The original character axes are shown by the frame of broken lines. The new axes are scaled in D units, which are effectively within-taxon average standard deviations in the transformed space.

that partly overlap; good examples are those of Giles and Elliot (1962, 1963) on human skulls of different sexes and racial groups. Sokal (1965) lists examples of their use in taxonomy. Hill et al. (1965) obtain a type of discriminant function from their gradient factor analysis. DuPraw (1965a) achieved excellent discrimination of wings of honeybees of different geographical origin by multiple discriminant analysis. Blackith and Reyment (1971) describe numerous applications of D^2, discriminant functions, and canonical variates. New discriminant methods have been suggested by Hall (1968) and Saila and Flowers (1969).

Among references dealing with more complex methods than discriminant functions and D^2, and reviewing various parts of the field of discrimination, are Rao (1952), Sebestyen (1962), Reyment (1963), Kossack (1963), Sokal (1965), and Chaddha and Marcus (1968). Related work is that of Birnbaum and Maxwell (1961). Cavalli (1949) discusses the "mean correlation coefficient," defined as the mean of the $n(n - 1)/2$ correlation coefficients between n pairs of characters, and discusses its relation to Gini's synthetic coefficient and the work of Zarapkin. Penrose (1954) found that it is possible to obtain, in a simple manner, good approximations to D^2 by using an average measure of the correlations between characters.

Sebestyen (1962) points out that measures that reduce the general size factor may make discrimination more difficult, but the point at issue is whether the difference in size in the particular case is indeed a reliable discriminant or an artifact of sampling, for example. Sebestyen considers that the most powerful methods are those giving equiprobable envelopes of clusters, but they have been little developed and require very large numbers of individuals. He also discusses some nonlinear methods, as does Rohlf (1970). Williams and Lance (1968) also discuss this general problem under the heading of extrinsic criteria of patterns; they conclude that nonlinear multiple regression may sometimes be a suitable technique but that the area is in need of deeper study.

It may often be useful to employ the simple method of Lubischew (1962) for testing single characters as discriminators. He calculates his coefficient of discrimination $K = (\bar{X}_{iA} - \bar{X}_{iB})^2/2s_i^2$ where s_i^2 is the pooled variance for character i from taxa **A** and **B** (that is an average weighted by the numbers of individuals). The greater K is, the better i is as a discriminator. This takes no account of correlation between characters. The probability of misclassification is approximately the probability that a normal deviate exceeds $\sqrt{K/2}$, so that, for example, with $K = 7.68$, 95 percent of identifications will be correct. This requires the distributions of X in **A** and **B** to be approximately normal and of equal variance.

General Conclusions on Identification and Discrimination

Recommendations in this area are somewhat tentative because of the small experience with alternative numerical methods. We suggest that for large studies with well-separated taxa the sequential methods are best. Simultaneous keys are useful for highly polythetic groups without much overlap, but discriminant analysis is indicated where there are a few close groups in which identification must be as certain as possible. The addition of simple probabilistic values to the traditional methods should prove rewarding.

9

Implications for Nomenclature

Although it is not yet possible to foresee the changes in nomenclatural procedures to be brought about by numerical taxonomy, some implications for nomenclature are evident now. A few constructive proposals will be discussed below and suggestions will be made about some of the lines along which nomenclature may develop. The student of numerical taxonomy may require a guide to the application of the present rules of nomenclature; for zoology, standard texts are Blackwelder (1967a) and Mayr (1969a) and for botany Lawrence (1951) or Core (1955). In microbiology and mycology a concise guide is given in Appendix I of Ainsworth and Sneath (1962, pp. 454–463). Savory (1962) has provided a useful discussion and comparison of the several International Codes of Nomenclature.

A new development in nomenclature is the use of punched cards and computers to handle the "book-keeping" of taxonomic names, keys, bibliographies, and so on. Although this is not part of numerical taxonomy as treated here, it is a parallel development that merits notice.

Jahn (1961) points out that schemes of classification and nomenclature are being increasingly developed in many branches of science as an aid to efficient automatic processing of information. Once established, such schemes are difficult to alter; therefore it behooves taxonomists to see that these are of the sort they want, lest they find themselves faced with a fait accompli. In such applications, the use of "unnatural" taxa may have damaging consequences; once the schemes are compiled it is difficult to disentangle the information pertaining to the different entities that have been lumped together. Jahn (1962) has also noted that such schemes may well force some changes also in the codes of nomenclature; the separation between plant and animal kingdoms may be abandoned, with consequent alteration of many of the present rules (especially those allowing homonymy between animals and plants).

9.1   SOME GENERAL CONSIDERATIONS

An excellent discussion of the problems of nomenclature is that of Simpson (1961, pp. 28–34). Systems of nomenclature have three major objectives: to provide for names that are (1) universally applied, (2) unambiguous, and (3) stable. Other considerations, such as that names should indicate taxonomic rank, position, or relationship, or that they should be descriptive are of secondary importance. The codes that regulate taxonomic nomenclature (but not necessarily other codes) also suppress unnecessary names (e.g., synonyms) by the application of the rules of priority and the nomenclatural type method. Since these different requirements are often in conflict, the codes offer an uneasy compromise; present nomenclature attempts to serve many functions but does none of them very well.

Numerical taxonomic techniques have implications in particular for stability, which is a matter of some practical consequence. It may be argued, as has been done by Gilmour (1961), that numerical taxonomy may increase instability, and it must be admitted that it may do so at least during the first studies on a taxonomic group. Instability may be of at least three kinds, referring to (a) OTU's to be included in a taxon, (b) the rank to be accorded to such a taxon, and (c) the name this taxon should be given. We believe, however, that numerical taxonomies will in the end prove very stable. It is clear that one could, by raising or lowering the phenon level a little, produce considerable changes in the nomenclature. This we believe to be undesirable; in common with others (e.g., Walters, 1965; Watson, Williams, and Lance, 1966) we would not recommend the rigorous application of phenon lines if this severely disturbed the nomenclature without making any positive taxonomic contribution. For example, if a second study showed that the phenon level of the majority of subgenera of the first study now fell just below the line chosen to indicate genera, we would not rename them all on this account: a third study might well shift them again into the subgeneric level. However, changes that in the opinion of the taxonomist are major and significant should result in renaming. This is necessary if the nomenclature is to reflect reasonably well the "natural" taxonomic groupings. To do otherwise is to deny biologists the benefits of improved taxonomies. Eventually one would hope for a time when the International Commissions on Nomenclature would no longer permit name changes for reasons of priority and author citations would become unnecessary.

Rohlf (1962) has pointed out that in successive studies the least disturbance of nomenclature would occur if the phenograms are divided at points where the stems show the widest gaps between successive branchings. Sometimes these optimal levels would be easy to determine, but the temptation to let the rank lines wander up and down in their course across a single phenogram would introduce an element of subjectivity that is at variance with our hope for objective representation of the relationships. The phenon nomenclature described in Section 5.11 is suggested as a means of expressing finer details of taxonomic relationships without having to force them into a rigid and formal system of nomenclature.

A number of new proposals for nomenclature have been made in recent years, even to the extent of questioning the value of any rules at all (Oldroyd, 1966; Cowan, 1970). Some of these, such as proposals for new starting dates based on recent monographs (Howden, Evans, and Wilson, 1968) can be accommodated within the framework of the present codes, and may become feasible as advances in data processing make available complete and annotated lists of names that will lighten much of the present labor of taxonomists. Bacteriologists are already moving in this direction (Lessel, 1971). Other proposals, such as systems of virus nomenclature, are so radical that they involve intense debate (e.g., Gibbs et al., 1966).

Many of the new proposals revive the question of the extent to which names of taxa should be indicators of taxonomic relations rather than be simply labels for taxonomic groups. The former was an original intention of Linnean binominal nomenclature, though now it plays a minor function because most genus and species names are familiar to only a few specialists. Attempts to continue using names to indicate relationships bring disadvantages. They may cause, in particular, instability of names due to later taxonomic revisions. Biologists have given up the attempt to make names descriptive or to make nomenclatural types typical in the usual sense; it is therefore not surprising that there is increasing discussion about whether biological names can successfully serve any other function than as labels for taxa. It is not so much the indication of taxonomic rank that causes difficulty; the use of uniform endings for names of each rank category above genus is becoming increasingly fashionable, even though it leads to some changes of familiar names. Rather, problems are engendered by attempts to make a name carry an indication of the position of the taxon in classification, e.g., the class, order, and family to which it belongs.

The trend towards names as pure labels rather than as vehicles for taxonomic information is seen in proposals for uninominal nomenclature put forward by Cain (1959b) and Michener (1963, 1964). The genus and species name (taken from the most recent revision) would be hyphenated to make a uninomen that would thereafter never be broken up. On transfer to a new genus it would remain unchanged. A genus would then contain species whose names were entirely different; there would be no common generic part. Genera and higher taxa would receive the name of one of the contained species in Michener's scheme. Thus the Hymenoptera might be called Order *Apis-mellifera*, and would contain Family *Apis-mellifera* (for the Family Apidae) and Genus *Apis-mellifera* containing the species *Apis-mellifera*. The inconvenience of this plan could be ameliorated by adding uniform endings for each rank, and the resulting nomenclature would then be very similar to the present one if it were the custom to give names to no taxon below the level of genus. It would indeed offer numerous attractions if one required a nomenclature for entirely new groups of organisms that had no existing names hallowed by long usage, e.g., viruses, or as Michener mentions, creatures on another planet. It may be noted, however, that the functions (though not the form) of the names in such a system are essentially those given in current nomenclature by the original name (basionym).

Proposals in the other direction, to add information on the taxonomic position of a taxon, have a long history. Mayr (1969a, p. 345) discusses some early suggestions for supplementing the generic name with letters indicating the higher taxa. Thus *Papilio* would become *Ylpapilia* (*Y* for Insecta, *l* for Lepidoptera and *-a* for Invertebrata). Amadon (1966) has put forward an idea intended to strengthen the indication of taxonomic position without causing much instability: the rank of the category that receives the first name in a binominal would be raised from the genus to the family level (or thereabouts), and within each family no two species would have the same specific epithet. Taxonomic revisions resulting in change of family would be quite infrequent. The usual custom of adding the class and order in parentheses after first mention of a name is an informal scheme of a similar kind, and it would seem that no one has made the suggestion that it should be formalized and made obligatory, so that the higher taxon names would become part of the name of the species.

There have also been proposals to supplement taxonomic names and perhaps eventually supplant them by a system of numbers—a "numericlature" (Little, 1964). The reason for using numbers is twofold, to make the system readily handled by data-processing machinery, and to avoid difficulties with names owing to their associations, pronouncability, and so on. But it should be mentioned (as Hull, 1968b; Randal and Scott, 1967, and others have pointed out) that computers can handle names as easily as numbers. Also, names are easier to remember and to check for accuracy by eye. It is the ease with which numbers can serve as labels, as indicators of taxonomic rank or position, or even descriptors, that makes them attractive. For such purposes the numbers must have a system of rules, what Hull refers to as a syntax, and which in due course taxonomists will have to learn. It seems unlikely though that these rules will ever be as complicated as the present rules of nomenclature!

Suggested schemes of numericlature have been numerous. A pioneering paper is that of Gould (1958), who was one of the first to appreciate the potential of data-processing in taxonomy. He proposed that taxa should have a number to indicate taxonomic position that we may refer to as the *classification number*. Others (e.g., Michener, 1963; Little, 1964; Rivas, 1965) suggested adding a unique number (the *reference number*) as a label, and Hull (1966) proposes in addition a number indicating cladistic position (if known).

Stability is achieved by the reference number, which would never be changed, although union or division of taxa would require cross-references, and possibly supplementary numbers, to clarify the situation. The setting up of a system of reference numbers is simply a conversion of existing basionyms, because a basionym with its full citation is effectively a unique label in alphabetic characters.

The classification number would contain groups of digits for phylum, class, order, family, genus, and species. Taxonomists would soon become familiar with the main outlines of the system, as well as with the numbers for the taxa they specialized upon. Change in taxonomic position would involve change of the classification number only. Again suitable cross-references would be needed from time to time. Provision can be made for uncertainty of taxonomic position. Systems of numericlature can be applied at all taxonomic ranks, though they are often principally intended for species. The numbers would of course have to be international, and allocated from a central source. In a comprehensive system it would be easy to prevent homonyms and to identify many synonyms.

Finally, Jahn (1961) has suggested that modern data-processing equipment could allocate new names to newly discovered organisms, and we have noted (Sokal and Sneath, 1966) that the completely automatic renaming of taxa would be feasible.

Several schemes for codifying names are now being developed. In taxonomy the most ambitious of these is the International Plant Index (IPIx) at the Connecticut Agricultural Experiment Station, New Haven, Connecticut. Its outlines are described by Gould (1962). A comprehensive review of recent projects is provided by Crovello and MacDonald (1970).

The danger that the advent of electronic data processing may rigidify classification through its effects on nomenclature does not seem too serious. Any new nomenclatural system developed today should, of course, allow for automated information storage and retrieval, but taxonomic flexibility can also be provided. An important consideration of a nomenclatural system is the question of how information about the organisms is to be retrieved. Among the problems of document retrieval being discussed currently is the following: what is the optimum system of classification for a series of documents so that a document can be retrieved with a minimum of searches? Documents must be indexed under those headings that will most frequently be employed, and storage should be arranged so that access to the more frequently required documents is easier than to those less often needed. These problems have clear relevance to taxonomy. The cross-indexing of taxa and their more salient properties is desirable. Research in this field is urgently needed.

9.2 NUMERICAL TAXONOMY AND NOMENCLATURAL PROBLEMS

It is now generally recognized that modern nomenclature does not concern itself with the limits of taxa but only with reference points to the taxonomic names. What is to be included in a taxon is left to the decision of the taxonomist. Bradley (1939) expressed this as follows. "Nomenclature is concerned with the nuclei of groups, never with their limits. Taxonomy is concerned with the limits of groups, not their nuclei. The limits are debatable, subjective, forever changeable, not amenable to decision by authority. The nuclei can be fixed by common consent, for they are objective, utilitarian, permanent." Numerical taxonomy will change this position, for it will be possible to determine the centers and boundaries of taxa by exact estimation of resemblances, so that what organisms should be placed in a taxon will no longer be simply a matter of opinion. The limits then, as well as the nuclei, may also be objective, utilitarian, permanent, and fixed by common consent.

Numerical taxonomy will sometimes be applied to groups in which there is no significant earlier taxonomy, or it may cause extensive revision of an existing taxonomy. In such cases it may be necessary to set up types for the names of the new taxa, or lectotypes or neotypes for old ones. Such types need not be phenetically typical of the taxon. Their function is expressed better by the term "nomenifer," or name bearer, suggested by Schopf (1960) than by the term "type," implying typicality. Nevertheless, there are advantages in choosing a nomenifer that is also reasonably typical, and the taxonomist can choose a typical specimen for a species from the results of numerical classification. Similarly, a typical species can be chosen as the type of a genus or higher ranked taxon. In general we require an OTU that is central in a geometrical sense in a cluster of OTU's in A-space; however, there may sometimes be practical considerations indicating the adoption of a noncentral OTU as the type of a taxon. Measures of the center of taxa have been discussed in Section 5.2.

The OTU nearest to the centroid (usually the centrotype) may be chosen as the most typical OTU, and in microbiology the organism closest to the hypothetical median organism is commonly selected. Of course, if a number of OTU's are equidistant from the geometric center the choice is arbitrary. Similarly, boundaries of taxa can be set by techniques discussed in Sections 5.2 and 5.11, and Niemalä and Gyllenberg (1968) have even suggested that several types per taxon could serve the same purpose. Descriptions and diagnoses, with precise indications of character variation, could also be produced automatically by numerical techniques.

There are a number of potential applications of numerical taxonomy to variation patterns not envisaged by orthodox nomenclature. One of these is terminology for intermediate forms. They may need a special terminology similar to that already used for hybrids and for intermediate forms in phylogenies, such as "$X - Y$ intermediates," or "X inter. Y." This might even take numerical form. The intermediate form $I(X, Y)$ could lie on what may, as a manner of speaking, be envisaged

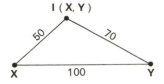

as the direct line in some appropriate taxonomic space between taxa **X** and **Y**. If it lies off this line, the sum of the distances $d_{X,I(X,Y)}$ and $d_{Y,I(X,Y)}$ will be greater than the distance $d_{X,Y}$. An intermediate could then be a "50 **X** − 70 **Y** intermediate," where $d_{X,Y} = 100$, $d_{X,I(X,Y)} = 50$, and $d_{Y,I(X,Y)} = 70$, as shown in Figure 9-1. The difference $|d_{X,Y} - (d_{X,I(X,Y)} + d_{Y,I(X,Y)})|$ gives an idea of how far **I(X, Y)** deviates from the straight line joining **X** and **Y**—an indication of epistasis or overdominance —regardless of the direction of the deviation. It should be noted that **X** and **Y** may be represented either by their most central or most typical members or by their nomenclatural types (which may not be central or typical), and this must be made clear.

Similar occasions may occur in phylogenetic studies, and we have already given methods for the construction of HTU's and their location in A-space.

A new development would be a nomenclature based on the volume occupied by a taxon in taxonomic hyperspace. Whether it would have advantages remains to be seen, but the principle would be to define a volume of a certain size as a generic volume, and so on, and to name the taxa within the corresponding volumes accordingly. This would lead directly to a nomenclature or numericlature based upon the coordinates of taxa in a suitable phenetic space by giving each taxon a measure of location and a measure of dispersion. It would probably be applied to ordinations to make the numericlature more concise, although some simple arithmetic would be needed when using it with actual specimens. Du Praw (1964, 1965a,b) has given some examples on these lines.

Should overlapping taxa prove useful, present nomenclature, which is by nature hierarchic, is not well suited to these (see Michener, 1963). Uninominals and nu-mericlature would be easy to apply however, as a species could be listed in two genera and readily cross-referenced by code numbers. Hierarchic nomenclature is also ill adapted to some other variation patterns (see Section 5.14), particularly the pattern in which dense clusters of OTU's are embedded in a sparse scattering of single OTU's. It is likely that this pattern is not uncommon at low taxonomic ranks in apomictic groups, where the OTU's are individuals or clones, but nomen-clatural treatment of apomicts is not at all uniform (see Davis and Heywood, 1963). If the clusters are named as species (in this case taxospecies) how should the scattered OTU's be named? Would each one be named as a species on the grounds that a little search would surely turn up a tight cluster around it? This would often be quite impracticable because of their number, and these aberrant OTU's might indeed be unique. If one raised the species level to include all the OTU's then the clusters could be named as subspecies, which might be the simplest solution. But any course adopted might place considerable strain on Linnean nomenclature, partly because the variation could not be well represented as a hierarchy, and partly because of different definitions of the species category.

The requirement that every organism (with few exceptions) must have a binominal name is also a disadvantage in naming apomicts; it would often be more convenient to refer some forms simply to a genus or section of a genus rather than force them into the nearest species. This practice is finding increasingly favor in botany and microbiology (see Hill, 1959; Davis and Heywood, 1963). Whether similar problems occur with variation patterns at higher ranks is not yet clear, but the increasing number of phenetic studies will improve the knowledge of variation patterns at all levels and enable taxonomists to develop more appropriate systems of names or numbers.

A Critical Examination of Numerical Taxonomy

Having completed our discussion of the various aspects of taxonomic procedure in biology from the point of view of numerical taxonomy, we now look to the task of critically evaluating the alleged and real weaknesses of the philosophy and methodology of numerical taxonomic procedure. Of course we wish also to stress positive contributions that numerical taxonomy has made and is making to biology, and the promise it holds for the future. In the first section we shall review briefly some of the more recent criticisms that have been leveled against one aspect or other of such work. In the second section we summarize unsolved problems of numerical taxonomy. Finally in Section 10.3 we shall point out how numerical taxonomy has led to a reevaluation of taxonomic principles and is leading to new knowledge of taxonomic facts as well as of evolutionary and ecological principles.

10.1 CRITICISMS OF NUMERICAL TAXONOMY

Ever since numerical taxonomy was first presented to the scientific community it has been subjected to criticisms on a variety of grounds. These criticisms have on occasion led to acrimonious and emotional controversy, much of which has fortunately subsided in recent years. It is important, however, for the reader of this book to be acquainted with a representative sample of the criticisms aimed at the methods and philosophy of numerical taxonomy and we have therefore already discussed a variety of views contrary to ours at several places throughout the text. In this section we shall mention some fundamental criticisms that may not have been adequately covered earlier, many of which are put forth in several comprehensive critical reviews of numerical taxonomy to which we shall refer.

Before we embark on this discussion it is important for the reader in the early 1970's to have some historical perspective regarding the development of our views and of the ensuing controversy. When numerical taxonomy was first propounded in the middle 1950's, a well-developed and relatively stable theory of systematics was held by most biologists over the world. These views reflected the maturing of the important conceptual contributions that had been made during the 1930's and 1940's by the "New Systematics," an evolutionary view of biological systematics based on an integration of genetic and biosystematic knowledge. This impressive body of knowledge had resulted from the work of numerous outstanding scientists such as J. S. Huxley, E. Mayr, T. Dobzhansky, G. G. Simpson, G. L. Stebbins, Jr., B. Rensch, and E. Anderson.

The development at that time of numerical and phenetic taxonomy was not a sudden event, but a gradually increasing questioning of some of the established assumptions and techniques. Nor were the views we have propounded in the present book arrived at suddenly; they were, of course, developed gradually over a period of time and were modified in response to constructive criticism received from our colleagues. To illustrate the development, we may outline the following steps in the evolution of present day numerical taxonomy: (1) the discovery (or in historical perspective, rediscovery) that the measurement of similarity could be approximated (Sneath, 1957a; Michener and Sokal, 1957); (2) the realization that such a procedure in taxonomy was contrary to established philosophical tenets (e.g., Inger, 1958; Sokal, 1959); (3) the discovery (rediscovery may again be the more appropriate term) of the looseness, possible circularity, and nonoperational nature of most of the concepts of phylogenetic systematics (Sneath, 1961; Sokal, 1962b; Sneath and Sokal, 1962); (4) the further realization that purported and actual conventional practice in taxonomy were widely disparate (e.g., Michener, 1963; Sokal, 1964a); (5) the redefinition of taxonomic principles in terms of an empirical and operational science (Sokal and Sneath, 1963; Sokal and Camin, 1965); (6) the initiation of the heuristic phase in which the new insights gained through phenetic taxonomy were applied toward the development of new theory (Ehrlich and Holm, 1962; Sokal and Sneath, 1963).

The controversy surrounding numerical taxonomy increased subsequent to the publication of the *Principles of Numerical Taxonomy* (Sokal and Sneath, 1963) and is reflected in the records of numerous symposia held during the first half of the 1960's. The critical views were summarized by Mayr (1965) in an important article, the main arguments of which were answered by Sokal et al., (1965). Because these publications are widely known and accessible and also because of changes in the points of view adopted by numerical taxonomists in the time span from the middle 1950's to the beginning of the 1970's, some of the early criticisms and controversies are no longer relevant today and are therefore omitted in this book. Persons interested in these historical aspects are referred to Sokal and Sneath (1963) and the cited papers, where they are discussed in detail. We should point out, however, that possibly because of linguistic barriers, arguments that had been debated and resolved, and that to some degree are no longer even relevant to modern numerical taxonomy, continued to be debated by some, e.g., Janetschek (1967) and Ziswiler (1967).

Even among some English speaking authors there remain misunderstandings of important points of our philosophy of classification. The book by Crowson (1970) is an example. Our views on unit characters, equal numbers of characters for all OTU's, and the implications of R studies are all misinterpreted. Another curious trend is illustrated by Crowson's book. No more than five references to numerical taxonomy are included in a bibliography of over 220 papers. Only one of these papers is dated 1967, the others are from 1963 or earlier. Many other important recent papers on taxonomic philosophy and methodology are not considered. It would be unfair to single out Crowson's book in this respect when some other recent works on systematics have been similarly deficient. Yet we should deplore a tendency among systematicists to think of their field as unchanged and unchanging. What modern textbook of ecology or evolution (not to mention fields such as molecular genetics or neurophysiology) could appear in 1970 and not discuss the important concepts and contributions engendered during the 1960's?

Phyletics versus Phenetics in Taxonomy

This remains a matter for continued dispute in taxonomic circles. We have already discussed this topic in some depth in Section 2.6. Some authors dismiss phenetic taxonomy out of hand; Janetschek (1967) and Gisin (1964) apparently do not consider the possibility of a nonphyletic approach to taxonomy worthy of discussion. Ghiselin (1966, 1969) criticizes the concept of overall similarity severely. Whether similarity is "over all" or "over some" has already been discussed earlier (Section 3.8). But the fact that numerical taxonomists are able to measure phenetic similarity and to do so quite successfully should be obvious even to the casual reader of the literature in systematics. Ghiselin's numerous other criticisms in these philosophically obscure papers have been answered sufficiently by Hull (1969) and Farris (1967c). L. A. S. Johnson (1968), in what is undoubtedly the most incisive and knowledgeable critical review of numerical taxonomy published to date, states that "... floral adaptations for pollination by long-tongued insects or by birds cannot have occurred in geological periods before such animals existed. Any phenetic classification which grouped organisms in a manner inconsistent with such a fact (taking all other relevant information into account) would not be acceptable as even roughly consistent with phylogeny, and there is no reason why we should be asked to accept it simply because of claims of repeatability, objectivity, precision, or stability." The key point here is that one would not ask a systematist to accept a phenetic classification as the true phylogeny of the organisms. But one would ask him to accept it as the *phenetic* classification. Perhaps the crux of the matter is: Is Johnson prepared to accept a phenetic classification? He does not seem to recognize the general usefulness of phenetic classifications although he does not furnish cogent arguments why only phylogenetic classifications should be employed in taxonomy. This is an argument scarcely heard outside biology and not always very coherent within it: Constance (1964) in a revealing footnote, comments that many biologists appear to use the term "phyletic" to mean phenetic.

It is easy for phylogenetically oriented taxonomists to fall into the error of expecting phenetic taxonomies to yield the phylogenetic conclusions that they anticipate. As an example we may cite Kendrick and Weresub (1966), who, when classifying orders of basidiomycetes, obtained a phenetic classification that did not agree with the traditional orders. They reject numerical taxonomy on that account because implicit in their argument (although not clearly expressed) is that they wish a phylogenetic (possibly cladistic) classification. However, it should be obvious that a phenetic classification will not necessarily give them a cladogram or even a phylogenetic tree, whatever that may mean. Nor do they state anywhere on what evidence, other than their opinions as taxonomists, they base the correctness of the established classification. The reasons why the phenetic classification was so unsatisfactory have not been explained, although convergence may have been a cause. It is questionable whether the traditional higher categories in this group are as solidly established as Kendrick and Weresub seem to imply.

Another case in point is two statements by Cracraft (1967) discussing the merits of phylogenetic versus operational homology: "... rather it suggests that a non-evolutionary approach will fail to give us as accurate a picture of historical (biological) events as the evolutionary definition will." This statement is almost tautological. Obviously, a nonevolutionary approach need not provide an accurate picture of past events. Another quotation from the same author: "... there is the danger of both methods calling homologous structures (in the sense of common ancestry) that have changed radically with time nonhomologous. Furthermore, structures exist that are almost identical morphologically but that cannot be traced back to a common ancestor; here either method might incorrectly reach a decision of homology." Since operational homology is simply based on identity or similarity among characters and their relationships, the question of deciding whether these data are homologous in the sense of common ancestry and possible errors in such decisions are simply not relevant.

L. A. S. Johnson (1968) seems to feel that homologies must be worked out phylogenetically before any phenetic comparison can be useful (although he does not show us how such phylogenetic analyses could be done). He does not think much of attempts at operational homology published up to the time of his own publication. However, as we have seen in Section 3.4 the work of Jardine and others has made impressive progress here. Also the evidence from falsifying homologies (Fisher and Rohlf, 1969) and from the pseudoscanning experiment (Rohlf and Sokal, 1967) indicate that the reliability of similarity estimates based on characters of dubious homology is much better than one would anticipate.

Mayr (1969a, p. 82) realizes that both cladistic and phenetic elements must enter taxonomic theory. Yet his fundamental contradiction is the statement "it is therefore completely legitimate to define taxonomic category in evolutionary (largely phylogenetic) terms, but to use evidence (comparative character analysis) that, as such, is almost entirely nonphylogenetic." He claims (p. 208) that a theoretical weakness of the purely phenetic method is its "inability to distinguish between phena and taxa." This is not true by the definitions that Mayr gives for phenon and taxon, since according to him a taxon is a matter of opinion (p. 4). It would be very difficult for any taxonomist, conventional or numerical, to decide how to turn one into the other but we would maintain that if anyone could do it, a numerical taxonomist could.

Criticisms of the Philosophical Bases of Numerical Taxonomy

Mayr's (1969a) discussion of the philosophical bases of classification suffers from overemphasis on categorizing ideas and from an eclectic use of citations to support his views. He attacks (1) essentialism and typology in their old fashioned manifestations, (2) nominalism, or the point of view that only individuals exist and all universals are artifacts of the human mind, (3) empiricism, and (4) cladism. He advocates evolutionary classification in which "the biologist classifies populations, not individuals or phena." The lower taxa are not "arbitrary aggregates, but reproductive communities tied together by courtship responses and separated from other similar units not by arbitrary decisions of the classifier but by isolating mechanisms encoded in the genetic program of the organism... The higher taxa, likewise, are characterized by the joint possession of components of an ancestral genetic program... Organisms have another unique property which distinguishes them from inanimate objects: they have a phenotype and a genotype..." Such a categorization into "good" and "bad" schools of taxonomy misses the important point that rather than conform rigidly to an orthodox taxonomic philosophy the new taxonomy must borrow and combine ideas from all. We have already rejected old-fashioned idealistic taxonomy (see Section 3.4) but a statistical typology (Sokal, 1962b) that seeks to discover the nature of patterns in the system of nature would surely be one of the goals of taxonomy. Mayr himself (1969a, p. 76) in describing the goals of his evolutionary classification says that "the taxonomist

no longer 'makes' taxa, he becomes a 'discoverer' of groups made by evolution."
Yet this discovery must be based on phenetic evidence.

Although an exclusively nominalist point of view would be extreme, it seems to
us desirable to proceed cautiously from entities that we *can* hope to define (individuals) to classes that we *may* hope to define. Mayr feels that numerical taxonomists
err in their misinterpretation of causal relation between similarity and relationship.
As Sokal (1969) has already pointed out, Mayr uses a poorly chosen example to
illustrate his case, for he states (p. 68) that "it is exactly as with identical twins; two
brothers are not identical twins because they are similar, but they are similar
because they are both derived from a single zygote, that is, because they are
identical twins." Mayr misses the point that the statement about the two being
identical twins is a hypothesis derived from our observation of the very great
similarity between any two individuals and our knowledge of genetics and
cytology. On this point Mayr enlists Hull's (1967) statement of the view
that phylogenetic inferences and classificatory procedure are not circular,
but Hull's ensuing remark that most inferences of this sort are unwarranted is
not quoted. Whatever the philosophical merits of empiricism it should be evident
from the account in the present volume and in the by now very large literature
of numerical taxonomy that numerical methods do work and taxa do "emerge
automatically."

In summary the basic difficulty between our philosophy and the evolutionary
classification of Mayr is that we try to distinguish between the process of classifying and the subsequent process of generating hypotheses about evolution, and
Mayr wishes to combine these two. Yet in his entire volume he fails to tell us
how to make such a combination. Our dissatisfaction with vague philosophies
and methods has led to the principles and procedures propounded in our
book.

The ideas by Gisin (1964) critical of numerical taxonomy (quoted approvingly
by Mayr, 1969a) are based on quite obvious misreadings of the aims and methods
of numerical taxonomy. Thus he does not, for example, understand that numerical
taxonomists do study character correlations (R matrices). His later writings aiming
at a quantum theory of evolution grope towards a synthetic view of cladistic and
phenetic systematics, but in his later paper (Gisin, 1967) before his untimely death
he had not as yet provided us with an operational method for achieving these goals.
There seem to us two flaws in Gisin's system. The first is the lack of a quantitative
approach to a task that must essentially be mathematical (although in his 1967
paper he begins to approach this goal). The second is the assumption that there
are in fact distinct nonarbitrary categories such as species, genus, family, order,
class, and phylum that, being there, must therefore be definable. Thus much of
his theory (Gisin, 1966) is concerned with finding objective ways of defining these
categories.

Criticisms of Methodology

Some writers feel that biological classification is so different from that of inanimate objects that few if any general principles and methods may be derived from a study of the latter. Mayr (1969a), surely no vitalist, still maintains a strong emphasis on the difference in principle between classifying inanimate objects and organisms. Burtt (1966), Rollins (1965a), and Gisin (1964) are others who claim that the classifying of living organisms is not necessarily ruled by the logic appropriate to the classification of inanimate objects. We feel that the areas of common ground in classificatory theory are far greater than these critics allow.

An important point stressed by Mayr (1969a, pp. 205, 208) is whether the chosen sample of characters on which a similarity coefficient is based is sufficient for making a classification. He cites sibling species and sexually dimorphic groups (birds of paradise) as two phenetic extremes. In the former, phenetic similarity would underestimate their "drastic genetic differences"; in the latter, phenetics would exaggerate the differences. The pheneticist's answer to this problem is that (1) if overall phenetic differences correspond to the assumptions implied by these extremes, then this would be an important biological fact worth recording and measuring and would be useful in the generation of ecological hypotheses having to do with width of niches, evolutionary rates, and similar phenomena; (2) we really do not have any hard data measuring the similarity of either sibling species or highly sexually dimorphic species by appropriate methods of phenetic analysis and until such measurements are made, a statement such as Mayr's is clearly misleading. The criticism of equal weighting of characters as not conforming to evolutionary patterns (Mayr, 1969a, p. 209) would be valid if evolutionary patterns were the primary aim of phenetic taxonomy. Numerical phyletics has of necessity begun to weight characters (see Farris, 1969b).

Other criticisms illustrate the difficulty of thinking in quantitative terms even when discussing an inherently quantitative subject. An example is provided by the recent critique by Inglis (1970) on the nexus, nonspecificity, and matches asymptote hypotheses. Such hypotheses are not simply true or false: they hold in some places to a varying degree. We need to discover where and how far they hold, and even if they did not hold at all, it would not destroy the bases of numerical phenetics as Inglis implies. It would simply mean that our material was more difficult than we had thought.

It is easy today to point out the weaknesses of the mathematical and theoretical assumptions of the earliest papers in the field. L. A. S. Johnson (1968) is quite correct when he states that much more work in various areas of mathematics, especially topology, is needed to gain further insight into the problems of taxonomy. Several numerical taxonomists have also pointed out these problems (among others, Williams and Dale, 1965; Estabrook, 1966; Gower, 1967a; Jardine, Jardine, and

Sibson, 1967; Jardine, 1969b; Jardine, van Rijsbergen, and Jardine, 1969; Lerman, 1970; Jardine and Sibson, 1971), and have begun to attack them imaginatively. We must agree with L. A. S. Johnson (1968) that there probably is no one optimal classification, but his dictum that taxonomy is necessarily inexact and that compromise solutions are what we should settle for is surely too pessimistic. We may also wish to designate inherent principles of optimality that are mathematically defensible rather than those that imitate human psychological processes.

Burtt (1966) and numerous others mention the importance of experience in taxonomy, but this can be countered by the work of Sokal and Rohlf (1970) as well as by the pseudoscanning study of Rohlf and Sokal (1967). In addition, we know of numerous unpublished studies by our students and others, who in term projects in numerical taxonomy courses have come up with acceptable classifications over a wide spectrum of organized nature, often with minimal (or no prior) knowledge of the particular group being classified.

Restrictions of Numerical Taxonomy to Special Data Sets

Some authors, though granting a measure of merit to numerical taxonomic techniques, would wish to restrict their application to special kinds of organisms or special ranks only, although no obvious limitations of this kind are evident in the existing literature (see Appendix A). Oldroyd (1966) feels that numerical taxonomy should be used exclusively at the species level but not above it. Throckmorton (1965) expresses similar views. Yet others, such as Mayr (1969a) have felt very strongly that the species, more than any other category, should be defined by biological rather than phenetic criteria. It seems difficult to us to find any logical grounds for restricting numerical phenetic studies to any one categorical level. Sokal and Crovello (1970) have recently emphasized that the supposedly biological criteria for species are dependent on phenetic considerations. And even though modern evolutionary theory uses the population as its basic unit, there is considerable difficulty in defining such a concept, as Sokal and Crovello (1970) point out at the end of their critique of the biological species concept.

Burtt, Hedge, and Stevens (1970) state that computer taxonomy will make feasible "an attack on large compact groups of world-wide distribution which have just never been tackled as a whole." Michener (1963) maintains that numerical taxonomy should not be used in an exploratory way for a group of organisms to be classified for the first time, but should be used for refined analysis of a well established group, i.e., revision of a genus from the point of view of efficient use of a researcher's time. There have been so far few attempts at making actual time and efficiency estimates in classification and this question cannot be fully resolved until such work is carried out, but in microbiology, in some botanical groups, and also in entomology (Michener's own field), numerical taxonomy has been used precisely for these purposes, i.e., as an exploratory tool in several groups of organisms that previously had not been studied. The inevitable development of automatic character recording devices may also outdate this particular argument.

It is significant that there have been rather few criticisms of numerical taxonomy applied to bacteria. The main ones are by Adams (1964) and Leifson (1966). Adams criticizes equal weighting of characters but proposes no practicable alternative. Leifson's main objection is that certain classes of characters, particularly morphological ones, were overemphasized in some published studies, an assertion that appears extremely dubious. The great majority of studies have given acceptable taxonomic results. Whey then do some critics maintain that numerical taxonomy does not work in higher organisms? One might anticipate that bacteria were particularly difficult organisms to classify. The reasons given by Blackwelder (1967b) that the characters are those of bacterial populations and are inherently quantitative seem to have little force: the characters of single cells (if available) would doubtless be much the same, while the nature of bacterial characters cannot be said to be inherently either qualitative or quantitative (the qualitative characters are due in the main to dichotomizing quantitative characters at some level convenient in chemical assay). It is true that homology in bacteria is mostly based on external criteria, and the implication of this for other organisms has not yet been explored. It is also true that most characters are not morphological but biochemical or physiological, yet it would be hard to argue cogently that their significance was very different, particularly since no very obvious peculiarity in taxonomic pattern of chemotaxonomic characters has been noted in higher organisms. If dichotomizing characters avoids some problems (like that of the general size factor), it may be profitable to use this method in higher organisms. But it would seem to us more likely that the main reason for the general acceptability of numerical phenetics in bacteria is that the user is expecting only phenetic groups and is satisfied when he obtains them (see Sneath, 1971b).

Criticisms of Empirical Findings

Typical of such publications is a critique of the paper by El-Gazzar et al. (1968) in Burtt et al. (1970), in which I. C. Hedge points out that the classification comes out in an unexpected way. Yet no reasons other than those of preconceived notions regarding the genus are given: "the underlying reason for this is not hard to find. The characters used were unsatisfactory for the genus... In any future study of *Salvia* ways would have to be found to codify the features of the corolla form in which much of the diversity of the genus is expressed..." These statements summarize the preconceived notion that "experienced taxonomists" know what characters are useful for describing species and genera, and that any studies are obviously wrong which do not use these characters or which use these characters intermixed with others that these experts have not found to be equally useful. Numerical taxonomy and its philosophy have attempted to escape this kind of dogmatism. The critique of a study of Ericales by Watson, Williams, and Lance (1967) by P. F. Stevens in Burtt et al. (1970) is of greater interest because some of the comments are clearer and more relevant. The criticism of the choice of characters remains philosophically unacceptable to us, but Stevens engages in a cogent discussion of the delimitation of character states and of errors of observation (see also Stevens, 1970). As we have discussed elsewhere in this book (Sections 5.10 and 5.13) errors in such work will necessarily result in classificatory errors of possibly predictable magnitude.

The criticisms also raise another point: what are the reasons for the failure of the numerical taxonomic method with certain OTU's when with others the phenetic groups were quite acceptable? What characters would have to be recoded, differentially weighted, and so on, to give results that would be acceptable with all the OTU's? If we could find this out we would be better able to avoid such errors, or the alterations in the data would have to be so extreme as to cast doubt on our preconceived ideas. Indeed, Brown (1965) illustrates this last point in asking whether the impressive list of common characters of Strepsiptera and rhipiphorid beetles is due to convergence or close common ancestry; the problem is seen to turn on whether it is more probable that there has been a gain rather than a loss in the number of tarsal segments, a point that is evidently very hard to decide. Similar comments are made by Blackith and Blackith (1968) who say, in speaking of the results of a numerical taxonomic study, "there do not seem to be any grounds for accepting only those parts of the picture which are easily assimilated and rejecting those parts which are less so. To select features in that way would be to condemn numerical taxonomy for being wrong when it disagreed with accepted ideas and superfluous whenever it agreed."

An important point to be stressed here is that the explicit procedures of numerical taxonomy make it possible to engage in criticism of this sort, which is not generally possible in orthodox taxonomy. Many if not all of the criticisms of methodology and results of taxometric studies apply to conventional classifications as well but cannot be stated as clearly in that branch of taxonomy. In other words, the very same criticisms could be directed at the treatment of the problem by conventional taxonomy, except that the problem did not become obvious until it was treated by numerical taxonomy. As just one of many possible examples we cite Steyskal (1968), who feels that many more than the usual number of characters may be required for numerical taxonomy in certain insect groups in which taxa can only be recognized on details of genitalia. Because the nonspecificity hypothesis does not hold in its entirety, it is possible that inadequate sampling of characters may distort phenetic relationships among such groups of insects, but this danger would affect conventional and numerical studies equally.

L. A. S. Johnson (1968) states that the effort expended in numerical phenetics is not worthwhile in view of the published results. We believe he underrates the significance of the findings that have been made and that we have illustrated in our present book. There is hardly a numerical taxonomic study that has not resulted in some new insight and understanding and there are many that have been of considerable heuristic value.

We share with L. A. S. Johnson (1968) his desire for a heuristic element in taxonomy and we shall show in Section 10.3 that numerical taxonomy can help pose some important biological questions and provide some answers, however automated the procedure of character gathering, defining, and classifying eventually becomes. We should clarify our view and state that we do not thereby imply that *only* numerical taxonomy has these heuristic properties. There are numerous ways of looking at biological systematics and using it to generate hypotheses about a variety of biological problems. Many of these can and will be quantified, but not all of them can be or need to be.

10.2 SHORTCOMINGS OF NUMERICAL TAXONOMY

Throughout the text we have pointed out the various problems encountered when one attempts to make taxonomic procedures operational. As in all sciences, the solution of such problems is a never-ending procedure. Having satisfactorily overcome the obstacles of one aspect of the work, new difficulties emerge at another level. It may be useful to bring together in one place the various problems discussed in the preceding sections that still await solution.

Perhaps the foremost challenge is the development of a generally acceptable system for coding and scaling characters. The problems involved have been outlined in Section 4.8. Possibly the best hope lies in analyzing in some way the information content provided by various characters, followed by their reduction into binary unit characters. The difficulties posed by logical dependence and biological correlation are manifold. Until such time as a general system for describing characters is available, workers must be content with the present methodology, which in spite of the lack of elegance of the theory is fortunately robust in its results. Uniform coding of characters will of its own accord lead to a type of automatic weighting. Characters conveying more information will contribute more toward the estimate of resemblance between OTU's.

Clearly more work is needed on defining the character sets that can and should be used in taxometric work, and more precise and profound analyses of the nature of phenetic similarity are required. We have already departed from the concept of a parametric overall similarity (Section 3.8). The difficulties inherent in such a concept have been enunciated very cogently by L. A. S. Johnson (1968). He furnishes the following list of subjective decisions that still must be made when a measure of similarity is adopted:

(i) the set of objects considered usefully comparable;
(ii) the domain of attributes which we consider relevant to our interest in the objects;
(iii) the "fineness" with which we analyse the features into elementary attributes ($\equiv$ states);
(iv) the establishment of equivalences or homologies between parts of the objects under comparison; and the consequent grouping of the elementary attributes into two- or multi-state sets, thus specifying what we usually term "the attributes" or "the characters" ("multi-state" here includes "continuously-varying");
(v) the method and intensity of sampling of the objects;
(vi) the method and intensity of sampling of the acceptable sets of "relevant" attributes;
(vii) the quantitative or qualitative measures to be used in expressing the "states" of each attribute (involving an often arbitrary assignment of working commensurability);
(viii) the measure of similarity (or distance) to be adopted.

Similar problems arise with apparently attractive alternatives in the study of proteins or DNA sequences. Much work lies ahead in these areas. It should be added that most of these problems face the traditional taxonomist as well.

Tests of significance of similarity coefficients (discussed in Section 4.10) and of classifications (Section 5.10) will undoubtedly progress considerably in the next few years and will, as a consequence, bring about profound changes in present day procedures.

The discussion of taxonomic structure in Chapter 5, especially optimality criteria (Section 5.10), has pointed out the difficulties that still beset any definition of an optimal classification. Such a definition will comprise decisions on the kind of structure needed by taxonomists, desirable mathematical properties of optimality criteria, and suitable algorithms for computer handling.

In this connection we may again cite L. A. S. Johnson (1968), who provides a large and impressive list of problems to be solved by numerical taxonomists. He furnishes this list as an example of the tasks that should occupy numerical taxonomists, with the implication that such activities are aside from and possibly have no relevance to the problems of taxonomy. Yet we feel that the cogent questions posed by him are the very questions to which any taxonomist needs answers if he is to understand the natural world that he wishes to classify.

What we consider the primary shortcomings of numerical taxonomy have been listed above, but we might round out our account with some problems of secondary importance. As should be evident from the extended discussion in Section 3.4, we still lack a consistent system of operational homology. It appears that phenetic analyses can be carried out in spite of this drawback, but nevertheless it is philosophically embarrassing to rest character coding on such shaky foundations. The importance of correctly defining homologies becomes especially relevant in cladistic work and in evolutionary studies based on protein sequences, and recent work has emphasized this problem. We still know little of the causes of incongruence; methods for partitioning the variation into components of biological interest, and accounting for sampling error are thus much needed. Some patterns of variation have not been looked for by numerical pheneticists, such as that in which two taxa are sharply separated on a few characters only, so that they form, in phenetic space, two flat plates separated by a small gap. This might occur with pairs of sibling species, and the existence of two such taxa would not be readily detected by the usual methods of finding taxonomic structure. In Section 6.3 we have stressed the importance of correct assumptions about evolutionary processes in order to make cladistic inferences for numerical cladistics and numerical phyletics. In order to construct explicitly analytical algorithms we must have some understanding of how to reconstruct ancestral character states and the degree of confidence we can have in such work. We are in a period of rapid change in our techniques in this field at the moment and readers are referred to Kluge and Farris (1973) and to the numerous articles appearing on this subject in the technical literature.

10.3 HEURISTIC ASPECTS OF NUMERICAL TAXONOMY

The term "heuristic" is defined as "revealing new knowledge." Specific examples have been given in other chapters of the new information that can be gained from numerical taxonomy, so we are more concerned here to point out the kinds of new knowledge that can be thus obtained, rather than details. The discussion includes some aspects of related multivariate techniques, because new knowledge is often obtained from a combination of these with the more narrowly taxometric methods.

Several authors have noted the potential of numerical taxonomy for forming new hypotheses. Thus Silvestri and Hill (1964) point out that the discovery of well-separated phenons implies that in nature intermediate forms will be rare, which can be tested by further work. They foresee that taxonomy will cease to be a purely descriptive science but will also be used for formulating hypotheses that can be tested experimentally. Similar views are expressed by Williams and Dale (1965), Williams and Lance (1965), and Goodall (1966e). This has, of course, always been true of taxonomic work to a certain extent, but its power will be greatly increased by quantitative techniques. Such work may have important economic consequences, especially in such fields as agriculture and public health. For example, the association between streptomycete taxa and the antibiotics they produce is of importance to the pharmaceutical industry (e.g., Gilardi et al., 1960; Silvestri et al., 1962). The observation of clusters of OTU's or of characters that are empirically correlated may lead to a search for a causal explanation. Thus floral character correlations may be causally associated with pollination mechanisms. Blackwelder (1967a) and Mayr (1969a) give numerous other examples of the practical importance of taxonomy.

We may first consider what general taxonomic information may be obtained from numerical studies. The most obvious is that the numerical taxonomy confirms the broad outlines of existing orthodox taxonomy. Although this may come as no surprise, and may sometimes be categorized as "self-evident information," such confirmation should not be despised. It increases the general credibility of the taxonomy of the group and of the numerical techniques appropriate for such study. Numerous examples could be cited, e.g., Michener and Sokal (1957) on bees; Sims (1966) on earthworms; Wilkinson (1967, 1968) on lepidoptera; Ackermann (1967) on birds; Watson, Williams, and Lance (1966), Rhodes et al. (1968), Stearn (1968, 1969), Ivimey-Cook (1969a,b), Prance, Rogers, and White (1969), and Rowley (1969) on flowering plants; and Seyfried (1968) and Stevens (1969) on bacteria. More to the point is that in many cases there have previously been several conflicting orthodox taxonomies, and the numerical analyses have often provided strong support for one of these. The studies just mentioned contain several such examples.

It is very common to find minor points of dispute in the previous taxonomy of a group, affecting only a few of the smaller taxa. Numerical studies have been extremely useful in giving more convincing evidence of the phenetic relationships and the better placement of such taxa (for examples see Hudson et al., 1966; Thornton and Wong, 1967; Rowley, 1967; McNeill, Parker, and Heywood, 1969a). Numerical taxonomy occasionally suggests quite unsuspected relationships (e.g., Barnes and Goldberg, 1968, on anaerobic bacteria). Again, some accepted groupings have been shown to be of dubious validity; for example, Stephenson, Williams, and Lance (1968) were unable to substantiate the existence of some "twin species" that had been proposed in orthodox classifications.

Numerical studies have been particularly informative in groups whose current taxonomy is very confused or is refractory to orthodox methods. Most often this has suggested major revisions of the taxonomy. Such confusion is most widespread in microbiology, where many examples can be found (reviewed in *Principles of Numerical Taxonomy* and in Sneath, 1964a). In higher organisms we know of fewer examples, but the studies of Ducker, Williams, and Lance (1965), Wilkinson (1970a), El-Gazzar and Watson (1970a,b), and El-Gazzar et al. (1968) may be cited. In the last study, on the genus *Salvia*, of the existing subdivisions only one of five subgenera and only one of twelve sections appeared valid, though this study has been criticized by Burtt, Hedge, and Stevens (1970); the numerical results yielded new and more convincing phenons in close accord with geographic distribution and with several additional characters. An even more drastic reassessment of current views is posed by the study of Young and Watson (1970) on the families of dicotyledons. An allied problem is to extract some sense out of data of very low quality, and Parker-Rhodes and Jackson (1969) give an example. In bacteriology numerical taxonomy has been of great value in the initial grouping together of individual strains to yield taxa of low rank conventionally regarded as species, and most of the work on bacteria deals in part with this problem. One common result has been to simplify greatly the taxonomy of bacterial taxa by reducing many dubiously named species to synonymy, but sometimes numerical methods have revealed subgroups considered to be valid new species within apparently homogeneous groups of strains. There have been a few studies in higher organisms where numerical taxonomy has been used to find new taxa without first attempting an orthodox taxonomy (e.g., Imaizumi, 1967, on a new genus of cat).

Although numerical taxonomy has occasionally increased the number of rank categories that may be profitably recognized, we do not feel that it has resulted in any marked tendency toward either "lumping" or "splitting" when all kinds of organisms are considered. It has however greatly improved our knowledge of the finer points of relationship.

The results of numerical taxonomy have often been confirmed by subsequent work. This is particularly true in bacteriology (reviewed by Jones and Sneath, 1970) and some such confirmatory work on higher organisms has been mentioned in Section 5.12. Additional examples may be found in the studies of Hudson et al. (1966), Drury and Randal (1969), and McNeill et al. (1969a) to mention only a few. But although confirmatory evidence has not always been presented, it is noteworthy that cases where clear and apparently acceptable numerical results have been contradicted by later work seem to be uncommon. We find it difficult to cite good examples of this. In most cases the discrepancies appear to be due to inadequate techniques, or methods inappropriate to the desired taxonomic aims. Some of these are discussed in Sections 10.1 and 10.2. We have noted, however, in common with others (e.g., Moss, 1968b; Rhodes et al., 1968; Huber, 1968; Johnson and Holm, 1968), that many minor discrepancies may well be due to size factors (see Section 4.11), which therefore deserve careful attention.

More specific information that may be obtained from numerical studies includes the prediction of states of characters that were not included in the original survey. A good example is given by Williams (1967a). In the study of Ericales by Watson, Williams, and Lance (1967) the genus *Epigaea* (always something of a puzzle) was placed in the Phyllodoceae. But members of this tribe have an unusual feature, the presence of viscid threads among the pollen grains, which was not recorded for *Epigaea*. On further examination of *Epigaea* the threads were found. This is, of course, a special case of the general congruence principle that good phenons are likely to be relatively homogeneous for states of new characters, of which Drury and Randal (1969) give an example. Numerical studies may also reveal the insufficiency of the available characters for adequate taxonomic work, e.g., as revealed by large errors of sampling or measurement or because unsatisfactory groupings emerge (see Sections 3.8, 4.10, 6.4).

Congruence and incongruence between phenetic relationships based on different life forms is another field where quantitative techniques are indispensable. Such work could give new knowledge of the probable identity of larval stages and adult stages (e.g., in parasites) or of the sexual and asexual forms of fungi (e.g., Leth-Bak and Stenderup, 1969). The grouping of fossil fragments into associations representing different species has been noted in Section 6.5. More important, the measurement of incongruence requires new hypotheses to explain it. Thus larval-adult incongruence will some day illuminate aspects of the evolution of these stages, the selective pressures and adaptations involved in the habitats of larvae and adults, etc. (Rohlf, 1963a). Similarly incongruence between floral and vegetative characters (Crovello, 1969; Johnson and Holm, 1968) must have biological meaning. Pernes, Combes, and Réné-Chaume (1970) proposed a hypothesis on inheritance mechanisms in millet as a result of observing incongruence. Incongruence often seems to affect mainly a few of the OTU's (Moss, 1967; Ehrlich and Ehrlich, 1967; Johnson and Holm, 1968), a finding which is still poorly understood.

Other details obtainable from numerical studies include patterns of phenetic variation. Thus phenons may be found to be hyperspherical or elongated, they may show trends or clines in ordination studies, or chains of intermediate forms may be revealed (see, in particular, Sections 5.14 and 7.5). It is also to be noted that satisfactory schemes of discrimination may require meticulous descriptions of phenons in numerical form; this can be a critical step in difficult situations (e.g., Lapage et al., 1970). The use of phenetics in nomenclature, for choosing types and limits of taxa, has been discussed in Chapter 9. New knowledge on morphogenesis, growth, functional analysis, and the interaction of genotype and phenotype can also be obtained from phenetic studies (Blackith, Davies, and Moy, 1963; Sneath, 1967a, 1968a; Soulé, 1967b; Oxnard and Neely, 1969). Borgelt (1968) discovered by numerical methods a new form in the solitary stage of the life cycle of a salp that corresponded to a known form in the aggregate stage.

In the field of evolution the growth of numerical cladistics has opened up a large new area. Numerical phenetics and cladistics together can allow the measurement of rates and patterns of evolution, selective pressures, the study of convergence, parallelism, and reticulate and mosaic evolution. These points are fully discussed in Chapter 6, and here we need merely add that studies of this kind can only be accomplished by numerical methods. Broader information includes generalizations about the way evolution operates, concordance between geological time and divergence (both morphological and biochemical, as in serology and protein sequences), and the logic of cladistic inference.

Phenetic studies of various kinds have been valuable in genetics. Thus Whitehouse (1971) has used multivariate methods to predict the parents most likely to yield desired hybrid crop plants. Rhodes, Carmer, and Courter (1969) have employed numerical taxonomy to select the widest range of variation for programs of plant breeding and Rhodes and Carmer (1966) discovered errors in pedigree records as a result of unexpected phenetic relationships. Incongruence between phenetics and ability to hybridize may be an interesting field (Heiser, Soria, and Burton, 1965). Numerical phenetics can give new information about the effect of environment on phenotype (an unexplored field); Williams and Lance (1969) mention the use of cluster analysis in elucidating a complex problem of this kind.

Geographic and ecological variation has been discussed in Chapter 7, but we may summarize this by saying that phenetics gives abundant information on the pattern of geographical variation, the reality (or otherwise) of subspecies and races, and the occurrence of temporal phenetic changes (Mason, Ehrlich, and Emmel, 1968). Numerical taxonomy may provide information of especial value in epidemiology. Thus Ibrahim and Threlfall (1966a) found a very close phenetic resemblance between isolates of a fungus from oats and from rye grass, and comment that this appears to solve the problem of the source of secondary infection of oats. Talbot and Sneath (1960) noted a correspondence between phenetics and pathogenicity of the hemorrhagic septicemia bacillus: all the strains from cats and from internal lesions in humans were extremely similar but differed from most dog strains; this suggests that human infections with this bacillus are usually contracted from cats. Roberts (1968) found certain subclusters of *Corynebacterium pyogenes* associated with different hosts, and more weakly with the organ infected and with geography. Jewsbury (1968) discovered complex relationships between host, geography, and possibly also snail vectors in a study of schistosomes.

Applications of numerical methods outside systematics are discussed at length in Chapter 11. We mention here only a few examples that illustrate the kind of new knowledge that may be obtained. Most work has been done in ecology. This may reveal quite unsuspected information. A floristic study may uncover a discontinuity due to a change in soil, a difference in agriculture, or the effects of burning of vegetation. Detailed ecological study of forests is giving an entirely new view of a complex habitat. These developments are reviewed by Williams (1967), Williams and Lance (1969), and Webb et al., (1967a,b). Major biogeographical patterns may be well shown by numerical methods, and invasion routes may be indicated (e.g., Holloway and Jardine, 1968; Berry, 1969). Climate and the geological history may be reflected in ecological resemblances (Proctor, 1967; Sneath, 1967c). Other fields where phenetic studies are providing new knowledge are also discussed in Chapter 11 (e.g., stratigaphy).

A major contribution to evolutionary and ecological theory has been the study of biological diversity initiated by Margalef (1958) and widely applied by numerous authors in recent years (e.g., MacArthur and MacArthur, 1961; Pianka, 1967; Johnson, 1970). Diversity indices have so far been based on numbers and frequencies of species—the traditional species of conventional taxonomy. The tacit assumption of these formulations is that each species is ecologically equivalent. Yet to weight such species occupying the ecological space in these faunas and floras equally in a diversity formula is clearly unrealistic. One looks forward therefore to a new field of research in which the study of ecological diversity will be combined with numerical taxonomy and in which diversity measures will include indices of the hypervolumes and the density of the various species concerned. Initial attempts at numerical taxonomic definitions of niches have been made by Hutchinson (1968),

Fujii (1969), and Maguire (1967). Hurlbutt (1968) made an attempt to illustrate Gause's principle by means of phenetic differences between species coexisting in the same habitat. Other work in this field is in progress (e.g., Findley, 1973).

Finally, numerical taxonomy has a very important part to play in scientific methodology. Indeed this alone would justify its study as a scientific discipline (as admitted by its critics). It has given deeper insight into the nature of the taxonomic process and the nature of taxonomic judgment. Concepts such as size and shape have only become susceptible to effective study as a result of numerical methods. Many questions, often awkward for taxonomists, are raised: what is a taxon?; what is a character?; what is resemblance? Whether or not the answers of numerical taxonomy have been useful, its questions certainly have been. It will contribute toward the psychology of perception. It seems from the little present work (see Hodson, Sneath, and Doran, 1966; Sneath, 1968b) that trends, clines, or progressions are much more difficult to detect by eye than are discrete clusters when variation is complex. There are some suggestions (from examples in Sneath, 1964b) that the eye is often misled to overvalue an attribute that is large or affects a large area (such as prominent coloration), while position, stance, and orientation are less often misinterpreted. Several studies currently underway (Moss, 1971; and the TAXOCRIT experiment by Sokal and several associates) are beginning to answer questions about what taxonomists actually see in making their judgments. An interesting essay on the methodology of classification and perception is that of Ornstein (1965). The nature and value of taxonomic experience has been investigated in a first study by Sokal and Rohlf (1970). The study of automatic scanning methods (Sokal and Rohlf, 1966) is a logical extension of work on perception. Numerical taxonomy contributes also to the study of homology, where work is being actively pursued both on gross morphology and protein sequences.

In conclusion we must emphasize that though many of the kinds of information discussed in this section may seem to be self-evident from the crude data, this is not by any means so in most instances. We may cite here a comment from an anonymous contributor to the discussion on the paper of Willmott and Grimshaw (1969), who states: "One simple way of checking on the production of 'self-evident' results, which I have myself used with ecologists in the application of association analysis is to ask the ecologist to write down his interpretation and place his description in a sealed envelope before starting the analysis. The interesting result is that in every case the analysis gives the information which a trained ecologist will derive from his knowledge of factors external to the data, and that in most cases it will find something which that ecologist, despite his extra knowledge, has missed." Our personal experience in a variety of other fields is virtually identical.

Numerical Taxonomy in Fields Other than Biological Systematics

In this chapter we give an account of the application of taxometric and closely related methods to fields outside systematic biology. We have been guided by two considerations: to show the range of subjects in which such methods are useful, and to describe the special problems that arise when they are applied to these other fields.

There is some overlap in the subject matter of the various sections, and the reader may find it useful to consult several of these for borderline topics.

11.1 ECOLOGY AND BIOGEOGRAPHY

In the last few years, there has been a swift growth of taxometric methods in ecology. Notable early work was that of Sørensen (1948), who employed both coefficients of association and cluster analysis. The modern phase can be said to have been initiated with the development of the monothetic method of Goodall (1953). In recent years W. T. Williams and his colleagues have contributed significantly to this field and there has been a considerable interchange of ideas with taxonomy. Reviews and critical discussion are given by Grieg-Smith (1964) and Gimingham (1969) and Whittaker (1972). These methods have had a powerful impact on descriptive and community ecology, especially in areas where ecology and systematics overlap (see Sections 7.4 and 7.5). Numerical taxonomy is widely used in plant ecology for vegetation studies, in which "taxa" such as woodland, prairie, or moorland can be distinguished and analyzed.

These types of vegetation are sometimes suitably represented by hierarchies analogous to those in systematics. For OTU's the basic data matrix has stands or quadrats; the species found in these units are used as characters. Those species are recorded as present or absent, or by some measure of their abundance. As well as grouping together areas (stands, quadrats) similar in their species composition, one can also group those species that are most usually associated, and Williams and Lambert (1961a) refer to these, respectively, as "normal" and "inverse" analyses, which are, of course, the Q and R analyses of the basic data matrix identified by terms less likely to cause confusion. Williams and Lambert also suggest simultaneous normal and inverse analysis, with the object of producing "noda," that is, joint clusters of both quadrats and species (Williams and Lambert, 1961b; Lambert and Williams, 1962; see also Webb et al., 1967a; Moore, Benninghoff, and Dwyer, 1967).

Because of the continuous nature of the variation in most vegetation, ordination methods are much used. The earlier ordination methods are now being superseded by various types of factor analysis (e.g., Austin and Orloci, 1966; Ivimey-Cook and Proctor, 1967; and Whittaker, 1972), but multidimensional scaling might be worth trial, particularly for such difficult studies as those on cyclic changes in vegetation (e.g., Anderson, 1967). Grieg-Smith, Austin, and Whitmore (1967) found some support for the view that cluster analysis would be most appropriate for studying major vegetation types, because it is here that sharp discontinuities are most likely.

Monothetic clustering methods have been used a good deal in ecology, particularly the association analysis of Williams and Lambert (1959). In this monothetic method a quadrat may be occasionally far removed from quadrats that are highly similar in overall floristic content, because of the chance presence or absence of the species that served to define the monothetic group. Such misplacements are evidently not uncommon (Lange, Stenhouse, and Offler, 1965; Lambert and Williams, 1966; Grieg-Smith, Austin, and Whitmore, 1967), both for normal and inverse analysis. Polythetic methods are better if detailed ecological relationships are to be studied. Nevertheless, because of their speed and ability to handle large numbers of OTU's, monothetic methods may be convenient for preliminary grouping (e.g., Crawford and Wishart, 1967; Crawford, Wishart, and Campbell, 1970).

A number of comparisons have been made in ecology between different polythetic methods, particularly between the usual clustering methods and those employing information statistics (Lambert and Williams, 1966; Williams, Lambert, and Lance, 1966; Hall, 1967b; Webb et al., 1967a; Orloci, 1968a,c, 1969a; Kikkawa, 1968). Information statistics appear promising for ecology, though they seem best suited to nonprobabilistic studies on unique sets of OTU's (Hall, 1967b). Webb et al. (1967a) found association analysis and information analysis superior to factor analytic methods in isolating the kinds of association that ecologists recognize. Williams and Lambert (1966) reported that centroid clustering with their "nonmetric coefficient" was almost as good as information analysis, but in ecology taxonomic distances were unsatisfactory, and standardization of 0,1 data was disadvantageous. Criteria of optimality present difficulties in ecological classification as in biological systematics, and "informed opinion" has most often been the test of success. However, independent criteria are sometimes available from the environment (e.g., clusters may agree closely with soil type or rainfall); thus Williams and Lambert (1960) found agreement between agricultural drainage history and subtle differences of vegetation type, and Sundman (1970) showed congruence of bacterial populations and soil type.

A considerable body of new knowledge is coming from the numerical studies of Williams and his colleagues on tropical rain forests (Webb et al., 1967a,b; Williams and Webb, 1968; Williams et al., 1969a,b; Webb et al., 1970), much of which has wide implications for ecology. This includes investigations on the extent to which certain classes of plants reflect the total floristic composition, the relation between floristic composition and microclimate, the detection of small scale vegetational pattern, and the rate and direction (in ecological hyperspace) of plant successions during the regeneration of forest.

Some special problems of ecological classifications may be noted. The number of characters may be unavoidably small; thus, there may be only ten species of flowering plants in the area studied. Williams, Dale, and McNaughton-Smith (1963) have suggested a method of effectively weighting characters in proportion to their average correlation with all other characters if the number of characters is small. The significance of the absence of a plant may be uncertain: the plant may not be able to grow, or it may not have had the opportunity to colonize the area studied, or it was by chance not in the sampled quadrat. Therefore the association coefficients employed in ecology commonly exclude negative matches. Mountford (1962) considers coefficients that are independent of quadrat size (see Section 4.4). Mixed qualitative and quantitative data are discussed by Williams and Dale (1962, 1965) and Williams and Lance (1965). Goodall (1969) proposes a method for testing the significance of rare floristic associations. A list of similarity coefficients used in ecology is given by Peters (1968). Trend surface analysis has been employed by Gittins (1968) and Fisher (1968).

Most of the papers cited above have emphasized methodology. Some descriptive studies are those of West (1966) on clusters of vegetation types in forests, Schmid (1965) on underwater vegetation, and Gyllenberg (1965), Gyllenberg and Rauramaa (1966), and Sundman and Carlberg (1967) in bacterial ecology. Rasnitsyn (1965) used Smirnov's similarity coefficient in describing the ecology of aquatic larvae. Hurlbutt (1964, 1968), by phenetic studies of mites, found some evidence for

Gause's law that species with identical ecological requirements do not occupy the same spatial niche. Ebeling et al. (1970) were unable to find support for this law in a marine habitat, but consider that they may not have had suitable material. There is considerable promise in using taxometric methods to cluster similar environments and correlate these with phenetic patterns in populations of organisms. The whole area of species diversity versus environmental diversity in which so much recent work has been done (see our discussion in Sections 7.4 and 10.3) is based on conventional interpretation of species. It is here that a quantitative phenetic interpretation of diversity could be of great heuristic value.

Studies on a larger geographic scale include those of Hagmeier and Stults (1964), Hagmeier (1966), and Kikkawa and Pearse (1969), in which zoogeographic provinces were constructed by cluster analysis. Holloway and Jardine (1968) and Holloway (1969) also distinguished faunal elements with different distribution patterns. Sakai (1971) obtained faunal regions of Dermaptera from cluster analysis. Fisher (1968), studying amphibia, reptiles, and mammals, noted good correspondence between environment (rainfall, vegetation) and trends in factors summarizing the occurrence of these species. Proctor (1967) made a principal component analysis of the liverwort flora of the British Isles. The major division was into highlands and lowlands, but the richness of the flora and oceanic influence were also reflected in the factors. Heatwole and MacKenzie (1967) noted that insular faunal similarity decreased evenly with increasing geographic distance. Sneath (1967c) applied principal component analysis to the world distribution of conifers in relation to evidence of continental drift. Holloway and Jardine (1968) applied multidimensional scaling to relate faunal distributions to geography in the Indo-Australian area, and clearly showed the different boundaries of zoogeographic regions based on different taxonomic groups (birds, butterflies, and bats). Indications of past routes of spread were also obtained. Tobler, Mielke, and Detwyler (1970) have carried this type of work further by partitioning the variation into factors representing latitude, longitude, and mobility of organisms.

Taxometric methods are also used in hydrobiology (e.g., Williamson, 1961; Fager and McGowan, 1963; Mello and Buzas, 1968) to characterize ecological associations of plankton. Neushal (1967) analyzed the effect of depth of water on the associations and Howarth and Murray (1969) found species clusters that were strongly related to environmental variables, such as salinity and ionic composition. A point raised by the work of Roback, Cairns, and Kaesler (1969) is whether multivariate methods are more sensitive indicators of pollution than the usual techniques based on a few indicator species. Valentine (1966) observed excellent concordance between geographic latitude and faunal similarity on the west coast of North America; he found that the faunal provinces given by the dendrogram were more informative than the conventional ones.

Similar ecological work has been done on fossil communities (for reviews see Buzas, 1970). Thus Johnson (1962) used linkage diagrams for clustering species of fossils in different localities and also demonstrated satisfactory results upon modern marine habitats. Kaesler (1966) made both normal and inverse studies on ostracods and foraminifers employing S_{SM}, S_J, and average linkage clustering; he found substantial agreement with earlier groups based on intuitive analyses. Maddocks (1966) and Rucker (1967) have made similar studies. Valentine and Peddicord (1967) used cluster analysis on fossil mollusk assemblages; they noted subsidiary resemblances in the similarity matrix between localities that were not evident in the dendrogram. These subsidiary resemblances are similar to ones noted by Rayner (1966) in his study on soil, though their significance is not yet clear. Park (1968) and others have studied fossil assemblages by ordination and interpreted main vectors as salinity, depth of water, and rate of sedimentation.

Soil classification has received increased interest lately. Leeper (1954) criticized the use of postulated origins of the soil rather than their observed properties, and this trend toward phenetic classifications is reviewed by Mulcahey and Humphries (1967), Webster (1968), and Rayner (1969).

Soils are similar to vegetation in that it is difficult to know the relative merits of ordination and clustering (see Russell and Moore, 1967; Cuanalo and Webster, 1970). But in addition there are problems of homology. It is not easy to decide which soil horizons are comparable when the resemblance between soil profiles is to be estimated. There is an arbitrary element in defining the conventional horizons. Rayner (1966, 1969) has shown that there is in general higher resemblance between the same conventional horizons than between different ones, but exceptions are not unusual. The best results (as judged against the traditional soil classifications) were obtained by first homologizing the horizons on the basis of their greatest resemblances from one soil to another and computing an average resemblance of the homologies thus determined. Quite acceptable results were obtained by employing only one horizon from each of a pair of soils, choosing this pair of horizons on the basis of the greatest degree of homology. Ordination of separate horizons was particularly instructive. Similar work has been done by Moore and Russell (1966), who examined in more detail the degree of homology between samples from different depths of a single soil profile. They noted difficulty in finding sharp distinctions between horizons, and in one case the analysis indicated subzones that were not evident before. Russell and Moore (1968) suggest a method of summing resemblances between set depths in two soil profiles to obtain an overall resemblance between the profiles. Lance and Williams (1967c) make a similar suggestion to average the character values over different depths.

Bidwell and his colleagues (Bidwell and Hole, 1964a,b; Cipra, Bidwell, and Rohlf, 1970) have shown that numerical taxonomy of soils is quite workable. The polythetic method placed together some soils that were very similar overall, but which were in separate "Great Groups" based monothetically on color. Moore and Russell (1967) note that conventional soil classifications appear to contain a substantial "general size" factor. Earlier work is that of Hughes and Lindley (1955) and

Hole and Hironaka (1960). Sarkar, Bidwell, and Marcus (1966) have investigated the minimum number of characters required to give satisfactory classifications, but their method of retaining only uncorrelated characters has serious drawbacks (Sneath, 1967b), and factor analysis sensu lato would seem preferable if the effect of correlation is to be excluded. Grigal and Arneman (1970) have compared soils and their associated vegetation by numerical methods. Webster and Wong (1969) report that soil boundaries in soil survey maps are more readily found from changes in the first vector of factor analyses than from changes in any single property.

11.2　MEDICINE

A main emphasis in taxometric work in medicine has been computer diagnosis of disease. It is thus aimed at problems of identification, and an extensive literature exists on this (for a review see Anderson and Boyle, 1968). But before discussing this it is necessary to consider the earlier stages in the classificatory process. For diagnosis one must assume prior knowledge of the disease entities and the validity of these is of critical importance. If disease taxa are unsound, the diagnostic schemes will inevitably be unsatisfactory. It is now being realized that the classification of diseases (nosology) warrants more attention than it has received; that also, like any taxonomic process, it can be carried out by taxometric methods (Sokal, 1964b; Sneath, 1965, 1966a; Baron and Fraser, 1965, 1968). Disease entities—clusters of similar individual cases of disease—can be made by Q studies, and syndromes (clusters of signs and symptoms) by R studies, although the distinction between Q and R clusters is not always made clear. Indeed diseases are in this respect analogous to the noda of ecology (see the previous section).

It is, of course, very probable that most well-known diseases are valid groupings in some sense, but there has been little study on this point; it is doubtful whether most of the obscurer diseases have been satisfactorily classified. It has been conventional to define a disease by its cause (etiology). Yet, despite the great influence of etiology on treatment, it may not always be suitable for classification. The argument here harks back to the phenetic-phyletic controversy in biological taxonomy. Some clinical entities may have a varied etiology: for example, meningitis caused by very different bacteria can be clinically indistinguishable; conversely, one etiology may produce varied signs, symptoms, and pathology: for example, syphilis. It is by no means clear how one should define health or what attributes should be used to classify diseases. It is true that there is an external criterion against which to test many disease classifications—all cases belonging to the same disease should respond to an appropriate remedy. But such a concept can permit disease groupings that may be unnatural in other respects. There is, in fact, no unifying concept of disease (see Engle, 1963; Engle and Davis, 1963; Scadding, 1963). Diseases may be defined on etiological, clinical, or anatomical bases, and many diseases are polythetic groupings.

The primary data are the individual patient and his signs and symptoms: case histories that are initially used to construct disease classifications and are the logical OTU's. Engle (1963) notes the disturbing tendency to ignore atypical cases, and the repeated redefinition of a disease monothetically according to the latest diagnostic test causes much confusion. Atypical cases point to another important reason for proper classifications: unrecognized subgroups of diseases might be found that respond to different treatments, thereby improving the overall success of treatment.

Hayhoe, Quaglino, and Doll (1964) have studied cases of acute leukemia numerically. The classification of these has been much disputed. The authors found four groups with some overlap and nearly all the cases fitted well into these. One group had distinctive clinical features. Although the authors weighted the characters in proportion to their rarity, a reexamination without such weighting showed almost identical results apart from a scaling factor (Sneath, 1965). A similar study by Leech (1963) on milk fever in cattle showed one homogeneous cluster; Zinsser (1964) found four rather indistinct subgroups in pyelonephritis patients. Manning and Watson (1966) classified heart disease using taxonomic distance and average linkage clustering; they found three main phenons that were in reasonable agreement with clinical assessments. Goldstein and Mackay (1967) were able to distinguish two main clusters in lupoid hepatitis reflecting different pathologies. Jones et al. (1970) were able to separate, by cluster analysis, most cases of two colonic diseases with a very variable and overlapping symptomatology. Zinsser (1964) has used both factor and cluster analysis on cases of pyelonephritis.

Classification in psychiatry has many of the attributes of classification in medicine. Although some of the data employed in psychiatric classification are the same as or similar to those in psychology, the aims of psychiatry, like medicine, are etiology, therapy, and prognosis. Because psychiatry is related to both medicine and psychology, psychiatric classification encounters the taxonomic problems of both disciplines, including difficulties in validation, reliability of data, and selection of variables, which are discussed in some detail in Section 11.3, dealing with classification in psychology. Classification in psychiatry is often unreliable. Sandifer, Green, and Carr-Harris (1966) investigated the reliability of diagnosis in mental disorders by tests on case histories that were examined by different clinicians on several occasions. The cases were also classified to give a phenogram that showed six main groups of mental disorder (such as depression or schizophrenia). They found that in diagnosis the clinicians nearly always agreed as to the main group, but within these groups they very frequently disagreed about the variety or subgroup. In addition, the same clinician frequently placed a case in a different subgroup on a second occasion. This paper is an interesting contrast to that of Overall (1963), who used discriminant analysis, in which much sharper groupings were assumed.

Because of the unreliability and limited validity of present day psychiatric classification (see also Bannister, 1968), there has recently been renewed interest in evaluating current diagnostic systems and developing new ones. Numerical taxonomy has been used to evaluate depressive syndromes (Pilowski, Levine, and Boulton, 1969; Paykel, 1971), borderline psychotic states (Grinker, Werble, and Drye, 1968), the general area of psychosis (Lorr, 1966), and the severe psychiatric disorders in general (Strauss et al., 1972). In some instances classical diagnostic concepts have been upheld by these methods; in other instances new classifications have been suggested (as by Lorr). The increasing attention to more careful data collection in psychiatry and clarification of the properties of clustering algorithms as they relate to the underlying structure of the input data promise more fruitful classification systems in this field.

There are, however, a number of special problems in disease classification, as Jacquez (1964b) points out. Manifestations of a disease change during its course (a point recently taken up by Thompson and Woodbury, 1970, but one we believe needs much deeper investigation) and may become increasingly variable and difficult to summarize by classification. The normal grades into the abnormal. Again we have the parallel with biological taxonomy: the disease must be considered through its entire ontogeny (as in Hennig's "holomorph"). There are also problems in making classifications based on signs and symptoms agree with those based on etiology and other clinical evidence. This point has been emphasized in particular by Baron and Fraser (1968) and Fraser and Baron (1968) in studies of liver disease. These authors also found considerable chaining during clustering; so did Wishart (1969d) in a study of thyroid disease. For these reasons cluster analysis may be less suitable than ordination methods, though these too may present difficulties (e.g., Freer and Adkins, 1968, studying dental malocclusion, found abnormals difficult to differentiate from normals). Capon and Jellett (1968) have pointed out, in a study of polycytemia, that the effect of different coding schemes can be quite large, particularly on clusters derived from correlation coefficients. This is of more than commonplace importance in medicine, where appropriate coding and scaling is not always obvious. Capon and Jellett found that taxonomic distances gave clusters in better accord with clinical diagnoses than did correlation coefficients, so their work raises the question of what part size factors (usually representing clinical severity) should play in taxometric work in medicine. Another area requiring investigation is the observational error of clinical findings and its effect on resemblance values; this introduces problems similar to those discussed in Section 4.10.

Current views on diagnosis are well summarized in the volume edited by Jacquez (1964a). As in other fields both Bayesian and discriminant function methods are popular, varying from simple mechanical devices (for example, Nash, 1960) to those that are on-line to a computer. In the more thorough studies there is now evidence that when a straightforward question is posed (such as the chance of survival or the choice between two diagnoses), and adequate data are furnished, a good statistical method can be even more accurate than the clinical specialist. Thus, Warner et al. (1961) were able by computer to discriminate at least as well as

experienced clinicians between congenital heart cases, and the discriminant function of Hughes et al. (1963) was better at prognosis than the best clinicians. Discriminant methods are now being actively applied in electrocardiography (Caceres,
1963; Klingeman and Pipberger, 1967).

Successful diagnostic techniques would seem to need provision for (1) deciding
on what signs, symptoms, and diseases are pertinent (and conversely, what is considered "normal" or "healthy"); (2) the satisfactory construction of disease
entities or the corroboration thereof; (3) the weighting of attributes for diagnosis
and discrimination; (4) the questioning of the physician to confirm signs he may
have misread or overlooked; and (5) the correction of the disease entities by new
data. There will be many opportunities for fruitful collaboration between those
working on medical diagnosis and those working on taxonomic identification
methods; Card (1967) presents a good discussion of this. Many of the problems
may be beyond our present capabilities, but attempts to solve them may yield
important byproducts. Thus it is already clear that better clinical data are
needed, which may lead to improved medical textbooks and methods for earlier
diagnosis.

In conclusion, some miscellaneous studies may be briefly mentioned. Gyllenberg
et al. (1963b) applied numerical taxonomy to food hygiene in delimiting bacterial
flora associated with good and bad keeping qualities of milk. Sneath (1966d)
studied the relationship between chemical structure and biological activity in some
peptides of pharmacological interest, the results suggesting that one might make
some predictions on activity from the chemical structure. Izzo and Coles (1962)
described a polythetic optical recognition method for identifying abnormal blood
cells, and recent taxometric applications to cytopathology are discussed in some
detail by Bartels, Bahr, and Wied (1968) and Wied, Bahr, and Bartels (1970). Some
numerical taxonomic work was done by Ornstein (1965) on serum proteins, and by
James (1964) on nutrition. Selwood (1969) used the expected similarity value of
leucocyte antisera on cell panels to identify allelic antigens in tissue typing.

11.3 THE SOCIAL SCIENCES

Methods similar to those in numerical taxonomy have been used for many years in
the social sciences. Factor analysis and clustering methods were pioneered in
psychometrics, and in social anthropology some farsighted work was done by
Boas at the end of the last century and by Kroeber in the first half of this century
(reviewed by Driver, 1962). Early quantitative classifications in political science
were carried out by Rice and Beyle (see Grumm, 1965). General reviews of taxometric methods include those of Driver (1965) on social and physical anthropology,
Cattell (1966a) in psychology, and Cowgill (1967) and Clarke (1968) in archaeology.
Alker's (1969) review in social science is particularly useful because of its wide
range.

Classification work in psychology has been dominated by factor analytic methods, and there is an extensive literature on this. Much less has been done on cluster analysis (this is reviewed by Cattell and Coulter, 1966, and Bromley, 1966). This emphasis on ordination rather than class making has the empirical justification that psychological traits show continuous distributions rather than discrete ones. Cluster analyses have nevertheless proved interesting in the study of emotions and verbal concepts (Stringer, 1967; Miller, 1969).

A serious problem, one constantly argued about since the first attempts to measure intelligence, is the validity of the character sets on which resemblances are to be based. Thus tests that discriminate between the sexes are excluded from intelligence tests because intelligence is intended to be uncorrelated with sex; but tests that discriminate between races may not be excluded. Many tests are greatly influenced by social environment. In addition, tests administered to subjects on successive occasions do not elicit exactly the same responses, so that workers in psychology are much concerned with the reliability of the variates used in their computations. These questions have been discussed at length by Hawkins (1964) and it is clear that it is difficult to choose objective sets of characters for psychometrics.

Numerical taxonomic tests of the ease and accuracy by which intuitive classifications and ordinations can be made by eye (see Section 5.10) relate to the psychology of perception and should have interest for workers in this field (see, for example, Micko and Fischer, 1970; Marr, 1970).

In social anthropology and linguistics, problems of evolution have attracted much interest; this has been well discussed by Kroeber (1960). Such studies are particularly difficult because of the reticulate nature of cultural evolution. Again we meet with the problem of the validity of sets of characters. Some early papers employing special resemblance coefficients are those of Klimek (1935) and Clements (1954). Milke (1949) and Howells (1966) made important observations on the numerical relation between increasing geographic distance and increasing cultural dissimilarity. Driver and Schuessler (1967) have followed up the suggestion of Leach (1962) and examined a world ethnographic sample of cultural traits by factor analysis to see what support is given to traditional ways of dividing up cultures. This is evidently a complex matter, for only two of the five main factors were readily interpreted. These two were evidently related to patricentered and agricultural societies, respectively. Hudson (1967) made a numerical taxonomy of social attitudes to artists and scientists.

Clausen (1967) reviews numerical taxonomy in political science. In this field, Grumm (1965) studied the voting behavior of legislators by average linkage clustering; he found, besides the main political parties, subgroups of a few individuals who followed a distinctive voting policy. Alker (1969) has used numerical cladistics to make an interesting study of the evolution of political systems.

In the study of languages we may mention the work of Dyen (1962, 1965) and Carroll and Dyen (1962) on computer classification of languages, and Ross (1950) on philology. When vocabularies are used as the character sets of language studies it is usual to measure resemblance as the proportion of "plausible cognates" in the word list. Thus "decem" (Latin) and "deka" (Greek) are obviously cognate, but are these cognate with "ten"? Homologies may therefore be difficult to decide. There may be incongruence between different parts of the vocabulary, or between vocabulary and grammar, when languages of mixed origin like English are studied. It may not be clear whether low resemblance is due to interdependence (see Ellegård, 1959, on this and the significance of negative matches). Computing problems also arise with the very big matrices of word lists. Dyen, James, and Cole (1967) have attempted numerical cladistics based on the rate of replacement of words in different branches of a family of languages, but reticulation (due to word borrowing) poses formidable problems. Some work on grammar has been done by Svartvik (1966) and Carvell and Svartvik (1968), in which clusters of prepositional phrases and idioms were found. An important study relating language to culture is the taxometric study of the dialects of Salish Indians of the American Northwest by Jorgensen (1969).

In physical anthropology taxometric methods are being slowly introduced, though the application of multivariate statistics in this field dates back several decades. Mention has been made in Sections 4.9 and 8.5 of studies on growth and on discriminant functions in man. There have been a few attempts to measure resemblance between human populations (e.g., Rao, 1952; Cavalli-Sforza and Edwards, 1964; Vyas et al., 1958; Neel and Ward, 1970; Jardine, 1971), though Huizinga (1962, 1965) has some pertinent comments on resemblance coefficients. Much disagreement exists on the pattern of variation in humans, despite the large body of data obtained by earlier methods (Bielicki, 1962; Wierciński, 1962), and there is urgent need for investigating racial differences by the study of individuals as the OTU's. Numerical taxonomic methods have recently been applied to social geography (Goddard, 1968; Krueckeberg, 1969; Willmott and Grimshaw, 1969).

In archaeology the OTU's can be of many kinds, from individual artifacts like axes to collections of many kinds of artifacts that together represent a cultural assemblage. The OTU's may be from many sites and periods. The choice of appropriate OTU's is therefore a critical step, depending on the sort of study planned (well discussed by Tugby, 1965). Material may however be scanty, and with simple objects (e.g., flint arrowheads) there may be very few characters that can be considered meaningful. There are a number of special resemblance coefficients used for cultural assemblages and Driver (1965) reviews these. Neither factor analysis nor cluster analysis has been employed widely, though clusters of artifact types or traits have been recognized by eye from Q or R kinds of resemblance matrices (see Tugby, 1958; Lewis and Kneberg, 1959; Clarke, 1962). Polythetic methods appear to be superior to monothetic ones (Doran and Hodson, 1966). A detailed study of Iron Age brooches (Hodson, Sneath, and Doran, 1966) using details of style and manufacture as characters showed that cluster analysis was satisfactory in some respects: near-duplicates were clustered together and atypical brooches were confirmed as such. In this study no large clusters were found, evidently because the designs belonged to a single cultural tradition, though as noted below fashions slowly changed with time. Hodson (1969) has shown the power of taxometrics to reveal structure in a collection of bronze tools analyzed for trace elements. True and Matson (1970) report on cluster analysis of archaeological sites in Chile.

The problem of finding time trends in fashions is one of special importance to archaeology and has its parallel in evolutionary biology. Frequently no independent dating is available, so that the trends must be discovered from changes in pattern of the artifacts and assemblages; the aim is to arrange these into chronological sequence. The classic conventional study is that of predynastic Egypt by Flinders Petrie (see Kendall, 1963). This was first attempted numerically by Brainerd (1951) and Robinson (1951) and has been discussed in Section 5.6 under the heading of seriation. Freeman and Brown (1964) illustrate with a statistical study the danger of assuming temporal changes on slender evidence. A study with a nonlinear scaling method has been made by Hodson et al. (1966), and this showed a clear correlation between the major axis of variation and the date (from independent criteria). This trend was quite difficult to detect by eye. The problem is one of ordination, and, if nonlinear methods are not essential, factor analysis sensu lato may be useful and can handle much larger numbers of OTU's, as pointed out by Cowgill (1968) and illustrated by Hodson (1970).

11.4   THE EARTH SCIENCES

Applications to the earth sciences are becoming numerous. Paleontology has already been discussed in Section 6.5, paleoecology and soil studies have been described in Section 11.1. Useful general references to numerical taxonomic methods in geology are those by Miller and Kahn (1962), Krumbein and Graybill (1965), and Harbaugh and Merriam (1968). These works also cover factor analysis and discriminant functions, topics covered in more detail in the review by Reyment (1963). Clustering methods in particular are discussed by Jizba (1964) and Parks (1966). Much work is summarized in Merriam (1966).

Classification of rock types is now an active field. When such classifications are based on specimens with few petrographic components, scaling is likely to present problems. Procedures like standardization of the composition percentages may have marked effects on the relationships, and it is therefore important to consider with some care the purpose of these procedures. In some specimens the proportions of minor constituents may be correlated, and it may then be useful to treat the analogues of size and shape separately; Imbrie and Purdy (1962) have suggested the use of the vector angle (Section 4.11). More usually distances, correlations, and association coefficients have been used (see Harbaugh and Demirmen, 1964;

Behrens, 1965; Bonham-Carter, 1965). Another problem arises when the composition is expressed as percentages that necessarily add up to 100. This introduces effects on correlations, which have been discussed by Krumbein (1962).

Studies on rocks include those of Varty and White (1964) on the classification of clays, which the authors found to form a single cluster, whose extreme members had been given separate names, although a reexamination by Rayner (1965) gave some evidence of subclusters. Silicious rocks were studied by Howd (1964), who used the method of Rogers and Tanimoto (1960). Obial (1970) applied numerical methods to classifying stream sediments from a mineralogical survey and obtained clusters that closely reflected the bedrocks.

Imbrie and Purdy (1962), Purdy (1963), and Bonham-Carter (1965) examined the same data on modern carbonate sediments by factor analysis and cluster analysis. The factors and clusters were highly congruent. Parks (1966) also found agreement between factors and clusters in Q and R studies of rocks. Behrens (1965) used correlations and cluster analysis in a study of the composition of limestones; he found four main clusters of the constituent materials, which appeared to relate to the environments in which these facies were deposited. Similar studies were made by Bonham-Carter (1967a) and Veevers (1968). Kaesler and McElroy (1966) examined clusters of different sandstones and the localities from which they came; members of the same cluster usually occurred close together on the map. Chave and Mackenzie (1961) have used linkage diagrams (Section 5.9) for work of this kind. As the result of a faunal distance study, Hecht (1969) obtained evidence that in the Miocene there was no sharp change in sea temperature between Florida and New Jersey, unlike the present day.

A special problem that is similar in many ways to finding homologous positions in protein sequences (for which similar methods are used) is that of matching up the strata in different rock sections and revealing cyclic phenomena of deposition (cyclothems). Cross-association (Sackin, Sneath and Merriam, 1965; Sackin and Merriam, 1969) has been used for such studies. Some illustrative examples and discussions of problems are given by Merriam and Sneath (1967) and Sneath (1967e, 1969c). This exploratory work suggests that the method has considerable power for finding the best matches and can tolerate a limited number of gaps in the successions of strata. Although it is also capable of demonstrating cyclothems where successive cycles of rock types occur (such as shale, limestone, sandstone, shale, limestone, sandstone,...), its sensitivity for this is less certain. A problem awaiting solution is to allow for differences in thickness of corresponding strata.

Another, more difficult subject is the measuring of resemblance between geological or geographic surfaces. Some exploratory work has been reported, most of it based on results of trend surface analysis (Merriam and Lippert, 1966; Merriam and Sneath, 1966; Sneath, 1966c). This may employ either the trend coefficients or the residuals from fitted surfaces. Comparison between maps is also pertinent in geography (see Haggett, 1965), and numerical taxonomic methods are now being actively used in several branches of geography (Ahmad, 1965; Berry, 1968; Johnston, 1968; Cole and King, 1968; Spence, 1968), and in the field of tree-ring research (e.g., Fritts, 1963). Skaggs (1968) applied cluster analysis to tornados using meteorologic variables, and demonstrated several types of tornados characteristic of different geographic regions. A review of applications to land surveys is given by Williams and Lance (1969).

11.5   OTHER SCIENCES AND TECHNOLOGY

Most of the applications of taxometric methods in other sciences have been in the fields of library science and pattern recognition; an interest common to both is information theory. The close connections between taxonomy and information theory may be seen by reading the articles of Good (1958, 1962). He discusses clumps and clusters and how to find them (under the name of "botryology") and also takes up some of the philosophical points of polythetic classes. An example is the concept of a cow cited in lectures by the Cambridge philosopher John Wisdom. A cow has four legs and gives milk—but it may have three legs and may not supply milk. No one property may be essential to its "cowness." Estrin, quoted by Good (1958), suggests information retrieval by asking for k out of n index words before selecting a document as relevant. But the group "cows" is a polythetic taxon. So are Estrin's document clusters. This is clearly similar to the philosophical considerations of Beckner (1959, 1964), discussed in Section 2.2. A perennial problem in library science is the classification of knowledge. Subject headings such as "chemistry" or "biology" are evidently polythetic concepts, with much overlap. They also change with time, as new disciplines arise. The conventional problems of classification are thus formidable. A less formidable problem is now being explored by taxometric methods. This is the construction by computer of clusters of words that occur together in documents, which is reviewed in Stevens, Giuliano, and Heilprin (1965), Needham (1967) and Sparck Jones and Jackson (1970). The aim is to find the best key words for indexing documents. Words that consistently appear together in documents are good candidates because they indicate a cluster of related concepts. For example, "magnetic," "transistor," "voltage" would point toward electronics as the subject of the document. These studies face two major problems: the matrices of word occurrences are very large, although most entries will be zeros; and the required clusters are overlapping and perhaps of rather special kinds. Methods will therefore develop along somewhat different lines from those in biological numerical taxonomy. Work on the classification of knowledge is relevant to language translation research (Parker-Rhodes, 1961). And Deutsch (1966) has attempted to formalize the performance of a classification as a way of establishing communication codes for information retrieval.

Numerical taxonomy can assist in pattern recognition, which Lusted (1960) refers to as one of the challenging problems to be found in many fields of science.

Pattern recognition usually means the identification of written or printed characters, also sounds, pictures, etc., with patterns already described and stored in the
memory of a computer (see Nagy, 1968, for a general review, and Casey and Nagy,
1971, for a popular account). It is therefore mainly concerned with techniques of
identification, with the special requirement that incorrect identifications must be
very few. A great variety of identification methods have been proposed, but probably most are simultaneous polythetic ones and in consequence various resemblance measures have been devised, often paralleling closely those in numerical
taxonomy (Unger, 1959; Bonner, 1962; Firschein and Fischler, 1963; Mattson and
Dammann, 1965). A modified numerical taxonomy was used by Sneath (1964b) to
solve a simple jigsaw puzzle.

A recent development is the proposal of mixed systems for classification and
identification. In these each unknown is in turn incorporated into the classification
and new classes of pattern are set up when examples are received that do not fit the
previous classes (Sebestyen, 1962; Ornstein, 1965; Rosen, 1967). Although this
work is directed mainly toward making machines that can learn to read the handwriting of different people, clearly it has potential for many fields.

Applications of numerical taxonomy in scattered fields include economics
(Möller, 1964; Fisher, 1969; Goronzy, 1969; Fisher, Williams, and Lance, 1967);
chemistry (Oyama and Carle, 1967); naval studies (Cattell and Coulter, 1966);
market research (Joyce and Channon, 1966; Green, Frank, and Robinson, 1967;
Frank and Green, 1968); and genetic analysis of mutants (Gillie and Peto, 1969).
Such methods have also been used in studies on biological activity in relation to
chemical structure (Sneath, 1966d; Simon, 1968). Using a clustering method,
Harrison (1968) was able to obtain considerably better prediction of activity of new
chemical compounds than would be expected by chance, and such techniques may
well have advantages over the more usual regression analysis approaches.

11.6 THE ARTS AND HUMANITIES

Most applications of numerical taxonomic methods in fields of the arts and the
humanities have been directed toward questions of disputed authorship, with a
little work on literary style. These are mostly problems of identification rather than
classification. A general survey is the volume edited by Leed (1966). There has been
much controversy over the kinds of traits needed to reach accurate conclusions
about authorship, particularly with ancient texts (for example, see Morton and
Winspear, 1967). From the *known* works of specific authors, it can be determined
which traits are easily imitated, or which vary much with the author's period or
subject matter. Such traits are obviously undesirable for identification. Less
attention has been given to the best discriminatory statistics or to the number of
traits that are needed. A milestone is the work of Mosteller and Wallace (1964) on
the authorship of *The Federalist* papers. Another detailed study is that by Ellegård

(1962) on the authorship of the letters, also published in the eighteenth century, signed by "Junius." Meier-Ewert and Gibbs (1970) demonstrate by cluster analysis the great sensitivity of doublets of letters in distinguishing not merely different languages (as is well known) but, less expectedly, different authors using the same language.

A different approach is illustrated by the elegant study of Blackith (1963) on Latin poets, employing ordination. He was able to isolate several components of style that appear to be reliable indicators of authorship. Of particular interest was the finding that the mental stress of Ovid's exile to Tomis was clearly reflected in stylistic features of his poetry written during his exile.

Less attention has been given in arts and humanities to quantifying relationships or making classifications. Buechley (1967) has made cluster analyses of family names in different geographic localities and Griffith (1967, 1968) has applied numerical taxonomy to medieval manuscripts by Juvenal; his ultimate aim is the more difficult task of determining their cladistic, as well as phenetic, relationships. In a similar study of New Testament manuscripts Griffith (1969) found that their dates were clearly reflected in a seriation, with the Codex Bezae (a puzzling manuscript of uncertain origin) consistently aberrant. Wishart and Leach (1970) used several methods of clustering and ordination to examine the works of Plato and to determine the probable chronology; they obtained an arrangement identical with that in an earlier independent mathematical study by Cox and Brandwood (1959). Such methods are now being applied to music; an example is the project of Bernstein (1967) to quantitate the properties of musical style. The proceedings of a recent symposium on mathematical methods in the humanities have been edited by Hodson, Kendall and Tăutu (1971).

12

The Future of Systematics

The future of systematics as viewed by numerical taxonomists was the subject of a section in *Principles of Numerical Taxonomy* (Sokal and Sneath, 1963). More detailed discussions of this subject can be found in Ehrlich (1961), and Sokal (1964a, 1970), discussions that (admittedly) contain a strong propagandistic element in an effort to convince the generally conservative taxonomic establishment of the merits of computer processing for classificatory purposes as well as for information storage and retrieval purposes. Although the views proposed by these authors are far from generally accepted—witness the several replies to the cited prognostications (e.g., Rollins, 1965a,b; Kalkman, 1966; L. A. S. Johnson, 1968) as well as the tone of some recent symposia (Sibley, 1969) or the antiquated outlook of a report by a committee of the U.S. National Academy of Sciences (Handler, 1970)—it would appear that many of the developments forecast by these early "prophets" are well under way and will continue with their own momentum.

It seems appropriate therefore to discuss only those aspects of systematics upon which the development of numerical and computer methods will impinge in one way or another.

The ever increasing application of numerical methods to all kinds of organisms and all types of characters or descriptors must by now be evident even to the casual reader and is documented in detail in the taxonomic lists of Appendix A.

The main impetus for the development of computer methods of handling taxonomic data will come from the great number of taxonomists in various groups experimenting with new types of taxonomic information. Most important among these are the various kinds of chemical characters being obtained by ever more sophisticated and automated technology. These include data on proteins and molecular biology discussed in detail with numerous references in Sections 3.5 and 5.12, but they will also include much information on secondary chemical products found especially in plants (for a review see Alston, 1967 or B. L. Turner, 1969). The need for numerical taxonomy to integrate these many different kinds of characters into the overall body of descriptive taxonomic knowledge is discussed by Sokal and Sneath (1966) and Heywood (1968).

In this connection we should mention various instruments that will automatically yield large data sets about organisms. These include autoanalyzers that automatically obtain the results of numerous chemical analyses, amino-acid sequencers that break down proteins into their constituent parts, and optical scanners that will give morphological descriptors of macroscopic and microscopic structures of various organisms. If an electron microscope is developed that can read molecular structure directly this would open the way to cladistic studies on a large scale from protein sequences by methods discussed in Chapter 6, and even molecular paleontology would then be possible if undegraded proteins could be found in fossil material. Instruments such as scanning electron microscopes coupled to computers (see Heywood, 1969, for just one illustration of the profusion of new information that will result from scanning electron microscopy), or automatic scanning by holography (Gabor, 1965) are already opening up an undreamed-of wealth of new taxonomic information. These developments will interact with those in pattern recognition discussed in Section 11.5. Unless classical taxonomists will categorically rule out evidence obtained by the means of such devices (and there is nothing in the history of systematics or in biological theory to justify such a parochial view), it will be necessary to process this information by computer, regardless of whether the aim of the taxonomist is a phenetic or a cladistic classification.

Machinery of this type and computer processing will become accepted in time as a routine part of taxonomic work. Systematists should therefore have sufficient training in mathematical methods to understand the rationale behind some of the techniques beeing applied, and also training in operating the various devices that yield this information. For a discussion of minimal educational requirements for systematists and ecologists of the coming generation, see Sokal (1970) and Turner (1971). Automatic data gathering machines and computer terminals will be employed not only in description and classification but also in identification. Machines that classify and identify (usually by means of optical scanners coupled to computers) are now available for cytological and pathological research. Their justification for routine identifications of economically and medically important organisms must be investigated. The degree of their employment will largely depend on the economics of the computer field in the years to come. Although heavy capital investment is necessary for computer installations, the overall cost of computing for specific tasks has steadily decreased in recent years by orders of magnitude. The use of electronic data processing (EDP) for the description of organisms, for the preparation of geographic distribution maps, for taxonomic keys, monographs and other taxonomic endeavors has been described by many persons and we need merely cite reviews by Soper and Perring (1967) on map preparation in botany, and by Crovello (1967) on data storage and retrieval. Cutbill (1971) presents articles on various aspects of EDP as it affects biological systematics.

These various developments will clearly affect museum operations, as Sokal (1964a), Sokal and Sneath (1966), Ehrlich (1964), Raven and Holm (1967), and Williams (1967a) have all pointed out. To what degree museum procedures should be modified is an open question that requires considerable study and discussion. A recent report (Steere, 1971) from a committee of natural history museum directors strongly supports computerization of many of the operations in the great systematic biology collections of the United States. A number of authors (Sokal and Sneath, 1966, and Williams, 1967a, among others) have suggested the establishment of international taxonomic centers in which data storage and retrieval will be carried out on a large scale and made available to taxonomic users at remote locations by various means of computerized interfacing.

The realization of these many developments will depend partly on the pressure exerted by the load of material to be processed, partly on technical developments (the speed of which cannot always be accurately predicted), and partly on the training received by the current generation of taxonomists. We cannot know how many of these innovations will be adopted and how rapidly this will take place. We can, however, say that the changes in taxonomic practice are going to be profound and that taxonomists entering the field at this time must acquaint themselves with these techniques and be conversant with them in order to be competent workers in the field. It has become obvious that taxonomy is as dynamic and as changing a field as other aspects of biology and there is little doubt that exciting realms of new endeavor are in store for the taxonomists of the next decade.

Appendixes
Bibliography
Indexes

Applications of Numerical Taxonomy
to Biological Systematics

In this Appendix we list the publications known to us of numerical taxonomy as applied to the systematics of organisms. We have included a number of borderline papers because they contain significant taxonomic implications, but we have restricted the list of multivariate studies to those directed mainly toward problems of classification. A few references have also been included mainly because they consider problems, such as the choice of material or the coding of characters, that are especially important in numerical work on a particular taxonomic group. Papers before 1956 have not been listed, as these are largely of historical interest; they have been summarized by Sokal and Sneath (1963), Davis and Heywood (1963), and Hubac (1967).

The lists are arranged as follows into major taxonomic groups: studies spanning several major groups of organisms, vertebrates, arthropods, other invertebrates, dicotyledons, monocotyledons, other eucaryote plants, bacteria, and viruses. Brief annotations have been added when the paper cited covers major problems beside the classification of the organisms mentioned. From the extensive literature on numerical analysis of protein sequences we have made a short selection of those that are most relevant to taxonomy. Some of these cover organisms from several of the major groups; we have therefore included them under a separate heading. The higher groupings within the bacteria are not yet stabilized, so these references are listed under somewhat arbitrary divisions commonly employed by bacteriologists—Actinomycetales, and Gram positive and Gram negative genera—with a separate list for those studies that cover more than one of these divisions. Paleontological applications may be identified by the inclusion of the word "fossil" in the annotations, and a † symbol has been added to the citation to assist in finding them.

STUDIES OF VERTEBRATA (*continued*)

Delany (1965)	Field-mice (*Apodemus sylvaticus*): populations, ordination
Doolittle and Blombäck (1964)	Mammals (Artiodactyla): proteins, cladistics
Edwards and Cavalli-Sforza (1964)	Man: blood groups, races, and cladistics
Fitch (1966b)	Hemoglobin evolution
Fitch and Neel (1969)	Man: tribes, cladistics
Forman, Baker, and Gerber (1968)	Bat genera: serology
Géry (1965)	Fish (Characidae): comparison of several genera
Goodman and Moore (1971)	Primates: serology, cladistics
Goodman et al. (1971)	Primates: serology, cladistics
Gould (1967)†	Fossil reptiles (Pelycosauria): ordination, Q and R analyses
Hedges (1969)	Field-mice (*Apodemus*): populations, geography
Hendrickson (1967)†	Fossil Equidae: cladistics
Hudson and Lanzillotti (1964)	Birds (Galliformes): musculature
Hudson, Lanzillotti, and Edwards (1959)	Birds (Galliformes): musculature
Hudson et al. (1966)	Birds (Galliformes): musculature, also congruence between character suites
Hudson et al. (1969)	Birds (Lari and Alcae): musculature
Horne (1967)	Primates: proteins, cladistics
Hureau (1967)	Fish (Nototheniidae)
Imaizumi (1967)	Mammals (Felidae)
Jardine (1969a)†	Fossil fish (Rhipidistia): homology
Jardine (1969c)†	Fossil fish (Rhipidistia): homology
Jardine (1971)	Man: populations
Jardine and Jardine (1967)	Skulls: homology
Johnston (1969)	Sparrows (*Passer domesticus*): populations, geography
Kirsch (1968)	Marsupialia: serology
Kluge and Eckardt (1969)	Lizards (*Hemidactylus garnotii*): populations
Kluge and Farris (1969)	Amphibia (Anura): cladistics
Lerman (1965b)†	Fossil Equidae and Oreodontidae: evolution rates

(*continued*)

STUDIES OF VERTEBRATA (*continued*)

Mainardi (1963)	Birds: serology
McAllister (1966)	Fish (Osmeridae)
Minkoff (1965)	Primates: skulls, teeth
Mohagheghpour and Leone (1969)	Primates: serology
Mross and Doolittle (1967)	Mammals (Artiodactyla): proteins, cladistics
Olson (1964)†	Fossil Oreodontidae: skulls, Q and R analyses
Petras (1967)	Mice (*Mus musculus*): phenetics, geography
Porter and Porter (1967)	Amphibia (*Bufo*): chemotaxonomy
Power (1970)	Birds (*Aegelaius phoeniceus*): populations, geography
Rees (1969a)	Deer (*Odocoileus virginianus*): phenetics, geography
Rees (1969b)	Deer (*Odocoileus virginianus*): phenetics, geography
Rising (1968)	Birds (*Parus*): hybridization, geography
Sarich (1969a)	Carnivora: serology, cladistics
Sarich (1969b)	Carnivora: serology, cladistics
Schnell (1969)	Birds (Lari)
Schnell (1970a)	Birds (Lari)
Schnell (1970b)	Birds (Lari)
Selander, Hunt, and Yang (1969)	Mice (*Mus musculus*): populations
Sneath (1961)†	Fossil fish (*Knightia*)
Sneath (1967a)	Hominoidia: skulls
Soulé (1966)	Lizards (*Uta stansburiana*): phenetics, geography
Soulé (1967a)	Lizards (*Uta stansburiana*): phenetics, geography
Wallace and Bader (1967)	Mice (*Mus musculus*): ordination

STUDIES OF ARTHROPODA

Bächli, G. (1971)	*Leucophenga, Paraleucophenga* (Diptera): congruence
Basford et al. (1968)	Coleoptera: serology
Blackith and Blackith (1968)	Orthoptera and allied orders

STUDIES OF ARTHROPODA (*continued*)

Blackith and Kevan (1967)	*Chrotogonus* (Orthoptera): canonical analysis
Blair, Blackith, and Boratyński (1964)	*Coccus hesperidum*: R analysis
Boratyński (1971)	Coccoidea (Homoptera): males
Boratyński and Davies (1971)	Diaspididae (Homoptera): male scale insects
Butler and Leone (1967)	Coleoptera: serology
Chillcot (1960)	Fanniinae (Diptera)
Chui (1969)	Psocoptera: Hawaiian complexes
da Cunha (1969)	Meliponinae (Hymenoptera)
DuPraw (1964)	*Apis*: bee wings
DuPraw (1965a)	*Apis mellifera*: bee populations, ordination, discrimination, wings
DuPraw (1965b)	*Apis mellifera*: bee populations; ordination, discrimination, wings
Eades (1970)	Orthoptera: comparison of methods
Ehrlich (1961b)	*Euphydryas* (Lepidoptera)
Ehrlich and Ehrlich (1967)	Papilionoidea and Hesperioidea (Lepidoptera): congruence between character suites
Eickwort (1969)	Augochlorini (Hymenoptera)
Fisher and Rohlf (1969)	Culicidae (Diptera): homology
Fry (1964)	*Ammothea* (Pycnogonida): allometry
Fry and Hedgpeth (1969)	*Ammothea* and *Achelia* (Pycnogonida)
Fujii (1969)	*Callosobruchus* (Coleoptera): populations and ecology
Funk (1964)	Euzerconidae (Acari)
Giles (1963)	Dermaptera and allied orders
Hendrickson (1967)	*Scellus* (Diptera): cladistics
Hendrickson and Sokal (1968)	*Psorophora* (Diptera)
Herrin (1970)	*Hirstionyssus* (Acari)
Hubby and Throckmorton (1965)	*Drosophila* (Diptera): chemotaxonomy
Huber (1968)	Blattaria (Dictyoptera): congruence between character suites
Huber (1969)	Blattaria (Dictyoptera)
Hurlbutt (1968)	*Veigaia* and *Asca* (Acari): phenetics and ecology
Ihm et al. (1967)	*Epilachna* (Coleoptera)

(*continued*)

STUDIES OF ARTHROPODA (*continued*)

Jago (1967)	Callimptaminae (Orthoptera)
Jago (1969)	Gomphocerinae (Orthoptera)
Jago (1971)	Gomphocerinae (Orthoptera)
Kaesler (1969a)	Ostracoda: Q and R analyses
Kaesler (1969b)	Ostracoda
Kathirithamby (1971)	Cicadellidae (Homoptera)
Kerr, Pisani, and Aily (1967)	*Melipona* (Hymenoptera)
Kim, Brown, and Cook (1966)	*Hoplopleura* (Anoplura): discriminant functions
Kistner (1967a)	*Termitodiscus* (Coleoptera)
Kistner (1967b)	*Aenictonia* and *Anommatochara* (Coleoptera)
Kistner (1968a)	Termitopaedini (Coleoptera)
Kistner (1968b)	Corotocini (Coleoptera)
Kistner and Pasteels (1970)	Coptotermoeciina (Coleoptera)
Klimaszewski (1967)	*Trioza* (Homoptera)
Kovalev (1968)	*Drapetis* and allied genera (Diptera)
Kovalev and Shatalkin (1969)	*Platypalpus* (Diptera)
Le Quesne (1969)	*Argodrepana* (Lepidoptera): cladistics
Louis and Lefebvre (1968)	*Apis mellifera*: honey bee colonies
Manischewitz (1971)	Ixodorhynchidae (Acari)
Mason, Ehrlich, and Emmel (1967)	*Euphydryas editha* (Lepidoptera)
Mason, Ehrlich, and Emmel (1968)	*Euphydryas editha* (Lepidoptera): R analysis and temporal change
Michener and Sokal (1957)	Megachilidae (Hymenoptera)
Michener and Sokal (1966)	Megachilidae (Hymenoptera): congruence
Moss (1967)	Dermanyssidae (Acari)
Moss (1968a)	*Dermanyssus* (Acari): extraction of characters for keys
Moss (1968b)	Dermanyssidae (Acari): comparison of methods
Petrova (1967)	Parholaspidae (Acari)
Pisani et al. (1969)	*Melipona* (Hymenoptera)
Procaccini (1966)	*Protesilaus* and allied genera (Lepidoptera)

STUDIES OF ARTHROPODA (*continued*)

Reyment (1965)†	Cytherellidae (Ostracoda): ordination of fossils
Rohlf (1962)	*Aedes* (Diptera): congruence in mosquitos
Rohlf (1963a)	*Aedes* (Diptera): congruence in mosquitos
Rohlf (1963b)	*Aedes*: (Diptera)
Rohlf (1965)	*Aedes* (Diptera), Megachilidae (Hymenoptera), and Papilionoidea and Hesperioidea (Lepidoptera): congruence
Rohlf (1967)	Culicidae (Diptera): effect of character correlations
Rohlf (1968)	Megachilidae (Hymenoptera): stereograms
Rohlf (1970)	Culicidae (Diptera)
Rohlf and Sokal (1962)	Megachilidae (Hymenoptera)
Rohlf and Sokal (1965)	Megachilidae (Hymenoptera) and *Aedes* (Diptera): comparison of methods
Rohlf and Sokal (1967)	Culicidae (Diptera): image scanning
Rowell (1970)†	Pterocephaliidae: cladistics of fossil trilobites
Rubin (1966)	Megachilidae (Hymenoptera)
Sakai (1970)	Dermaptera: revision of order, zoogeography
Sakai (1971)	Dermaptera: revision of order, zoogeography
Scudder (1963)	Lygaeoidea and Coreidae (Heteroptera)
Selander and Mathieu (1969)	*Epicauta* (Coleoptera)
Sheals (1964)	Laelaptoidea (Acari)
Sheals (1969)	Phthiracaroidea (Acari)
Shepard (1971)	Luciliini (Diptera)
Smirnov (1969)	*Meromyza*, *Chlorops*, Simuliidea (Diptera); *Dermestes* (Coleoptera); Parholaspidae (Acari); families of Araneida: illustrative examples
Smirnov and Fedoseeva (1967)	*Meromyza* (Diptera)
Sokal (1958a)	Megachilidae (Hymenoptera)
Sokal (1962b)	Syrphidea (Diptera)

(*continued*)

STUDIES OF ARTHROPODA (*continued*)

Sokal and Michener (1958)	Megachilidae (Hymenoptera)
Sokal and Michener (1967)	Megachilidae (Hymenoptera): comparison of methods
Sokal and Rinkel (1963)	*Pemphigus populi-transversus* (Homoptera): geographic variation
Sokal and Rohlf (1966)	*Aedes* (Diptera): image scanning
Sokal and Rohlf (1970)	Megachilidae (Hymenoptera): congruence
Sokal and Thomas (1965)	*Pemphigus populi-transversus* (Homoptera): geographic variation
Stallings and Turner (1957)	Megathymidae (Lepidoptera)
Stephenson, Williams, and Lance (1968)	Portunidea (Malacostraca)
Steward (1968)	*Aedes* (Diptera): species in Canada
Styron (1969)	*Lirceus fontinalis* (Isopoda): populations
Thomas (1968a)	*Haemaphysalis leporispalustris* (Acari): geographic variation
Thomas (1968b)	*Haemaphysalis leporispalustris* (Acari): R studies, geographic variation
Thornton and Wong (1967)	Peripsocidae (Psocoptera): congruence between character suites
Throckmorton (1968)	*Drosophila*: phenetics and cladistics
Wainstein (1968)	Phytoseiidae (Acari)
Wilkinson (1967)	*Teldenia* and *Argodrepana* (Lepidoptera)
Wilkinson (1968)	*Ditrigona* (Lepidoptera)
Wilkinson (1970a)	Drepanidae (Lepidoptera)
Wilkinson (1970b)	Drepanidae (Lepidoptera): ordination methodology
Willis (1971)	*Cicindela* (Coleoptera): cladistics
Wrenn (1972)	*Euschoengastia* (Acari)
Zhantiev (1967)	*Dermestes* (Coleoptera)

STUDIES OF OTHER INVERTEBRATA

Bird (1967)	Nematoda (*Trichodorus*): ordination
Bird and Mai (1967)	Nematoda (*Trichodorus christiei*)
Borgelt (1968)	Tunicata (*Thalia democratica*): subspecies
Camin and Sokal (1965)†	Fossil Fusulinidae: cladistics

STUDIES OF OTHER INVERTEBRATA (*continued*)

Cheetham (1968)†	Fossil Bryozoa (*Metrarabdotos*): phenetics and evolution
Eldredge (1968)†	Fossil Mollusca (*Worthenia* and *Glabrocingulum*): evolution
Evans and Fisher (1966)	Mollusca (*Penitella*)
Fry (1970)	Sponges (*Ophlitaspongia seriata*): populations
Ghiselin et al. (1967)	Molluscan genera: shell amino acids
Gould (1969a)†	Mollusca (*Poecilozonites*), Recent and fossil
Gould (1969b)	Mollusca (*Cerion uva* and *Tudora megacheilos*): geography, subspecies
Hendrickson (1967)†	Fossil Fusulinidae: cladistics
Jamieson (1968)	Annelida (Alluroididae)
Jewsbury (1968)	Trematoda (*Schistosoma haematobium*): egg morphology
Kaesler (1970b)†	Fossil Fusulinidae (*Pseudoschwagerina*): cladistics
Kesling and Sigler (1969)†	Fossil Crinoidea: cladistics
Kohut (1969)†	Fossil conodont groups
Lerman (1965a)†	Fossil Mollusca (*Exogyra*)
Lerman (1965b)†	Fossil Mollusca (*Exogyra* and *Kosmoceras*): evolution rates
Lima (1965)	Nematoda (*Xiphinema*)
Lima (1968)	Nematoda (*Xiphinema*)
Little (1963)	Sponges (*Cliona*)
Moss and Webster (1969)	Nemotoda (Strongylidae)
Moss and Webster (1970)	Nematoda (Strongylidae): ordination
Pitcher (1966)†	Fossil Fusulinidae: ordination
Powers (1970)	Corals of the Hawaiian reef
Reyment and Naidin (1962)†	Fossil Belemnitidae (*Actinocamax*): ordination
Rowell (1967)†	Fossil Brachiopoda (Chonetacea)
Sims (1966)	Earthworms (Megascolecidae)
Sims (1969a)	Earthworms (Megascolecidae)
Sims (1969b)	Earthworms (Megascolecidae)
Ukoli (1967)	Trematoda (*Apharyngostrigea*)

Balbach (1965)	*Apocynum* (Apocynaceae)
Banks and Hillis (1969)	*Eucalyptus camaldulensis* (Myrtaceae): chemotaxonomy and geography
Beals (1968)	*Antennaria* (Compositae): ordination
Bemis et al. (1970)	*Cucurbita* (Cucurbitaceae) and hybrids
Bisby (1970a)	*Crotalaria* (Leguminosae)
Bisby (1970b)	*Crotalaria* (Leguminosae): Q and R analyses
Crovello (1966)	*Salix* (Salicaceae)
Crovello (1968a)	*Salix* (Salicaceae)
Crovello (1968b)	Limnanthaceae
Crovello (1968d)	*Arabidopsis thaliana* (Cruciferae): cultivated races
Crovello (1968e)	*Salix* (Salicaceae)
Crovello (1968f)	*Salix* (Salicaceae)
Crovello (1968g)	*Salix* (Salicaceae): different sources of data
Crovello (1968h)	*Salix* (Salicaceae)
Crovello (1968i)	Limnanthaceae
Crovello (1969)	*Salix* (Salicaceae)
Dass and Nybom (1967)	*Brassica* (Cruciferae) and hybrids: chemotaxonomy
Davidson (1963)	*Cirsium* (Compositae): R analysis
Davidson and Dunn (1967)	*Froelichia* (Amaranthaceae): R analysis
Davidson and Dunn (1968)	*Froelichia* (Amaranthaceae)
Drury and Randal (1969)	*Erechtites* (Compositae)
Edye, Williams, and Pritchard (1970)	*Glycine wightii* (Leguminosae): populations
El-Gazzar and Watson (1970a)	Labiatae, Verbenaceae, and allied families
El-Gazzar et al. (1968)	*Salvia* (Labiatae)
Ernst (1967)	Platystemonoideae (Papaveraceae)
Eshbaugh (1964)	*Capsicum* (Solanaceae)
Eshbaugh (1970)	*Capsicum baccatum* (Solanaceae): wild and cultivated forms
Grant (1969)	*Betula* (Betulaceae): chemotaxonomy

STUDIES OF ANGIOSPERMAE: DICOTYLEDONES (*continued*)

Grant and Zandstra (1968) — *Lotus* (Leguminosae): chemotaxonomy

Hawksworth, Estabrook, and Rogers (1968) — *Arceuthobium* (Viscaceae)

Heiser, Soria, and Burton (1965) — *Solanum* (Solanaceae)

Hickman and Johnson (1969) — *Menziesia* (Ericaceae): geography

Hubac (1964) — *Campanula rotundifolia* (Campanulaceae): populations and keys

Hubac (1967) — *Campanula rotundifolia* (Campanulaceae): populations

Hubac (1969) — *Campanula rotundifolia* (Campanulaceae): hybrids

Irwin and Rogers (1967) — *Cassia* (Leguminosae)

Ivimey-Cook (1969a) — *Ononis* (Leguminosae)

Ivimey-Cook (1969b) — *Ononis* (Leguminosae)

Jancey (1966b) — *Phyllota phylicoides* (Leguminosae): populations

Jardine and Sibson (1968b) — *Sagina apetala* (Caryophyllaceae): populations

Jardine and Sibson (1971) — *Silene* (Caryophyllaceae)

Jaworska and Nybom (1967) — *Saxifraga* (Saxifragaceae): hybridization, chemotaxonomy

Johnson and Holm (1968) — *Sarcostemma* (Asclepiadaceae)

Johnson and Thien (1970) — *Gossypium* (Malvaceae): chemotaxonomy

Katz and Torres (1965) — *Zinnia* (Compositae)

Klotz (1967) — *Cotoneaster* (Rosaceae)

Kowal and Kuźniewski (1959) — *Chenopodium* and *Atriplex* (Chenopodiaceae)

Levin and Schaal (1970) — *Phlox* (Polemoniaceae): chemotaxonomy and hybrids

't Mannetje (1967b) — *Trifolium* (Leguminosae): susceptibility to strains of *Rhizobium*

't Mannetje (1969) — *Stylosanthes* (Leguminosae): morphology and susceptibility to *Rhizobium* strains

McNeill, Parker, and Heywood (1969a) — Caucalideae (Umbelliferae)

McNeill, Parker, and Heywood (1969b) — Caucalideae (Umbelliferae)

(*continued*)

STUDIES OF ANGIOSPERMAE: DICOTYLEDONES (*continued*)

Menitskii (1966)	*Quercus* (Fagaceae)
Mooney and Emboden (1968)	*Bursera* (Burseraceae): populations, geography, and chemotaxonomy
Moore, Harborne, and Williams (1970)	*Empetrum rubrum* (Empetraceae): chemotaxonomy, geography
Olsson (1967)	*Mentha* (Labiatae): hybrids, chemotaxonomy
Orloci (1968b)	*Phyllodoce* (Ericaceae)
Orloci (1970)	*Phyllodoce* (Ericaceae)
Ornduff and Crovello (1968)	Limnanthaceae and hybrids
Parups et al. (1966)	*Trifolium* (Leguminosae)
Prance, Rogers, and White (1969)	Chrysobalanaceae
Ramon (1968)	*Haplopappus* (Compositae) and hybrids
Rhodes, Carmer, and Courter (1969)	*Armoracia rusticana* (Cruciferae): cultivars
Rhodes et al. (1968)	*Cucurbita* (Cucurbitaceae)
Rhodes et al. (1970)	*Mangifera* (Anacardiaceae): cultivars and hybrids
Rogers (in IBM, 1959)	*Manihota* (Euphorbiaceae)
Rogers and Tanimoto (1960)	*Manihota* (Euphorbiaceae)
Rostánski (1968)	*Oenothera* (Onagraceae)
Rostánski (1969)	*Oenothera* (Onagraceae)
Shmidt (1962)	*Odontites* (Scrophulariaceae)
Simon and Goodall (1968)	*Medicago* (Leguminosae): chemotaxonomy
Smith (1969)	*Vaccinium* (Ericaceae) and hybrids
Sneath (1968a)	*Laburnocytisus* (Leguminosae): graft chimera
Soria and Heiser (1961)	*Solanum* (Solanaceae)
Stearn (1968)	*Columnea* and *Alloplectus* (Gesneriaceae)
Stearn (1969)	*Columnea* and *Alloplectus* (Gesneriaceae): species in Jamaica
Stone, Adrouny, and Flake (1969)	*Carya* (Juglandaceae): chemotaxonomy
Taylor (1966)	*Lithophragma* (Saxifragaceae)
Taylor (1971)	*Tiarella* (Saxifragaceae): chemotaxonomy, seasonal variation

---

(*continued*)

STUDIES OF ANGIOSPERMAE: MONOCOTYLEDONES (*continued*)

Goodman (1968b)	Maize cultivars
Hall (1965b)	*Eulophia* (Orchidaceae)
Hall (1967b)	*Eulophia* and *Satyrium* (Orchidaceae)
Hall (1969a)	*Iris* (Iridaceae): discriminant analysis
Hamann (1961)	Monocotyledon families
Ising and Fröst (1969)	*Cyrtanthus* (Amaryllidaceae): chemotaxonomy
Kaltsikes and Dedio (1970a)	*Triticum* and *Aegilops* (Gramineae): chemotaxonomy
Kaltsikes and Dedio (1970b)	*Triticum* and *Aegilops* (Gramineae): chemotaxonomy
Kaltsikes, Dedio, and Larter (1969)	*Secale* (Gramineae): chemotaxonomy
Liang and Casady (1966)	*Sorghum* (Gramineae)
Morishima and Oka (1960)	*Oryza* (Gramineae)
Morishima (1969b)	*Oryza perennis* (Gramineae): cultivars
Oka (1964)	*Oryza* (Gramineae)
Pernes, Combes, and Réné-Chaume (1970)	*Panicum maximum* (Gramineae): cultivars
Rhodes and Carmer (1966)	Maize cultivars
Rowley (1967)	Aloineae (Liliaceae)
Rowley (1969)	Aloineae (Liliaceae)
Shahi, Morishima, and Oka (1969)	*Oryza* (Gramineae): chemotaxonomy, ordination
Stant (1964)	Alismataceae
Stant (1967)	Butomaceae
Takakura (1962)	*Oryza* (Gramineae): ordination
de Wet and Huckabay (1967)	*Sorghum* (Gramineae)
Whitehouse (1971)	Barley cultivars and hybrids
Wirth, Estabrook, and Rogers (1966)	Oncidiinae (Orchidaceae)

STUDIES OF GYMNOSPERMAE

Adams and Turner (1970)	*Juniperus ashei* (Cupressaceae): populations
Flake (1969)	*Juniperus virginiana* (Cupressaceae): populations
Flake, von Rudloff, and Turner (1969)	*Juniperus virginiana* (Cupressaceae): clines, chemotaxonomy
Gambaryan (1965)	*Pinus* (Pinaceae)

STUDIES OF GYMNOSPERMAE (*continued*)

Jeffers and Black (1963)	*Pinus contorta* (Pinaceae): subspecies, ordination
Lange, Stenhouse, and Offler (1965)†	Fossil Coniferales
La Roi and Dugle (1968)	*Picea* (Pinaceae): chemotaxonomy, geography
Thielges (1969)	*Pinus* (Pinaceae): chemotaxonomy, geography
Young and Watson (1969)	Coniferales: wood structure

STUDIES OF OTHER EUCARYOTE PLANTS

Bischler and Joly (1969)	Lichens (*Calypogeia*)
Campbell (1969)	Yeasts (*Saccharomyces*): serology
Cullimore (1969)	Algae (*Chlorella vulgaris*)
Ducker, Williams, and Lance (1965)	Algae (*Chlorodesmis*)
Ibrahim (1963)	Fungi (*Helminthosporium*)
Ibrahim and Threlfall (1966a)	Fungi (*Helminthosporium*)
Ibrahim and Threlfall (1966b)	Fungi (*Helminthosporium*)
Joly (1969)	Fungi (*Alternaria*)
Kendrick (1964)	Fungi (*Verticicladiella*)
Kendrick and Proctor (1964)	Fungi (*Verticicladiella* and *Phialocephala*)
Kendrick and Weresub (1966)	Orders of Basidiomycetes: character weighting
Kockovà-Kratochvílová (1969a)	Yeasts (*Saccharomyces*)
Kockovà-Kratochvílová (1969b)	Yeasts (*Saccharomyces*)
Kockovà-Kratochvílová et al. (1968)	Yeasts (*Saccharomyces*)
Kockovà-Kratochvílová, Šandula, and Vojtková-Lepšíková (1969)	Yeasts (*Candida*)

Landau, Shechter, and Newcomer
    (1968)
	Fungi (Dermatophyta):
	    chemotaxonomy

Lellinger (1964)	Ferns (cheilanthoids): cladistics
Lellinger (1965)	Ferns (adiantoids): cladistics
Levin and Rogers (1964)	Algae (Nemalionales)
Lichtwardt et al. (1969)	Fungi (*Smittium*): chemotaxonomy, serology
McGuire (1969)	Algae (*Chlorococcum*)
Mickel (1962)	Ferns (*Anemia*): cladistics
Morishima (1969a)	Fungi (*Piricularia oryzae*)

(*continued*)

STUDIES OF OTHER EUCARYOTE PLANTS (*continued*)

Pokorná (1969)	Yeasts (*Candida*)
Poncet (1967a)	Yeasts (*Pichia*): ordination
Poncet (1967b)	Yeasts (*Pichia*): ordination
Poncet (1970)	Yeasts (*Hansenula*)
Proctor (1966)	Fungi (*Verticicladiella*)
Proctor and Kendrick (1963)	Fungi (*Haplobasidion* and allied genera)
Rogers and Fleming (1964)	Algae (*Halimeda*)
Seki (1968)	Mosses (Sematophyllaceae): ordination
Shipton and Fleischmann (1969)	Fungi (*Puccinia*): chemotaxonomy
Whitney, Vaughan, and Heale (1968)	Fungi (*Fusarium* and *Verticillium*): chemotaxonomy

WIDE RANGE STUDIES OF BACTERIA

Allen and Pelczar (1967)	Bacteria from fish, including a new pathogen
Bean and Everton (1969)	Bacteria from cannery environments
Beers et al. (1962)	Pseudomonadaceae, Enterobacteriaceae, and *Streptococcus*
Brisbane and Rovira (1961)	Soil bacteria
Davis and Newton (1969)	Coryneform bacteria, and Actinomycetales
Focht and Lockhart (1965)	Genera of several orders
Goodfellow (1967)	Genera of several orders
Goodfellow (1969)	Soil bacteria
Graham (1964)	*Rhizobium* and allied genera, and *Bacillus*
Gyllenberg (1967)	Soil bacteria: R ordination
Gyllenberg and Rauramaa (1966)	Soil bacteria: cluster parameters
Hayashi (1968)	Various Gram positive genera and Gram negative cocci
Johnson, Katarski, and Weisrock (1968)	Marine bacteria
Litchfield, Colwell, and Prescott (1969)	*Pseudomonas* and allied genera, and *Bacillus*
Lockhart and Hartman (1963)	Pseudomonadaceae, Enterobacteriaceae, and *Streptococcus*

WIDE RANGE STUDIES OF BACTERIA (*continued*)

't Mannetje (1967a) | *Rhizobium* and allied genera, and *Bacillus*

Melchiorri-Santolini (1968) | Marine bacteria

Pfister and Burkholder (1965) | Marine bacteria

Quadling and Hopkins (1966) | Pseudomonadaceae, Enterobacteriaceae, and *Bacillus*

Quadling and Hopkins (1967) | Pseudomonadaceae, Enterobacteriaceae, and *Bacillus*: two-stage ordination

Rouatt et al. (1970) | Coryneform bacteria, and Actinomycetales: chemotaxonomy

Rovira and Brisbane (1967) | Soil bacteria: Q and R analyses

Skyring and Quadling (1969b) | Soil bacteria

Sneath and Cowan (1958) | Genera of several orders

Sundman and Gyllenberg (1967) | Soil bacteria: R analysis

STUDIES OF BACTERIA: ACTINOMYCETALES

Bogdanescu and Racotta (1967) | *Mycobacterium*

Bojalil and Cerbón (1961) | *Mycobacterium*

Bojalil, Cerbón, and Trujillo (1962) | *Mycobacterium*

Cerbón and Bojalil (1961) | *Mycobacterium*

Gilardi et al. (1960) | *Streptomyces* and allied genera

Goodall (1966a) | *Mycobacterium*

Goodfellow (1971) | *Nocardia* and allied genera

Gyllenberg (1970) | *Streptomyces*: R analysis

Gyllenberg, Woźnicka, and Kurylowicz (1967) | *Streptomyces*: ordination

Hill and Silvestri (1962) | *Streptomyces* and allied genera: probabilistic keys

Hill et al. (1961) | *Streptomyces* and allied genera

Jones and Bradley (1964) | *Mycobacterium*, *Nocardia*, and allied genera

Kazda (1966) | *Mycobacterium*

Kazda (1967) | *Mycobacterium*

Kazda, Vrubel, and Dornetzhuber (1967) | *Mycobacterium*

Kestle, Abbott, and Kubica (1967) | *Mycobacterium*

Kubica et al. (1970) | *Mycobacterium*

Kurylowicz et al. (1970) | *Streptomyces*   (*continued*)

STUDIES OF BACTERIA: ACTINOMYCETALES (*continued*)

Melville (1965)	*Actinomyces*: congruence under different conditions of test
Möller (1962c)	*Streptomyccs* and allied genera: probabilistic keys
Nakayama, Nakayama, and Takeya (1970)	*Mycobacterium fortuitum* and *M. runyonii*
Saito, Tasaka, and Takei (1968)	*Mycobacterium*
Silvestri et al. (1962)	*Streptomyces* and allied genera
Takeya, Nakayama, and Nakayama (1967)	*Mycobacterium*: immunology
Tsukamura (1966)	*Mycobacterium*
Tsukamura (1967a)	*Mycobacterium*
Tsukamura (1967b)	*Mycobacterium chitae*
Tsukamura (1967c)	*Mycobacterium terrae* and *M. novum*
Tsukamura (1967d)	*Mycobacterium*: the distinctness of taxa
Tsukamura (1968)	*Mycobacterium*
Tsukamura (1969)	*Nocardia*
Tsukamura and Mizuno (1968)	*Mycobacterium*: cluster parameters
Tsukamura and Mizuno (1969)	*Mycobacterium*
Tsukamura and Tsukamura (1966)	*Mycobacterium*
Tsukamura, Mizuno, and Tsukamura (1967)	*Mycobacterium*
Tsukamura, Tsukamura, and Mizuno (1967)	*Mycobacterium fortuitum*
Wayne (1966)	*Mycobacterium*
Wayne (1967)	*Mycobacterium*
Wayne, Doubek, and Diaz (1967)	*Mycobacterium*
Williams, Davies, and Hall (1969)	*Streptomyces*
Woźnicka (1967)	*Streptomyces*

STUDIES OF BACTERIA: GRAM POSITIVE GROUPS

Antila and Gyllenberg (1963)	*Propionibacterium*
Barre (1969)	*Lactobacillus*: strains from wine
Blondeau (1961)	*Streptococcus faecalis*
Bonde (1965)	*Bacillus*: marine strains
Carlsson (1968)	*Streptococcus*: oral forms
Chatelain and Second (1966)	*Brevibacterium* and allied genera

STUDIES OF BACTERIA: GRAM POSITIVE GROUPS (*continued*)

Cheeseman and Berridge (1959)	*Lactobacillus*
Colman (1968)	*Streptococcus*
Colobert and Blondeau (1962)	*Streptococcus faecalis*
Davis (1964)	*Lactobacillus*
Davis et al. (1969)	*Listeria, Streptococcus*, and other lactic acid and coryneform bacteria
Defayolle and Colobert (1962)	*Streptococcus faecalis*: ordination
Defayolle et al. (1968)	*Bacillus*: ordination
Drucker and Melville (1969a)	*Streptococcus*: oral forms
Drucker and Melville (1969b)	*Streptococcus*
Gray (1969)	*Arthrobacter*
Harrington (1966)	Coryneform bacteria
Hauser and Smith (1964)	*Lactobacillus*: strains from cheese
Hayashi, Mimura, and Nakabe (1968a)	Halophilic micrococci
Hayashi, Mimura, and Nakabe (1968b)	Lactobacillaceae
Hayashi et al. (1965)	*Micrococcus* and *Sarcina*
Hayashi et al. (1966a)	*Streptococcus* and allied genera
Hesser, Hartman, and Saul (1967)	*Lactobacillus*: strains from silage
Hester and Weeks (1969)	*Brevibacterium*
Hester and Weeks (1970)	Coryneform bacteria
Hill (1959)	*Micrococcus* and *Staphylococcus*
Hill et al. (1965)	*Micrococcus* and *Staphylococcus*
Hubálec (1969)	Micrococcaceae
Jarvis and Annison (1967)	*Ruminococcus*
Lowe (1969)	Soil bacteria
Lysenko (1962)	*Bacillus cereus*: insect pathogenic forms
Lysenko (1963b)	*Bacillus cereus*: insect pathogenic forms
Malik, Reinbold, and Vedamuthu (1968)	*Propionibacterium*
Masuo and Nakagawa (1968)	Coryneform bacteria
Masuo and Nakagawa (1969a)	Coryneform bacteria
Masuo and Nakagawa (1969b)	Various genera
Masuo and Nakagawa (1969c)	Various genera and DNA data

(*continued*)

STUDIES OF BACTERIA: GRAM POSITIVE GROUPS (*continued*)

Masuo and Nakagawa (1970)	*Corynebacterium*: serology
Mullakhanbhai and Bhat (1967)	*Arthrobacter*
Nakamura and Nishida (1970)	*Clostridium tetani* and allied species
Pike (1965a)	Micrococcaceae: R analysis
Pike (1965b)	Micrococcaceae: R analysis
Pohja (1960)	*Micrococcus*
Pohja and Gyllenberg (1962)	*Micrococcus*
Raj and Colwell (1966)	*Streptococcus*
Raj, Colwell, and Liston (1964)	*Streptococcus*
Roberts (1968)	*Corynebacterium pyogenes*: host origin, geography
Rosypal, Rosypalová, and Hořejš (1966)	*Micrococcus* and *Staphylococcus*: congruence with DNA base ratios
Seyfried (1968)	*Streptococcus, Lactobacillus*, and *Propionibacterium*
Silva and Holt (1965)	Coryneform bacteria
Silvestri and Hill (1965)	*Micrococcus* and *Staphylococcus*: congruence with DNA base ratios
Skyring and Quadling (1969a)	Coryneform bacteria: two-stage ordination
Skyring and Quadling (1970)	Coryneform bacteria
Sneath (1962)	*Bacillus*
Splittstoesser, Mautz, and Colwell (1968)	*Streptococcus* and allied genera from vegetables
Splittstoesser et al. (1967)	Coryneform bacteria from vegetables
Wang (1968)	Coryneform bacteria, glutamate-accumulating

STUDIES OF BACTERIA: GRAM NEGATIVE GROUPS

Baptist, Shaw, and Mandel (1969)	Enterobacteriaceae: chemotaxonomy
Barnes and Goldberg (1968)	Bacteroidaceae
Baumann, Doudoroff, and Stanier (1968)	*Acinetobacter*
Carmichael and Sneath (1969)	*Pasteurella, Yersinia*, and allied genera
Colwell (1964)	*Pseudomonas aeruginosa*
Colwell (1969)	Myxobacterales and allied genera
Colwell (1970a)	*Vibrio* and allied genera

Colwell (1970b)	*Vibrio*: allied genera and DNA data
Colwell and Chapman (1966)	*Vibrio* and allied genera: marine strains
Colwell, Citarella, and Ryman (1965)	*Pseudomonas*: congruence with DNA base ratios
Colwell and Gochnauer (1963)	*Pseudomonas* and *Vibrio*: marine strains
Colwell and Liston (1961a)	*Pseudomonas* and *Xanthomonas*
Colwell and Liston (1961b)	Pseudomonadales and *Vibrio*
Colwell and Liston (1961c)	*Pseudomonas* and *Xanthomonas*
Colwell and Liston (1961d)	*Pseudomonas* and *Xanthomonas*
Colwell and Mandel (1964)	*Pseudomonas* and allied genera: enterobacteria and congruence with DNA base ratios
Colwell and Mandel (1965)	*Serratia*: congruence with DNA base ratios
Colwell, Mandel, and Gochnauer (1964)	*Serratia marcescens*
Colwell, Moffett, and Sutton (1968)	*Xanthomonas* and plant-pathogenic strains of *Pseudomonas*
Colwell, Morita, and Gochnauer (1964)	*Vibrio*: marine strains
Colwell and Yuter (1965)	*Vibrio*
Eddy and Carpenter (1964)	*Aeromonas*
Evans and Falkow (1969)	*Escherichia coli*: congruence with DNA pairing
Fager (1969)	Myxobacterales and allied genera
Floodgate and Hayes (1963)	*Flavobacterium* and *Cytophaga*
Goodall (1966a)	*Chromobacterium* and Enterobacteriaceae
Grimont (1969)	*Serratia*
Gyllenberg and Eklund (1967)	Pseudomonadaceae: Q and R analyses
Gyllenberg et al. (1963a)	*Pseudomonas* in milk
Hansen and Weeks (1964)	*Flavobacterium piscicida* (*Pseudomonas piscicida*)
Hansen, Weeks, and Colwell (1965)	*Pseudomonas*, including *P. piscicida*
Hayashi et al. (1966b)	*Neisseria* and allied forms
Heberlein, De Ley, and Tijtgat (1967)	*Rhizobium* and allied organisms: DNA pairing
Hodgkiss and Shewan (1968)	*Aeromonas* and *Vibrio* (*continued*)

STUDIES OF BACTERIA: GRAM NEGATIVE GROUPS (*continued*)

Hutchinson and White (1964)	*Thiobacillus*
Hutchinson, Johnstone, and White (1965)	*Thiobacillus*: distinctness of phenons
Hutchinson, Johnstone, and White (1966)	*Thiobacillus*: cluster parameters
Hutchinson, Johnstone, and White (1967)	*Thiobacillus*
Hutchinson, Johnstone, and White (1969)	*Thiobacillus* and allied organisms
Komagata, Tamagawa, and Iizuka (1968)	*Erwinia*
Krantz, Colwell, and Lovelace (1969)	*Vibrio parahaemolyticus*
Kreig and Lockhart (1966a)	Enterobacteriaceae
Kreig and Lockhart (1966b)	Enterobacteriaceae
Lewin (1969)	Myxobacterales and genera in allied orders
De Ley (1968)	*Acinetobacter* and allied generain
De Ley et al. (1966a)	*Pseudomonas, Xanthomonas*, and allied genera: congruence with DNA pairing
De Ley et al. (1966b)	*Agrobacterium*
Liston (1960)	*Pseudomonas, Achromobacter*, and allied genera
Liston and Colwell (1960)	Pseudomonadales
Liston, Weibe, and Colwell (1963)	*Pseudomonas*: cluster parameters
Lockhart (1967)	Enterobacteriaceae: effect of experimental errors
Lockhart and Holt (1964)	*Salmonella* serotypes
Lockhart and Koenig (1965)	*Erwinia*: different coding methods
Lysenko (1961)	*Pseudomonas*
Lysenko and Sneath (1959)	*Chromobacterium* and Enterobacteriaceae: ordination
McCurdy and Wolf (1967)	Myxobacterales
McDonald, Quadling, and Chambers (1963)	*Cytophaga* and allied genera: bacteria found in Artic sediments
Moffett and Colwell (1967)	*Rhizobium* and allied genera
Moffett and Colwell (1968)	*Rhizobium* and allied genera
Papacostea, Missirliu, and Preda (1965)	*Pseudomonas*
Pintér and Bende (1967)	*Acinetobacter, Moraxella*, and allied organisms

STUDIES OF BACTERIA: GRAM NEGATIVE GROUPS (*continued*)	
Pintér and Bende (1968)	*Acinetobacter, Moraxella*, and allied organisms
Pintér and De Ley (1969)	*Acinetobacter*: similarity of strains, DNA base ratios
Poindexter (1964)	Caulobacteraceae
Quadling and Colwell (1963)	*Pseudomonas, Vibrio.* and *Cytophaga*: Arctic bacteria
Quadling, Cook, and Colwell (1964)	*Cytophaga*
Quigley and Colwell (1968a)	*Pseudomonas* and allied genera: marine strains
Quigley and Colwell (1968b)	*Pseudomonas bathycetes*
Reich et al. (1966)	*Mycoplasma*: DNA pairing
Rhodes (1961)	*Pseudomonas*
Sakazaki, Gomez, and Sebald (1967)	*Vibrio* and *Aeromonas*
Sands, Schroth, and Hildebrand (1970)	*Pseudomonas*: plant pathogens
Shewan, Hobbs, and Hodgkiss (1960)	Pseudomonadaceae
Smith (1963)	*Aeromonas*
Smith and Thal (1965)	*Pasteurella* and *Yersinia*
Sneath (1957b)	*Chromobacterium*
Sneath (1964b)	Pseudomonadaceae
Sneath (1968d)	*Chromobacterium*: vigor and pattern differences
Stevens (1969)	*Pasteurella, Yersinia*, and allied genera
Talbot and Sneath (1960)	*Pasteurella multocida*
Thornley (1960)	*Pseudomonas* and *Achromobacter*
Thornley (1967)	*Acinetobacter* and allied genera
Thornley (1968)	*Acinetobacter* and allied genera
Véron (1966a)	*Vibrio* and allied genera
Véron (1966b)	*Vibrio* and allied genera

STUDIES OF VIRUSES	
Andrewes and Sneath (1958)	Animal
Bellett (1967a)	Animal: nucleic acid data
Bellett (1967b)	Animal: nucleic acid data
Bellett (1967c)	Animal: nucleic acid data
Bellett (1969)	Animal

(*continued*)

STUDIES OF VIRUSES (*continued*)

Dowdle et al. (1969)	Influenza: serology
Gibbs (1969)	Plant
Lee (1967)	Influenza: serology
Lee (1968)	Influenza: serology
Lee and Tauraso (1968)	Influenza: serology
Meier-Ewert, Gibbs, and Dimmock (1970)	Influenza: serology
Sneath (1962)	Animal
Tremaine (1970)	Plant: amino acid data
Tremaine and Argyle (1970)	Plant: amino acid data
Varma, Gibbs, and Woods (1969)	Plant: nucleic acid data

Some Hints on Techniques, Sources, and References

The rapid advance of computer techniques has made it inappropriate for us to present an appendix of worked examples such as was featured in our *Principles of Numerical Taxonomy*. There are, however, still practical points that may be conveniently brought together here. These come under the following headings: preparing material for numerical taxonomic studies; computational methods and strategy; and sources of computer programs, and bibliographies and reviews of special areas of numerical taxonomy. A full treatment of practical and computational aspects will appear in a forthcoming volume by F. J. Rohlf and P. M. Neely.

In selecting material for numerical taxonomic studies the investigator should have a clear idea of the kind of variation he wishes to explore, among both characters and OTU's. We believe that in most organisms it will be possible to find the necessary numbers of characters for analysis. In some difficult groups there may be few available characters, which in itself would show that previous taxonomies must have been based on inadequate data; any improvement in the classification, therefore, must first require new methods of study. This difficulty may also arise with studies at very high taxonomic ranks, as it may not be easy to know what characters can be selected for comparing, for example, an insect with an echinoderm; in such cases chemical data, especially protein sequences, are useful.

The selection of OTU's also requires attention, particularly if space-distorting clustering methods are employed, because of the sensitivity of these methods to the sample of OTU's chosen to represent the taxa under study. Ideally one would study all the species of a genus, all the genera of a family, and so on, but although this is often not feasible, efforts should be made to make the set of OTU's as complete as possible. A more difficult problem is that of incomplete data for the characters of the OTU's. Again the worker may have to undertake extensive studies to obtain the missing data (which again implies the inadequate bases of previous taxonomies). Our present advice on when to exclude OTU's or characters because of incompleteness of the data has been given in Section 4.12.

Substantial saving of time and effort can come from a planned approach to collecting the data. In some applications (e.g., microbiology) it is inconvenient to add OTU's to a partly completed study; in others (e.g., when using highly specialized techniques in chemistry or electron microscopy) it is inconvenient to reexamine OTU's for further characters. It is therefore useful to make preliminary lists of OTU's and characters early in the study, and this will also ensure that obvious kinds of information are looked for and recorded. After all the data have been recorded, unnecessary and laborious copying should be avoided. This is facilitated by first inspecting all the OTU's and characters and clearing up any ambiguities. A checklist of OTU's should then be made. Next, the specification of the computer program should be studied to see in what format the data should be presented. It should now be possible to code the character state values in the form required for punching. Some characters may be rejected at this stage for various reasons, and it is best to place the characters in some logical order to avoid accidental repetitions and inconsistences. It usually does not matter what order the OTU's are in, because ties in resemblance values are uncommon when large numbers of characters are employed. Further details can be found in Sneath (1967d).

One aspect of computational strategy is the writing of convenient computer programs for carrying out the computations. There are technical aspects that, though extremely important for the success of numerical taxonomic work, go beyond the scope of this book and are discussed in detail in a forthcoming volume by Rohlf and Neely. Among other problems, Rohlf and Neely concern themselves with questions such as how data matrices should be read in and stored, which particular algorithms should be used for various matrix computations, and what particular combination of central processor and peripheral equipment such as tape drives, disks, and drums permits an optimal analysis of a data matrix of given dimensions. The successful solution of these problems by several programming groups (e.g., the team headed by F. J. Rohlf, first at the University of Kansas and subsequently at the State University of New York at Stony Brook, or that of G. N. Lance and W. T. Williams at Canberra, Australia, among others) has greatly aided the advance of numerical taxonomic work all around the world. Several taxonomic program packages have been developed (e.g., those of Rohlf, Kishpaugh, and Kirk, 1971; of Wishart 1969e, 1970; of Gower and Ross of Rothamsted Experimental Station; of Sackin at Leicester University; and of Lance and Williams at Canberra); these permit a wide choice of resemblance coefficients and methods for displaying taxonomic structure.

The time required for computation depends greatly on the numerical method employed. For most methods, in which the full similarity matrix is computed, the time is roughly proportional to nt^2, so that increasing the number of OTU's has more effect than increasing the number of characters. For association analysis the reverse holds, as the time is proportional to n^2t. Any method for which the time rises too steeply with increasing t or n is impracticable for realistic taxonomic problems. Examples are the methods of Edwards and Cavalli-Sforza (1965), and complete searches for seriation (Kendall, 1963) or rooted trees (Fitch and Margoliash, 1968), for which the times are roughly proportional to 2^{t-1}, $t!$, and $(2t)!/(t+1)!2^{t+1}$ respectively. Some methods require only the computation of part of the S matrix (e.g., that of Rose, 1964, and use of a graph-theoretical approach to single linkage clustering by Gower and Ross, 1969). Special combinatorial algorithms (e.g., Wishart, 1969a, 1970) can reduce the time appreciably. Češka (1968) gives a method whereby the mean values of association coefficients within and between sets of OTU's may be computed directly from the $n \times t$ matrix without calculating all similarity values (provided the sets are already known). Further information on computing times is given by W. T. Williams (1964) and Lance and Williams (1966b).

For most present computer installations, with more conventional methods, capacity is limited by t with an upper limit of usually about 400. In some applications very large numbers of OTU's, perhaps many thousands, must be processed, and special methods are then needed. Although experience with such methods is still limited, they have been described by a number of authors. These include methods of Lockhart and Hartman (1963), Crawford and Wishart (1967; 1968), Ross (1969d), Kaminuma, Takekawa, and Watanabe (1969), and Switzer (1970), as well as methods of Rose (1964) and of Gower and Ross (1969).

These fast methods may also be used to divide very large data sets into manageable subsets, but because many of them are monothetic there is the risk of misplacement of a few OTU's unless special reallocation facilities are provided (e.g., Crawford and Wishart, 1968). Other ways of handling more OTU's than can be accommodated at once have been mentioned in Section 3.1. With ordination methods it is usually possible to obtain a desired end result by algebraic manipulation of either Q- or R-type matrices, so one can choose the technique involving least computation according to whether n or t is the greater (see Gower, 1966a,b, 1967b; Orloci, 1967a). Some steps in orthodox taxonomy are analogous to using R analyses to break down very large sets of OTU's into smaller ones, and this strategy is also available.

An allied problem is to add a new point to an ordination, and a method has been described by Gower (1968) for which an example is given by Wilkinson (1970b). A general solution for n dimensions to the related problem of matching diagrams (see Section 3.3) has been derived by Gower (1971b), as follows. The n-dimensional coordinates of the h points are first referred to the centroids of their respective diagrams A and B, giving two $h \times n$ matrices ($h > n$), $\mathbf{A}$ and $\mathbf{B}$. If $\mathbf{A}$ and $\mathbf{B}$ have originally different numbers of columns, the smaller is filled out with zeros on the right to ensure that $\mathbf{A}$ and $\mathbf{B}$ are both $h \times n$ matrices. One then computes $\mathbf{R} = \mathbf{A'B} = \mathbf{USV'}$ where $\mathbf{U}$ and $\mathbf{V}$ are orthogonal (their elements scaled so the sum of squares of columns is unity) and $\mathbf{S}$ is diagonal. $\mathbf{U}, \mathbf{V}, \mathbf{S}$ are obtained as eigenvectors of the equations $\mathbf{RR'U} = \mathbf{US^2}$ and $\mathbf{R'RV} = \mathbf{VS^2}$ or directly by a singular value decomposition algorithm (Golub and Reinsch, 1970). The required orthogonal rotation matrix $\mathbf{H}$ that minimizes least square distances between corresponding h-points of A and B is $\mathbf{H} = \mathbf{VU'}$. This is not unique if $\mathbf{R}$ has rank less than $h - 1$, but this simply means that several possible rotations exist. The columns of $\mathbf{U}$ and $\mathbf{V}$ may each be changed in sign independently, without affecting the validity of the above solution, giving 2^n different results, each one associated with a different set of reflections of B relative to A. The reflection that gives the best fit is obtained by calculating $\mathbf{U'RV}$ for some solution $\mathbf{U}$ and $\mathbf{V}$ and then multiplying the ith column of $\mathbf{V}$ by the sign of the ith diagonal element of $\mathbf{U'RV}$. The new $\mathbf{V}$ is then used in subsequent calculations. These considerations are conveniently covered by calculating the matrix $\mathbf{G} = \mathbf{JH}$, where $\mathbf{J}$ is the reflection matrix, a diagonal matrix of $+1$ and -1 elements whose signs are derived as explained above. $\mathbf{G}$ gives the best fit after allowing for reflection. The coordinates of B referred to A are given by $\delta\mathbf{BG}$, where δ is a scaling factor of B. The value of δ that gives the minimum sum of squared distances between corresponding pairs of points, $\Sigma\Delta^2$, is obtained as $\delta = \text{trace}(\mathbf{BGA'})/\text{trace}(\mathbf{BB'})$. After scaling $\mathbf{B}$ by multiplying by δ, $\Sigma\Delta^2$ is then trace $(\mathbf{AA'})$, which equals δ^2 trace $(\mathbf{BB'})$. However, because of the nonreciprocal scaling this gives when fitting B to A compared with A to B, other methods of scaling may be preferred. A simple method is to scale $\mathbf{A}$ and $\mathbf{B}$ to have the same total sums of squares, say unity; this amounts to dividing $\mathbf{A}$ and $\mathbf{B}$ by the square roots of the traces of $\mathbf{A'A}$ and $\mathbf{B'B}$ respectively. Sometimes it is clear that both A and B are on the same scale so no additional scaling is required.

Many computer programs now have facilities for graphic output, ranging from diagrams constructed on the line printer to cathode ray displays, but most commonly on a graph plotter (for example the GRAFPAC subroutines of Rohlf, 1969). Subroutines for producing phenograms or cladograms include those of Bartcher (1966), Bonham-Carter (1967b), and McCammon and Wenniger (1970). Rearranged similarity matrices are also often provided (sometimes with only the first significant digit and no spaces between digits, giving much the same impression as shaded similarity matrices). Reordered $n \times t$ matrices (e.g., Bonham-Carter, 1967b) are useful for selecting diagnostic characters. Ordination plots can also be provided, though problems occur if plotted points overlap. Stereograms are assuming increasing importance, and here exact positioning is very important, as pointed out by Rohlf (1968). Formulae for stereograms are given by Fraser and Kovats (1966) and by Rohlf (1968). Rohlf's formulae can be calculated on desk calculators, and are given below.

For any given OTU with coordinates X_1, X_{II}, and X_{III} on three ordination axes I, II, and III, one calculates its position for the stereoimages on axes X (horizontal) and Y (vertical). First the coordinates are suitably scaled by subtracting the minimum value for any OTU in the study, and dividing by a constant M that is conveniently taken as rather larger than the greatest of the ranges of the values on I, II, and III. The scaled values are indicated by primes

$$X_I' = (X_I - X_{I,min})/M$$
$$X_{II}' = (X_{II} - X_{II,min})/M$$
$$X_{III}' = (X_{III} - X_{III,min})/M$$

It is most convenient to choose the axis with the greatest range as I, and that with the least as III.

Next, viewing points are chosen, where the left viewing point has the coordinates L_I, L_{II}, and H, and for the right they are R_I, R_{II}, and H. Rohlf recommends for general use $L_I = L_{II} = R_{II} = 1/2$, $R_I = 2/3$, and $H = 3$. The position for the left stereoimage is then

$$X_L = (HX_I' - L_I X_{III}')/(H - X_{III}')$$
$$Y_L = (HX_{II}' - L_{II} X_{III}')/(H - X_{III}')$$

and for the right stereoimage

$$X_R = (HX_I' - R_I X_{III}')/(H - X_{III}')$$
$$Y_R = (HX_{II}' - R_{II} X_{III}')/(H - X_{III}')$$

The X, Y coordinates are calculated for each OTU and may also be calculated for the corners of a rectangular box that acts as a viewing frame, which can then be drawn in with straight lines. Because of the need for accuracy, enlarged drawings should be made and reduced photographically to the size appropriate to the viewing device to be used. If an oblique view is preferred, convenient viewing points are given by $L_I = L_{II} = R_{II} = 1.5$, $R_I = 1.7$, $H = 3$.

For unusually difficult jobs, consultation with computer experts in processing multivariate data is essential. In a study of the suborder Blattaria by Huber (1968) involving 177 OTU's representing 37 species scored for 446 characters, a major problem was storing the original data matrix on tape for subsequent computer processing. In spite of competent assistance it took the better part of a month before the data were successfully converted from cards to tape. The subsequent handling of the data was relatively routine by the NT-SYS system at The University of Kansas, although on some computers such large numbers of characters would also present a problem.

Another aspect of computational strategy is how much time a given investigator should devote to learning how to program by himself, whether he should attempt to implement numerical taxonomy programs at his own computation center or use larger, remote facilities where these programs are already implemented if he has the opportunity to do so. There are still many problems in transferring programs to another machine, or even to the same machine at another installation. Recent tendency has been to develop sophisticated systems at several large computation centers and have these used by clients outside the institution, because such a strategy would generally be considerably less expensive and less time-consuming for a potential user than to attempt to implement even a simple numerical taxonomy program at his own computation center. The availability of long distance, time-shared computing makes this approach even more attractive. Our present advice therefore would be that for "one shot jobs" it is simplest and most economical to send the data to a center that is equipped to handle them. However, persons who wish to do numerical taxonomy on a routine basis should establish at their centers a series of programs that would carry out these computations for them. The recent development, by various teams of workers, of libraries of basic numerical taxonomy routines not interrelated as a computing system, but standing independent of each other and written in a simple, widely compatible style, has made this strategy more feasible than before.

Programs that are written for numerical taxonomy should have detailed write-ups to enable machine operators unfamiliar with the computations and taxonomists unfamiliar with computers to do the work with maximum facility. Information on operating instructions should include how many OTU's and characters can be processed, the exact format of input and output, estimates for execution, and restrictions on the kinds of characters permitted. It is particularly important that the write-up should not only describe the general idea of what the program will do, but should also give in detail the actual algebra used (unless it is a very standard procedure for which reference to a publication would be adequate). This information is essential for the user to enable him to be certain that the program carries out the kind of analysis that he requires. It is also extremely useful for the write-up to contain a small worked example with input data and results for checking the program if it is implemented on a new machine.

In *Principles of Numerical Taxonomy* a grouping of numerical taxonomic programs was outlined, based on suggestions by Sneath and Rohlf in *Taxometrics*, **2**, December 1962. Many programs incorporate several groups, so that these are not always convenient in practice, but they illustrate the logical arrangement of subprograms in a computer program package and may therefore still be of use in planning program layout. They are briefly described below with minor modifications.

GROUP 1
Control programs control the subsequent programs of Groups 2–9. Control programs call up different subprograms as required and direct the flow of operations.

GROUP 2
Translation programs take in descriptions of OTU's in words or diagrams and convert them into appropriate numerical codes. These programs may eventually be able to remove much of the tedium of coding characters from the shoulders of the taxonomist, and are now being used particularly in key-making programs.

GROUP 3
Character conversion programs convert data to the form necessary for computing resemblance coefficients, such as standardization (or other transformations), transposition of rows and columns, and augmentation (or deletion) of OTU's (or characters).

GROUP 4
Resemblance coefficient programs use the output of Group 3 programs and calculate matrices of resemblance between OTU's. Some special procedures in cladistic analysis (e.g., character compatibility) may be conveniently included here.

GROUP 5
Programs for analysis of taxonomic structure, a group originally restricted to programs for cluster analysis, can now conveniently be extended to include (a) cluster analysis programs,

(b) ordination programs, (c) programs for cladograms, and (d) cophenetic programs. This group takes resemblance matrices or their equivalent and yield, as output, dendrograms, cluster parameters, ordination plots, distortion measures, etc.

GROUP 6

Data extraction programs extract data from earlier steps, answering specific questions addressed to the study. They include (a) programs that compute average resemblances within and between specified clusters of OTU's and (b) identification programs. They require additional input to indicate specified phenons (sometimes single OTU's).

GROUP 7

Interstudy coordination programs store and sort out previous studies, establish reference taxa and their characters, and correlate different studies (e.g., by computing distortion measures).

GROUP 8

Publication programs convert results into forms that are legible and publishable, such as diagnostic keys and graphic outputs of quality suitable for use as phenograms and ordination plots.

GROUP 9

Miscellaneous programs; some programs do not readily fit into the other classes.

There are now many computer programs for numerical taxonomy. Most of these are unpublished, and workers must contact those who have written them; however, there are several sources through which these may be traced, in addition to standard sources of papers on applications to numerical taxonomy. The following periodicals among others contain descriptions and sometimes full program listings: the *Computer Journal, Applied Statistics, Behavioral Science*, and the *Kansas Geological Survey Computer Contributions*. Two newsletters, *Taxometrics* (issued by the National Collection of Type Cultures, Colindale, London N.W. 9) and the *Classification Programs Newsletter* (issued by the M.R.C. Microbial Systematics Unit, University of Leicester) contain lists of programs and their sources.

The series of computer contributions of the Kansas Geological Survey is especially valuable because these contain full descriptions and examples of input and output as well as the programs themselves. Numbers of special interest include the following: Bartcher (1966) for cladistic relationships by the Camin–Sokal method; Bonham-Carter (1967b) for Q cluster analysis of binary WPGM or UPGM; Wahlstedt and Davis (1968) for principal components (and a form equivalent to principal coordinates); Wishart (1969e) for numerous resemblance coefficients and cluster methods; Ondrick and Srivastava (1970) for correlations and R and Q factor analysis with varimax rotation; Demirmen (1969) for iterative reallocation in cluster analysis; Reyment, Ramden, and Wahlstedt (1969) for Mahalanobis distance; Reyment and Ramden (1970) for canonical variates; and McCammon and Wenniger (1970) for a special form of phenogram called the dendrograph. An extensive compilation of programs for environmental sciences (Tarrant, 1972) lists many taxometric ones. Many useful multivariate statistical programs are given in Cooley and Lohnes (1971). Sokal and Rohlf (1969) include programs for basic statistical procedures. Certain algorithms of use in graphs and trees are given by Gower and Ross (1969), Ross (1969a,b,c), and Farris (1970). Wishart (1970) gives many useful combinatorial formulae for clustering methods that can considerably reduce time and programming. Other papers that consider improved algorithms for various methods are Proctor (1966), Jensen (1969), Vinod (1969), Wishart (1969a), and Cole and Wishart (1970)). Estabrook and Brill (1969) describe a program for taxonomic data processing.

There are now numerous publications reviewing special areas in numerical taxonomy. Our earlier volume (Sokal and Sneath, 1963) is available in French translation from Laboratoire Central, Compagnie Française de Petrole, Bordeaux, and a summary is available in Russian (Sokal, 1968). The proceedings of a numerical taxonomy symposium have been edited by Cole (1969). Publications mainly on methodology in special groups of organisms include Lockhart and Liston (1970) and Sneath (1972) in microbiology, and J. McNeill (in preparation) in botany. Reviews of numerical taxonomy in special groups include the following: in zoology Funk (1963, acarology), Johnston (1964, acarology), Moss and Webster (1970, nematology); in botany Williams (1967b), Gilmartin (1967b), and Sneath (1969d); in microbiology Sneath (1962, 1964a, 1968c), Lysenko (1963a), Véron (1969), and Colwell (1971). A more general review is that of Rogers, Fleming, and Estabrook (1967). Similar reviews on applications in paleontology and subjects outside systematics have been listed in Section 6.5 and in Chapter 11.

Certain methodological fields are also covered by recent or forthcoming works: in mathematics, Fernandez de la Véga (1965), Lerman (1970), and Jardine and Sibson (1971); on computational aspects, Rohlf and Neely (in preparation); on cluster analysis, Spence and Taylor (1970) and Wishart (1969b, 1970); on phyletics, Kluge and Farris (1973).

Useful bibliographies may be found in various issues of *Taxometrics* (a KWIC index to this has been issued) and the *Classification Society Bulletin*. Journals of special interest to numerical taxonomists include *Systematic Zoology*, *Taxon*, the *Classification Society Bulletin*, the *Journal of General Microbiology*, *Biometrics*, and *Computers and the Humanities*.

Introductory texts for statistics and mathematics for numerical taxonomy include Sokal and Rohlf (1969), Simpson, Roe, and Lewontin (1960), Schwartz (1961), Searle (1966), Graybill (1969), and Cooley and Lohnes (1971); more advanced treatments of multivariate methods are found in Rao (1952), Anderson (1958), Seal (1964), Morrison (1967), Harman (1967), and Van de Geer (1971).

Bibliography

Abelson, P. H. (1957). Some aspects of paleobiochemistry. *Ann. New York Acad. Sci.*, **69**, 276–285.

Abou-El-Fittouh, H. A., J. O. Rawlings, and P. A. Miller (1969). Classification of environments to control genotype by environment interactions with an application to cotton. *Crop Sci.*, **9**, 135–140.

Ackermann, A. (1967). Quantitative Untersuchungen an körnerfressenden Singvögeln. *J. Ornithol.*, **108**, 430–473.

Adams, J. N. (1964). A critical evaluation of Adansonian taxonomy. *Devel. Ind. Microbiol.*, **5**, 173–179.

Adams, R. P., and B. L. Turner (1970). Chemosystematic and numerical studies of natural populations of *Juniperus ashei* Buch. *Taxon*, **19**, 728–751.

Adanson, M. (1757). *Histoire naturelle du Sénégal. Coquillages. Avec la relation abrégée d'un voyage fait en ce pays, pendant les années 1749, 50, 51, 52 et 53.* Coquillages. Préface, pp. xi, xx, xxix–lxxxviii. Bauche, Paris. 190 + xcvi + 175 pp.

Adanson, M. (1763). *Familles des plants.* Vol. 1. Préface, pp. cliv et seq., clxiii, clxiv. Vincent, Paris. cccxxv + 190 pp.

Ahmad, Q. (1965). Indian cities: characteristics and correlates. *Univ. Chicago Dept. Geogr. Res. Pap.*, No. 102. 184 pp.

Ainsworth, G. C. (1941). A method for characterizing smut fungi exemplified by some British species. *Trans. Brit. Mycol. Soc.*, **25**, 141–147.

Ainsworth, G. C., and P. H. A. Sneath (eds.) (1962). The proposal and selection of scientific names for microorganisms. Appendix I, pp. 456–463. In *Microbial Classification.* 12th Symposium of the Society for General Microbiology. Cambridge University Press, Cambridge. 483 pp.

Alker, H. R., Jr. (1969). Statistics and politics: the need for causal data analysis. In S. M. Lipset (ed.), *Politics and the Social Sciences*, pp. 244–313. Oxford University Press, New York. 328 pp.

Allen, N., and M. J. Pelczar, Jr. (1967). Bacteriological studies on the white perch, *Roccus americanus. Chesapeake Sci.*, **8**, 135–154.

Alpatov, W. W., and A. M. Boschko-Stepanenko (1928). Variation and correlation in serially situated organs in insects, fishes and birds. *Amer. Natur.*, **62**, 409–424.

Alston, R. E. (1967). Biochemical systematics. *Evolut. Biol.*, **1**, 197–305.

Alston, R. E., and B. L. Turner (1963). *Biochemical Systematics*. Prentice-Hall, Englewood Cliffs, N. J. 404 pp.

Amadon, D. (1949). The seventy-five per cent rule for subspecies. *Condor*, **51**, 250–258.

Amadon, D. (1966). Another suggestion for stabilizing nomenclature. *Systematic Zool.*, **15**, 54–58.

Anderson, A. J. B. (1971). Similarity measure for mixed attribute types. *Nature*, **232**, 416–417.

Anderson, D. J. (1967). Studies on structure in plant communities. IV. Cyclical succession in *Dryas* communities from north-west Iceland. *J. Ecol.*, **55**, 629–635.

Anderson, E. (1936). Hybridization in American Tradescantias. *Ann. Missouri Botan. Gard.*, **23**, 511–525.

Anderson, E., and E. C. Abbe (1934). A quantitative comparison of specific and generic differences in the Betulaceae. *J. Arnold Arboretum*, **15**, 43–49.

Anderson, E., and R. P. Owenbey (1939). The genetic coefficients of specific difference. *Ann. Missouri Botan. Gard.*, **26**, 325–348.

Anderson, E., and T. W. Whitaker (1934). Speciation in *Uvularia. J. Arnold Arboretum*, **15**, 28–42.

Anderson, H. E. (1966). Regression, discriminant analysis and a standard notation for basic statistics. In R. B. Cattell (ed.), *Handbook of Multivariate Experimental Psychology*, pp. 153–173. Rand McNally, Chicago, 959 pp.

Anderson, J. A., and J. A. Boyle (1968). Computer diagnosis: statistical aspects. *Brit. Med. Bull.*, **24**, 230–235.

Anderson, N. G. (1970). Evolutionary significance of virus infection. *Nature*, **227**, 1346–1347.

Anderson, T. W. (1958). *An Introduction to Multivariate Statistical Analysis*. Wiley, New York. 374 pp.

Andrewes, C. H., and P. H. A. Sneath (1958). The species concept among viruses. *Nature*, **182**, 12–14.

Anstey, R. L., and T. G. Perry (1970). Biometric procedures in taxonomic studies of Paleozoic bryozoans. *J. Paleontol.*, **44**, 383–398.

Antila, M., and H. G. Gyllenberg (1963). A taxometric study of the propionic acid bacteria of dairy origin. *Milchwissenschaft*, **18**, 398–400.

Arnheim, N., Jr., E. M. Prager, and A. C. Wilson (1969). Immunological prediction of sequence differences among proteins. *J. Biol. Chem.*, **244**, 2085–2094.

Ashlock, P. D. (1971). Monophyly and associated terms. *Systematic Zool.*, **20**, 63–69.

Austin, M. P., and L. Orloci (1966). Geometric models in ecology. II. An evaluation of some ordination techniques. *J. Ecol.*, **54**, 217–227.

Bächli, G. (1971). *Leucophenga* and *Paraleucophenga* (Diptera, Brachycera) Fam. Drosophilidae. *Exploration du Parc National de l'Upemba, Bruxelles.* Fasc. 71. 192 pp.

Bailey, N. T. J. (1967). *The Mathematical Approach to Biology and Medicine*. Wiley, New York. 296 pp.

Balbach, H. E. (1965). Computer analysis of variation in the genus *Apocynum. Amer. J. Bot.*, **52**, 646–647.

Ball, G. H. (1965). Data analysis in the social sciences: what about the details? In W. R. Rector (Chairman), *Fall Joint Computer Conference*, AFIPS Conference Proceedings, vol. 27, part 1, 1965, pp. 533–559. Spartan Books, Washington, D.C. 1100 pp.

Ball, G. H. (1970). *Classification Analysis*. Stanford Research Institute, Menlo Park, Calif. 117 pp.

Ball, G. H., and D. J. Hall (1965). *ISODATA, a Novel Method of Data Analysis and Pattern Classification*. Stanford Research Institute, Menlo Park, Calif. 61 pp.

Ball, G. H., and D. J. Hall (1967). A clustering technique for summarizing multivariate data. *Behav. Sci.*, **12**, 153–155.

Banks, J. C. G., and W. E. Hillis (1969). The characterization of populations of *Eucalyptus camaldulensis* by chemical features. *Aust. J. Bot.*, **17**, 133–146.

Bannister, D. (1968). The logical requirements of research into schizophrenia. *Brit. J. Psychiatry*, **114**, 181–188.

Baptist, J. N., C. R. Shaw, and M. Mandel (1969). Zone electrophoresis of enzymes in bacterial taxonomy. *J. Bacteriol.*, **99**, 180–188.

Barnes, E. M., and H. S. Goldberg (1968). The relationships of bacteria within the family Bacteroidaceae as shown by numerical taxonomy. *J. Gen. Microbiol.*, **51**, 313–324.

Baron, D. N., and P. M. Fraser (1965). The digital computer in the classification and diagnosis of diseases. *Lancet*, **1965** (ii), 1066–1069.

Baron, D. N., and P. M. Fraser (1968). Medical applications of taxonomic methods. *Brit. Med. Bull.*, **24**, 236–240.

Barre, P. (1969). Taxonomie numérique de lactobacilles isolés du vin. *Arch. Mikrobiol.*, **68**, 74–86.

Bartcher, R. L. (1966). FORTRAN IV program for estimation of cladistic relationships using the IBM 7040. *Kansas Geol. Surv. Computer Contrib.*, No. 6. 54 pp.

Bartels, P. H., G. F. Bahr, D. W. Calhoun, and G. L. Wied (1970). Cell recognition by neighborhood grouping techniques in TICAS. *Acta Cytol.*, **14**, 313–324.

Bartels, P. H., G. F. Bahr, and G. L. Wied (1968). Cell recognition by cluster analysis in pure parameter hyperspaces. *Acta Cytol.*, **12**, 371–380.

Basford, N. L., J. E. Butler, C. A. Leone, and F. J. Rohlf (1968). Immunologic comparisons of selected Coleoptera with analyses of relationships using numerical taxonomic methods. *Systematic Zool.*, **17**, 388–406.

Basrur, P. K. (1959). The salivary gland chromosomes of seven segregates of *Prosimulium* (Diptera : Simuliidae) with a transformed centromere. *Can. J. Zool.*, **37**, 527–570.

Basrur, P. K. (1962). The salivary gland chromosomes of seven species of *Prosimulium* (Diptera : Simuliidae) from Alaska and British Columbia. *Can. J. Zool.*, **40**, 1019–1033.

Bather, F. A. (1927). Biological classification: past and future. *Quart. J. Geol. Soc. Lond.*, **83**, Proc. lxii–civ.

Baumann, P., M. Doudoroff, and R. Y. Stanier (1968). A study of the *Moraxella* group. II. Oxidative-negative species (genus *Acinetobacter*). *J. Bacteriol.*, **95**, 1520–1541.

Beals, E. W. (1968). A taxonomic continuum in the genus *Antennaria* in Wisconsin. *Amer. Midland Natur.*, **79**, 31–47.

Bean, P. G., and J. R. Everton (1969). Observations on the taxonomy of chromogenic bacteria isolated from cannery environments. *J. Appl. Bacteriol.*, **32**, 51–59.

Beckner, M. (1959). *The Biological Way of Thought*. Columbia University Press, New York. 200 pp.

Beckner, M. (1964). Metaphysical presuppositions and the description of biological systems. In J. R. Gregg and F. T. C. Harris (eds.), *Form and Strategy in Science*. Studies Dedicated to Joseph Henry Woodger on the Occasion of His Seventieth Birthday, pp. 15–29. D. Reidel Publ. Co., Dordrecht, The Netherlands. 476 pp.

Beermann, W., and U. Clever (1964). Chromosome puffs. *Sci. Amer.*, **210**(4), 50–58.

Beers, R. J., J. Fisher, S. Megraw, and W. R. Lockhart (1962). A comparison of methods for computer taxonomy. *J. Gen. Microbiol.*, **28**, 641–652.

Beers, R. J., and W. R. Lockhart (1962). Experimental methods in computer taxonomy. *J. Gen. Microbiol.*, **28**, 633–640.

Behrens, E. W. (1965). Environment reconstruction for a part of the Glen Rose limestone, central Texas. *Sedimentology*, **4**, 65–111.

Bellett, A. J. D. (1967a). The use of computer-based quantitative comparisons of viral nucleic acids in the taxonomy of viruses: a preliminary classification of some animal viruses. *J. Gen. Virol.*, **1**, 583–585.

Bellett, A. J. D. (1967b). Numerical classification of some viruses, bacteria and animals according to nearest-neighbour base sequence frequency. *J. Mol. Biol.*, **27**, 107–112.

Bellett, A. J. D. (1967c). Preliminary classification of viruses based on quantitative comparisons of viral nucleic acids. *J. Virol.*, **1**, 245–259.

Bellett, A. J. D. (1969). Relationships among the polyhedrosis and granulosis viruses of insects. *Virology*, **37**, 117–123.

Bemis, W. P., A. M. Rhodes, T. W. Whitaker, and S. G. Carmer (1970). Numerical taxonomy applied to *Cucurbita* relationships. *Amer. J. Bot.*, **57**, 404–412.

Bendich, A. J., and B. J. McCarthy (1970). DNA comparisons among barley, oats, rye, and wheat. *Genetics*, **65**, 545–565.

Benson, R. H. (1967). Muscle-scar patterns of Pleistocene (Kansan) ostracodes. In C. Teichert and E. L. Yochelson (eds.), *Essays in Paleontology and Stratigraphy: R. C. Moore Commemorative Volume*, pp. 211–241. University of Kansas Press, Lawrence, Kan. 626 pp.

Bernstein, L. F. (1967). University of Chicago Chanson Project. *Computers and the Humanities*, **1**, 221.

Berry, B. J. L. (1968). A synthesis of formal and functional regions using a general field theory of spatial behavior. In B. J. L. Berry and D. F. Marble (eds.), *Spatial Analysis: A Reader in Statistical Geography*, pp. 419–428. Prentice-Hall, Englewood Cliffs, N. J. 512 pp.

Berry, R. J. (1963). Epigenetic polymorphism in wild populations of *Mus musculus*. *Genet. Res.*, **4**, 193–220.

Berry, R. J. (1969). History in the evolution of *Apodemus sylvaticus* (Mammalia) at one edge of its range. *J. Zool., Lond.*, **159**, 311–328.

Berry, R. J., and A. G. Searle (1963). Epigenetic polymorphism of the rodent skeleton. *Proc. Zool. Soc. Lond.*, **140**, 577–615.

Bhatt, G. M. (1970). Multivariate analysis approach to selection of parents for hybridization aiming at yield improvement in self-pollinated crops. *Aust. J. Agr. Res.*, **21**, 1–7.

Bhattacharya, C. G. (1967). A simple method of resolution of a distribution into Gaussian components. *Biometics*, **23**, 115–135.

Bidault, M. (1968). Essai de taxonomie expérimental et numérique sur *Festuca ovina L. s. l.* dans le sud-est de la France. *Rev. Cytol. Biol. Vég.*, **31**, 217–356.

Bidault, M., and J. M. Hubac (1967). Application des méthodes numériques de la taxinomie sur une série de populations de *Festuca ovina* L. spp. *eu-ovina* Hack. *Compt. Rend. Acad. Sci. Paris, Sér. D.*, **264**, 1785–1788.

Bidwell, O. W., and F. D. Hole (1964a). Numerical taxonomy and soil classification. *Soil Sci.*, **97**, 58–62.

Bidwell, O. W., and F. D. Hole (1964b). An experiment in the numerical classification of some Kansas soils. *Soil Sci. Soc. Amer. Proc.*, **28**, 263–268.

Bielicki, T. (1962). Some possibilities for estimating inter-population relationships on the basis of continuous traits. *Current Anthropology*, **3**, 3–8, 20–46.

Bigelow, R. S. (1956). Monophyletic classification and evolution. *Systematic Zool.*, **5**, 145–146.

Bird, G. W. (1967). Numerical analysis of the genus *Trichodorus*. *Phytopathology*, **57**, 804.

Bird, G. W., and W. F. Mai (1967). Morphometric and allometric variations of *Trichodorus christiei*. *Nematologia*, **13**, 617–632.

Birnbaum, A., and A. E. Maxwell (1961). Classification procedures based on Bayes's formula. *Appl. Stat.*, **9**, 152–168.

Bisby, F. A. (1970a). The application of numerical techniques to some problems in plant taxonomy. D.Phil. thesis, Oxford University.

Bisby, F. A. (1970b). The evaluation and selection of characters in angiosperm taxonomy: an example from *Crotalaria*. *New Phytol.*, **69**, 1149–1160.

Bischler, H., and P. Joly (1969). Essais d'application de méthodes de traitment numérique des informations systématiques. II. Etude des espèces européennes, africaines et sud-américaines de *Calypogeia*. *Rev. Bryol. Lichénol.*, **36**, 691–714.

Blackith, R. E. (1957). Polymorphism in some Australian locusts and grasshoppers. *Biometrics*, **13**, 183–196.

Blackith, R. E. (1963). A multivariate analysis of Latin elegiac verse. *Language and Speech*, **6**, 196–205.

Blackith, R. E. (1965). Morphometrics. In T. H. Waterman and H. J. Morowitz (eds.), *Theoretical and Mathematical Biology*, pp. 225–249. Blaisdell Publ. Co., New York. 426 pp.

Blackith, R. E., and R. M. Blackith (1968). A numerical taxonomy of orthopteroid insects. *Aust. J. Zool.*, **16**, 111–131.

Blackith, R. E., R. G. Davies, and E. A. Moy (1963). A biometric analysis of development in *Dysdercus fasciatus* Sign (Hemiptera: Pyrrhocoridae). *Growth*, **27**, 317–334.

Blackith, R. E., and D. K. McE. Kevan (1967). A study of the genus *Chrotogonus* (Orthoptera). VIII. Patterns of variation in external morphology. *Evolution*, **21**, 76–84.

Blackith, R. E., and R. A. Reyment (1971). *Multivariate Morphometrics*. Academic Press, London and New York. 412 pp.

Blackwelder, R. E. (1962). Animal taxonomy and the New Systematics. *Survey Biol. Progr.*, **4**, 1–57.

Blackwelder, R. E. (1967a). *Taxonomy: A Text and Reference Book*. Wiley, New York. 698 pp.

Blackwelder, R. E. (1967b). A critique of numerical taxonomy. *Systematic Zool.*, **16**, 64–72.

Blair, C. A., R. E. Blackith, and K. Boratyński (1964). Variation in *Coccus hesperidum* L. (Homoptera: Coccidae). *Proc. Roy. Entomol. Soc. Lond. A*, **39**, 129–134.

Blondeau, H. (1961). *Utilisation des ordinateurs électroniques pour l'étude de l'homogénéité de l'espèce* Streptococcus faecalis. *Application à la détermination de l'origine de la contamination des semi-conserves de viande*. Maurice Fabre, Lyon. 98 pp.

Bock, W. J. (1963). The role of preadaption and adaption in the evolution of higher levels of organization. *Proc. XVI Int. Congr. Zool., Wash., 1963*, **3**, 297–300.

Boeke, J. E. (1942). On quantitative statistical methods in taxonomy; subdivision of a polymorphous species: *Planchonella sandwicensis* (Gray) Pierre. *Blumea*, **5**, 47–65.

Bogdanescu, V., and R. Racotta (1967). Identification of mycobacteria by overall similarity analysis. *J. Gen. Microbiol.*, **48**, 111–126.

Bojalil, L. F., and J. Cerbón (1961). Taxonomic analysis of nonpigmented, rapidly growing mycobacteria. *J. Bacteriol.*, **8**, 338–345.

Bojalil, L. F., J. Cerbón, and A. Trujillo (1962). Adansonian classification of mycobacteria. *J. Gen. Microbiol.*, **28**, 333–346.

Bolton, E. T., C. A. Leone, and A. A. Boyden (1948). A critical analysis of the performance of the photronreflectometer in the measurement of serological and other turbid systems. *J. Immunol.*, **58**, 169–181.

Bonde, G. J. (1965). Classification of *Bacillus* spp. from marine sediments. *J. Gen. Microbiol.*, **41**, xxii.

Bonham-Carter, G. F. (1965). A numerical method of classification using qualitative and semi-quantitative data, as applied to the facies analysis of limestones. *Bull. Can. Petroleum Geol.*, **13**, 482–502.

Bonham-Carter, G. F. (1967a). An example of the analysis of semi-quantitative petrographic data. *Proc. VIIth World Petroleum Congr.*, **2**, 567–583.

Bonham-Carter, G. F. (1967b). FORTRAN IV program for Q-mode cluster analysis of non-quantitative data using IBM 7090/7094 computers. *Kansas Geol. Surv. Computer Contrib.*, No. 17. 28 pp.

Bonner, R. E. (1962). A "logical pattern" recognition program. *IBM J. Res. Develop.*, **6**, 353–360.

Bonner, R. E. (1964). On some clustering techniques. *IBM J. Res. Develop.*, **8**, 22–32.

Boratyński, K. (1971). Advances in our knowledge of Coccoidea (Homoptera) with reference to studies of males and the application of some numerical methods of classification. *Polish Congr. Contemporary Culture in Exile, London*, **1**, 585–595.

Boratyński, K., and R. G. Davies (1971). The taxonomic value of male Coccoidea (Homoptera) with an evaluation of some numerical techniques. *Biol. J. Linn. Soc.*, **3**, 57–102.

Borgelt, J. P. (1968). The subspecific differentiation of the salp *Thalia democratica* (Forskål, 1775), based on numerical taxonomical studies. *Trans. Roy. Soc. S. Afr.*, **38**, 45–64.

Boughey, A. S., K. W. Bridges, and A. G. Ikeda (1968). An automated biological identification key. *Mus. Syst. Biol. University of California, Irvine, Calif., Res. Ser.* No. 2, 36 pp.

Boulter, D., E. W. Thompson, J. A. M. Ramshaw, and M. Richardson (1970). Higher plant cytochrome *c*. *Nature*, **228**, 552–554.

Boulton, D. M., and C. S. Wallace (1970). A program for numerical classification. *Computer J.*, **13**, 63–69.

Boyce, A. J. (1964). The value of some methods of numerical taxonomy with reference to hominoid classification. In. V. H. Heywood and J. McNeill (eds.), *Phenetic and Phylogenetic Classification.*, pp. 47–65. Syst. Ass. Pub. 6. 164 pp.

Boyce, A. J. (1965). The methods of quantitative taxonomy with special reference to functional analysis. D.Phil. thesis, Oxford University. 190 pp.

Boyce, A. J. (1969). Mapping diversity: a comparative study of some numerical methods. In A. J. Cole (ed.), *Numerical Taxonomy*. Proceedings of the Colloquium in Numerical Taxonomy Held in the University of St. Andrews, September 1968, pp. 1–31. Academic Press, London. 324 pp.

Boyden, A. (1942). Systematic serology: a critical appreciation. *Physiol. Zool.*, **15**, 109–145.

Boyden, A. (1958). Comparative serology: aims, methods, and results. In W. Cole (ed.), *Serological and Biochemical Comparison of Proteins*. XIV Annual Protein Conference, pp. 3–24. Rutgers University Press, New Brunswick, N.J.

Boyden, A., R. J. DeFalco, and D. Gemeroy (1951). Parallelism in serological correspondence. *Bull. Serol. Mus.*, **6**, 6–7.

Bradley, J. C. (1939). The philosophy of biological nomenclature. *Verhandl. VII. Int. Kongr. Entomol.*, **1**, 531–534.

Brainerd, G. W. (1951). The place of chronological ordering in archaeological analysis. *Amer. Antiquity*, **16**, 301–313.

Brew, K., T. C. Vanaman, and R. L. Hill (1967). Comparison of the amino acid sequence of bovine α-lactalbumin and hens egg white lysozyme. *J. Biol. Chem.*, **242**, 3747–3749.

Brisbane, P. G., and A. D. Rovira (1961). A comparison of methods for classifying rhizosphere bacteria. *J. Gen. Microbiol.*, **26**, 379–392.

Bromley, D. B. (1966). Rank order cluster analysis. *Brit. J. Math. Statist. Psychol.*, **19**, 105–123.

Brown, W. L., Jr. (1965). Numerical taxonomy, convergence and evolutionary reduction. *Systematic Zool.*, **14**, 101–109.

Buck, R. C., and D. L. Hull (1966). The logical structure of the Linnaean hierarchy. *Systematic Zool.*, **15**, 97–111.

Buechley, R. W. (1967). Characteristic name sets of Spanish populations. *Names*, **15**, 53–69.

Buettner-Janusch, J., and R. H. Hill (1965). Molecules and monkeys. *Science*, **147**, 836–842.

Burnaby, T. P. (1966). Growth-invariant discriminant functions and generalized distances. *Biometrics*, **22**, 96–110.

Burnaby, T. P. (1970). On a method for character weighting a similarity coefficient, employing the concept of information. *J. Int. Ass. Math. Geol.*, **2**, 25–38.

Burr, E. J. (1968). Cluster sorting with mixed character types. I. Standardization of character values. *Aust. Computer J.*, **1**, 97–99.

Burt, C. (1937). Correlations between persons. *Brit. J. Psychol.*, **28**, 59–96.

Burtt, B. L. (1964). Angiosperm taxonomy in practice. In V. H. Heywood and J. McNeill (eds.), *Phenetic and Phylogenetic Classification*, pp. 5–16. Syst. Ass. Pub. 6. 164 pp.

Burtt, B. L. (1966). Adanson and modern taxonomy. *Notes Roy. Bot. Gard. Edinburgh*, **26**, 427–431.

Burtt, B. L., I. C. Hedge, and P. F. Stevens (1970). A taxonomic critique of recent numerical studies in Ericales and *Salvia*. *Notes Roy. Bot. Gard. Edinburgh*, **30**, 141–158.

Busacker, R. G., and T. L. Saaty (1965). *Finite Graphs and Networks*. McGraw-Hill, New York. 294 pp.

Butler, G. A. (1969). A vector field approach to cluster analysis. *Pattern Recognition*, **1**, 291–299.

Butler, J. E., and C. A. Leone (1967). Immunotaxonomic investigations of the Coleoptera. *Systematic Zool.*, **16**, 56–63.

Buzas, M. A. (1970). On the quantification of biofacies. *In* E. L. Yochelson (ed.), *Proceedings of the North American Paleontological Convention, Chicago, 1967*, vol. 1, part B, pp. 101–116. Allen Press, Lawrence, Kan. 2 vols.

Buzzati-Traverso, A. A. (1959). Quantitative traits, and polygenic systems in evolution. *Cold Spring Harbor Symp. Quant. Biol.*, **24**, 41–46.

Caceres, C. A. (1963). Electrocardiographic analysis by a computer system. *Arch. Internal Med.*, **111**, 196–202.

Cain, A. J. (1958). Logic and memory in Linnaeus's system of taxonomy. *Proc. Linn. Soc. Lond.*, 169th session, pp. 144–163.

Cain, A. J. (1959a). Deductive and inductive methods in post-Linnaean taxonomy. *Proc. Linn. Soc. Lond.*, 170th session, pp. 185–217.

Cain, A. J. (1959b). The post-Linnaean development of taxonomy. *Proc. Linn. Soc. Lond.*, 170th session, pp. 234–244.

Cain, A. J. (1959c). Taxonomic concepts. *Ibis*, **101**, 302–318.

Cain, A. J. (1962). The evolution of taxonomic principles. In G. C. Ainsworth and P. H. A. Sneath (eds.), *Microbial Classification*, 12th Symposium of the Society for General Microbiology, pp. 1–13, Cambridge University Press, Cambridge. 483 pp.

Cain, A. J., and G. A. Harrison (1958). An analysis of the taxonomist's judgement of affinity. *Proc. Zool. Soc. Lond.*, **131**, 85–98.

Cain, A. J., and G. A. Harrison (1960a). The phenetic affinity of some hominoid skulls. *Proc. Soc. Study Human Biol.*, 27 May, 1960.

Cain, A. J., and G. A. Harrison (1960b). Phyletic weighting. *Proc. Zool. Soc. Lond.*, **135**, 1–31.

Camin, J. H., and R. R. Sokal (1965). A method for deducing branching sequences in phylogeny. *Evolution*, **19**, 311–326.

Camp, W. H., and C. L. Gilly (1943). The structure and origin of species. *Brittonia*, **4**, 323–385.

Campbell, B. (1964). Quantitative taxonomy and human evolution. In S.L. Washburn (ed.), *Classification and Human Evolution*, pp. 50–74. Methuren, London. 371 pp.

Campbell, I. (1969). Serological classification of *Saccharomyces* spp. *Antonie van Leeuwenhoek*, **35** (Suppl.), A19–A20.

Candolle, A. P. de (1813). *Théorie élémentaire de la botanique, ou exposition des principes de la classification naturelle et de l'art de décrire et d'étudier les végétaux.* Déterville, Paris. 528 pp.

Cantor, C. R. (1968). The occurrence of gaps in protein sequences. *Biochem. Biophys. Res. Commun.*, **31**, 410–416.

Capon, I. N., and L. B. Jellett (1968). Studies concerning a computer analysis of patients with primary, secondary or stress polycythaemia. *Australasian Ann. Med.*, **17**, 110–117.

Card, W. (1967). Towards a calculus of medicine. *Med. Ann.*, *1967*, pp. 9–21.

Carlsson, J. (1968). A numerical taxonomic study of human oral streptococci. *Odontologisk Revy*, **19**, 137–160.

Carmichael, J. W., J. A. George, and R. S. Julius (1968). Finding natural clusters. *Systematic Zool.*, **17**, 144–150.

Carmichael, J. W., R. S. Julius, and P. M. D. Martin (1965). Relative similarities in one dimension. *Nature*, **208**, 544–547.

Carmichael, J. W., and P. H. A. Sneath (1969). Taxometric maps. *Systematic Zool.*, **18**, 402–415.

Carroll, J. B., and I. Dyen (1962). High-speed computation of lexicostatistical indices. *Language*, **38**, 274–278.

Carvell, H. T., and J. Svartvik (1968). *Computational Experiments in Grammatical Classification.* Mouton, The Hague. 271 pp.

Casas, E., W. D. Hanson, and E. J. Wellhausen (1968). Genetic relationships among collections representing three Mexican race composites of *Zea mays* L. *Genetics*, **59**, 299–310.

Casey, R. G., and G. Nagy (1971). Advances in pattern recognition. *Sci. Amer.*, **224**(4), 56–71.

Cattell, R. B. (1949). r_p and other coefficients of pattern similarity. *Psychometrika*, **14**, 279–298.

Cattell, R. B. (1952). *Factor Analysis.* Harper, New York. 462 pp.

Cattell, R. B. (ed.) (1966a). *Handbook of Multivariate Experimental Psychology.* Rand McNally, Chicago. 959 pp.

Cattell, R. B. (1966b). The data box: its ordering of total resources in terms of possible relational systems. In R. B. Cattell (ed.), *Handbook of Multivariate Experimental Psychology*, pp. 67–128. Rand McNally, Chicago. 959 pp.

Cattell, R. B. (1966c). The meaning and strategic use of factor analysis. In R. B. Cattell (ed.), *Handbook of Multivariate Experimental Psychology*, pp. 174–243. Rand McNally, Chicago. 959 pp.

Cattell, R. B., and M. A. Coulter (1966). Principles of behavioural taxonomy and the mathematical basis of the taxonome computer program. *Brit. J. Math. Statist. Psychol.*, **19**, 237–269.

Cattell, R. B., M. A. Coulter, and B. Tsujioka (1966). The taxonometric recognition of types and functional emergents. In R. B. Cattell (ed.), *Handbook of Multivariate Experimental Psychology*, pp. 288–329. Rand McNally, Chicago. 959 pp.

Cavalli-Sforza, L. L. (1949). Sulla correlazione media fra più caratteri in relazione alla biometria. *Metron*, **15**, 1–16.

Cavalli-Sforza, L. L. (1966). Population structure and human evolution. *Proc. Roy. Soc. B.*, **164**, 362–379.

Cavalli-Sforza, L. L., I. Barrai, and A. W. F. Edwards (1964). Analysis of human evolution under random genetic drift. *Cold Spring Harbor Symp. Quant. Biol.*, **29**, 9–20.

Cavalli-Sforza, L. L., and A. W. F. Edwards (1964). Analysis of human evolution. In *Genetics Today*. Proc. XI Int. Congr. Genet., pp. 923–933.

Cavalli-Sforza, L. L., and A. W. F. Edwards (1967). Phylogenetic analysis: models and estimation procedures. *Evolution*, **21**, 550–570.

Cerbón, J., and L. F. Bojalil (1961). Physiological relationships of rapidly growing mycobacteria. Adansonian classification. *J. Gen. Microbiol.*, **25**, 7–15.

Ceška, A. (1968). Application of association coefficients for estimating the mean similarity between sets of vegetational relevés. *Folia Geobot. Phytotaxonom.*, **3**, 57–64.

Chaddha, R. L., and L. F. Marcus (1968). An empirical comparison of distance statistics for populations with unequal covariance matrices. *Biometrics*, **24**, 683–694.

Challinor, J. (1959). Palaeontology and evolution. In P. R. Bell (ed.), *Darwin's Biological Work*, pp. 50–100. Cambridge University Press, Cambridge. 343 pp.

Chatelain, R., and L. Second (1966). Taxonomie numérique de quelques *Brevibacterium*. *Ann. Inst. Pasteur*, **111**, 630–644.

Chave, K. E., and F. T. Mackenzie (1961). A statistical technique applied to the geochemistry of pelagic muds. *J. Geol.*, **69**, 572–582.

Cheeseman, G. C., and N. J. Berridge (1959). The differentiation of bacterial species by paper chromatography. VII. The use of electronic computation for the objective assessment of chromatographic results. *J. Appl. Bacteriol.*, **22**, 307–316.

Cheetham, A. H. (1968). Morphology and systematics of the bryozoan genus *Metrarabdotos*. *Smithsonian Misc. Coll.*, **153**(1). 121 pp.

Cheetham, A. H., and J. E. Hazel (1969). Binary (presence-absence) similarity coefficients. *J. Paleontol.*, **43**, 1130–1136.

Chillcot, J. G. (1960). A revision of the nearctic species of Fanniinae (Diptera: Muscidae). *Can. Entomol.*, **92** (suppl. 14), 1–295.

Chui, Wun Duen V. (1969). A study of the distribution and relationships of endemic complexes of Psocoptera in the Hawaiian Islands. Ph.D. thesis, University of Hong Kong.

Cipra, J. E., O. W. Bidwell, and F. J. Rohlf (1970). Numerical taxonomy of soils from nine orders by cluster and centroid-component analyses. *Soil. Sci. Soc. Amer. Proc.*, **34**, 281–287.

Clark, P. J. (1952). An extension of the coefficient of divergence for use with multiple characters. *Copeia*, **2**, 61–64.

Clarke, B. (1970). Selective constraints on amino-acid substitutions during the evolution of proteins. *Nature*, **228**, 159–160.

Clarke, D. L. (1962). Matrix analysis and archaeology with particular reference to British beaker pottery. *Proc. Prehist. Soc. Lond.*, **28**, 371–382.

Clarke, D. L. (1968). *Analytical archaeology*. Methuen, London. 684 pp.

Claude, C. (1970). Biometrie und Fortpflanzungsbiologie der Rötelmaus *Clethrionomys glareolus* (Schreber, 1780) auf verschiedenen Höhenstufen der Schweiz. *Rev. Suisse Zool.*, **77**, 435–480.

Clausen, A. R. (1967). The measurement of legislative group behavior. *Midwest J. Political Sci.*, **11**, 212–224.

Clausen, J., and W. M. Heisey (1958). Experimental studies on the nature of species. IV. Genetic structure of ecological races. *Carnegie Inst. Wash. Pub.*, No. 615. 312 pp.

Clausen, J., D. D. Keck, and W. M. Heisey (1940). Experimental studies on the nature of species. I. The effect of varied environments on Western North American plants. *Carnegie Inst. Wash. Pub.*, No. 520. 452 pp.

Clay, T. (1949). Some problems in the evolution of a group of ectoparasites. *Evolution*, **3**, 279–299.

Clayton, W. D. (1970). Studies in the Gramineae: XXI. *Coelorhachis* and *Rhytachne*: a study in numerical taxonomy. *Kew Bull.*, **24**, 309–314.

Clements, F. E. (1954). Use of cluster analysis with anthropological data. *Amer. Anthropol.*, **56**, 180–199.

Clifford, H. T. (1964). The systematic position of the grass genus *Micraira* F. Muell. *Univ. Queensland Pap. Bot.*, **4**, 87–94.

Clifford, H. T. (1965). The classification of the Poaceae: a statistical study. *Univ. Queensland Pap. Bot.*, **4**, 243–253.

Clifford, H. T. (1969). Attribute correlations in the Poaceae (grasses). *Bot. J. Linn. Soc.*, **62**, 59–67.

Clifford, H. T., and D. W. Goodall (1967). A numerical contribution to the classification of the Poaceae. *Aust. J. Bot.*, **15**, 499–519.

Clifford, H. T., W. T. Williams, and G. N. Lance (1969). A further numerical contribution to the classification of the Poaceae. *Aust. J. Bot.*, **17**, 119–131.

Colbert, E. H. (1963). Phylogeny and the dimension of time. *Amer. Natur.*, **97**, 319–331.

Cole, A. J. (ed.) (1969). *Numerical taxonomy*. Proceedings of the Colloquium in Numerical Taxonomy Held in the University of St. Andrews, September 1968. Academic Press, London, 324 pp.

Cole, A. J., and D. Wishart (1970). An improved algorithm for the Jardine-Sibson method of generating overlapping clusters. *Computer J.*, **13**, 156–163.

Cole, J. P., and C. A. M. King (1968). *Quantitative Geography: Techniques and Theories in Geography*. Wiley, London. 692 pp.

Cole, L. C. (1949). The measurement of interspecific association. *Ecology*, **30**, 411–424.

Cole, L. C. (1957). The measurement of partial interspecific association. *Ecology*, **38**, 226–233.

Colless, D. H. (1966). A note on Wilson's consistency test for phylogenetic hypotheses. *Systematic Zool.*, **15**, 358–359.

Colless, D. H. (1967a). An examination of certain concepts in phenetic taxonomy. *Systematic Zool.*, **16**, 6–27.

Colless, D. H. (1967b). The phylogenetic fallacy. *Systematic Zool.*, **16**, 289–295.

Colless, D. H. (1969a). The phylogenetic fallacy revisited. *Systematic Zool.*, **18**, 115–126.

Colless, D. H. (1969b). The interpretation of Hennig's "Phylogenetic Systematics"—a reply to Dr. Schlee. *Systematic Zool.*, **18**, 134–144.

Colless, D. H. (1969c). A note on the equivalence of characters and numbers of characters needed. *Systematic Zool.*, **18**, 455–456.

Colless, D. H. (1969d). The relationship of evolutionary theory to phenetic taxonomy. *Evolution*, **23**, 721–722.

Colless, D. H. (1970). The phenogram as an estimate of phylogeny. *Systematic Zool.*, **19**, 352–362.

Colman, G. (1968). The application of computers to the classification of streptococci. *J. Gen. Microbiol.*, **50**, 149–158.

Colobert, L., and H. Blondeau (1962). L'espèce *Streptococcus faecalis*. I. Etude de l'homogénéité par la méthode adansonnienne. *Ann. Inst. Pasteur*, **103**, 345–362.

Colwell, R. R. (1964). A study of features used in the diagnosis of *Pseudomonas aeruginosa*. *J. Gen. Microbiol.*, **37**, 181–194.

Colwell, R. R. (1969). Numerical taxonomy of the flexibacteria. *J. Gen. Microbiol.*, **58**, 207–215.

Colwell, R. R. (1970a). Polyphasic taxonomy of bacteria. In H. Iizuka and T. Hasegawa (eds.), *Culture Collections of Microorganisms*, Proceedings of the International Conference on Culture Collections, Tokyo, Oct. 7–11, 1968, pp. 421–436. University Park Press, Baltimore, Md. 625 pp.

Colwell, R. R. (1970b). Polyphasic taxonomy of the genus *Vibrio*: numerical taxonomy of *Vibrio cholerae*, *Vibrio parahaemolyticus*, and related *Vibrio* species. *J. Bacteriol.*, **104**, 410–433.

Colwell, R. R. (1971). The value of numerical taxonomy in bacterial systematics. In A. Peréz-Miravete and D. Peláez (eds.), *Recent Advances in Microbiology*, pp. 587–593. Asociacion Mexicana de Microbiologia, Mexico City. 623 pp.

Colwell, R. R., and G. B. Chapman (1966). Adansonian analysis and fine structure studies of marine psychrophiles. *Bacteriol. Proc.*, **1966**, 20.

Colwell, R. R., R. V. Citarella, and I. Ryman (1965). Deoxyribonucleic acid base composition and Adansonian analysis of heterotrophic, aerobic pseudomonads. *J. Bacteriol.*, **90**, 1148–1149.

Colwell, R. R., and M. B. Gochnauer (1963). The taxonomy of marine bacteria. *Bacteriol. Proc.*, **1963**, 40.

Colwell, R. R., and J. Liston (1961a). An electronic computer analysis of some *Xanthomonas* and *Pseudomonas* species. *Bacteriol. Proc.*, **1961**, 72.

Colwell, R. R., and J. Liston (1961b). Taxonomic relationships among the pseudomonads. *J. Bacteriol.*, **82**, 1–14.

Colwell, R. R., and J. Liston (1961c). Taxonomy of *Xanthomonas* and *Pseudomonas*. *Nature*, **191**, 617–619.

Colwell, R. R., and J. Liston (1961d). Taxonomic analysis with the electronic computer of some *Xanthomonas* and *Pseudomonas* species. *J. Bacteriol.*, **82**, 913–919.

Colwell, R. R., and M. Mandel (1964). Adansonian analysis and deoxyribonucleic acid base composition of some gram-negative bacteria. *J. Bacteriol.*, **87**, 1412–1422.

Colwell, R. R., and M. Mandel (1965). Adansonian analysis and deoxyribonucleic acid base composition of *Serratia marcescens*. *J. Bacteriol.*, **89**, 454–461.

Colwell, R. R., M. Mandel, and M. B. Gochnauer (1964). *Serratia marcescens*: correlation of numerical taxonomy studies with deoxyribonucleic acid composition. *Bacteriol. Proc.*, **1964**, 21.

Colwell, R. R., M. L. Moffett, and M. D. Sutton (1968). Computer analysis of relationships among phytopathogenic bacteria. *Phytopathology*, **58**, 1207–1215.

Colwell, R. R., R. Y. Morita, and M. B. Gochnauer (1964). Taxonomy of marine vibrios. *Bacteriol. Proc.*, **1964**, 37.

Colwell, R. R., and M. Yuter (1965). Adansonian analysis and deoxyribonucleic acid base composition studies of *Vibrio cholerae* and El Tor vibrios. *Bacteriol. Proc.*, **1965**, 18.

Constance, L. (1964). Systematic botany—an unending synthesis. *Taxon*, **13**, 257–273.

Cooley, W. W., and P. R. Lohnes (1971). *Multivariate Data Analysis*. Wiley, New York. 364 pp.

Core, E. L. (1955). *Plant Taxonomy*. Prentice-Hall, Englewood Cliffs, N. J. 459 pp.

Corlett, D. A., Jr., J. S. Lee, and R. O. Sinnhuber (1965). Application of replica plating and computer analysis for rapid identification of bacteria in some foods. I. Identification scheme. *Appl. Microbiol.*, **13**, 808–817.

Cormack, R. M. (1971). A review of classification. *J. Roy. Statist. Soc. A*, **134**, 321–367.

Cousin, G. (1956a). Intérêt de la zoologie quantitative pour la systématique. *Bull. Soc. Zool. Fran.*, **81**, 2–8.

Cousin, G. (1956b). Biometrie et definitions de morphologie quantitative des espèces et de leurs hybrides. *Bull. Soc. Zool. Fran.*, **81**, 247–289.

Cowan, S. T. (1965). Development of coding schemes for microbial taxonomy. *Advanc. Appl. Microbiol.*, **7**, 139–167.

Cowan, S. T. (1970). Heretical taxonomy for bacteriologists. *J. Gen. Microbiol.*, **61**, 145–154.

Cowan, S. T., and K. J. Steel (1960). A device for the identification of microorganisms. *Lancet*, **1960**(i), 1172–1173.

Cowan, S. T., and K. J. Steel (1965). *Manual for the Identification of Medical Bacteria*. Cambridge University Press, Cambridge. 217 pp.

Cowgill, G. L. (1967). Computer applications in archaeology. *Computers and the Humanities*, **2**, 17–23.

Cowgill, G. L. (1968). Archaeological applications of factor, cluster, and proximity analysis. *Amer. Antiquity*, **33**, 367–375.

Cowie, D. B. (1967). Genetic relationships among viruses and bacteria. *Science*, **156**, 537.

Cox, D. R., and L. Brandwood (1959). On a discriminatory problem connected with the works of Plato. *J. Roy. Statist. Soc. B.*, **21**, 195–200.

Cracraft, J. (1967). Comments on homology and analogy. *Systematic Zool.*, **16**, 355–359.

Crain, I. K. (1970). Computer interpolation and contouring of two-dimensional data: a review. *Geoexploration*, **8**, 71–86.

Crawford, R. M. M. (1969). The use of graphical methods in classification. In A. J. Cole (ed.), *Numerical Taxonomy*. Proceedings of the Colloquium in Numerical Taxonomy Held in the University of St. Andrews, September 1968, pp. 32–41. Academic Press, London. 324 pp.

Crawford, R. M. M., and D. Wishart (1967). A rapid multivariate method for the detection and classification of groups of ecologically related species. *J. Ecol.*, **55**, 505–524.

Crawford, R. M. M., and D. Wishart (1968). A rapid classification and ordination method and its application to vegetation mapping. *J. Ecol.*, **56**, 385–404.

Crawford, R. M. M., D. Wishart, and R. M. Campbell (1970). A numerical analysis of high altitude scrub vegetation in relation to soil erosion in the Eastern Cordillera of Peru. *J. Ecol.*, **58**, 173–191.

Craytor, W. B., and L. Johnson, Jr. (1968). Refinements in computerized item seriation. *Bull. Mus. Natur. Hist. Univ. Oregon*, No. 10, 22 pp.

Crick, F. H. C. (1958). On protein synthesis. In *The Biological Replication of Macromolecules.* XIIth Symposium of the Society for Experimental Biology, pp. 138–163. Cambridge University Press, Cambridge. 255 pp.

Crovello, T. J. (1966). Quantitative taxonomic studies in the genus *Salix*. Ph.D. thesis, University of California, Berkeley. 251 pp.

Crovello, T. J. (1967). Problems in the use of electronic data processing in biological collections. *Taxon*, **16**, 481–494.

Crovello, T. J. (1968a). A numerical taxonomic study of the genus *Salix*, section *Sitchenses*. *Univ. California Pub. Bot.*, **44**, 1–61.

Crovello, T. J. (1968b). Key communality cluster analysis as a taxonomic tool. *Taxon*, **17**, 241–258.

Crovello, T. J. (1968c). Different concepts of relevance in a numerical taxonomic study. *Nature*, **218**, 492.

Crovello, T. J. (1968d). The value of numerical taxonomy for *Arabidopsis* research. *Arabidopsis Information Service Bull.*, **5**, 7–9.

Crovello, T. J. (1968e). The effect of alteration of technique at two stages in a numerical taxonomic study. *Univ. Kansas Sci. Bull.*, **47**, 761–786.

Crovello, T. J. (1968f). The effect of change of number of OTU's in a numerical taxonomic study. *Brittonia*, **20**, 346–367.

Crovello, T. J. (1968g). The effect of missing data and of two sources of character values on a phenetic study of the willows of California. *Madroño*, **19**, 301–315.

Crovello, T. J. (1968h). Effect of change of characters in numerical taxonomy. *Amer. J. Bot.*, **55**, 733.

Crovello, T. J. (1968i). Numerical taxonomy: its value to biosystematics. *Amer. J. Bot.*, **55**, 735.

Crovello, T. J. (1969). Effects of change of characters and of number of characters in numerical taxonomy. *Amer. Midland Natur.*, **81**, 68–86.

Crovello, T. J. (1971). The Fourth Annual Numerical Taxonomy Conference. *Systematic Zool.*, **20**, 233–238.

Crovello, T. J., and R. D. MacDonald (1970). Index of EDP-IR projects in systematics. *Taxon*, **19**, 63–76.

Crowson, R. A. (1958). Darwin and classification. In S. A. Barnett (ed.), *A Century of Darwin*, pp. 102–129. Heinemann, London. 376 pp.

Crowson, R. A. (1970). *Classification and Biology.* Atherton, New York. 350 pp.

Cuanalo de la C., H. E., and R. Webster (1970). A comparative study of numerical classification and ordination of soil profiles in a locality near Oxford. Part I. Analysis of 85 sites. *J. Soil Sci.*, **21**, 340–352.

Cullimore, D. R. (1969). The Adansonian classification using the heterotrophic spectra of *Chlorella vulgaris* by a simplified procedure involving a desk-top computer. *J. Appl. Bacteriol.*, **32**, 439–447.

Cunha, R. A. da (1969). Contribuição ao estudo da taxonomia dos Meliponinae (Hymenoptera—Apidae). D.Sc. thesis, University of Rio Claro, Brazil. 83 pp.

Curtis, J. T. (1959). *The Vegetation of Wisconsin: An Ordination of Plant Communities.* University of Wisconsin Press, Madison, Wis. 657 pp.

Cutbill, J. L. (ed.) (1971). *Data Processing in Biology and Geology.* Proceedings of a Symposium Held at the Department of Geology, University of Cambridge, 24–26 September, 1969. Academic Press, London. 346 pp.

Czekanowski, J. (1909). Zur Differentialdiagnose der Neandertalgruppe. *Korrespondenzblatt Deutsch. Ges. Anthropol. Ethnol. Urgesch.*, **40**, 44–47.

Czekanowski, J. (1932). "Coefficient of racial likeness" und "durchschnittliche Differenz." *Anthrop. Anz.*, **9**, 227–249.

Daget, J. (1966). Taxonomie numérique des Citharininae (Poissons, Characiformes). *Bull. Mus. Nat. Hist. Natur.*, **38**, 376–386.

Daget, J., and J. C. Hureau (1968). Utilisation des statistiques d'ordre en taxonomie numérique. *Bull. Mus. Nat. His. Natur.*, **40**, 465–473.

Dagnelie, P. (1960). Contribution à l'étude des communautés végétales par l'analyse factorielle. *Bull. Serv. Carte Phytogéogr., B*, **5**, 7–71, 93–195.

Danser, B. H. (1950). A theory of systematics. *Bibl. Biotheoret.*, **4**, 113–180.

Dass, H., and N. Nybom (1967). The relationships between *Brassica nigra, B. campestris, B. oleracea* and their amphidiploid hybrids studied by means of numerical chemotaxonomy. *Can. J. Genet. Cytol.*, **9**, 880–890.

Davidson, R. A. (1963). Initial biometric survey of morphological variation in the *Cirsium altissum-C. discolor* complex. *Brittonia*, **15**, 222–241.

Davidson, R. A., and R. A. Dunn (1967). A correlation approach to certain problems of population-environment relations. *Amer. J. Bot.*, **54**, 529–538.

Davidson, R. A., and R. A. Dunn (1968). Computer simulation of certain forms of evolutionary change: a preliminary report. *Taxon*, **17**, 1–10.

Davis, G. H. G. (1964). Notes on the phylogenetic background to lactobacillus taxonomy. *J. Gen. Microbiol.*, **34**, 177–184.

Davis, G. H. G., L. Fomin, E. Wilson, and K. G. Newton (1969). Numerical taxonomy of *Listeria*, streptococci and possibly related bacteria. *J. Gen. Microbiol.*, **57**, 333–348.

Davis, G. H. G., and K. G. Newton (1969). Numerical taxonomy of some named coryneform bacteria. *J. Gen. Microbiol.*, **56**, 195–214.

Davis, P. H., and V. H. Heywood (1963). *Principles of Angiosperm Taxonomy*. Oliver & Boyd, Edinburgh. 556 pp.

Dayhoff, M. O. (ed.) (1969a). *Atlas of Protein Sequence and Structure 1969*. National Biomedical Research Foundation, Silver Spring, Md. 315 pp.

Dayhoff, M. O. (1969b). Computer analysis of protein evolution. *Sci. Amer.*, **221**(1), 86–94.

Dayhoff, M. O., and R. V. Eck (1968). *Atlas of Protein Sequence and Structure 1967–68*. National Biomedical Research Foundation, Silver Spring, Md. 356 pp.

Dayhoff, M. O., and R. V. Eck (1969). Evolution of the globins. In M. O. Dayhoff (ed.), *Atlas of Protein Sequence and Structure 1969*, pp. 17–24. National Biomedical Research Foundation, Silver Spring, Md. 315 pp.

Dayhoff, M. O., R. V. Eck, and C. M. Park (1969). A model of evolutionary change in proteins. In M. O. Dayhoff (ed.), *Atlas of Protein Sequence and Structure 1969*, pp. 75–83. National Biomedical Research Foundation, Silver Spring, Md. 315 pp.

Dayhoff, M. O., and P. J. McLaughlin (1969) Evolution of other protein families. In M. O. Dayhoff (ed.), *Atlas of Protein Sequence and Structure 1969*, pp. 33–38. National Biomedical Research Foundation, Silver Spring, Md. 315 pp.

Dayhoff, M. O., M. R. Sochard, and P. J. McLaughlin (1969). The immunoglobulins: relationships and evolution. In M. O. Dayhoff (ed.), *Atlas of Protein Sequence and Structure 1969*, pp. 25–32. National Biomedical Research Foundation, Silver Spring, Md. 315 pp.

De Beer, G. R. (1951). *Embryos and Ancestors*, 2nd ed. Clarendon Press, Oxford. 159 pp.

De Beer, G. R. (1954). *Archaeopteryx* and evolution. *Advance. Sci.*, **11**, 160–170.

Dedio, W., P. J. Kaltsikes, and E. N. Larter (1969a). Numerical chemotaxonomy in the genus *Secale. Can. J. Bot.*, **47**, 1175–1180.

Dedio, W., P. J. Kaltsikes, and E. N. Larter (1969b). A thin-layer chromatographic study of the phenolics of *Triticale* and its parental species. *Can. J. Bot.*, **47**, 1589–1593.

Deegener, P. (1918). *Die Formen der Vergesellschaftung im Tierreiche. Ein systematisch-soziologischer Versuch*. Veit, Leipzig. 420 pp.

Defayolle, M., and L. Colobert (1962). L'espèce *Streptococcus faecalis*. II. Étude de le'homogénéité par l'analyse factorielle. *Ann. Inst. Pasteur*, **103**, 505–522.

Defayolle, M., L. Colobert, P. Poncet, J. Buissière, and J. Pontier (1968). Application de l'analyse factorielle à la taxonomie des microorganismes. *Biometrie-Praximetrie*, **9**, 14–51.

Delany, M. J. (1965). The application of factor analysis to the study of variation in the long-tailed field-mouse *(Apodemus sylvaticus* (L.)) in north-west Scotland. *Proc. Linn. Soc. Lond.*, **176**, 103–111.

Demirmen, F. (1969). Multivariate procedures and FORTRAN IV program for evaluation and improvement of classifications. *Kansas Geol. Surv. Computer Contrib.*, No. 31. 51 pp.

Dessauer, H. C. (1969). Molecular data in animal systematics. In C. G. Sibley (Chairman), *Systematic Biology*. Proceedings of an International Conference Conducted at the University of Michigan, Ann Arbor, Michigan, June 14–16, 1967, pp. 325–357. Nat. Acad. Sci., Wash., Pub. 1692. 632 pp.

Deutsch, K. W. (1966). On theories, taxonomies, and models as communication codes for organising information. *Behav. Sci.*, **11**, 1–17.

De Wet, J. M. J., and J. P. Huckabay (1967). The origin of *Sorghum bicolor*. II. Distribution and domestication. *Evolution*, **21**, 787–802.

Dice, L. R. (1945). Measures of the amount of ecologic association between species. *Ecology*, **26**, 297–302.

Dobzhansky, T. (1970). *The Genetics of the Evolutionary Process*. Columbia University Press, New York, 505 pp.

Doolittle, R. F., and B. Blombäck (1964). Amino-acid sequence investigations of fibrinopeptides from various mammals: evolutionary implications. *Nature*, **202**, 147–152.

Doran, J. E., and F. R. Hodson (1966). A digital computer analysis of palaeolithic flint assemblages. *Nature*, **210**, 688–689.

Doty, P., J. Marmur, J. Eigner, and C. Schildkraut (1960). Strand separation and specific recombination in deoxyribonucleic acids: physical chemical studies. *Proc. Nat. Acad. Sci. U.S.A.*, **46**, 461–476.

Dowdle, W. R., M. T. Coleman, E. C. Hall, and V. Knez (1969). Properties of the Hong Kong influenza virus. 2. Antigenic relationship of the Hong Kong virus haemagglutinin to that of other human influenza A viruses. *Bull. World Health Organ.*, **41**, 419–424.

Downe, A. E. R. (1963). Mosquitoes: comparative serology of four species of *Aedes (Ochlerotatus)*. *Science*, **139**, 1286–1287.

Downs, T. (1961). A study of variation and evolution in miocene *Merychippus*. *Contrib. Sci. Los Angeles County Mus.*, **45**, 1–75.

Driver, H. E. (1962). The contribution of A. L. Kroeber to culture area theory and practice. *Int. J. Amer. Linguistics*, **28** (Suppl., Memoir No. 18). 28 pp.

Driver, H. E. (1965). Survey of numerical classification in anthropology. In D. Hymes (ed.), *The Use of Computers in Anthropology*, pp. 301–344. Mouton, La Hague. 558 pp.

Driver, H. E., and K. F. Schuessler (1967). Correlational analysis of Murdock's 1957 ethnographic sample. *Amer. Anthropol.*, **69**, 332–352.

Drucker, D. B., and T. H. Melville (1969a). Computer classification of streptococci, mostly of oral origin. *Nature*, **221**, 664.

Drucker, D. B., and T. H. Melville (1969b). Streptococcal taxonomy. *J. Gen. Microbiol.*, **58**, x–xi.

Drury, D. G., and J. M. Randal (1969). A numerical study of the variation in the New Zealand *Erechtites arguta-scaberula* complex (Senecioneae-Compositae). *New Zealand J. Bot.*, **7**, 56–75.

Ducker, S. C., W. T. Williams, and G. N. Lance (1965). Numerical classification of the Pacific forms of *Chlorodesmis* (Chlorophyta). *Aust. J. Bot.*, **13**, 489–499.

Dunn, O. J., and P. D. Varady (1966). Probabilities of correct classification in discriminant analysis. *Biometrics*, **22**, 908–924.

DuPraw, E. J. (1964). Non-Linnean taxonomy. *Nature*, **202**, 849–852.

DuPraw, E. J. (1965a). The recognition and handling of honeybee specimens in non-Linnean taxonomy. *J. Apicult. Res.*, **4**, 71–84.

DuPraw, E. J. (1965b). Non-Linnean taxonomy and the systematics of honeybees. *Systematic Zool.*, **14**, 1–24.

Dybowski, W., and D. A. Franklin (1968). Conditional probability and the identification of bacteria: a pilot study. *J. Gen. Microbiol.*, **54**, 215–229.

Dybowski, W., D. A. Franklin, and L. C. Payne (1963). Computer for bacteriological diagnosis. *Lancet*, **1963**(ii), 866.

Dyen, I. (1962). The lexicostatistically determined relationship of a language group. *Int. J. Amer. Linguistics*, **28**, 153–161.

Dyen, I. (1965). A lexicostatistical classification of the Austronesian languages. *Int. J. Amer. Linguistics*, **31**(1, Suppl.), 1–64.

Dyen, I., A. T. James, and J. W. L. Cole (1967). Language divergence and estimated word retention rate. *Language*, **43**, 150–171.

Eades, D. C. (1965). The inappropriateness of the correlation coefficient as a measure of taxonomic resemblance. *Systematic Zool.*, **14**, 98–100.

Eades, D. C. (1970). Theoretical and procedural aspects of numerical phyletics. *Systematic Zool.*, **19**, 142–171.

Ebeling, A. W., R. M. Ibara, R. J. Lavenberg, and F. J. Rohlf (1970). Ecological groups of deep-sea animals off Southern California. *Bull. Los Angeles Mus. Natur. Hist. Sci.*, No. 6, 43 pp.

Eck, R. V., and M. O. Dayhoff (1966). *Atlas of Protein Sequence and Structure 1966*. National Biomedical Research Foundation, Silver Spring, Md. 215 pp.

Eddy, B. P., and K. P. Carpenter (1964). Further studies on *Aeromonas*. II. Taxonomy of *Aeromonas* and C27 strains. *J. Appl. Bacteriol.*, **27**, 96–109.

Edwards, A. W. F. (1970). Estimation of the branch points of a branching diffusion process. *J. Roy. Statist. Soc. B.*, **32**, 155–174.

Edwards, A. W. F., and L. L. Cavalli-Sforza (1964). Reconstruction of evolutionary trees. In V. H. Heywood and J. McNeill (eds.), *Phenetic and Phylogenetic Classification*, pp. 67–76. Syst. Ass. Pub. 6. 164 pp.

Edwards, A. W. F., and L. L. Cavalli-Sforza (1965). A method for cluster analysis. *Biometrics*, **21**, 362–375.

Edye, L. A., W. T. Williams, and A. J. Pritchard (1970. A numerical analysis of variation patterns in Australian introductions of *Glycine wightii (G. javanica)*. *Aust. J. Agr. Res.*, **21**, 57–69.

Ehrlich, P. R. (1961a). Systematics in 1970: some unpopular predictions. *Systematic Zool.*, **10**, 157–158.

Ehrlich, P. R. (1961b). Has the biological species concept outlived its usefulness? *Systematic Zool.*, **10**, 167–176.

Ehrlich, P. R. (1964). Some axioms of taxonomy. *Systematic Zool.*, **13**, 109–123.

Ehrlich, P. R., and A. H. Ehrlich (1967). The phenetic relationships of the butterflies. I. Adult taxonomy and the nonspecificity hypothesis. *Systematic Zool.*, **16**, 301–317.

Ehrlich, P. R., and R. W. Holm (1962). Patterns and populations. *Science*, **137**, 652–657.

Eickwort, G. C. (1969). A comparative morphological study and generic revision of the augochlorine bees (Hymenoptera: Halictidea). *Univ. Kansas Sci. Bull.*, **48**, 325–524.

Eldredge, N. (1968). Convergence between two Pennsylvanian gastropod species: a multivariate mathematical approach. *J. Paleontol.*, **42**, 186–196.

El-Gazzar, A., and L. Watson (1970a). A taxonomic study of Labiatae and related genera. *New Phytol.*, **69**, 451–486.

El-Gazzar, A., and L. Watson (1970b). Some economic implications of the taxonomy of Labiatae. Essential oils and rusts. *New Phytol.*, **69**, 487–492.

El-Gazzar, A., L. Watson, W. T. Williams, and G. N. Lance (1968). The taxonomy of *Salvia*: a test of two radically different numerical methods. *J. Linn. Soc. Lond. Bot.*, **60**, 237–250.

Ellegård, A. (1959). Statistical measurement of linguistic relationship. *Language*, **35**, 131–156.

Ellegård, A. (1962). *Who Was Junius?* Almqvist & Wiksell, Stockholm. 159 pp.

Ellison, W. L., R. E. Alston, and B. L. Turner (1962). Methods of presentation of crude biochemical data for systematic purposes, with particular reference to the genus *Bahia* (Compositae). *Amer. J. Bot.*, **49**, 599–604.

Elsasser, W. M. (1958). *The Physical Foundation of Biology*. Pergamon Press, London, New York, Paris, Los Angeles. 219 pp.

Emerson, A. E. (1945). Taxonomic categories and population genetics. *Entomol. News*, **56**, 14–19.

Engelman, L., and J. A. Hartigan (1969). Percentage points of a test for clusters. *J. Amer. Statist. Ass.*, **64**, 1647–1648.

Engle, R. L., Jr. (1963). Medical diagnosis: present, past, and future. III. Diagnosis in the future, including a critique on the use of electronic computers as diagnostic aids to the physician. *Arch. Internal Med.*, **112**, 530–543.

Engle, R. L., Jr., and B. J. Davis (1963). Medical diagnosis: present, past, and future. I. Present concepts of the meaning and limitations of medical diagnosis. *Arch. Internal Med.*, **112**, 512–519.

Ernst, W. R. (1967). Floral morphology and systematics of *Platystemon* and its allies *Hesperomecon* and *Meconella* (Papaveraceae: Platsytemonoideae). *Univ. Kansas Sci. Bull.*, **47**, 25–70.

Eshbaugh, W. H. (1964). A numerical taxonomic and cytogenetic study of certain species of the genus *Capsicum*. Ph.D. thesis, Indiana University. 112 pp.

Eshbaugh, W. H. (1970). A biosystematic and evolutionary study of *Capsicum baccatum* (Salanaceae). *Brittonia*, **22**, 31–43.

Estabrook, G. F. (1966). A mathematical model in graph theory for biological classification. *J. Theoret. Biol.*, **12**, 297–310.

Estabrook, G. F. (1967). An information theory model for character analysis. *Taxon*, **16**, 86–97.

Estabrook, G. F. (1968). A general solution in partial orders for the Camin-Sokal model in phylogeny. *J. Theoret. Biol.*, **21**, 421–438.

Estabrook, G. F., and R. C. Brill (1969). The theory of the TAXIR accessioner. *Math. Biosci.*, **5**, 327–340.

Evans, J. J., and S. Falkow (1969). Evolutionary divergence among strains of *Escherichia coli*. *Bacteriol. Proc.*, **1969**. 40.

Evans, J. W., and D. Fisher (1966). A new species of *Penitella* (family Pholadidae) from Coos Bay, Oregon. *Veliger*, **8**, 222–224.

Fager, E. W. (1969). Recurrent group analysis in the classification of flexibacteria. *J. Gen. Microbiol.*, **58**, 179–187.

Fager, E. W., and J. A. McGowan (1963). Zooplankton species groups in the North Pacific. *Science*, **140**, 453–460.

Farchi, G. (1966). Un metodo per la classificazione automatica in gruppi. *Ann. Ist. Super Sanità (Roma)*, **2**, 717–721.

Farris, J. S. (1966). Estimation of conservatism of characters by constancy within biological populations. *Evolution*, **20**, 587–591.

Farris, J. S. (1967a). The meaning of relationship and taxonomic procedure. *Systematic Zool.*, **16**, 44–51.

Farris, J. S. (1967b). Definitions of taxa. *Systematic Zool.*, **16**, 174–175.

Farris, J. S. (1967c). Comment on psychologism. *Systematic Zool.*, **16**, 345–347.

Farris, J. S. (1968). Categorical ranks and evolutionary taxa in numerical taxonomy. *Systematic Zool.*, **17**, 151–159.

Farris, J. S. (1969a). On the cophenetic correlation coefficient. *Systematic Zool.*, **18**, 279–285.

Farris, J. S. (1969b). A successive approximations approach to character weighting. *Systematic Zool.*, **18**, 374–385.

Farris, J. S. (1970). Methods for computing Wagner trees. *Systematic Zool.*, **19**, 83–92.

Farris, J. S. (1971). The hypothesis of nonspecificity and taxonomic congruence. *Ann. Rev. Ecol. Systematics*, **2**, 277–302.

Farris, J. S. (1972). Estimating phylogenetic trees from distance matrices. *Amer. Natur.*, **106**, 645–668.

Farris, J. S., A. G. Kluge, and M. J. Eckardt (1970). A numerical approach to phylogenetic systematics. *Systematic Zool.*, **19**, 172–189.

Feldman, S., D. F. Klein, and G. Honigfeld (1969). A comparison of successive screening and discriminant function techniques in medical taxonomy. *Biometrics*, **25**, 725–734.

Fernandez de la Véga, W. F. (1965). *Techniques de classification automatique utilisant un indice de ressemblance*. Maison des sciences de l'homme, Paris. 134 pp.

Findley, J. S. (1973). Phenetic packing as a measure of faunal diversity. *Amer. Natur.*, **107** (in press).

Firschein, O., and M. Fischler (1963). Automatic subclass determination for pattern-recognition applications. *I.E.E.E. Trans. Electronic Computers*, **EC–12**, 137–141.

Fisher, D. R. (1968). A study of faunal resemblance using numerical taxonomy and factor analysis. *Systematic Zool.*, **17**, 48–63.

Fisher, D. R. (1970). A comparison of the various techniques of multiple factor analysis applied to biosystematic data. Ph.D. thesis, Kansas University. 122 pp.

Fisher, D. R., and F. J. Rohlf (1969). Robustness of numerical taxonomic methods and errors in homology. *Systematic Zool.*, **18**, 33–36.

Fisher, R. A. (1936). The use of multiple measurements in taxonomic problems. *Ann. Eugen.*, **7**, 179–188.

Fisher, R. G., W. T. Williams, and G. N. Lance (1967). An application of techniques of numerical taxonomy to company information. *Econ. Rec.*, **43**, 566–587.

Fisher, W. D. (1969). *Clustering and Aggregation in Economics*. Johns Hopkins Press, Baltimore, Md. 195 pp.

Fitch, W. M. (1966a). An improved method of testing for evolutionary homology. *J. Mol. Biol.*, **16**, 9–16.

Fitch, W. M. (1966b). Evidence suggesting a partial, internal duplication in the ancestral gene for heme-containing globins. *J. Mol. Biol.*, **16**, 17–27.

Fitch, W. M. (1969). Locating gaps in amino acid sequences to optimize the homology between two proteins. *Biochem. Genet.*, **3**, 99–108.

Fitch, W. M. (1970a). Further improvements in the method of testing for evolutionary homology among proteins. *J. Mol. Biol.*, **49**, 1–14.

Fitch, W. M. (1970b). Distinguishing homologous from analogous proteins. *Systematic Zool.*, **19**, 99–113.

Fitch, W. M. (1970c). A method for estimating the probability that a specific frameshift mutation was selected in the course of evolution. *J. Mol. Biol.*, **49**, 15–21.

Fitch, W. M., and E. Margoliash (1967). Construction of phylogenetic trees. *Science*, **155**, 279–284.

Fitch, W. M., and E. Margoliash (1968). The construction of phylogenetic trees. II. How well do they reflect past history? *Brookhaven Symp. Biol.*, **21**, 217–242.

Fitch, W. M., and E. Margoliash (1970). The usefulness of amino acid and nucleotide sequences in evolutionary studies. *Evolut. Biol.*, **4**, 67–109.

Fitch, W. M., and J. V. Neel (1969). The phylogenetic relationships of some Indian tribes of Central and South America. *Amer. J. Hum. Genet.*, **21**, 384–397.

Flake, R. H. (1969). Numerical classification techniques for population studies at the infraspecific level. *Abstr. Pap. XI Int. Bot. Congr.*, Seattle, p. 60.

Flake, R. H., E. von Rudloff, and B. L. Turner (1969). Quantitative study of clinical variation in *Juniperus virginiana* using terpenoid data. *Proc. Nat. Acad. Sci. U.S.A.*, **64**, 487–494.

Floodgate, G. D., and P. R. Hayes (1963). The Adansonian taxonomy of some yellow pigmented marine bacteria. *J. Gen. Microbiol.*, **30**, 237–244.

Florek, K., J. Łukaszewicz, J. Perkal, H. Steinhaus, and S. Zubrzycki (1951a). Sur la liason et la division des points d'un ensemble fini. *Colloquium Math.*, **2**, 282–285.

Florek, K., J. Łukaszewicz, J. Perkal, H. Steinhaus, and S. Zubrzycki (1951b). Taksonomia Wrocławska. *Przegl. Antropol.*, **17**, 193–211.

Florkin, M. (1962). Isologie, homologie, analogie et convergence en biochimie comparée. *Bull. Acad. Roy. Belg. Classe de Sci.*, Sér. 5, **48**, 819–824.

Florkin, M. (1966). *A Molecular Approach to Phylogeny*. Elsevier Publ. Co., Amsterdam. 176 pp.

Focht, D. D., and W. R. Lockhart (1965). Numerical survey of some bacterial taxa. *J. Bacteriol.*, **90**, 1314–1319.

Forbes, W. T. M. (1933). A grouping of the Agrotine genera. *Entomol. Amer. N.S.*, **14**, 1–40.

Forman, G. L., R. J. Baker, and J. D. Gerber (1968). Comments on the systematic status of vampire bats (family Desmodontidae). *Systematic Zool.*, **17**, 417–425.

Frank, R. E., and P. E. Green (1968). Numerical taxonomy in marketing analysis. A review article. *J. Marketing Res.*, **5**, 83–94.

Fraser, A. R., and M. Kovats (1966). Stereoscopic models of multivariate statistical data. *Biometrics*, **22**, 358–367.

Fraser, P. M., and D. N. Baron (1968). Taxonomic procedures applied to liver disease. *Proc. Roy. Soc. Med.*, **61**, 1043–1046.

Freeman, L. G., Jr., and J. A. Brown (1964). Statistical analysis of Carter Ranch Pottery, *Fieldiana, Anthropology*, **55**, 126–154.

Freer, T. J., and B. L. Adkins (1968). New approach to malocclusion and indices. *J. Dent. Res.*, **47**, 1111–1117.

Friedman, H. P., and J. Rubin (1967). On some invariant criteria for grouping data. *J. Amer. Statist. Ass.*, **62**, 1159–1178.

Fritts, H. C. (1963). Computer programs for tree-ring research. *Tree-Ring Bull.*, **25**, (3–4), 2–7.

Fry, W. G. (1964). The Pycnogonida and the coding of biological information for numerical taxonomy. *Systematic Zool.*, **13**, 32–41.

Fry, W. G. (1970). The sponge as a population: a biometric approach. *Symp. Zool. Soc. Lond.*, **25**, 135–162.

Fry, W. G., and J. W. Hedgpeth (1969). The fauna of the Ross Sea. Part 7. Pycnogonida, 1, Colossendeidae, Pycnogonidae, Endeidae, Ammotheidae. *Bull. New Zealand Dep. Sci. Ind. Res.*, No. 198. 139 pp.

Fujii, K. (1969). Numerical taxonomy of ecological characteristics and the niche concept. *Systematic Zool.*, **18**, 151–153.

Funk, R. C. (1963). The application of numerical taxonomy to acarology. In J. A. Naegele (ed.), *Advances in Acarology*, vol. 1, pp. 374–378. Cornell University Press, Ithaca, N. Y. 492 pp.

Funk, R. C. (1964). An investigation of the Euzerconidae (Mesostigmata: Celaenopsoidea) based on the procedures of numerical taxonomy. *Acarologia fasc. h.s.*, **1964**, 127–132.

Gabor, D. (1965). Character recognition by holography. *Nature*, **208**, 422–423.

Gabriel, K. R., and R. R. Sokal (1969). A new statistical approach to geographic variation analysis. *Systematic Zool.*, **18**, 259–278.

Gambaryan, P. (1964). A mathematical method of taxonomy. *Izvest. Akad. Nauk Armen. SRR, Biol. Nauki.*, **17**(12), 47–53. In Russian, NLLST transl. RTS 3101.

Gambaryan, P. (1965). Taxonomic analysis of the genus *Pinus* L. *Izvest. Akad. Nauk Armen. SSR, Biol. Nauki.*, **18**(8), 75–81. In Russian, NLLST transl. RTS 3547.

Gasking, D. (1960). Clusters. *Australasian J. Phil.*, **38**, 1–36.

George, T. N. (1933). Palingenesis and palaeontology. *Biol. Rev.*, **8**, 107–135.

Géry, J. R. (1965). A new genus from Brazil—*Brittanichthys*, a new, sexually-dimorphic characid genus with peculiar caudal ornament, from the Rio Negro, Brazil, with a discussion of certain cheirodontin genera and a description of two new species, *B. axelrodi* and *B. myersi. Tropical Fish Hobbyist*, **Feb. 1965**, 13–24, 61–69.

Ghiselin, M. T. (1966). On psychologism in the logic of taxonomic controversies. *Systematic Zool.*, **15**, 207–215.

Ghiselin, M. T. (1969). The principles and concepts of systematic biology. In C. G. Sibley (Chairman), *Systematic Biology*. Proceedings of an International Conference Conducted at the University of Michigan, Ann Arbor, Michigan, June 14–16, 1967, pp. 45–55. Nat. Acad. Sci., Wash., Pub. 1692. 632 pp.

Ghiselin, M. T., E. T. Degens, D. W. Spencer, and R. H. Parker (1967). A phylogenetic survey of molluscan shell matrix proteins. *Breviora*, No. 262. 35 pp.

Giacomelli, F., J. Wiener, J. B. Kruskal, J. V. Pomeranz, and A. V. Loud (1971). Subpopulations of blood lymphocytes demonstrated by quantitative cytochemistry. *J. Histochem. Cytochem.*, **19**, 426–433.

Gibbs, A. (1969). Plant virus classification. *Advance. Virus Res.*, **14**, 263–328.

Gibbs, A. J., B. D. Harrison, D. H. Watson, and P. Wildy (1966). What's in a virus name. *Nature*, **209**, 450–454.

Gilardi, E., L. R. Hill, M. Turri, and L. G. Silvestri (1960). Quantitative methods in the systematics of Actinomycetales. I. *Giorn. Microbiol.*, **8**, 203–218.

Gilbert, E. N., and H. O. Pollak (1968). Steiner minimal trees. *SIAM J. Appl. Math.*, **16**, 1–29.

Gilbert, E. S. (1968). On discrimination using qualitative variables. *J. Amer. Statist. Ass.*, **63**, 1399–1412.

Giles, E., and O. Elliot (1962). Race identification from cranial measurements. *J. Forensic Sci.*, **7**, 147–157.

Giles, E., and O. Elliot (1963). Sex determination by discriminant function analysis of crania. *Amer. J. Phys. Anthropol.*, **21**, 53–68.

Giles, E. T. (1963). The comparative external morphology and affinities of the Dermaptera. *Trans. Roy. Entomol. Soc. Lond.*, **115**, 95–164.

Gillett, S. (1968). Airborne factor affecting the grouping behaviour of locusts. *Nature*, **218**, 782–783.

Gillham, N. W. (1956). Geographic variation and the subspecies concept in butterflies. *Systematic Zool.*, **5**, 110–120.

Gillie, O. J., and R. Peto (1969). The detection of complementation map clusters by computer analysis. *Genetics*, **63**, 329–347.

Gilmartin, A. J. (1967a). Variance of phenetic affinities with different randomly selected character sets in an alpha-numerical taxonomic study of the Bromeliaceae. *Amer. J. Bot.*, **54**, 655.

Gilmartin, A. J. (1967b). Numerical taxonomy—an eclectic viewpoint. *Taxon*, **16**, 8–12.

Gilmartin, A. J. (1969a). The quantification of some plant-taxa circumscriptions. *Amer. J. Bot.*, **56**, 654–663.

Gilmartin, A. J. (1969b). Numerical phenetic samples of taxonomic circumscriptions in the Bromeliaceae. *Taxon*, **18**, 378–392.

Gilmour, J. S. L. (1937). A taxonomic problem. *Nature*, **139**, 1040–1042.

Gilmour, J. S. L. (1940). Taxonomy and philosophy. In J. Huxley (ed.), *The New Systematics*, pp. 461–474. Clarendon Press, Oxford. 583 pp.

Gilmour, J. S. L. (1951). The development of taxonomic theory since 1851. *Nature*, **168**, 400–402.

Gilmour, J. S. L. (1961). Taxonomy. In A. M. MacLeod and L. S. Cobley (eds.), *Contemporary Botanical Thought*, pp. 27–45. Oliver & Boyd, Edinburgh, and Quadrangle Books, Chicago. 197 pp.

Gilmour, J. S. L., and J. W. Gregor (1939). Demes: a suggested new terminology. *Nature*, **144**, 333.

Gilmour, J. S. L., and J. Heslop-Harrison (1954). The deme terminology and the units of micro-evolutionary change. *Genetica*, **27**, 147–161.

Gilmour, J. S. L., and S. M. Walters (1963). Philosophy and classification. In W. B. Turrill (ed.), *Vistas in Botany*, vol. 4, pp. 1–22. Pergamon Press, London. 328 pp.

Gimingham, C. H. (1969). The interpretation of variation in North European dwarf-shrub heath communities. *Vegetatio*, **17**, 89–108.

Gisin, H. (1964). Synthetische Theorie der Systematik. *Z. Zool. Syst. Evolutionsforsch.*, **2**, 1–17.

Gisin, H. (1966). Signification des modalités de l'évolution pour la théorie de la systématique. *Z. Zool. Syst. Evolutionsforsch.*, **4**, 1–12.

Gisin, H. (1967). La systématique idéale. *Z. Zool. Syst. Evolutionsforsch.*, **5**, 111–128.

Gittins, R. (1968). Trend-surface analysis of ecological data. *J. Ecol.*, **56**, 845–869.

Gittins, R. (1969). The application of ordination techniques. In I. H. Rorison (ed.), *Ecological Aspects of the Mineral Nutrition of Plants*. British Ecological Society Symposium 9, pp. 37–66. Blackwell Scientific Publications, Oxford. 506 pp.

Goddard, J. (1968). Multivariate analysis of office location patterns in the city centre: a London example. *Regional Stud.*, **2**, 69–85.

Goldstein, G., and I. R. Mackay (1967). Lupoid hepatitis: computer analysis defining "Hepatitis" and "Cirrhosis" phases and relationships between hepatocellular damage and immune reactions in the liver. *Australasian Ann. Med.*, **16**, 62–69.

Golub, G. H., and C. Reinsch (1970). Handbook series linear algebra. Singular value decomposition and least squares solutions. *Numer. Math.*, **14**, 403–420.

Good, I. J. (1958). How much science can you have at your fingertips? *IBM J. Res. Develop.*, **2**, 282–288.

Good, I. J. (1962). Botryological speculations. In I. J. Good (ed.), *The Scientist Speculates: An Anthology of Half-Baked Ideas*, pp. 120–132. Heinemann, London. 413 pp.

Goodall, D. W. (1953). Objective methods for the classification of vegetation. I. The use of positive interspecific correlation. *Aust. J. Bot.*, **1**, 39–63.

Goodall, D. W. (1964). A probabilistic similarity index. *Nature*, **203**, 1098.

Goodall, D. W. (1966a). Numerical taxonomy of bacteria—some published data re-examined. *J. Gen. Microbiol.*, **42**, 25–37.

Goodall, D. W. (1966b). Deviant index: a new tool for numerical taxonomy. *Nature*, **210**, 216.

Goodall, D. W. (1966c). A new similarity index based on probability. *Biometrics*, **22**, 882–907.

Goodall, D. W. (1966d). Classification, probability and utility. *Nature*, **211**, 53–54.

Goodall, D. W. (1966e). Hypothesis-testing in classification. *Nature*, **211**, 329–330.

Goodall, D. W. (1967). The distribution of the matching coefficient. *Biometrics*, **23**, 647–656.

Goodall, D. W. (1968a). Identification by computer. *BioScience*, **18**, 485–488.

Goodall, D. W. (1968b). Affinity between an individual and a cluster in numerical taxonomy. *Biometrie-Praximetrie*, **9**, 52–55.

Goodall, D. W. (1969). A procedure for recognition of uncommon species combinations in sets of vegetation samples. *Vegetatio*, **18**, 19–35.

Goodfellow, M. (1967). Numerical taxonomy of some named bacterial cultures. *Can. J. Microbiol.*, **13**, 1365–1374.

Goodfellow, M. (1969). Numerical taxonomy of some heterotrophic bacteria isolated from a pine forest soil. In J. C. Sheals (ed.), *The Soil Ecosystem*, pp. 83–104. Syst. Ass. Pub. 8. 247 pp.

Goodfellow, M. (1971). Numerical taxonomy of some nocardioform bacteria. *J. Gen. Microbiol.*, **69**, 33–80.

Goodman, L. A., and W. H. Kruskal (1954). Measures of association for cross classifications. *J. Amer. Statist. Ass.*, **49**, 732–764.

Goodman, L. A., and W. H. Kruskal (1959). Measures of association for cross classifications. II. Further discussion and references. *J. Amer. Statist. Ass.*, **54**, 123–163.

Goodman, L. A., and W. H. Kruskal (1963). Measures of association for cross classifications. III. Approximate sampling theory. *J. Amer. Statist. Ass.*, **58**, 310–364.

Goodman, M., J. Barnabas, G. Matsuda and G. W. Moore (1971). Molecular evolution in the descent of man. *Nature*, **233**, 604–613.

Goodman, M., and G. W. Moore (1971). Immunodiffusion systematics of the primates. I. The Catarrhini. *Systematic Zool.*, **20**, 19–62.

Goodman, M. M. (1967a). The races of maize. I. The use of Mahalanobis' Generalized Distances to measure morphological similarity. *Fitotecnia Latinoamer.*, **4**, 1–22.

Goodman, M. M. (1967b). The identification of hybrid plants in segregating populations. *Evolution*, **21**, 334–340.

Goodman, M. M. (1968a). A measure of 'overall variability' in populations. *Biometrics*, **24**, 189–192.

Goodman, M. M. (1968b). The races of maize. II. Use of multivariate analysis of variance to measure morphological similarity. *Crop. Sci.*, **8**, 693–698.

Goodman, M. M. (1969). Measuring evolutionary divergence. *Jap. J. Genet.*, **44** (Suppl. 1), 310–316.

Goodman, M. M., and E. Paterniani (1969). The races of maize. III. Choices of appropriate characters for racial classification. *Econ. Bot.*, **23**, 265–273.

Goronzy, F. (1969). A numerical taxonomy on business enterprises. In A. J. Cole (ed.), *Numerical Taxonomy*. Proceedings of the Colloquium in Numerical Taxonomy Held in the University of St. Andrews, September 1968, pp. 42–52. Academic Press, London. 324 pp.

Gould, S. J. (1966). Allometry and size in ontogeny and phylogeny. *Biol. Rev.*, **41**, 587–640.

Gould, S. J. (1967). Evolutionary patterns in pelycosaurian reptiles: a factor analytic study. *Evolution*, **21**, 385–401.

Gould, S. J. (1969a). An evolutionary microcosm: Pleistocene and Recent history of the land snail *P. (Poecilozonites)* in Bermuda. *Bull. Mus. Comp. Zool. Harvard Univ.*, **138**, 407–532.

Gould, S. J. (1969b). Character variation in two land snails from the Dutch Leeward Islands: geography, environment and evolution. *Systematic Zool.*, **18**, 185–200.

Gould, S. J., and R. F. Johnston (1972). Geographic variation. *Ann. Rev. Ecol. Syst.*, **3**, 457–498.

Gould, S. W. (1958). Punched cards, binomial names and numbers. *Amer. J. Bot.*, **45**, 331–339.

Gould, S. W. (1962). *Family Names of the Plant Kingdom*, vol. 1. International Plant Index, New Haven and New York. 111 pp.

Gower, J. C. (1966a). Some distance properties of latent root and vector methods used in multivariate analysis. *Biometrika*, **53**, 325–338.

Gower, J. C. (1966b). A Q-technique for the calculation of canonical variates. *Biometrika*, **53**, 588–589.

Gower, J. C. (1967a). A comparison of some methods of cluster analysis. *Biometrics*, **23**, 623–637.

Gower, J. C. (1967b). Multivariate analysis and multidimensional geometry. *Statistician*, **17**, 13–28.

Gower, J. C. (1968). Adding a point to vector diagrams in multivariate analysis. *Biometrika*, **55**, 582–585.

Gower, J. C. (1970). A note on Burnaby's character-weighted similarity coefficient. *J. Int. Ass. Math. Geol.*, **2**, 39–45.

Gower, J. C. (1971a). A general coefficient of similarity and some of its properties. *Biometrics*, **27**, 857–871.

Gower, J. C. (1971b). Statistical methods of comparing different multivariate analyses of the same data. In F. R. Hodson, D. G. Kendall, and P. Tăutu (eds.), *Mathematics in the Archaeological and Historical Sciences*, pp. 138–149. Edinburgh University Press, Edinburgh. 565 pp.

Gower, J. C. (1972). Measures of taxonomic distance and their analysis. In J. S. Weiner and J. Huizinga (eds.), *The Assessment of Population Affinities in Man*, pp. 1–24. Clarendon Press, Oxford. 224 pp.

Gower, J. C., and G. J. S. Ross (1969). Minimum spanning trees and single linkage cluster analysis. *Appl. Statist.*, **18**, 54–64.

Graham, P. H. (1964). The application of computer techniques to the taxonomy of the root-nodule bacteria of legumes. *J. Gen. Microbiol.*, **35**, 511–517.

Grant, V. E. (1957). The plant species in theory and practice. In E. Mayr (ed.), *The Species Problem*, pp. 39–80. Amer. Ass. Advance. Sci. Pub. 50. 395 pp.

Grant, W. F. (1969). A numerical chemotaxonomic study of the *Betula caerulea* complex. *Abstr. Pap. XI Int. Bot. Congr.*, Seattle, p. 76.

Grant, W. F., and I. I. Zandstra (1968). The biosystematics of the genus *Lotus* (Leguminosae) in Canada. II. Numerical chemotaxonomy. *Can. J. Bot.*, **46**, 585–589.

Gray, T. R. G. (1969). The identification of soil bacteria. In J. G. Sheals (ed.), *The Soil Ecosystem*, pp. 73–81. Syst. Ass. Pub. 8. 247 pp.

Graybill, F. A. (1969). *Introduction to Matrices with Applications in Statistics*. Wadsworth Publ. Co., Belmont, Calif. 372 pp.

Green, P. E., and F. J. Carmone (1970). *Multidimensional Scaling and Related Techniques in Marketing Analysis*. Allyn & Bacon, Boston, Mass. 203 pp.

Green, P. E., R. E. Frank, and P. J. Robinson (1967). Cluster analysis in test market selection. *Management Sci.*, **13**, B387–B400.

Gregg, J. R. (1954). *The Language of Taxonomy*. Columbia University Press, New York. 71.

Gregg, J. R. (1967). Finite Linnaean structures. *Bull. Math. Biophys.*, **29**, 191–206.

Gregg, J. R. (1968). Buck and Hull: a critical rejoinder. *Systematic Zool.*, **17**, 342–344.

Greig-Smith, P. (1964). *Quantitative Plant Ecology*, 2nd ed. Butterworth, London. 256 pp.

Greig-Smith, P., M. P. Austin, and T. C. Whitmore (1967). The application of quantitative methods to vegetation survey. I. Association-analysis and principal component ordination of rain forest. *J. Ecol.*, **55**, 483–503.

Grewal, M. S. (1962). The rate of genetic divergence of sublines in the C57BL strain of mice. *Genet. Res.*, **3**, 226–237.

Griffith, J. G. (1967). Numerical taxonomy and textual criticism of a classical Latin author. *Nature*, **215**, 326.

Griffith, J. G. (1968). A taxonomic study of the manuscript tradition of Juvenal. *Mus. Helveticum*, **25**, 101–138.

Griffith, J. G. (1969). Numerical taxonomy and some primary manuscripts of the Gospels. *J. Theol. Stud.*, n.s. **20**, 389–406.

Grigal, D. F., and H. F. Arneman (1970). Quantitative relationships among vegetation and soil classifications from northeastern Minnesota. *Can. J. Bot.*, **48**, 555–566.

Grimont, P. A. D. (1969). *Les Serratia: étude taxométrique*. Editions Bergeret, Bordeaux. 141 pp. M.D. thesis, Université de Bordeaux.

Grinker, R. R., Sr., B. Werble, and R. C. Drye (1968). *The Borderline Syndrome. A Behavioral Study of Ego Functions*. Basic Books, New York. 274 pp.

Grumm, J. G. (1965). The systematic analysis of blocs in the study of legislative behavior. *Western Political Quart.*, **18**, 350–362.

Guédès, M. (1967). La méthode taxonomique d'Adanson. *Rev. Hist. Sci.*, **20**, 361–386.

Guttman, L. (1954). A new approach to factor analysis: the radex. In P. F. Lazarsfeld (ed.), *Mathematical Thinking in the Social Sciences*, pp. 258–348. Free Press, Glencoe, Ill. 444 pp.

Guttman, L. (1966). Order analysis of correlation matrices. In R. B. Cattell (ed.), *Handbook of Multivariate Experimental Psychology*, pp. 439–458. Rand McNally, Chicago. 959 pp.

Guttman, L. (1968). A general nonmetric technique for finding the smallest coordinate space for a configuration of points. *Psychometrika*, **33**, 469–506.

Guttman, R., and L. Guttman (1965). A new approach to the analysis of growth patterns: the simplex structure of intercorrelations of measurements. *Growth*, **29**, 219–232.

Gyllenberg, H. G. (1963). A general method for deriving determination schemes for random collections of microbial isolates. *Ann. Acad. Sci. Fenn., Ser. A. IV. Biol.*, No. 69. 23 pp.

Gyllenberg, H. G. (1964). An approach to numerical description of microbial populations. *Ann. Acad. Sci. Fenn., Ser. A. IV. Biol.*, No. 81. 23 pp.

Gyllenberg, H. G. (1965a). Character correlations in certain taxonomic and ecologic groups of bacteria. A study based on factor analysis. *Ann. Med. Exp. Biol. Fenn.*, No. 43, 82–90.

Gyllenberg, H. G. (1965b). A model for computer identification of microorganisms. *J. Gen. Microbiol.*, **39**, 401–405.

Gyllenberg, H. G. (1967). Significance of the Gram strain in the classification of soil bacteria. In T. R. G. Gray and D. Parkinson (eds.), *The Ecology of Soil Bacteria. An International Symposium*, pp. 351–359. Liverpool University Press, Liverpool. 681 pp.

Gyllenberg, H. G. (1970). Factoranalytical evaluation of patterns of correlated characteristics in streptomycetes. In H. Prauser (ed.), *The Actinomycetales*, pp. 101–105. Gustav Fischer, Jena. 439 pp.

Gyllenberg, H. G., and E. Eklund (1967). Application of factor analysis in microbiology. 2. Evaluation of character correlation patterns in psychrophilic pseudomonads. *Ann. Acad. Sci. Fenn., Ser. A. IV. Biol.*, No. 113. 19 pp.

Gyllenberg, H. G., E. Eklund, M. Antila, and U. Vartiovaara (1963a). Contamination and deterioration of market milk. V. Taxometric classification of pseudomonads. *Acta Agr. Scand.*, **13**, 157–176.

Gyllenberg, H. G., E. Eklund, G. Carlberg, M. Antila, and U. Vartiovaara (1963b). Contamination and deterioration of market milk. VI. Application of taxometrics in order to evaluate relationships between microbiological characteristics and keeping quality of market milk. *Acta Agr. Scand.*, **13**, 177–194.

Gyllenberg, H. G., and V. Rauramaa (1966). Taxometric models of bacterial soil populations. *Acta Agr. Scand.*, **16**, 30–38.

Gyllenberg, H. G., W. Woźnicka, and W. Kurylowicz (1967). Application of factor analysis in microbiology. 3. A study of the "yellow series" of streptomycetes. *Ann. Acad. Sci. Fenn., Ser. A. IV. Biol.*, No. 114. 15 pp.

Haas, O. (1962). Comment on numerical taxonomy. *Systematic Zool.*, **11**, 186.

Haas, O., and G. G. Simpson (1946). Analysis of some phylogenetic terms, with attempts at redefinition. *Proc. Amer. Phil. Soc.*, **90**, 319–349.

Haggett, P. (1965). *Locational Analysis in Human Geography*. Edward Arnold, London. 339 pp.

Hagmeier, E. M. (1966). A numerical analysis of the distribution patterns of North American Mammals. II. Re-evaluation of the provinces. *Systematic Zool.*, **15**, 279–299.

Hagmeier, E. M., and C. D. Stults (1964). A numerical analysis of the distribution patterns of North American mammals. *Systematic Zool.*, **13**, 125–155.

Haigh, J. (1970). The recovery of the root of a tree. *J. Appl. Probability*, **7**, 79–88.

Haigh, J. (1971). The manuscript linkage problem. In F. R. Hodson, D. G. Kendall and P. Tăutu (eds.), *Mathematics in the Archaeological and Historical Sciences*, pp. 396–400. Edinburgh University Press, Edinburgh. 565 pp.

Haldane, J. B. S. (1949). Suggestions as to quantitative measurement of rates of evolution. *Evolution*, **3**, 51–56.

Hall, A. V. (1965a). The pecularity index, a new function for use in numerical taxonomy. *Nature*, **206**, 952.

Hall, A. V. (1965b). Studies of the South African species of *Eulophia*. *J. S. Afr. Bot.*, suppl. vol. 5. 248 pp.

Hall, A. V. (1967a). Methods for demonstrating resemblance in taxonomy and ecology. *Nature*, **214**, 830–831.

Hall, A. V. (1967b). Studies in recently developed group-forming procedures in taxonomy and ecology. *J. S. Afr. Bot.*, **33**, 185–196.

Hall, A. V. (1968). Methods for showing distinctness and aiding identification of critical groups in taxonomy and ecology. *Nature*, **218**, 203–204.

Hall, A. V. (1969a). Group forming and discrimination with homogeneity functions. In A. J. Cole (ed.), *Numerical Taxonomy*. Proceedings of the Colloquium in Numerical Taxonomy Held in the University of St. Andrews, September 1968, pp. 53–68. Academic Press, London, 324 pp.

Hall, A. V. (1969b). Automatic grouping programs: the treatment of certain kinds of properties. *Biol. J. Linn. Soc.*, **1**, 321–325.

Hall, A. V. (1969c). Avoiding informational distortion in automatic grouping programs. *Systematic Zool.*, **18**, 318–329.

Hall, A. V. (1970). A computer-based system for forming identification keys. *Taxon*, **19**, 12–18.

Haltenorth, T. (1937). Die verwandtschaftliche Stellung der Grosskatzen zueinander. *Z. Säugetierkunde*, **12**, 97–240.

Hamann, U. (1961). Merkmalbestand und Verwandtschaftsbeziehungen der Farinosae. Ein Beitrag zum System der Monokotyledonen. *Willdenowia*, **2**, 639–768.

Hanan, M. (1966). On Steiner's problem with rectilinear distance. *SIAM J. Appl. Math.*, **14**, 255–265.

Handler, P. (ed.) (1970). *Biology and the Future of Man*. Oxford University Press, New York. 936 pp.

Hansen, A. J., and O. B. Weeks (1964). Taxonomy of *Flavobacterium piscicida* Bein. *Bacteriol. Proc.*, **1964**, 37.

Hansen, A. J., O. B. Weeks, and R. R. Colwell (1965). Taxonomy of *Pseudomonas piscicida* (Bein) Buck, Meyers, and Leifson. *J. Bacteriol.*, **89**, 752–761.

Hansen, B., and K. Rahn (1969). Determination of angiosperm families by means of a punched-card system. *Dansk. Bot. Ark.*, **26**, 1–45, and 172 cards.

Harary, F. (1964). A graph theoretic approach to similarity relations. *Psychometrika*, **29**, 143–151.

Harbaugh, J. W., and F. Demirmen (1964). Application of factor analysis to petrologic variations of Americus limestone (Lower Permian), Kansas and Oklahoma. *Kansas Geol. Surv. Spec. Distrib. Pub.*, No. 15. 40 p..

Harbaugh, J. W., and D. F. Merriam (1968). *Computer Applications in Stratigraphic Analysis*. Wiley, New York. 282 pp.

Harding, J. P. (1949). The use of probability paper for the graphical analysis of polymodal frequency distributions. *J. Marine Biol. Ass. U.K.*, **28**, 141–153.

Harman, H. H. (1967). *Modern Factor Analysis*, 2nd ed. University of Chicago Press, Chicago. 474 pp.

Harrington, B. J. (1966). A numerical taxonomical study of some corynebacteria and related organisms. *J. Gen. Microbiol.*, **45**, 31–40.

Harrison, P. J. (1968). A method of cluster analysis and some applications. *Appl. Statist.*, **17**, 226–236.

Hartigan, J. A. (1967). Representation of similarity matrices by trees. *J. Amer. Statist. Ass.*, **62**, 1140–1158.

Hauser, M. M., and R. E. Smith (1964). The characterization of lactobacilli from Cheddar cheese. II. A numerical analysis of the data by means of an electronic computer. *Can. J. Microbiol.*, **10**, 757–762.

Hawkes, J. G. (ed.) (1968). *Chemotaxonomy and Serotaxonomy*. Academic Press, London. 299 pp.

Hawkins, D. (1964). *The Language of Nature*. W. H. Freeman and Company, San Francisco. 372 pp.

Hawksworth, F. G., G. F. Estabrook, and D. J. Rogers (1968). Application of an information theory model for character analysis in the genus *Arceuthobium* (Viscaceae). *Taxon*, **17**, 605–619.

Hayashi, C. (1956). Theory and examples of quantification. II. *Proc. Inst. Statist. Math. Jap.*, **4**, 19–30. In Japanese.

Hayashi, K. (1964). Studies on a new concept of "center species" and classification based on the cross-relationship of microorganisms. *Jap. J. Bacteriol.*, **19**, 175–180. In Japanese.

Hayashi, K. (1968). New concept of center species and a new genus *Halococcus* induced theoretically. In *Taxonomy of Microorganisms*. Proceedings of the 10th I.A.M. Symposium on Microbiology, pp. 59–88. Inst. of Appl. Microbiology, University of Tokyo, Tokyo.

Hayashi, K., T. Kodaira, K. Baba, and K. Kikuchi (1965). Adansonian taxonomy and relationship of microorganisms based on the concepts of similarity value and center species. I. Studies on the tribes Micrococceae and Sarcineae in the family Coccaceae. *Jap. J. Bacteriol.*, **20**, 528–533. In Japanese.

Hayashi, K., T. Kodaira, K. Baba, and K. Kikuchi (1966a). Adansonian taxonomy and relationship of microorganisms based on the concepts of similarity value and center species. III. Studies on the tribe Neisserieae in the family Coccaceae and isolation and certification of species in a new genus *Halococcus* induced theoretically. *Jap. J. Bacteriol.*, **21**, 633–639. In Japanese.

Hayashi, K., T. Kodaira, K. Kikuchi, and K. Baba (1966b). Adansonian taxonomy and relationship of microorganisms based on the concepts of similarity value and center species. II. Studies on the tribe Streptococceae in the family Coccaceae. *Jap. J. Bacteriol.*, **21**, 336–340. In Japanese.

Hayashi, K., M. Mimura, and Y. Nakabe (1968a). Adansonian taxonomy and relationship of microorganisms based on the concepts of similarity value and center species. IV. Studies on the relationship of tribes Streptococceae, Lactobacilleae and Propionibacterieae. *Jap. J. Bacteriol.*, **23**, 184–190. In Japanese.

Hayashi, K., M. Mimura, and Y. Nakabe (1968b). Adansonian taxonomy and relationship of microorganisms based on the concepts of similarity value and center species. V. Biological characteristics of the genus *Halococcus* with special reference to the correlation of experimental findings and key features induced theoretically. *Jap. J. Bacteriol.*, **23**, 381–391. In Japanese.

Hayhoe, F. G. J., D. Quaglino, and R. Doll (1964). The cytology and cytochemistry of acute leukaemias: a study of 140 cases. *Med. Res. Council, Spec. Rep. Ser.*, No. 304. 105 pp.

Hazel, J. E. (1970). Binary coefficients and clustering in biostratigraphy. *Geol. Soc. Amer. Bull.*, **81**, 3237–3252.

Heatwole, H., and F. MacKenzie (1967). Herpetogeography of Puerto Rico. IV. Paleogeography, faunal similarity and endemism. *Evolution*, **21**, 429–438.

Heberlein, G. T., J. De Ley, and R. Tijtgat (1967). Deoxyribonucleic acid homology and taxonomy of *Agrobacterium*, *Rhizobium*, and *Chromobacterium*. *J. Bacteriol.*, **94**, 116–124.

Hecht, A. D. (1969). Miocene distribution of molluscan provinces along the East Coast of the United States. *Geol. Soc. Amer. Bull.*, **80**, 1617–1620.

Hedges, S. R. (1969). Epigenetic polymorphism in populations of *Apodemus sylvaticus* and *Apodemus flavicollis* (Rodentia, Muridea). *J. Zool., Lond.*, **159**, 425–442.

Hedrick, P. W. (1971). A new approach to measuring genetic similarity. *Evolution*, **25**, 276–280.

Hegenauer, R. (1962–69). *Chemotaxonomie der Pflanzen: eine Übersicht über die Verbreitung und die systematische Bedeutung der Pflanzenstoffe*. Birkhäuser, Basel & Stuttgart. 5 vols.

Heincke, F. (1898). Naturgeschichte des Herings. I. Die Lokalformen und die Wanderungen des Herings in den europäischen Meeren. *Abh. Deutsch. Seefischerei-Vereins*, **2**, i–cxxxvi, 1–223.

Heiser, C. B., Jr., J. Soria, and D. L. Burton (1965). A numerical taxonomic study of *Solanum* species and hybrids. *Amer. Natur.*, **99**, 471–488.

Hendrickson, J. A., Jr. (1967). A methodological analysis of numerical cladistics. Ph.D. thesis, University of Kansas. 88 pp.

Hendrickson, J. A., Jr. (1968). Clustering in numerical cladistics: a minimum-length directed tree problem. *Math. Biosci.*, **3**, 371–381.

Hendrickson, J. A., Jr., and R. R. Sokal (1968). A numerical taxonomic study of the genus *Psorophora* (Diptera: Culicidae). *Ann. Entomol. Soc. Amer.*, **61**, 385–392.

Hennig, W. (1950). *Grundzüge einer Theorie der phylogenetischen Systematik*. Deutscher Zentral-verlag, Berlin. 370 pp.

Hennig, W. (1957). Systematik und Phylogenese. *Ber. Hundertjahrfeier Deutsch. Entomol. Ges.*, pp. 50–70.

Hennig, W. (1966). *Phylogenetic Systematics*. University of Illinois Press, Urbana. 263 pp.

Henningsmoen, G. (1964). Zig-zag evolution. *Norsk Geol. Tidsskr.*, **44**, 341–352.

Herrin, C. S. (1970). A systematic revision of the genus *Hirstionyssus* (Acari: Mesostigmata) of the Nearctic region. *J. Med. Entomol.*, **7**, 391–437.

Heslop-Harrison, J. (1962). Purposes and procedures in the taxonomic treatment of higher organisms. In G. C. Ainsworth and P. H. A. Sneath (eds.), *Microbial Classification*. 12th Symposium of the Society for General Microbiology, pp. 14–36. Cambridge University Press, Cambridge. 483 pp.

Hesser, J. M., P. A. Hartman, and R. A. Saul (1967). Lactobacilli in ensiled high-moisture corn. *Appl. Microbiol.*, **15**, 49–54.

Hester, D. J., and O. B. Weeks (1969). Taxonometric study of the genus *Brevibacterium* Breed. *Bacteriol. Proc.*, **1969**, 19.

Hester, D. J., and O. B. Weeks (1970). Taxonometric studies of gram-positive, pigmented bacteria. *Abstr. X Int. Congr. Microbiol.*, Mexico City, p. 192.

Heywood, V. H. (1968). Plant taxonomy today. In V. H. Heywood (ed.), *Modern Methods in Plant Taxonomy*, pp. 3–12. Academic Press, London. 312 pp.

Heywood, V. H. (1969). Scanning electron microscopy in the study of plant materials. *Micron*, **1**, 1–14.

Heywood, V. H., and J. McNeill (eds.) (1964). *Phenetic and Phylogenetic Classification*. Syst. Ass. Pub. 6. 164 pp.

Hickman, J. C., and M. P. Johnson (1969). An analysis of geographical variation in Western North American *Menziesia* (Ericaceae). *Madroño*, **20**, 1–11.

Hiernaux, J. (1965). Une nouvelle mésure de distance anthropologique entre populations, utilisant simultanément des fréquences géniques, des pourcentages de traits descriptifs et des moyennes métriques. *Compt. Rend. Acad. Sci. Paris*, **260**, 1748–1750.

Hill, L. R. (1959). The Adansonian classification of the staphylococci. *J. Gen. Microbiol.*, **20**, 277–283.

Hill, L. R. (1966). An index to deoxyribonucleic acid base composition of bacterial species. *J. Gen. Microbiol.*, **44**, 419–437.

Hill, L. R., and L. Silvestri (1962). Quantitative methods in the systematics of Actinomycetales. III. The taxonomic significance of physiological-biochemical characters and the construction of a diagnostic key. *Giorn. Microbiol.*, **10**, 1–28.

Hill, L. R., L. G. Silvestri, P. Ihm, G. Farchi, and P. Lanciani (1965). Automatic classification of staphylococci by principal-component analysis and a gradient method. *J. Bacteriol.*, **89**, 1393–1401.

Hill, L. R., M. Turri, E. Gilardi, and L. G. Silvestri (1961). Quantitative methods in the systematics of Actinomycetales. II. *Giorn. Microbiol.*, **9**, 56–72.

Hodgkiss, W., and J. M. Shewan (1968). Problems and modern principles in the taxonomy of marine bacteria. In D. R. Droop and F. J. Ferguson Wood (eds.), *Advances in Microbiology of the Sea*, pp. 127–166. Academic Press, London. 239 pp.

Hodson, F. R. (1969). Searching for structure within multivariate archaeological data. *World Archaeol.*, **1**, 90–105.

Hodson, F. R. (1970). Cluster analysis and archaeology: some new developments and applications. *World Archaeol.*, **1**, 299–320.

Hodson, F. R., D. G. Kendall, and P. Tăutu (eds.) (1971). *Mathematics in the Archaeological and Historical Sciences*. Proceedings of the Anglo-Romanian Conference, Mamaia, 1970. Edinburgh Univ. Press, Edinburgh. 565 pp.

Hodson, F. R., P. H. A. Sneath, and J. E. Doran (1966). Some experiments in the numerical analysis of archaeological data. *Biometrika*, **53**, 311–324.

Hole, F., and M. Shaw (1967). Computer analysis of chronological seriation. *Rice Univ. Stud.*, **53**(3). 166 pp.

Hole, F. D., and M. Hironaka (1960). An experiment in ordination of some soil profiles. *Soil Sci. Soc. Amer. Proc.*, **24**, 309–312.

Holloway, J. D. (1969). A numerical investigation of the biogeography of the butterfly fauna of India, and its relation to continental drift. *Biol. J. Linn. Soc.*, **1**, 373–385.

Holloway, J. D., and N. Jardine (1968). Two approaches to zoogeography: a study based on the distributions of butterflies, birds and bats in the Indo-Australian area. *Proc. Linn. Soc. Lond.*, **179**, 153–188.

Holttum, R. E. (1949). The classification of ferns. *Biol. Rev.*, **24**, 267–296.

Hopkins, J. W. (1966). Some considerations in multivariate allometry. *Biometrics*, **22**, 747–760.

Horne, S. L. (1967). Comparisons of primate catalase tryptic peptides and implications for the study of molecular evolution. *Evolution*, **21**, 771–786.

Howarth, R. J., and J. W. Murray (1969). The foraminiferida of Christchurch Harbour, England: a reappraisal using multivariate techniques. *J. Paleontol.*, **43**, 660–675.

Howd, F. H. (1964). The taxonomy program—a computer technique for classifying geologic data. *Quart. Colorado Sch. Mines*, **59**, 207–222.

Howden, H. F., H. E. Evans, and E. O. Wilson (1968). A suggested revision of nomenclatural procedure in animal taxonomy. *Systematic Zool.*, **17**, 188–191.

Howells, W. W. (1966). Population distances: biological, linguistic, geographical, and environmental. *Current Anthropol.*, **7**, 531–540.

Hoyer, B. H., E. T. Bolton, B. J. McCarthy, and R. B. Roberts (1965). The evolution of polynucleotides. In V. Bryson and H. J. Vogel (eds.), *Evolving Genes and Proteins*, pp. 581–590. Academic Press, New York. 629 pp.

Hubac, J. M. (1964). Application de la taxonomie de Wraclaw (technique des dendrites) à quelques populations du *Campanula rotundifolia* L., s.l., et utilization de cette technique pour l'établissement des clés de détermination. *Bull. Soc. Bot. Fran.*, **111**, 331–346.

Hubac, J. M. (1967). Étude comparée de deux méthodes de taxinomie numérique appliquées à la systématique du *Campanula rotundifolia* L., s.l. *Compt. Rend. Acad. Sci. Paris, Sér. D*, **264**, 577–580.

Hubac, J. M. (1969). Premier essai d'étude de croisements expérimentaux à l'aide des méthodes numériques de la taxonomie chez *Campanula rotundifolia* L., s.l. *Compt. Rend. Soc. Biol. Paris*, **163**, 336–344.

Hubálek, Z. (1969). Numerical taxonomy of genera *Micrococcus* Cohn and *Sarcina* Goodsir. *J. Gen. Microbiol.*, **57**, 349–363.

Hubby, J. L., and L. H. Throckmorton (1965). Protein differences in *Drosophila*. II. Comparative species genetics and evolutionary problems. *Genetics*, **52**, 203–215.

Huber, I. (1968). Numerical taxonomic studies of cockroaches (Blattaria). Ph.D. thesis, Kansas University. 270 pp.

Huber, I. (1969). Numerical taxonomic studies of cockroaches. *Bull. New Jersey Acad. Sci.*, **14**, 61.

Hudson, G. E., K. M. Hoff, J. Vanden Berge, and E. C. Trivette (1969). A numerical study of the wing and leg muscles of Lari and Alcae. *Ibis*, **111**, 459–524.

Hudson, G. E., and P. J. Lanzillotti (1964). Muscles of the pectoral limb in galliform birds. *Amer. Midland Natur.*, **71**, 1–113.

Hudson, G. E., P. J. Lanzillotti, and G. D. Edwards (1959). Muscles of the pelvic limb in galliform birds. *Amer. Midland Natur.*, **61**, 1–67.

Hudson, G. E., R. A. Parker, J. Vanden Berge, and P. J. Lanzillotti (1966). A numerical analysis of the modifications of the appendicular muscles in various genera of gallinaceous birds. *Amer. Midland Natur.*, **76**, 1–73.

Hudson, L. (1967). Arts and sciences: the influence of stereotypes on language. *Nature*, **214**, 968–969.

Hughes, N. F. (1971). Remedy for the general data handling failure of palaeontology. In J. L. Cutbill (ed.), *Data Processing in Biology and Geology*, pp. 321–330. Academic Press, London and New York. 346 pp.

Hughes, R. E., and D. V. Lindley (1955). Application of biometric methods to problems of classification in ecology. *Nature*, **175**, 806–807.

Hughes, W. L., J. M. Kalbfleisch, E. N. Brandt, Jr., and J. P. Costiloe (1963). Myocardial infarction prognosis by discriminant analysis. *Arch. Internal Med.*, **111**, 338–345.

Huizinga, J. (1962). From DD to D^2 and back: the quantitative expression of resemblance. *Proc. Kon. Ned. Akad. Wet. Amsterdam, Ser. C*, **65**(4), 1–12.

Huizinga, J. (1965). Some more remarks on the quantitative expression of resemblance (distance coefficients). *Proc. Kon. Ned. Akad. Wet. Amsterdam, Ser. C*, **68**(1), 69–80.

Hull, D. L. (1964). Consistency and monophyly. *Systematic Zool.*, **13**, 1–11.

Hull, D. L. (1965). The effect of essentialism on taxonomy—two thousand years of stasis. *Brit. J. Phil. Sci.*, **15**, 314–326 and **16**, 1–18.

Hull, D. L. (1966). Phylogenetic numericlature. *Systematic Zool.*, **15**, 14–17.

Hull, D. L. (1967). Certainty and circularity in evolutionary taxonomy. *Evolution*, **21**, 174–189.

Hull, D. L. (1968a). The operational imperative: sense and nonsense in operationism. *Systematic Zool.*, **17**, 438–457.

Hull, D. L. (1968b). The syntax of numericlature. *Systematic Zool.*, **17**, 472–474.

Hull, D. L. (1969). The natural system and the species problem. In C. G. Sibley (Chairman), *Systematic Biology*. Proceedings of an International Conference Conducted at the University of Michigan, Ann Arbor, Michigan, June 14–16, 1967, pp. 56–61. Nat. Acad. Sci. Wash., Pub. 1692. 632 pp.

Hureau, J. C. (1967). Taxonomie numérique des Nototheniidae (Poissons, Perciformes). *Bull. Mus. Nat. Hist. Natur. Paris*, **39**, 488–500.

Hurlbutt, H. W. (1964). Structure and distribution of *Veigaia* (Mesostigmata) in forest soils. *Acarologia 4 fasc. h.s.*, **1964**, 150–152.

Hurlbutt, H. W. (1968). Coexistence and anatomical similarity in two genera of mites, *Veigaia* and *Asca*. *Systematic Zool.*, **17**, 261–271.

Hutchinson, G. E. (1957). Concluding remarks. *Cold Spring Harbor Symp. Quant. Biol.*, **22**, 415–427.

Hutchinson, G. E. (1968). When are species necessary? In R. C. Lewontin (ed.), *Population Biology and Evolution*, pp. 177–186. Syracuse University Press, Syracuse, N. Y. 205 pp.

Hutchinson, M., K. I. Johnstone, and D. White (1965). The taxonomy of certain thiobacilli. *J. Gen. Microbiol.*, **41**, 357–366.

Hutchinson, M., K. I. Johnstone, and D. White (1966). Taxonomy of the acidophilic thiobacilli. *J. Gen. Microbiol.*, **44**, 373–381.

Hutchinson, M., K. I. Johnstone, and D. White (1967). Taxonomy of anaerobic thiobacilli. *J. Gen. Microbiol.*, **47**, 17–23.

Hutchinson, M., K. I. Johnstone, and D. White (1969). Taxonomy of the genus *Thiobacillus*: the outcome of numerical taxonomy applied to the group as a whole. *J. Gen. Microbiol.*, **57**, 397–410.

Hutchinson, M., and D. White (1964). The types and distribution of thiobacilli in biological systems treating carbonization effluents. *J. Appl. Bacteriol.*, **27**, 244–251.

Huxley, J. S. (1932). *Problems of Relative Growth*. Methuen, London. 276 pp.

Huxley, J. S. (ed.) (1940). *The New Systematics*. Clarendon Press, Oxford. 583 pp.

Huxley, J. S. (1957). The three types of evolutionary process. *Nature*, **180**, 454–455.

Huxley, J. S. (1958). Evolutionary processes and taxonomy with special reference to grades. *Uppsala Univ. Arssks.*, **1958**, 21–39.

Hyvärinen, L. (1962). Classification of qualitative data. *Nord. Tidskr. Informations-Behandling*, **2**, 83–89.

IBM (1959). *The IBM 704 taxonomy application—an experimental procedure for classification and prediction purposes*. Parts I and II. IBM 704 Program IB CLF. Mathematics and Applications Department, Data Systems Division, International Business Machines Corporation, New York.

Ibrahim, F. M. (1963). Experimental taxonomy of fungi. Ph.D. thesis, University of London. 197 pp.

Ibrahim, F. M., and R. J. Threlfall (1966a). The application of numerical taxonomy to some graminicolous species of *Helminthosporium*. *Proc. Roy. Soc. B*, **165**, 362–388.

Ibrahim, F. M., and R. J. Threlfall (1966b). The application of numerical taxonomy to the genus *Helminthosporium*. *J. Gen. Microbiol.*, **42**, vi–vii.

Ihm, P., G. Himmelmann, U. Hinz, and H. Fürsch (1967). Taxometrische Untersuchungen an *Epilachna*-Stichproben aus Zentralafrika. *Biometr. Z.*, **9**, 159–179.

Imaizumi, Y. (1967). A new genus and species of cat from Iriomote, Ryukyu Islands. *J. Mammalol. Soc. Jap.*, **3**, 74–105.

Imbrie, J., and E. G. Purdy (1962). Classification of modern Bahamian carbonate sediments. In W. E. Ham (ed.), *Classification of Carbonate Rocks, a Symposium*, pp. 253–272. Amer. Ass. Petroleum Geol. Mem. 1. 279 pp.

Inger, R. F. (1958). Comments on the definition of genera. *Evolution*, **12**, 370–384.

Inglis, W. G. (1966). The observational basis of homology. *Systematic Zool.*, **15**, 219–228.

Inglis, W. G. (1970). The purpose and judgements of biological classifications. *Systematic Zool.*, **19**, 240–250.

Ingram, V. M. (1961). Gene evolution and the haemoglobins. *Nature*, **189**, 704–708.

Irwin, H. S., and D. J. Rogers (1967). Monographic studies in *Cassia* (Leguminosae-Caesalpinioideae). II. A taximetric study of Section Apoucouita. *Mem. New York Bot. Gard.*, **16**, 71–120.

Ising, G., and S. Fröst (1969). Thin-layer chromatographic studies in *Cyrtanthus*. I. *Cyrtanthus breviflorus* and *Cyrtanthus luteus*. *Hereditas*, **63**, 385–414.

Ivimey-Cook, R. B. (1969a). Investigations into the phenetic relationships between species of *Ononis* L. *Watsonia*, **7**, 1–23.

Ivimey-Cook, R. B. (1969b). The phenetic relationships between species of *Ononis*. In A. J. Cole (ed.), *Numerical Taxonomy*. Proceedings of the Colloquium in Numerical Taxonomy Held in the University of St. Andrews, September 1968, pp. 69–90. Academic Press, London. 324 pp.

Ivimey-Cook, R. B., and M. C. F. Proctor (1967). Factor analysis of data from an East Devon heath: a comparison of principal component and rotated solutions. *J. Ecol.*, **55**, 404–413.

Izzo, N. F., and W. Coles (1962). Blood-cell scanner identifies rare cells. *Electronics*, **35**(17), 52–57.

Jaccard, P. (1908). Nouvelles recherches sur la distribution florale. *Bull. Soc. Vaud. Sci. Nat.*, **44**, 223–270.

Jackson, D. M. (1970). The stability of classifications of binary attribute data. *Classification Soc. Bull.*, **2**(2), 40–46.

Jacob, F., and E. L. Wollman (1959). The relationship between the prophage and the bacterial chromosome in lysogenic bacteria. In G. Tunevall (ed.), *Recent Progress in Microbiology*. Symposia Held at the VIIth International Congress of Microbiology, pp. 15–30. Almqvist & Wiksell, Stockholm. 453 pp.

Jacobs, M. (1966). Adanson—the first neo-adansonian? *Taxon*, **15**, 51–55.

Jacquez, J. A. (ed.) (1964a). *The Diagnostic Process*. The Proceedings of a Conference Sponsored by the Biomedical Data Processing Training Program of the University of Michigan, 1963. University of Michigan, Ann Arbor, Michigan. 391 pp.

Jacquez, J. A. (1964b). The diagnostic process: problems and perspectives. In J. A. Jacquez (ed.), *The Diagnostic Process*. The Proceedings of a Conference Sponsored by the Biomedical Data Processing Training Program of the University of Michigan, 1963, pp. 23–27. University of Michigan, Ann Arbor, Michigan. 391 pp.

Jago, N. D. (1967). A key, checklist and synonymy to the species formerly included in the genera *Caloptenopsis* I. Bolivar, 1889, and *Acorypha* Krauss, 1877 (Orth. Calliptaminae). *Eos, Revista Espanola de Entomología*, **42**(3–4), 397–462.

Jago, N. D. (1969). A revision of the systematics and taxonomy of certain North American gomphocerine grasshoppers (Gomphocerinae, Acrididae, Orthoptera). *Proc. Acad. Nat. Sci. Phila.*, **121**, 229–335.

Jago, N. D. (1971). A review of the Gomphocerinae of the world with a key to the genera (Orthoptera, Acrididae). *Proc. Acad. Nat. Sci. Phila.*, **123**, 205–343.

Jahn, T. L. (1961). Man versus machine: a future problem in protozoan taxonomy. *Systematic Zool.*, **10**, 179–192.

Jahn, T. L. (1962). The use of computers in systematics. *J. Parasitol.*, **48**, 656–663.

James, M. T. (1953). An objective aid in determining generic limits. *Systematic Zool.*, **2**, 136–137.

James, W. H. (1964). Use of electronic computers in searching out nutritional interrelationships. *Analytical Chem.*, **36**, 1848–1849.

Jamieson, B. G. M. (1968). A taxonometric investigation of the Alluroididae (Oligochaeta). *J. Zool., Lond.*, **155**, 55–86.

Jancey, R. C. (1966a). Multidimensional group analysis. *Aust. J. Bot.*, **14**, 127–130.

Jancey, R. C. (1966b). The application of numerical methods of data analysis to the genus *Phyllota* Benth. in New South Wales. *Aust. J. Bot.*, **14**, 131–149.

Janetschek, H. (1967). Numerische Taxonomie? Mit Bemerkungen zur Methode synbiologischer Systematik. *Beitr. Entomol.*, **17**, 109–126.

Jardine, C. J., N. Jardine, and R. Sibson (1967). The structure and construction of taxonomic hierarchies. *Math. Biosci.*, **1**, 173–179.

Jardine, N. (1967). The concept of homology in biology. *Brit. J. Phil. Sci.*, **18**, 125–139.

Jardine, N. (1969a). Studies in the theory of classification. Ph.D. thesis, University of Cambridge. 234 pp.

Jardine, N. (1969b). A logical basis for biological classification. *Systematic Zool.*, **18**, 37–52.

Jardine, N. (1969c). The observational and theoretical components of homology: a study based on the morphology of the dermal skull-roofs of rhipidistian fishes. *Biol. J. Linn. Soc.*, **1**, 327–361.

Jardine, N. (1971). Patterns of differentiation between human local populations. *Phil. Trans. Roy. Soc. Lond. B.*, **263**, 1–33.

Jardine, N., and C. J. Jardine (1967). Numerical homology. *Nature*, **216**, 301–302.

Jardine, N., and C. J. Jardine (1969). Is there a concept of homology common to several sciences? *Classification Soc. Bull.*, **2**(1), 12–18.

Jardine, N., C. J. van Rijsbergen, and C. J. Jardine (1969). Evolutionary rates and the inference of evolutionary tree forms. *Nature*, **224**, 185.

Jardine, N., and R. Sibson (1968a). A model for taxonomy. *Math. Biosci.*, **2**, 465–482.

Jardine, N., and R. Sibson (1968b). The construction of hierarchic and non-hierarchic classifications. *Computer J.*, **11**, 177–184.

Jardine, N., and R. Sibson (1970). Quantitative attributes in taxonomic descriptions. *Taxon*, **19**, 862–870.

Jardine, N., and R. Sibson (1971). *Mathematical Taxonomy*. Wiley, London. 286 pp.

Jarvis, B. D. W. (1967). Antigenic relations of cellulolytic cocci in the sheep rumen. *J. Gen. Microbiol.*, **47**, 309–319.

Jarvis, B. D. W., and E. F. Annison (1967). Isolation, classification and nutritional requirements of cellulolytic cocci in the sheep rumen. *J. Gen. Microbiol.*, **47**, 295–307.

Jaworska, H., and N. Nybom (1967). A thin-layer chromatographic study of *Saxifraga caesia*, *S. aizoides*, and their putative hybrid. *Hereditas*, **57**, 159–177.

Jeffers, J. N. R., and T. M. Black (1963). An analysis of variability in *Pinus contorta*. *Forestry*, **36**, 199–218.

Jensen, R. E. (1969). A dynamic programming algorithm for cluster analysis. *Operations Res.*, **17**, 1034–1057.

Jevons, W. S. (1877). *The Principles of Science: A Treatise on Logic and Scientific Method*, 2nd ed., rev. Macmillan, London and New York. 786 pp.

Jewsbury, J. M. (1968). Variation in egg shape in *Schistosoma haematobium*. *Exp. Parasitol.*, **22**, 50–61.

Jičín, R., Z. Pilous, and Z. Vašíček (1969). Grundlagen einer formalen Methode zur Zusammenstellung und Bewertung von Bestimmungsschlüsseln. *Preslia*, **41**, 71–85.

Jičín, R., and Z. Vašíček (1969). The problem of the similarity of objects in numerical taxonomy. *J. Gen. Microbiol.*, **58**, 135–139.

Jizba, Z. V. (1964). A contribution to statistical theory of classification. *Stanford Univ. Pub. Geol. Sci.*, **9**, 729–756.

Johnson, B. L., and O. Hall (1965). Analysis of phylogenetic affinities in the Triticinae by protein electrophoresis. *Amer. J. Bot.*, **52**, 506–513.

Johnson, B. L., and M. M. Thien (1970). Assessment of evolutionary affinities in *Gossypium* by protein electrophoresis. *Amer. J. Bot.*, **57**, 1081–1092.

Johnson, L., Jr. (1968). Item seriation as an aid for elementary scale and cluster analysis. *Bull. Mus. Natur. Hist. Univ. Oregon*, No. 15. 46 pp.

Johnson, L. A. S. (1968). Rainbow's end: the quest for an optimal taxonomy. *Proc. Linn. Soc. New South Wales*, **93**, 8–45; reprinted with additional comments in *Systematic Zool.*, **19**, 203–239 (1970).

Johnson, M. P., and R. W. Holm (1968). Numerical taxonomic studies in the genus *Sarcostemma* R. Br. (Asclepiadaceae). In V. H. Heywood (ed.), *Modern Methods in Plant Taxonomy*, pp. 199–217. Academic Press, London. 312 pp.

Johnson, R. G. (1962). Interspecific associations in Pennsylvanian fossil assemblages. *J. Geol.*, **70**, 32–55.

Johnson, R. G. (1970). Variations in diversity within benthic marine communities. *Amer. Natur.*, **104**, 285–300.

Johnson, R. M., M. E. Katarski, and W. P. Weisrock (1968). Correlation of taxonomic criteria for a collection of marine bacteria. *Appl. Microbiol.*, **16**, 708–713.

Johnson, S. C. (1967). Hierarchical clustering schemes. *Psychometrika*, **32**, 241–254.

Johnston, D. E. (1964). The principles of numerical taxonomy and their application to the systematics of Acari. *Acarologia 4 fasc. h.s.*, **1964**, 117–125.

Johnston, R. F. (1969). Character variation and adaptation in European sparrows. *Systematic Zool.*, **18**, 206–231.

Johnston, R. J. (1968). Choice in classification: the subjectivity of objective methods. *Ann. Ass. Amer. Geogr.*, **58**, 575–589.

Jolicoeur, P. (1963). The multivariate generalization of the allometry equation. *Biometrics*, **19**, 497–499.

Joly, P. (1969). Essais d'applications de méthodes de traitement numérique des informations systématiques. I. Etude du groupe des *Alternaria sensu lato. Bull. Soc. Mycol. Fran.*, **85**, 213–233.

Jones, D., and P. H. A. Sneath (1970). Genetic transfer and bacterial taxonomy. *Bacteriol. Rev.*, **34**, 40–81.

Jones, J. H., W. Card, M. Chapman, J. E. Lennard-Jones, B. C. Morson, M. J. Sackin, and P. H. A. Sneath (1970). Heterogeneity of disease. *Classification Soc. Bull.*, **2**(2), 33–38.

Jones, L. A., and S. G. Bradley (1964). Relationships among streptomycetes, nocardiae, mycobacteria and other actinomycetes. *Mycologia*, **56**, 505–513.

Jorgensen, J. G. (1969). *Salish language and culture.* Language Science Monographs 3, Indiana University Publications, Bloomington. 173 pp.

Joyce, T., and C. Channon (1966). Classifying market survey respondents. *Appl. Statist.*, **15**, 191–215.

Joysey, K. A. (1956). The nomenclature and comparison of fossil communities. In P. C. Sylvester-Bradley (ed.), *The Species Concept in Palaentology*, pp. 83–94. Syst. Ass. Pub. 2. 145 pp.

Joysey, K. A. (1959). The evolution of the Liassic oysters *Ostrea-Gryphaea. Biol. Rev.*, **34**, 297–332.

Kaesler, R. L. (1966). Quantitative re-evaluation of ecology and distribution of Recent Foraminifera and Ostracoda of Todos Santos Bay, Baja California, Mexico. *Univ. Kansas Paleontol. Contrib. Pap. Ser.*, No. 10. 50 pp.

Kaesler, R. L. (1967). Numerical taxonomy in invertebrate paleontology. In C. Teichert and E. L. Yochelson, (eds.), *Essays in Paleontology and Stratigraphy: Raymond C. Moore Commemorative Volume*, pp. 63–81. University of Kansas Press, Lawrence, Kan. 626 pp.

Kaesler, R. L. (1969a). Ordination and character correlations of selected Recent British Ostracoda. *J. Int. Ass. Math. Geol.*, **1**, 97–111.

Kaesler, R. L. (1969b). Numerical taxonomy of selected Recent British Ostracoda. In J. W. Neale (ed.), *Symposium on the Taxonomy, Morphology and Ecology of Recent Ostracoda*, pp. 21–47. Oliver & Boyd, Edinburgh. 553 pp.

Kaesler, R. L. (1970a). The cophenetic correlation coefficient in paleoecology. *Geol. Soc. Amer. Bull.*, **81**, 1261–1266.

Kaesler, R. L. (1970b). Numerical taxonomy in paleontology: classification, ordination and reconstruction of phylogenies. In E. L. Yochelson (ed.), *Proceedings of the North American Paleontological Convention, Chicago 1969*, vol. 1, part B, pp. 84–100. Allen Press, Lawrence, Kan. 2 vols.

Kaesler, R. L., and M. N. McElroy (1966). Classification of subsurface localities of the Reagan Sandstone (Upper Cambrian) of Central and Northwest Kansas. In D. F. Merriam (ed.), *Computer Applications in the Earth Sciences.* Colloquium on classification procedures, pp. 42–47. *Kansas Geol. Surv. Computer Contrib.*, No. 7. 79 pp.

Kalkman, C. (1966). Keeping up with the Joneses. *Taxon*, **15**, 177–179.

Kaltsikes, P. J., and W. Dedio (1970a). A thin-layer chromatographic study of the phenolics of the genus *Aegilops*. I. Numerical chemotaxonomy of the diploid species. *Can. J. Bot.*, **48**, 1775–1780.

Kaltsikes, P. J., and W. Dedio (1970b). A thin-layer chromatographic study of the phenolics of the genus *Aegilops*. II. Numerical chemotaxonomy of the polyploid species. *Can. J. Bot.*, **48**, 1781–1786.

Kaltsikes, P. J., W. Dedio, and E. N. Larter (1969). Numerical chemotaxonomy in the genus *Secale*. *Abstr. Pap. XI Int. Bot. Congr.*, Seattle, p. 106.

Kaminuma, T., T. Takekawa, and S. Watanabe (1969). Reduction of clustering problem to pattern recognition. *Pattern Recognition*, **1**, 195–205.

Kansky, K. J. (1963). Structure of transportation networks: relationship between network geometry and regional characteristics. *Univ. Chicago Dept. Geogr. Res. Pap.*, No. 84, 155 pp.

Kaplan, A., and H. F. Schott (1951). A calculus for empirical classes. *Methodos*, **3**, 165–190.

Kathirithamby, J. (1971). Taxonomy, development and morphology of the immature stages of Cicadellidae (Homoptera). Ph.D. thesis, Univ. of London.

Katz, M. W., and A. M. Torres (1965). Numerical analyses of cespitose zinnias. *Brittonia*, **17**, 335–349.

Kazda, J. (1966). Isolierung and Beschreibung einer *Mycobacterium* Species, des Erregers einer Parallergie gegenüber Tuberkulin bein Geflügel. *Zentralbl. Bakteriol. 1 Abt. Orig.*, **199**, 529–532.

Kazda, J. (1967). Mykobakterien in Trinkwasser als Ursache der Parallergie gegenüber Tuberkulinen bei Tieren. III. *Zentralbl. Bakteriol. 1 Abt. Orig.*, **203**, 199–211.

Kazda, J., F. Vrubel, and V. Dornetzhuber (1967). Course of infection induced in man by inoculating mycobacteria originating in water. *Amer. Rev. Respiratory Dis.*, **95**, 848–853.

Kendall, D. G. (1963). A statistical approach to Flinders Petrie's sequence-dating. *Bull. Int. Statist. Inst.*, 34th session: 657–680.

Kendall, D. G. (1969). Incidence matrices, interval graphs and seriation in archaeology. *Pacific J. Math.*, **28**, 565–570.

Kendrick, W. B. (1964). Quantitative characters in computer taxonomy. In V. H. Heywood and J. McNeill (eds.), *Phenetic and Phylogenetic Classification*, pp. 105–114. Syst. Ass. Pub. 6. 164 pp.

Kendrick, W. B. (1965). Complexity and dependence in computer taxonomy. *Taxon*, **14**, 141–154.

Kendrick, W. B., and J. R. Proctor (1964). Computer taxonomy in the fungi imperfecti. *Can. J. Bot.*, **42**, 65–88.

Kendrick, W. B., and L. K. Weresub (1966). Attempting neo-Adansonian computer taxonomy at the ordinal level in the basidiomycetes. *Systematic Zool.*, **15**, 307–329.

Kerr, W. E., J. F. Pisani, and D. Aily (1967). Aplicação de princípios modernos à sistemática do gênero *Melipona* Illiger, com a divisão em dois subgêneros (Hymenoptera, Apoidea). *Pap. Avuls. Dept. Zool. São Paulo*, **20**, 135–145.

Kesling, R. V., and J. P. Sigler (1969). *Cunctocrinus*, a new Middle Devonian calceocrinid crinoid from the Silica Shale of Ohio. *Contrib. Mus. Paleontol. Univ. Michigan*, **22**, 339–360.

Kestle, D. G., V. D. Abbott, and G. P. Kubica (1967). Differential identification of mycobacteria. II. Subgroups of Groups II and III (Runyon) with different clinical significance. *Amer. Rev. Respiratory Dis.*, **95**, 1041–1052.

Key, K. H. L. (1967). Operational homology. *Systematic Zool.*, **16**, 275–276.

Kidd, K. K., and L. L. Cavalli-Sforza (1971). Number of characters examined and error in reconstruction of evolutionary trees. In F. R. Hodson, D. G. Kendall, and P. Tăutu (eds.), *Mathematics in the Archaeological and Historical Sciences*, pp. 335–346. Edinburgh University Press, Edinburgh. 565 pp.

Kidwell, J. F., and H. B. Chase (1967). Fitting the allometric equation—a comparison of ten methods by computer simulation. *Growth*, **31**, 165–179.

Kikkawa, J. (1968). Ecological association of bird species and habitats in eastern Australia; similarity analysis *J. Anim. Ecol.*, **37**, 143–165.

Kikkawa, J., and K. Pearse (1969). Geographical distribution of land birds in Australia—a numerical analysis. *Aust. J. Zool.*, **17**, 821–840.

Kim, Ke Chung, B. W. Brown, Jr., and E. F. Cook (1966). A quantitative taxonomic study of the *Hoplopleura hesperomydis* complex (Anoplura, Hoplopleuridae), with notes on *a posteriori* taxonomic characters. *Systematic Zool.*, **15**, 24–45.

Kimura, M. (1968). Evolutionary rate at the molecular level. *Nature*, **217**, 624–626.

Kimura, M. (1969). The rate of molecular evolution considered from the standpoint of population genetics. *Proc. Nat. Acad. Sci. U.S.A.*, **63**, 1181–1188.

Kirsch, J. A. W. (1968). Prodromus of the comparative serology of Marsupialia. *Nature*, **217**, 418–420.

Kirsch, J. A. W. (1969). Serological data and phylogenetic inference: the problem of rates of change. *Systematic Zool.*, **18**, 296–311.

Kistner, D. H. (1967a). A revision of the termitophilous tribe Termitodiscini (Coleoptera: Staphylinidae). Part I. The genus *Termitodiscus* Wasmann; its systematics, phylogeny and behavior. *J. New York Entomol. Soc.*, **75**, 204–235.

Kistner, D. H. (1967b). Revision of the myrmecophilous species of the tribe Myrmedoniini. Part II. The genera *Aenictonia* and *Anommatochara*—their relationship and behavior. *Ann. Entomol. Soc. Amer.*, **61**, 971–986.

Kistner, D. H. (1968a). A taxonomic revision of the termitophilous tribe Termitopaedini, with notes on behavior, systematics, and post-imaginal growth (Coleoptera: Staphylinidae). *Misc. Pub. Entomol. Soc. Amer.*, **6**, 141–196.

Kistner, D. H. (1968b). Revision of the African species of the termitophilous tribe Corotocini (Coleoptera, Staphylinidae) II. The genera *Termitomimus* Tragardh and *Nasutimimus* new genus and their relationships. *Coleopterists' Bull.*, **22**, 65–93.

Kistner, D. H., and J. M. Pasteels (1970). Taxonomic revision of the termitophilous subtribe Coptotermoeciina (Coleoptera: Staphylinidae) with a description of some integumentary glands and a numerical analysis of their relationships. *Pacific Insects*, **12**, 85–115.

Klimaszewski, S. M. (1967). Stosunki pokrewieństwa środkowoeuropejskich gatunków z rodzaju *Trioza* Först. (Homoptera, Psyllodea) w świetle badań metodami taksonomii numerycznej. *Ann. Univ. Mariae Curie-Skłodowska, Sec. C, Biol.*, **22**, 1–20.

Klimek, S. (1935). Culture element distributions. I. The structure of Californian Indian culture. *Univ. Calif. Pub. Amer. Archaeol. Ethnol.*, **35**, 1–70.

Klingeman, J., and H. V. Pipberger (1967). Computer classifications of electrocardiograms. *Computers and Biomedical Res.*, **1**, 1–17.

Klotz, G. (1967). Numerische Taxonomie und moderne Verwandtschaftsforschung. *Feddes Repert.*, **75**, 115–130.

Kluge, A. G., and M. J. Eckardt (1969). *Hemidactylus garnotii* Duméril and Bibron, a triploid all-female species of gekkonid lizard. *Copeia*, **1969**, 651–664.

Kluge, A. G., and J. S. Farris (1969). Quantitative phyletics and the evolution of anurans. *Systematic Zool.*, **18**, 1–32.

Kluge, A. G., and J. S. Farris (1973). *Quantitative Phyletics*. Macmillan, New York. In preparation.

Kocková-Kratochvílová, A. (1969a). *Taxometric Study of the Genus* Saccharomyces *(Meyen) Reess*. Swets & Zeitlinger, Amsterdam, 190 pp.

Kocková-Kratochvílová, A. (1969b). The taxometric study of the genus *Saccharomyces* (Meyen) Reess. *Antonie van Leeuwenhoek*, **35** (Suppl.): A15–A16.

Kocková-Kratochvílová, A., J. Šandula, and A. Vojtková-Lepšíková (1969). The genus *Candida* Berkhout. X. *Candida parapsilosis* (Ashford) Langeron et Talice. *Folia Microbiol. Acad. Sci. Bohemoslov.*, **14**, 239–250.

Kocková-Kratochvílová, A., A. Vojtková-Lepšíková, J. Šandula, and M. Pokorná (1968). The genus *Saccharomyces* (Meyen) Reess. V. *Folia Microbiol. Acad. Sci. Bohemoslov.*, **13**, 300–309.

Kohut, J. J. (1969). Determination, statistical analysis, and interpretation of recurrent conodont groups in middle and upper Ordovician strata of the Cincinnati region (Ohio, Kentucky, and Indiana). *J. Paleontol.*, **43**, 392–412.

Komagata, K., Y. Tamagawa, and H. Iizuka (1968). Characteristics of *Erwinia herbicola*. *J. Gen. Appl. Microbiol.*, **14**, 19–37.

Kossack, C. F. (1963). Statistical classification techniques, *IBM Systems J.*, **2**, 136–151.

Kovalev, V. G. (1968). The use of taxonomic analysis in systematics of Diptera. *Zool. Zh.*, **47**, 720–731. In Russian.

Kovalev, V. G., and A. I. Shatalkin (1969). Some aspects of taxonomic analysis. *Zh. Obsch. Biol.*, **30**, 556–560. In Russian.

Kowal, T., and E. Kuźniewski (1959). Uogolnienie metody dendrytowej i zastosowanie jej do systematyki roślin na przykladzie rodzajow *Chenopodium* L. i *Atriplex* L. *Acta Soc. Bot. Polon.*, **28**, 249–262.

Krantz, G. E., R. R. Colwell, and E. Lovelace (1969). *Vibrio parahaemolyticus* from the Blue Crab *Callinectes sapidus* in Chesapeake Bay. *Science*, **164**, 1286–1287.

Krieg, R. E., and W. R. Lockhart (1966a). Numerical taxonomy and deoxyribunucleic acid base composition of enterobacteria. *Bacteriol. Proc.*, **1966**, 23.

Krieg, R. E., and W. R. Lockhart (1966b). Classification of enterobacteria based on overall similarity. *J. Bacteriol.*, **92**, 1275–1280.

Kroeber, A. L. (1960). Statistics, Indo-European, and taxonomy. *Language*, **36**, 1–21.

Krueckeberg, D. A. (1969). A multivariate analysis of metropolitan planning. *J. Amer. Inst. Planners*, **35**, 319–325.

Krumbein, W. C. (1962). Open and closed number systems in stratigraphic mapping. *Bull. Amer. Ass. Petroleum Geol.*, **46**, 2229–2245.

Krumbein, W. C., and F. A. Graybill (1965). *An Introduction to Statistical Models in Geology*. McGraw-Hill, New York. 475 pp.

Kruskal, J. B. (1956). On the shortest spanning subtree of a graph and the traveling salesman problem. *Proc. Amer. Math. Soc.*, **7**, 48–50.

Kruskal, J. B. (1964a). Multidimensional scaling by optimizing goodness of fit to a nonmetric hypothesis. *Psychometrika*, **29**, 1–27.

Kruskal, J. B. (1964b). Nonmetric multidimensional scaling: a numerical method. *Psychometrika*, **29**, 115–129.

Kubica, G. P., V. A. Silcox, J. O. Kilburn, R. W. Smithwick, R. E. Beam, W. D. Jones, Jr., and K. D. Stottmeier (1970). Differential identification of mycobacteria. VI. *Mycobacterium triviale* Kubica sp. nov. *Int. J. Syst. Bacteriol.*, **20**, 161–174.

Kullback, S. (1968). *Information Theory and Statistics*. Dover, New York. 399 pp.

Kurczynski, T. W. (1970). Generalized distance and discrete variables. *Biometrics*, **26**, 525–534.

Kurtén, B. (1958). A differentiation index, and a new measure of evolutionary rates. *Evolution*, **12**, 146–157.

Kurtén, B. (1959). Rates of evolution in fossil mammals. *Cold Spring Harbor Symp. Quant. Biol.*, **24**, 205–215.

Kurylowicz, W., W. Woźnicka, A. Paszkiewicz, and K. Malinowski (1970). Application of numeric taxonomy in streptomycetes. In H. Prauser (ed.), *The Actinomycetales*, pp. 107–122. Gustav Fischer, Jena. 439 pp.

Laird, C. D., B. L. McConaughy, and B. J. McCarthy (1969). Rate of fixation of nucleotide substitutions in evolution. *Nature*, **224**, 149–154.

Lambert, J. M., and W. T. Williams (1962). Multivariate methods in plant ecology. IV. Nodal analysis. *J. Ecol.*, **50**, 775–802.

Lambert, J. M., and W. T. Williams (1966). Multivariate methods in plant ecology. VI. Comparison of information-analysis and association-analysis. *J. Ecol.*, **54**, 635–664.

Lance, G. N., P. W. Milne, and W. T. Williams (1968). Mixed-data classificatory programs. III. Diagnostic systems. *Aust. Computer J.*, **1**, 178–181.

Lance, G. N., and W. T. Williams (1965). Computer programs for monothetic classification ("association analysis"). *Computer J.*, **8**, 246–249.

Lance, G. N., and W. T. Williams (1966a). A generalized sorting strategy for computer classifications. *Nature*, **212**, 218.

Lance, G. N., and W. T. Williams (1966b). Computer programs for hierarchical polythetic classification ("similarity analyses"). *Computer J.*, **9**, 60–64.

Lance, G. N., and W. T. Williams (1967a). Mixed-data classificatory programs. I. Agglomerative systems. *Aust. Computer J.*, **1**, 15–20.

Lance, G. N., and W. T. Williams (1967b). A general theory of classificatory sorting strategies. I. Hierarchical systems. *Computer J.*, **9**, 373–380.

Lance, G. N., and W. T. Williams (1967c). Note on the classification of multi-level data. *Computer J*, **9**, 381–382.

Lance, G. N., and W. T. Williams (1967d). A general theory of classificatory sorting strategies. II. Clustering systems. *Computer J.*, **10**, 271–277.

Lance, G. N., and W. T. Williams (1968a). Note on a new information-statistic classificatory program. *Computer J.*, **11**, 195.

Lance, G. N., and W. T. Williams (1968b). Mixed-data classificatory programs. II. Divisive systems. *Aust. Computer J.*, **1**, 82–85.

Landau, J. W., Y. Shechter, and V. D. Newcomer (1968). Biochemical taxonomy of the dermatophytes. II. Numerical analysis of electrophoretic protein patterns. *J. Invest. Dermatol.*, **51**, 170–176.

Lange, R. T., N. S. Stenhouse, and C. E. Offler (1965). Experimental appraisal of certain procedures for the classification of data. *Aust. J. Biol. Sci.*, **18**, 1189–1205.

Lanjouw, J. (ed.) (1950). *Synopsis of Proposals Concerning the International Rules of Botanical Nomenclature*. Submitted to the Seventh International Botanical Congress, Stockholm, 1950. Int. Comm. Taxonomy. IUBS, Utrecht, and Chronica Botanica, Waltham, Mass. 255 pp.

Lapage, S. P., and S. Bascomb (1968). Use of selenite reduction in bacterial classification. *J. Appl. Bacteriol.*, **31**, 568–580.

Lapage, S. P., S. Bascomb, W. R. Wilcox, and M. A. Curtis (1970). Computer identification of bacteria. In A. Baillie and R. J. Gilbert (eds.), *Automation, Mechanization and Data Handling in Microbiology*, pp. 1–22. Academic Press, London. 233 pp.

La Roi, G. H., and J. R. Dugle (1968). A systematic and genecological study of *Picea glauca* and *P. engelmannii*, using paper chromatograms of needle extracts. *Can. J. Bot.*, **46**, 649–687.

Lawrence, G. H. M. (1951). *Taxonomy of Vascular Plants*. Macmillan, New York. 832 pp.

Leach, E. R. (1962). Classification in social anthropology. *Aslib Proc.*, **14**, 239–242.

Lederberg, E. M. (1960). Genetic and functional aspects of galactose metabolism in *Escherichia coli* K-12. In W. Hayes and R. C. Clowes (eds.), *Microbial Genetics*. Xth Symposium of the Society for General Microbiology, pp. 115–131. Cambridge University Press, Cambridge. 300 pp.

Ledley, R. S., and L. B. Lusted (1959a). Reasoning foundations of medical diagnosis. *Science*, **130**, 9–21.

Ledley, R. S., and L. B. Lusted (1959b). The use of electronic computers to aid in medical diagnosis. *Proc. Inst. Radio Eng.*, **47**, 1970–1977.

Lee, A. M. (1967). A comparative study of some methods of numerical taxonomy with reference to classification of influenza virus. Ph.D. thesis, University of Michigan. 211 pp.

Lee, A. M. (1968). Numerical taxonomy and influenza *B* virus. *Nature*, **217**, 620–622.

Lee, A. M., and N. M. Tauraso (1968). A method for the formulation of influenza virus vaccine using numerical taxonomy. *Bull. World Health Organ.*, **39**, 261–270.

Leech, F. B. (1963). Electronic computers in relation to veterinary science. *Proc. Roy. Soc. Med.*, **56**, 563–564.

Leed, J. (ed.) (1966). *The Computer and Literary Style: Introductory Essays and Studies*. Kent State University Press, Kent, Ohio. 179 pp.

Leenhouts, P. W. (1966). Keys in biology. I. A survey and a proposal of a new kind. *Proc. Kon. Ned. Akad. Wet. Amsterdam, Ser. C.*, **69**, 571–596.

Leeper, G. W. (1954). The classification of soils—an Australian approach. *Trans. V. Int. Congr. Soil Sci.*, **4**, 217–226.

Leifson, E. (1966). Bacterial taxonomy: a critique. *Bacteriol. Rev.*, **30**, 257–266.

Lellinger, D. B. (1964). Quantitative assessment of evolutionary patterns in the cheilanthoid ferns. *Abstr. Pap. X Int. Bot. Congr.*, Edinburgh, pp. 414–415.

Lellinger, D. B. (1965). A quantitative study of generic delimitation in the adiantoid ferns. Ph.D. theis, University of Michigan. 270 pp.

Leone, C. A. (ed.) (1964). *Taxonomic Biochemistry and Serology*. Ronald Press, New York. 728 pp.

Le Quesne, W. J. (1969). A method of selection of characters in numerical taxonomy. *Systematic Zool.*, **18**, 201–205.

Lerman, A. (1965a). Evolution of *Exogyra* in the late Cretaceous of the southeastern United States. *J. Paleontol.*, **39**, 414–435.

Lerman, A. (1965b). On rates of evolution of unit characters and character complexes. *Evolution*, **19**, 16–25.

Lerman, I. C. (1970). *Les bases de la classification automatique.* Gauthier-Villars, Paris. 117 pp.

Leroy, J.-F. (1967). Adanson dans l'histoire de la pensée scientifique. *Rev. Hist. Sci.*, **20**, 349–360.

Lessel, E. F. (1971). Judicial Commission of the International Committee on Nomenclature of Bacteria. Minutes of Meeting, 5 August 1970. *Int. J. Syst. Bacteriol.*, **21**, 100–103.

Leth Bak, A., and A. Stenderup (1969). Deoxyribonucleic acid homology in yeasts. Genetic relatedness within the genus *Candida. J. Gen. Microbiol.*, **59**, 21–30.

Levin, D. A., and B. A. Schaal (1970). Reticulate evolution in *Phlox* as seen through protein electrophoresis. *Amer. J. Bot.*, **57**, 977–987.

Levin, M. H., and D. J. Rogers (1964). A taximetric analysis of genera and families of the Nemalionales (Rhodophyceae, Florideae). *Amer. J. Bot.*, **51**, 689.

Lewin, R. A. (1969). A classification of flexibacteria. *J. Gen. Microbiol.*, **58**, 189–206.

Lewis, T. M. N., and M. Kneberg (1959). The Archaic culture in the Middle South. *Amer. Antiquity*, **25**, 161–183.

Ley, J. De (1968). DNA base composition and taxonomy of some *Acinetobacter* strains. *Antonie van Leeuwenhoek*, **34**, 109–114.

Ley, J. De. (1969a). Compositional nucleotide distribution and the theoretical prediction of homology in bacterial DNA. *J. Theoret. Biol.*, **22**, 89–116.

Ley, J. De (1969b). Molecular data in microbial systematics. In C. G. Sibley (Chairman), *Systematic Biology.* Proceedings of an International Conference Conducted at the University of Michigan, Ann Arbor, Michigan, June 14–16, 1967, pp. 248–268. Nat. Acad. Sci., Wash., Pub. 1692. 632 pp.

Ley, J. De, M. Bernaerts, A. Rassel, and J. Guilmot (1966a). Approach to an improved taxonomy of the genus *Agrobacterium. J. Gen. Microbiol.*, **43**, 7–17.

Ley, J. De, and I. W. Park (1966). Molecular biological taxonomy of some free-living nitrogen-fixing bacteria. *Antonie van Leeuwenhock*, **32**, 6–16.

Ley, J. De, I. W. Park, R. Tijtgat, and J. van Ermengem (1966b). DNA homology and taxonomy of *Pseudomonas* and *Xanthomonas. J. Gen. Microbiol.*, **42**, 43–56.

Liang, G. H. L., and A. J. Casady (1966). Quantitative presentation of the systematic relationships among twenty-one *Sorghum* species. *Crop Sci.*, **6**, 76–79.

Lichtwardt, R. W., V. K. Sangar, J. A. W. Kirsch, and R. N. Lester (1969). Immunologic and electrophoretic studies on the endozooic fungal genus *Smittium* (Trichomycetes). *Abstr. Pap. XI Int. Bot. Congr.*, Seattle, p. 128.

Lima, M. B. (1965). Studies on species of the genus *Xiphinema* and other nematodes. Ph.D. thesis, University of London. 165 pp.

Lima, M. B. (1968). A numerical approach to the *Xiphinema americanum* complex. *Compt. Rend. 8th Int. Symp. Nématol, Antibes, 1965*, p. 30. Brill, Leiden.

Liston, J. (1960). Some results of a computer analysis of strains of *Pseudomonas* and *Achromobacter*, and other organisms. *J. Appl. Bacteriol.*, **23**, 391–394.

Liston, J., and R. R. Colwell (1960). Taxonomic relationships among the pseudomonads. *Bacteriol. Proc.*, **1960**, 78–79.

Liston, J., W. Weibe, and R. R. Colwell (1963). Quantitative approach to the study of bacterial species. *J. Bacteriol.*, **85**, 1061–1070.

Litchfield, C. D., R. R. Colwell, and J. M. Prescott (1969). Numerical taxonomy of heterotrophic bacteria growing in association with continuous-culture *Chlorella sorokiniana. Appl. Microbiol.*, **18**, 1044–1049.

Little, F. J., Jr. (1963). An experimental or tentative revision of the genus *Cliona* utilizing the principles of numerical taxonomy. Ph.D. thesis, University of Texas. 255 pp.

Little, F. J., Jr. (1964). The need for a uniform system of biological numericlature. *Systematic Zool.*, **13**, 191–194.

Lockhart, W. R. (1964). Scoring of data and group-formation in quantitative taxonomy. *Devel. Ind. Microbiol.*, **5**, 162–168.

Lockhart, W. R. (1967). Factors affecting reproducibility of numerical classifications. *J. Bacteriol.*, **94**, 826–831.

Lockhart, W. R. (1970). Coding data. In W. R. Lockhart and J. Liston (eds.), *Methods for Numerical Taxonomy*, pp. 22–33. Amer. Soc. Microbiol., Bethesda, Md. 62 pp.

Lockhart, W. R., and P. A. Hartman (1963). Formation of monothetic groups in quantitative bacterial taxonomy. *J. Bacteriol.*, **85**, 68–77.

Lockhart, W. R., and J. G. Holt (1964). Numerical classification of *Salmonella* serotypes. *J. Gen. Microbiol.*, **35**, 115–124.

Lockhart, W. R., and K. Koenig (1965). Use of secondary data in numerical taxonomy of the genus *Erwinia*. *J. Bacteriol.*, **90**, 1638–1644.

Lockhart, W. R., and J. Liston (eds.) (1970). *Methods for Numerical Taxonomy*. Amer. Soc. Microbiol., Bethesda, Md. 62 pp.

Loening, U. E., and J. Ingle (1967). Diversity of RNA components in green plant tissues. *Nature*, **215**, 363-367.

Long, C. A. (1966). Dependence in taxonomy. *Taxon*, **15**, 49–51.

Lorr, M. (ed.) (1966). *Explorations in Typing Psychotics*. Pergamon Press, New York, 241 pp.

Louis, J., and J. Lefebvre (1968). Étude quantitative de la divergence dans l'évolution morphologique de certaines entités infraspécifiques d'abeilles domestiques (*A. mellifica* L.). *Compt. Rend. Acad. Sci. Paris, Sér. D*, **266**, 1131–1133.

Lowe, W. E. (1969). An ecological study of coccoid bacteria in soil. Ph.D. thesis, University of Liverpool. 266 pp.

Lu, K. H. (1965). Harmonic analysis of the human face. *Biometrics*, **21**, 491–505.

Lubischew, A. A. (1962). On the use of discriminant functions in taxonomy. *Biometrics*, **18**, 455–477.

Lusted, L. B. (1960). Logical analysis in Roentgen diagnosis. *Radiology*, **74**, 178–193.

Lysenko, O. (1961). *Pseudomonas*—an attempt at a general classification. *J. Gen. Microbiol.*, **25**, 379–408.

Lysenko, O. (1962). Some thoughts on the taxonomy of *Bacillus thuringiensis* Berliner. *Coll. Int. Pathol. Insectes Paris 1962, Entomophaga*, h.s., pp. 239–244.

Lysenko, O. (1963a). The statistical approach to bacterial taxonomy. In N. A. Gibbons (ed.), *Rec. Progr. Microbiol.*, vol. 8, pp. 625–629. University of Toronto Press, Toronto. 721 pp.

Lysenko, O. (1963b). The taxonomy of entomogenous bacteria. In E. A. Steinhaus (ed.), *Insect Pathology, an Advanced Treatise*, vol. 2, pp. 1–20. Academic Press, New York. 2 vols., 622 and 689 pp.

Lysenko, O., and P. H. A. Sneath (1959). The use of models in bacterial classification. *J. Gen. Microbiol.*, **20**, 284–290.

McAllister, E. E. (1966). Numerical taxonomy and the smelt family, Osmeridae. *Can. Field Natur.*, **80**, 227–238.

MacArthur, R. H., and J. W. MacArthur (1961). On bird species diversity. *Ecology*, **42**, 594–598.

Maccacaro, G. A. (1958). La misura delle informazione contenuta nei criteri di classificazione. *Ann. Microbiol. Enzimol.*, **8**, 231–239.

McCammon, R. B. (1968). The dendrograph: a new tool for correlation. *Geol. Soc. Amer. Bull.*, **79**, 1663–1670.

McCammon, R. B., and G. Wenniger (1970). The dendrograph. *Kansas Geol. Surv. Computer Contrib.*, No. 48, 28 pp.

McCarthy, B. J., and E. T. Bolton (1963). An approach to the measurement of genetic relatedness among organisms. *Proc. Nat. Acad. Sci. U.S.A.*, **50**, 156–164.

McCurdy, H. D., and S. Wolf (1967). Studies on the taxonomy of fruiting Myxobacterales. *Bacteriol. Proc.*, **1967**, 39.

McDonald, I. J., C. Quadling, and A. K. Chambers (1963). Proteolytic activity of some cold-tolerant bacteria from Arctic sediments. *Can. J. Microbiol.*, **9**, 303–315.

McGuire, R. F. (1969). Attributes of *Chlorococcum* species: a numerical analysis *J. Phycol.*, **5**, 220–223.

McLaughlin, P. J., and M. O. Dayhoff (1969). Evolution of species and proteins: a time scale. In M. O. Dayhoff, *Atlas of Protein Sequence and Structure 1969*, pp. 39–46. National Biomedical Research Foundation, Silver Spring, Md. 315 pp.

Macnaughton-Smith, P. (1965). *Some Statistical and Other Numerical Techniques for Classifying Individuals.* Home Office Research Unit Report. Her Majesty's Stationery Office, London. 33 pp.

Macnaughton-Smith, P., W. T. Williams, M. B. Dale, and L. G. Mockett (1964). Dissimilarity analysis: a new technique of hierarchical sub-division. *Nature,* **202,** 1034–1035.

McNeill, J., P. F. Parker, and V. H. Heywood (1969a). A taximetric approach to the classification of the spiny-fruited members (tribe Caucalideae) of the flowering-plant family Umbelliferae. In A. J. Cole (ed.), *Numerical Taxonomy.* Proceedings of the Colloquium in Numerical Taxonomy Held in the University of St. Andrews, September 1968, pp. 129–147. Academic Press, London. 324 pp.

McNeill, J., P. F. Parker, and V. H. Heywood (1969b). The effect of sorting strategy and character selection in taximetric studies of the Umbelliferae, tribe Caucalidae. *Abstr. Pap. XI Int. Bot. Congr.,* Seattle, p. 135.

MacQueen, J. (1967). Some methods for classification and analysis of multivariate observations. In L. M. Le Cam and J. Neyman (eds.), *Proceedings of the Fifth Berkeley Symposium on Mathematical Statistics and Probability,* vol. 1, pp. 281–297. University of California Press, Berkeley, Calif. 5 vols.

McQuitty, L. L. (1967a). A mutual development of some typological theories and pattern-analytic methods. *Educ. Psychol. Measurement,* **27,** 21–46.

McQuitty, L. L. (1967b). A novel application of the coefficient of correlation in the isolation of both typal and dimensional constructs. *Educ. Psychol. Measurement,* **27,** 591–599.

McQuitty, L. L. (1968a). Improving the validity of crucial decisions in pattern analytic methods. *Educ. Psychol. Measurement,* **28,** 9–21.

McQuitty, L. L. (1968b). Multiple clusters, types, and dimensions from iterative intercolumnar correlational analysis. *Multivariate Behav. Res.,* **3,** 465–478.

McQuitty, L. L., and J. A. Clark (1968). Clusters from iterative, intercolumnar correlational analysis. *Educ. Psychol. Measurement,* **28,** 211–238.

Maddocks, R. F. (1966). Distribution patterns of living and subfossil podocopid ostracodes in the Nosy Bé area, northern Madagascar. *Univ. Kansas Paleontol. Contrib. Pap. Ser.,* No. 12, 72 pp.

Maguire, B., Jr. (1967). A partial analysis of the niche. *Amer. Natur.,* **101,** 515–526.

Mahalanobis, P. C. (1936). On the generalized distance in statistics. *Proc. Nat. Inst. Sci. India.,* **2,** 49–55.

Mainardi, D. (1958). Immunology and chromatography in taxonomic studies on gallinaceous birds. *Nature,* **182,** 1388–1389.

Mainardi, D. (1959). Immunological distances among some gallinaceous birds. *Nature,* **184,** 913–914.

Mainardi, D. (1963). Immunological distances and phylogenetic relationships in birds. *Proc. XIII Int. Ornithol. Congr.,* pp. 103–114.

Malik, A. C., G. W. Reinbold, and E. R. Vedamuthu (1968). An evaluation of the taxonomy of *Propionibacterium. Can. J. Microbiol.,* **14,** 1185–1191.

Mandelbaum, H. 1963. Statistical and geological implications of trend mapping with nonorthogonal polynomials. *J. Geophys. Res.,* **68,** 505–519.

Manischewitz, J. R. (1971). A numerical phenetic study of the snake mites of the family Ixodorhynchidae (Acari, Mesostigmata). Ph.D. Dissertation, Ohio State Univ. 122 pp.

Mannetje, L. 't. (1967a). A re-examination of the taxonomy of the genus *Rhizobium* and related genera using numerical analysis. *Antonie van Leeuwenhoek,* **33,** 477–491.

Mannetje, L. 't. (1967b). A comparison of eight numerical procedures applied to the classification of some African *Trifolium* taxa based on *Rhizobium* affinities. *Aust. J. Bot.,* **15,** 521–528.

Mannetje, L. 't. (1969). *Rhizobium* affinities and phenetic relationships within the genus *Stylosanthes. Aust. J. Bot.,* **17,** 553–564.

Manning, R. T., and L. Watson (1966). Signs, symptoms, and systematics. *J. Amer. Med. Ass.,* **198,** 1180–1184.

Marable, I. W., and W. G. Glenn (1964). Quantitative serologic correspondence of ungulates by gel diffusion. In C. A. Leone (ed.), *Taxonomic Biochemistry and Serology,* pp. 527–534. Ronald Press, New York. 728 pp.

Marcus, L. F. (1969). Measurement of selection using distance statistics in the prehistoric orang-utan *Pongo pygmaeus palaeosumatrensis*. *Evolution*, **23**, 301–307.

Marcus, L. F., and J. H. Vandermeer (1966). Regional trends in geographic variation. *Systematic Zool.*, **15**, 1–13.

Margalef, D. R. (1958). Information theory in ecology. *Gen. Systems*, **3**, 36–71.

Margoliash, E., and E. L. Smith (1965). Structural and functional aspects of cytochrome *c* in relation to evolution. In V. Bryson and H. J. Vogel (eds.), *Evolving Genes and Proteins*, pp. 221–242. Academic Press, New York. 629 pp.

Marr, D. (1970). A theory for cerebral neocortex. *Proc. Roy. Soc. Lond. B*, **176**, 161–234.

Maslin, T. P. (1952). Morphological criteria of phyletic relationships. *Systematic Zool.*, **1**, 49–70.

Mason, L. G., P. R. Ehrlich, and T. C. Emmel (1967). The population biology of the butterfly *Euphydryas editha*. V. Character clusters and asymmetry. *Evolution*, **21**, 85–91.

Mason, L. G., P. R. Ehrlich, and T. C. Emmel (1968). The population biology of the butterfly *Euphydryas editha*. VI. Phenetics of the Jasper Ridge Colony, 1965–66. *Evolution*, **22**, 46–54.

Masuo, E., and T. Nakagawa (1968). Numerical classification of bacteria. I. Computer analysis of coryneform bacteria. *Nippon Nogeikagaku Kaishi*, **42**, 627–632. In Japanese.

Masuo, E., and T. Nakagawa (1969a). Numerical taxonomy of bacteria. II. Analysis of "coryne-form bacteria" and related ones on the basis of overall similarity of phenotypic characters. *Ann. Rep. Shionogi Res. Lab., Osaka.*, **19**, 121–133.

Masuo, E., and T. Nakagawa (1969b). Numerical classification of bacteria. Part II. Computer analysis of coryneform bacteria (2), comparison of group-formations obtained on two different methods of scoring data. *Agr. Biol. Chem.*, **33**, 1124–1133.

Masuo, E., and T. Nakagawa (1969c). Numerical classification of bacteria. Part III. Computer analysis of "coryneform bacteria," (3) classification based on DNA base compositions. *Agr. Biol. Chem.*, **33**, 1570–1576.

Masuo, E., and T. Nakagawa (1970). Numerical classification of bacteria. Part IV. Relationships among some corynebacteria based on serological similarity alone. *Agr. Biol. Chem.*, **34**, 1375–1382.

Mattson, R. L., and J. E. Dammann (1965). A technique for determining and coding subclasses in pattern recognition problems. *IBM J. Res. Devel.*, **9**, 294–302.

Mayr, E. (1963). *Animal Species and Evolution*. Belknap Press, Harvard University Press, Cambridge, Mass. 797 pp.

Mayr, E. (1964). The new systematics. In C. A. Leone (ed.), *Taxonomic Biochemistry and Serology*, pp. 13–32. Ronald Press, New York. 728 pp.

Mayr, E. (1965). Numerical phenetics and taxonomic theory. *Systematic Zool.*, **14**, 73–97.

Mayr, E. (1966). The proper spelling of taxonomy. *Systematic Zool.*, **15**, 88.

Mayr, E. (1968). The role of systematics in biology. *Science*, **159**, 595–599.

Mayr, E. (1969a). *Principles of Systematic Zoology*. McGraw-Hill, New York. 428 pp.

Mayr, E. (1969b). The biological meaning of species. *Biol. J. Linn. Soc.*, **1**, 311–320.

Mayr, E., E. G. Linsley, and R. L. Usinger (1953). *Methods and Principles of Systematic Zoology*. McGraw-Hill, New York. 328 pp.

Medawar, P. B. (1945). Size, shape, and age. In W. E. Le Gros Clark and P. B. Medawar (eds.), *Essays on Growth and Form Presented to D'Arcy Wentworth Thompson*, pp. 157–187. Clarendon Press, Oxford. 408 pp.

Meier-Ewert, C., and A. J. Gibbs (1970). An Adansonian classification of various texts. *Aumla, J. Australasian Univ. Lang. Lit. Ass.*, **33**, 39–47.

Meier-Ewert, H., A. J. Gibbs, and N. J. Dimmock (1970). Studies on antigenic variations of the haemagglutinin and neuraminidase of swine influenza virus isolates. *J. Gen. Virol.*, **6**, 409–419.

Melchiorri-Santolini, U. (1968). Numerical taxonomy of pelagic bacteria from the Ligurian sea. *Ann. Microbiol. Enzimol.*, **18**, 67–83.

Mello, J. F., and M. A. Buzas (1968). An application of cluster analysis as a method of determining biofacies. *J. Paleontol.*, **42**, 747–758.

Meltzer, B., N. H. Searle, and R. Brown (1967). Numerical specification of biological form. *Nature*, **216**, 32–36.

Melville, T. H. (1965). A study of the overall similarity of certain actinomycetes mainly of oral origin. *J. Gen. Microbiol.*, **40**, 309–315.

Mendel, J. M., and K. S. Fu (eds.) (1970). *Adaptive Learning and Pattern Recognition*. Academic Press, New York. 444 pp.

Menitskii, Y. (1966). The use of the quantitative estimates of resemblance in taxonomy. *Bot. Zh.*, **51**, 354–371. In Russian.

Merriam, D. F. (ed.) (1966). Computer applications in the earth sciences: colloquium on classification procedures. *Kansas Geol. Surv. Computer Contrib.*, No. 7, 79 pp.

Merriam, D. F., and R. H. Lippert (1966). Geologic model studies using trend-surface analysis. *J. Geol.*, **74**, 344–357.

Merriam, D. F., and P. H. A. Sneath (1966). Quantitative comparison of contour maps. *J. Geophys. Res.*, **71**, 1105–1115.

Merriam, D. F., and P. H. A. Sneath (1967). Comparison of cyclic rock sequences using cross-association. In C. Teichert and E. L. Yochelson (eds.), *Essays in Paleontology and Stratigraphy: Raymond C. Moore Commemorative Volume*, pp. 523–538. University of Kansas Press, Lawrence, Kan. 626 pp.

Metcalf, Z. P. (1954). The construction of keys. *Systematic Zool.*, **3**, 38–45.

Michener, C. D. (1957). Some bases for higher categories in classification. *Systematic Zool.*, **6**, 160–173.

Michener, C. D. (1963). Some future developments in taxonomy. *Systematic Zool.*, **12**, 151–172.

Michener, C. D. (1964). The possible use of uninominal nomenclature to increase the stability of names in biology. *Systematic Zool.*, **13**, 182–190.

Michener, C. D. (1970). Diverse approaches to systematics. *Evolut. Biol.*, **4**, 1–38.

Michener, C. D., and R. R. Sokal (1957). A quantitative approach to a problem in classification. *Evolution*, **11**, 130–162.

Michener, C. D., and R. R. Sokal (1966). Two tests of the hypothesis of nonspecificity in the *Hoplitis* complex (Hymenoptera: Megachilidae). *Ann. Entomol. Soc. Amer.*, **59**, 1211–1217.

Mickel, J. T. (1962). A monographic study of the fern genus *Anemia*, subgenus *Ceratophyllum*. *Iowa State J. Sci.*, **36**, 349–482.

Micko, H. C., and W. Fischer (1970). The metric of multidimensional psychological spaces as a function of the differential attention to subjective attributes. *J. Math. Psychol.*, **7**, 118–143.

Milke, W. (1949). The quantitative distribution of cultural similarities and their cartographic representation. *Amer. Anthropol.*, **51**, 237–252.

Miller, G. A. (1969). A psychological method to investigate verbal concepts. *J. Math. Psychol.*, **6**, 169–191.

Miller, R. L., and J. S. Kahn (1962). *Statistical Analysis in the Geological Sciences*. Wiley, New York. 483 pp.

Minkoff, E. C. (1965). The effects on classification of slight alterations in numerical technique. *Systematic Zool.*, **14**, 196–213.

Moffett, M. L., and R. R. Colwell (1967). Adansonian analysis of the Rhizobiaceae. *Bacteriol. Proc.*, **1967**, 41.

Moffett, M. L., and R. R. Colwell (1968). Adansonian analysis of the Rhizobiaceae. *J. Gen. Microbiol.*, **51**, 245–266.

Mohagheghpour, N., and C. A. Leone (1969). An immunologic study of the relationships of non-human primates to man. *Comp. Biochem. Physiol.*, **31**, 437–452.

Möller, F. (1962a). La formazione di gruppi in base al contenuto d'informazione dei criteri di classificazione. *Boll. Centro Ricerca Operativa*, **6**, 22–36.

Möller, F. (1962b). Proposta di una chiave di classificazione. *Boll. Centro Ricerca Operativa*, **6**, 37–44.

Möller, F. (1962c). Quantitative methods in the systematics of Actinomycetales. IV. The theory and application of a probabilistic identification key. *Giorn. Microbiol.*, **10**, 29–47.

Möller, F. (1964). Il problema della classificazione statistica al livello nominale. *Riv. Int. Sci. Econ. Commerciali*, **11**, 1–36.

Mooney, H. A., and W. A. Emboden, Jr. (1968). The relationship of terpene composition, morphology, and distribution of populations of *Bursera microphylla* (Burseraceae). *Brittonia*, **20**, 44–51.

Moore, A. W., and J. S. Russell (1966). Potential use of numerical analysis and Adansonian concepts in soil science. *Aust. J. Sci.*, **29**, 141–143.

Moore, A. W., and J. S. Russell (1967). Comparison of coefficients and grouping procedures in numerical analysis of soil trace element data. *Geoderma*, **1**, 139–158.

Moore, D. M., J. B. Harborne, and C. A. Williams (1970). Chemotaxonomy, variation and geographical distribution of the Empetraceae. *Bot. J. Linn. Soc.*, **63**, 277–293.

Moore, G. W. (1971). A mathematical model for the construction of cladograms. Ph.D. thesis, North Carolina State University. 262 pp.

Moore, G. W., W. S. Benninghoff, and P. S. Dwyer (1967). A computer method for the arrangement of phytosociological tables. In *Proceedings of the ACM 22nd National Conference*, pp. 297–299. Thompson Book Co., Washington, D.C. 601 pp.

Moore, G. W., and M. Goodman (1968). A set theoretical approach to immunotaxonomy: analysis of species comparisons in modified Ouchterlony plates. *Bull. Math. Biophys.*, **30**, 279–289.

Morishima, H. (1969a). Differentiation of pathogenic races of *Piricularia oryzae* into two groups, "indica" and "japonica." *Sabrao Newsletter*, **1**, 81–94.

Morishima, H. (1969b). Phenetic similarity and phylogenetic relationships among strains of *Oryza perennis*, estimated by methods of numerical taxonomy. *Evolution*, **23**, 429–443.

Morishima, H., and H. Oka (1960). The pattern of interspecific variation in the genus *Oryza*: its quantitative representation by statistical methods. *Evolution*, **14**, 153–165.

Morrison, D. F. (1967). *Multivariate Statistical Methods*. McGraw-Hill, New York. 338 pp.

Morse, L. E. (1968). Construction of identification keys by computer. *Amer. J. Bot.*, **55**, 737.

Morse, L. E. (1971). Specimen identification and key construction with time-sharing computers. *Taxon*, **20**, 269–282.

Morton, A. Q., and A. D. Winspear (1967). The computer and Plato's *Seventh Letter*. *Computers and the Humanities*, **1**, 72–73.

Moss, W. W. (1967). Some new analytic and graphic approaches to numerical taxonomy, with an example from the Dermanyssidae (Acari). *Systematic Zool.*, **16**, 177–207.

Moss, W. W. (1968a). An illustrated key to the species of the acarine genus *Dermanyssus* (Mesostigmata: Laelapoidea: Dermanyssidae). *J. Med. Entomol.*, **5**, 67–84.

Moss, W. W. (1968b). Experiments with various techniques of numerical taxonomy. *Systematic Zool.*, **17**, 31–47.

Moss, W. W. (1971). Taxonomic repeatability: an experimental approach. *Systematic Zool.*, **20**, 309–330.

Moss, W. W., and W. A. Webster (1969). A numerical taxonomic study of a group of selected strongylates (Nematoda). *Systematic Zool.*, **18**, 423–443.

Moss, W. W., and W. A. Webster (1970). Phenetics and numerical taxonomy applied to systematic nematology. *J. Nematol.*, **2**, 16–25.

Mosteller, F., and D. L. Wallace (1964). *Inference and Disputed Authorship: The Federalist*. Addison-Wesley, Reading, Mass. 287 pp.

Mountford, M. D. (1962). An index of similarity and its application to classificatory problems. In P. W. Murphy (ed.), *Progress in Soil Zoology*, pp. 43–50. Butterworth, London. 398 pp.

Mountford, M. D. (1970). A test of the difference between clusters. In G. P. Patil, E. C. Pielou, and W. E. Waters (eds.), *Statistical Ecology*, Vol. 3, pp. 237–257. Pennsylvania State University Press, University Park, Penna.

Mross, G. A., and R. F. Doolittle (1967). Amino acid sequence studies on artiodactyl fibrinopeptides. II. Vicuna, elk, muntjak, pronghorn, antelope, and water buffalo. *Arch. Biochem. Biophys.*, **122**: 674–684.

Mulcahy, M. J., and A. W. Humphries (1967). Soil classification, soil surveys and land use. *Soils Fertilizers*, **30**, 1–8.

Mullakhanbhai, M. F., and J. V. Bhat (1967). A numerical taxonomical study of *Arthrobacter*. *Current Sci.*, **36**, 115–118.

Munroe, E. (1964). Problems and trends in systematics. *Can. Entomol.*, **96**, 368–377.

Myers, G. S. (1960). The endemic fish fauna of Lake Lanao, and the evolution of higher taxonomic categories. *Evolution*, **14**, 323–333.

Naef, A. (1919). *Idealistische Morphologie und Phylogenetik*. Gustav Fischer, Jena. 77 pp.

Nagy, G. (1968). State of the art in pattern recognition. *Proc. I.E.E.E.*, **56**, 836–862.

Nakamura, S., and S. Nishida (1970). Computer taxonomy of *Clostridium tetani*-like strains. *Med. Biol. (Tokyo)*, **80**, 141–144. In Japanese.

Nakayama, Y., H. Nakayama, and K. Takeya (1970). Studies of the relationship between *Mycobacterium fortuitum* and *Mycobacterium runyonii*. *Amer. Rev. Respiratory Dis.*, **101**, 558–568.

Nash, F. A. (1960). Diagnostic reasoning and the logoscope. *Lancet*, **1960**(ii), 1442–1446.

Needham, R. M. (1967). Automatic classification in linguistics. *Statistician*, **17**, 45–54.

Needleman, S. B., and C. D. Wunsch (1970). A general method applicable to the search for similarities in the amino acid sequence of two proteins. *J. Mol. Biol.*, **48**, 443–453.

Neel, J. V., and R. H. Ward (1970). Village and tribal genetic distances among American Indians, and the possible implications for human evolution. *Proc. Nat. Acad. Sci. U.S.A.*, **65**, 323–330.

Nelson, G. J. (1971). Paraphyly and polyphyly: redefinitions. *Systematic Zool.*, **20**, 471–472.

Neushal, M. (1967). Studies of subtidal marine vegetation in western Washington. *Ecology*, **48**, 83–94.

Niall, H. D., H. T. Keutmann, D. H. Copp, and J. T. Potts, Jr. (1969). Amino acid sequence of salmon ultimobranchial calcitonin. *Proc. Nat. Acad. Sci. U.S.A.*, **64**, 771–778.

Niemelä, S. I., and H. G. Gyllenberg (1968). Application of numerical methods to the identification of micro-organisms. *Spisy Prir. Fak. Univ. Purkyne Brne, Ser. K*, **43**, 279–289.

Niemelä, S. I., J. W. Hopkins, and C. Quadling (1968). Selecting an economical binary test battery for a set of microbial cultures. *Can. J. Microbiol.*, **14**, 271–279.

Obial, R. C. (1970). Cluster analysis as an aid in the interpretation of multi-element geochemical data. *Trans. Inst. Mining Met. Sec. B*, **79**, B175–B180.

Oka, H. (1964). Pattern of interspecific relationships and evolutionary dynamics in *Oryza*. In *Rice Genetics and Cytogenetics*, pp. 71–90. Elsevier Publ. Co., Amsterdam. 268 pp.

Oldroyd, H. (1966). The future of taxonomic entomology. *Systematic Zool.*, **15**, 253–260.

Olds, R. J. (1970). Identification of bacteria with the aid of an improved information sorter. In A. Baillie and R. J. Gilbert (eds.), *Automation, Mechanization and Data Handling in Microbiology*, pp. 85–89. Academic Press, London. 233 pp.

Olson, E. C. (1964). Morphological integration and the meaning of characters in classification systems. In V. H. Heywood and J. McNeill (eds.), *Phenetic and Phylogenetic Classification*, pp. 123–156. Syst. Ass. Pub. 6. 164 pp.

Olson, E. C., and R. L. Miller (1958). *Morphological Integration*. University of Chicago Press, Chicago. 317 pp.

Olsson, U. (1967). Chemotaxonomic analysis of some cytotypes in the *Mentha × verticillata* complex (Labiatae). *Bot. Notiser*, **120**, 255–267.

Ondrick, C. W., and G. S. Srivastava (1970). CORFAN-FORTAN IV computer program for correlation, factor analysis (R- and Q-mode) and varimax rotation. *Kansas Geol. Surv. Computer Contrib.*, No. 42, 92 pp.

Ore, O. (1963). *Graphs and Their Uses*. Random House, New York. 131 pp.

Orloci, L. (1967a). Data centering: a review and evaluation with reference to component analysis. *Systematic Zool.*, **16**, 208–212.

Orloci, L. (1967b). An agglomerative method for classification of plant communities. *J. Ecol.*, **55**, 193–206.

Orloci, L. (1968a). Definitions of structure in multivariate phytosociological samples. *Vegetatio*, **15**, 281–291.

Orloci, L. (1968b). A model for the analysis of structure in taxonomic collections. *Can. J. Bot.*, **46**, 1093–1097.

Orloci, L. (1968c). Information analysis in phytosociology: partition, classification and prediction. *J. Theoret. Biol.*, **20**, 271–284.

Orloci, L. (1969a). Information analysis of structure in biological collections. *Nature*, **223**, 483–484.

Orloci, L. (1969b). Information theory models for hierarchic and non-hierarchic classifications. In A. J. Cole (ed.), *Numerical Taxonomy*. Proceedings of the Colloquium in Numerical Taxonomy Held in the University of St. Andrews, September 1968, pp. 148–164. Academic Press, London. 324 pp.

Orloci, L. (1970). Automatic classification of plants based on information content. *Can. J. Bot.*, **48**, 793–802.

Ornduff, R., and T. J. Crovello (1968). Numerical taxonomy of Limnanthaceae. *Amer. J. Bot.*, **55**, 173–182.

Ornstein, L. (1965). Computer learning and the scientific method: a proposed solution to the information theoretical problem of meaning. *J. Mount Sinai Hosp.*, **32**, 437–494.

Osborne, D. V. (1963a). A numerical representation for taxonomic keys. *New Phytol.*, **62**, 35–43.

Osborne, D. V. (1963b). Some aspects of the theory of dichotomous keys. *New Phytol.*, **62**, 144–160.

Overall, J. E. (1963). A configural analysis of psychiatric diagnostic stereotypes. *Behav. Sci.*, **8**, 211–219.

Oxnard, C. E. (1969a). Mathematics, shape and function: a study in primate anatomy. *Amer. Sci.*, **57**, 75–96.

Oxnard, C. E. (1969b). The combined use of multivariate and clustering analyses in functional morphology. *J. Biomechanics*, **2**, 73–88.

Oxnard, C. E., and P. M. Neely (1969). The descriptive use of neighborhood limited classification in functional morphology: an analysis of the shoulder in primates. *J. Morphol.*, **129**, 127–148.

Oyama, V. I., and G. C. Carle (1967). Pyrolysis–gas chromatography: application to life detection and chemotaxonomy. *J. Gas Chromatography*, **5**, 151–154.

Pankhurst, R. J. (1970a). Key generation by computer. *Nature*, **227**, 1269–1270.

Pankhurst, R. J. (1970b). A computer program for generating diagnostic keys. *Computer J.*, **12**, 145–151.

Papacostea, P., E. Missirliu, and C. Preda (1965). Actual problems of bacterial classification. In *Symposium on Methods in Soil Biology, Bucharest, December 1965*, pp. 47–80. Rumanian National Society of Soil Science, Bucharest.

Park, I. W., and J. De Ley (1967). Ancestral remnants in deoxyribonucleic acid from *Pseudomonas* and *Xanthomonas*. *Antonie van Leeuwenhoek*, **33**, 1–16.

Park, R. A. (1968). Paleoecology of *Venericardia sensu lato* (Pelecypoda) in the Atlantic and Gulf Coastal Province: an application of paleosynecologic methods. *J. Paleontol.*, **42**, 955–986.

Parker-Rhodes, A. F. (1961). *Contributions to the Theory of Clumps*. Cambridge Language Res. Unit, 20 Millington Road, Cambridge, England. 34 pp.

Parker-Rhodes, A. F., and D. M. Jackson (1969). Automatic classification in the ecology of the higher fungi. In A. J. Cole (ed.), *Numerical Taxonomy*. Proceedings of the Colloquium in Numerical Taxonomy Held in the University of St. Andrews, September 1968, pp. 181–215. Academic Press, London. 324 pp.

Parks, J. M. (1966). Cluster analysis applied to multivariate geologic problems. *J. Geol.*, **74**, 703–715.

Parups, E. V., J. R. Proctor, B. Meredith, and J. M. Gillett (1966). A numerotaxonomic study of some species of *Trifolium*, section *Lupinaster*. *Can. J. Bot.*, **44**, 1177–1182.

Paykel, E. S. (1971). Classification of depressed patients. A cluster analysis derived grouping. *Brit. J. Psychiatry.*, **118**, 275–288.

Pearson, K. (1926). On the coefficient of racial likeness. *Biometrika*, **18**, 105–117.

Peng, K. C. (1967). *The Design and Analysis of Scientific Experiments*. Addison-Wesley, Reading, Mass. 252 pp.

Penrose, L. S. (1954). Distance, size and shape. *Ann. Eugenics*, **18**, 337–343.

Perkal, J. (1951). Taksonomia Wrocławska. *Przegląd Antropol.*, **17**, 82–96.

Pernes, J., D. Combes, and R. Réné-Chaume (1970). Différenciation des populations naturelles du *Panicum maximum* Jacq. en Côte-d'Ivoire par acquisition de modifications transmissibles, les unes par grains apomictiques, d'autres par multiplication végétative. *Compt. Rend. Acad. Sci. Paris, Sér. D*, **270**, 1992–1995.

Peters, J. A. (1968). A computer program for calculating degree of biogeographic resemblance between areas. *Systematic Zool.*, **17**, 64–69.

Petras, M. L. (1967). Studies of natural populations of *Mus*. IV. Skeletal variations. *Can. J. Genet. Cytol.*, **9**, 575–588.

Petrova, A. D. (1967). On the taxonomical structure of the family Parholaspidae Krantz 1960 (Parasitiformes: Gamasoidea). *Biol. Nauki*, **10**, 15–26. In Russian.

Pfister, R. M., and P. R. Burkholder (1965). Numerical taxonomy of some bacteria isolated from antarctic and tropical seawater. *J. Bacteriol.*, **90**, 863–872.

Phipps, J. B. (1971). Dendrogram topology. *Syst. Zool.*, **20**, 306–308.

Pianka, E. R. (1967). On lizard species diversity: North American flatland deserts. *Ecology*, **48**, 333–351.

Pielou, E. C. (1969a). Association tests versus homogeneity tests: their uses in subdividing quadrats into groups. *Vegetatio*, **18**, 4–18.

Pielou, E. C. (1969b). *An Introduction to Mathematical Ecology*. Wiley-Interscience, New York. 286 pp.

Pike, E. B. (1965a). A trial of association methods for selecting determinative characters from a collection of Micrococcaceae isolates. *J. Gen. Microbiol.*, **41**, xix.

Pike, E. B. (1965b). A trial of statistical methods for selection of determinative characters from Micrococcaceae isolates. *Spisy Prir. Fak. Univ. Purkyne Brne, Ser. K*, **35**, 316–317.

Pilowsky, I., S. Levine, and D. M. Boulton (1969). The classification of depression by numerical taxonomy. *Brit. J. Psychiat.*, **115**, 937–945.

Pimentel, R. A. (1959). Mendelian infraspecific divergence levels and their analysis. *Systematic Zool.*, **8**, 139–159.

Pintér, M., and I. Bende (1967). Computer analysis of *Acinetobacter lwoffii (Moraxella lwoffii)* and *Acinetobacter anitratus (Moraxella glucidolytica)* strains. *J. Gen. Microbiol.*, **46**, 267–272.

Pintér, M., and I. Bende (1968). Biochemical similarity of *Acinetobacter lwoffii* and *Acinetobacter anitratus*. *Pathol. Microbiol.*, **31**, 41–50.

Pintér, M., and J. De Ley (1969). Overall similarity and DNA base composition of some *Acinetobacter* strains. *Antonie van Leeuwenhoek*, **35**, 209–214.

Pisani, J. F., W. E. Kerr, B. Crestana, D. Aily, and M. L. Lorenzetti (1969). Estudo sôbre a estrutura interspecifica de um grupo de espécies do gênero *Melipona* (Apidae, Hymenoptera). *Ann. Acad. Brasil. Cienc.*, **41**, 97–107.

Pitcher, M. (1966). A factor analytic scheme for grouping and separating types of fossils. *Kansas Geol. Surv. Computer Contrib.*, No. 7, pp. 30–41. 79 pp.

Pohja, M. S. (1960). Micrococci in fermented meat products. Classification and description of 171 different strains. *Suomen Maataloust. Seur. Julkais. (Acta Agralia Fenn.) Bull.*, No. 96, 80 pp.

Pohja, M. S., and H. G. Gyllenberg (1962). Numerical taxonomy of micrococci of fermented meat origin. *J. Appl. Bacteriol.*, **25**, 341–351.

Poindexter, J. S. (1964). Biological properties and classification of the *Caulobacter* group. *Bacteriol. Rev.*, **28**, 231–295.

Pokorná, M. (1969). Statistical analysis of micromorphological dimensions and inhibition of growth by antibiotics and acids in pathogenic species of the genus *Candida*. *Folia Microbiol. Acad. Sci. Bohemoslov*, **14**, 544–553.

Poncet, S. (1967a). Étude taxométrique du genre *Pichia* Hansen (Ascomycètes, Saccharomycetaceae). *Compt. Rend. Acad. Sci. Paris, Sér. D*, **264**, 43–46.

Poncet, S. (1967b). A numerical classification of yeasts of the genus *Pichia* Hansen by a factor analysis method. *Antonie van Leeuwenhoek*, **33**, 345–358.

Poncet, S. (1970). Le genre *Hansenula*, H. et P. Sydow (Ascomycètes, Endomycetaceae). Application d'une méthode d'analyse factorielle a la taxinomie de ce groupe. *Ann. Inst. Pasteur*, **119**, 232–248.

Porter, K. R., and W. F. Porter (1967). Venom comparisons and relationships of twenty species of New World toads (genus *Bufo*). *Copeia, 1967*, 298–307.

Power, D. M. (1969). Evolutionary implications of wing and size variation in the red-winged blackbird in relation to geography and climatic factors: a multiple regression analysis. *Systematic Zool., 18*, 363–373.

Power, D. M. (1970). Geographic variation of red-winged blackbirds in Central North America. *Univ. Kansas Pub. Mus. Natur. Hist., 19*, 1–83.

Powers, D. A. (1970). A numerical taxonomic study of Hawaiian reef corals. *Pacific Sci., 24*, 180–186.

Prance, G. T., D. J. Rogers, and F. White (1969). A taximetric study of an angiosperm family: generic delimination in the Chrysobalanceae. *New Phytol., 68*, 1203–1234.

Prim, R. C. (1957). Shortest connection networks and some generalizations. *Bell System Tech. J., 36*, 1389–1401.

Procaccini, D. J. (1966). A numerical phenetic study of the *Protesilaus* group (Lepidoptera: Papilionidae). Ph.D. thesis, Fordham University. 164 pp.

Proctor, J. R. (1966). Some processes of numerical taxonomy in terms of distance. *Systematic Zool., 15*, 131–140.

Proctor, J. R., and W. B. Kendrick (1963). Unequal weighting in numerical taxonomy. *Nature, 197*, 716–717.

Proctor, M. C. F. (1967). The distribution of British liverworts: a statistical analysis. *J. Ecol., 55*, 119–135.

Purdy, E. G. (1963). Recent calcium carbonate facies of the Great Bahama Bank. I. Petrography and reaction groups. *J. Geol., 71*, 334–355.

Quadling, C., and R. R. Colwell (1963). Taxonomic studies on Gram-negative bacteria from the Arctic. *Bacteriol. Proc., 1963*, 40.

Quadling, C., and R. R. Colwell (1964). The use of numerical methods in characterizing unknown isolates. *Devel. Ind. Microbiol., 5*, 151–161.

Quadling, C., F. D. Cook, and R. R. Colwell (1964). Taxonomy of newly isolated *Cytophaga* strains, *Bacteriol. Proc., 1964*, 28

Quadling, C., and J. W. Hopkins (1966). Two-stage geometric approach to numerical classification. *Bacteriol. Proc., 1966*, 23.

Quadling, C., and J. W. Hopkins (1967). Evaluation of tests and grouping of cultures by a two-stage principal component method. *Can. J. Microbiol., 13*, 1379–1400.

Quigley, M. M., and R. R. Colwell (1968a). Properties of bacteria isolated from deep-sea sediments. *J. Bacteriol., 95*, 211–220.

Quigley, M. M., and R. R. Colwell (1968b). Proposal of a new species *Pseudomonas bathycetes*. *Int. J. Syst. Bacteriol., 18*, 241–252.

Rahman, N. A. (1962). On the sampling distribution of the studentized Penrose measure of distance. *Ann. Human Genet, 26*, 97–106.

Raj, H., and R. R. Colwell (1966). Taxonomy of enterococci by computer analysis. *Can. J. Microbiol, 12*, 353–362.

Raj, H., R. R. Colwell, and J. Liston (1964). A taxonomic study of enterococci by computer analysis. *Bacteriol. Proc., 1964*, 21.

Rajski, C. (1961). Entropy and metric spaces. In C. Cherry (ed.), *Information Theory*. Papers Read at a Symposium on 'Information Theory' Held at the Royal Institution, London, August 29 to September 2, 1960, pp. 41–45. Butterworth, London. 476 pp.

Ramon, S. (1968). A numerical taxonomic study of certain taxa of *Haplopappus* section *Blepharodon*. *Univ. Kansas Sci. Bull., 47*, 863–900.

Randal, J. M., and G. H. Scott (1967). Linnaean nomenclature: an aid to data processing. *Systematic Zool., 16*, 278–281.

Rao, C. R. (1948). The utilization of multiple measurements in problems of biological classification. *J. Roy. Statist. Soc., Ser. B, 10*, 159–193.

Rao, C. R. (1952). *Advanced Statistical Methods in Biometric Research*. Wiley, New York. 390 pp.

Rasnitsyn, S. P. (1965). The use of taxonomic analysis to compare biotopes on the basis of their fauna and population. *Zh. Obshch. Biol.*, **26**, 335–340. In Russian: English transl. NLLST RTS 3693.

Raup, D. M. (1961). The geometry of coiling in gastropods. *Proc. Nat. Acad. Sci. U.S.A.*, **47**, 602–609.

Raup, D. M. (1966). Geometric analysis of shell coiling: general problems. *J. Paleontol.*, **40**, 1178–1190.

Raven, P. H., and R. W. Holm (1967). Systematics and the levels-of-organization approach. *Systematic Zool.*, **16**, 1–5.

Ravin, A. W. (1963). Experimental approaches to the study of bacterial phylogeny. *Amer. Natur.*, **97**, 307–318.

Rayner, J. H. (1965). Multivariate analysis of montmorillonite. *Clay Minerals*, **6**, 59–70.

Rayner, J. H. (1966). Classification of soils by numerical methods. *J. Soil Sci.*, **17**, 79–92.

Rayner, J. H. (1969). The numerical approach to soil systematics. In J. G. Sheals (ed.), *The Soil Ecosystem*, pp. 31–38. Syst. Ass. Pub. 8. 247 pp.

Read, D. W., and P. E. Lestrel (1970). Hominid phylogeny and immunology: a critical appraisal. *Science*, **168**, 578–580.

Rees, J. W. (1969a). Morphologic variation in the cranium and mandible of the white-tailed deer (*Odocoileus virginianus*): a comparative study of geographical and four biological distances. *J. Morphol.*, **128**, 95–112.

Rees, J. W. (1969b). Morphologic variation in the mandible of the white-tailed deer (*Odocoileus virginianus*): a study of populational skeletal variation by principal component and canonical analyses. *J. Morphol.*, **128**, 113–130.

Reeve, E. C. R. (1940). Relative growth in the snout of anteaters. A study in the application of quantitative methods to systematics. *Proc. Zool. Soc. Lond. Ser. A*, **110**, 47–80.

Reich, P. R., N. L. Somerson, C. J. Hybner, R. M. Chanock, and S. M. Weissman (1966). Genetic differentiation by nucleic acid homology. I. Relationships among *Mycoplasma* species of man. *J. Bacteriol.*, **92**, 302–310.

Remane, A. (1956). *Die Grundlagen des natürlichen Systems, der vergleichenden Anatomie und der Phylogenetik. Theoretische Morphologie und Systematik*. Vol. I, 2nd ed. Akademische Verlagsgesellschaft Geest & Portig, Leipzig. 364 pp.

Rensch, B. (1947). *Neuere Probleme der Abstammungslehre. Die Transspezifische Evolution*. Ferdinand Enke, Stuttgart. 407 pp.

Rescigno, A., and G. A. Maccacaro (1961). The information content of biological classifications. In C. Cherry (ed.), *Information Theory*. Papers Read at a Symposium on 'Information Theory' Held at the Royal Institution, London, August 29 to September 2, 1960, pp. 437–446. Butterworth, London. 476 pp.

Reyment, R. A. (1963). Paleontological applicability of certain recent advances in multivariate statistical analysis. *Geol. Föreningens Stockholm Förhandlingar*, **85**, 236–265.

Reyment, R. A. (1965). Quantitative morphologic variation and classification of some Nigerian Paleocene Cytherellidae. *Micropaleontology*, **11**, 457–465.

Reyment, R. A., and D. P. Naidin (1962). Biometric study of *Actinocamax verus* s.l. from the Upper Cretaceous of the Russian Platform. *Stockholm Contrib. Geol.*, **9**, 147–206.

Reyment, R. A., and H.-A. Ramden (1970). FORTRAN IV program for canonical variates analysis for the CDC 3600 computer. *Kansas Geol. Surv. Computer Contrib.*, No. 47, 40 pp.

Reyment, R. A., H.-A. Ramden, and W. J. Wahlstedt (1969). FORTRAN IV program for the generalized statistical distance and analysis of covariance matrices for the CDC 3600 computer. *Kansas Geol. Surv. Computer Contrib.*, No. 39, 42 pp.

Reynolds, K. A. (1965). Numerical taxonomy and comparative elaborateness, with a speculation on unused genes. *Nature*, **206**, 166–168.

Rhodes, A. M., W. P. Bemis, T. W. Whitaker, and S. G. Carmer (1968). A numerical taxonomic study of *Cucurbita*. *Brittonia*, **20**, 251–266.

Rhodes, A. M., and S. G. Carmer (1966). Classification of sweet corn inbreds by methods of numerical taxonomy. *Proc. Amer. Soc. Hort. Sci.*, **88**, 507–515.

Rhodes, A. M., S. G. Carmer, and J. W. Courter (1969). Measurement and classification of genetic variability in horseradish. *J. Amer. Soc. Hort. Sci.*, **94**, 98–102.

Rhodes, A. M., C. Campbell, S. E. Malo, and S. G. Carmer (1970). A numerical taxonomic study of the mango *Mangifera indica* L. *J. Amer. Soc. Hort. Sci.*, **95**, 252–256.

Rhodes, M. E. (1961). The characterization of *Pseudomonas fluorescens* with the aid of an electronic computer. *J. Gen. Microbiol.*, **25**, 331–345.

Rickett, H. W. (1958). So what is a taxon? *Taxon*, **7**, 37–38.

Rijsbergen, C. J. van (1970). A clustering algorithm. *Computer J.*, **13**, 113–115.

Rinkel, R. C. (1965). Microgeographic variation and covariation in *Pemphigus populi-transversus*. *Univ. Kansas Sci. Bull.*, **46**, 167–200.

Rising, J. D. (1968). A multivariate assessment of interbreeding between the chickadees *Parus atricapillus* and *P. carolinensis*. *Systematic Zool.*, **17**, 160–169.

Rivas, L. R. (1965). A proposed code system for storage and retrieval of information in systematic zoology. *Systematic Zool.*, **14**, 131–132.

Roback, S. S., J. Cairns, Jr., and R. L. Kaesler (1969). Cluster analysis of occurrence and distribution of insect species in a portion of the Potomac River. *Hydrobiologia*, **34**, 484–502.

Roberts, R. J. (1968). A numerical taxonomic study of 100 isolates of *Corynebacterium pyogenes*. *J. Gen. Microbiol.*, **53**, 299–303.

Robinson, W. S. (1951). A method for chronologically ordering archaeological deposits. *Amer. Antiquity*, **16**, 293–301.

Rogers, D. J. (1963). Taximetrics, new name, old concept. *Brittonia*, **15**, 285–290.

Rogers, D. J., and S. G. Appan (1969). Taximetric methods for delimiting biological species. *Taxon*, **18**, 609–624.

Rogers, D. J., and H. S. Fleming (1964). A computer program for classifying plants. II. A numerical handling of non-numerical data. *BioScience*, **14**, 15–28.

Rogers, D. J., H. S. Fleming, and G. Estabrook (1967). Use of computers in studies of taxonomy and evolution. *Evolut. Biol.*, **1**, 169–196.

Rogers, D. J., and T. T. Tanimoto (1960). A computer program for classifying plants. *Science*, **132**, 1115–1118.

Rogoff, M. (1957). Automatic analysis of infrared spectra. *Ann. New York Acad. Sci.*, **69**, 27–37.

Rohlf, F. J. (1962). A numerical taxonomic study of the genus *Aedes* (Diptera: Culicidae) with emphasis on the congruence of larval and adult classifications. Ph.D. thesis, University of Kansas. 98 pp.

Rohlf, F. J. (1963a). Congruence of larval and adult classifications in *Aedes* (Diptera: Culcidae). *Systematic Zool.*, **12**, 97–117.

Rohlf, F. J. (1963b). Classification of *Aedes* by numerical taxonomic methods (Diptera: Culicidae). *Ann. Entomol. Soc. Amer.*, **56**, 798–804.

Rohlf, F. J. (1964). Methods for checking the results of a numerical taxonomic study. *Systematic Zool.*, **13**, 102–104.

Rohlf, F. J. (1965). A randomization test of the nonspecificity hypothesis in numerical taxonomy. *Taxon*, **14**, 262–267.

Rohlf, F. J. (1967). Correlated characters in numerical taxonomy. *Systematic Zool.*, **16**, 109–126.

Rohlf, F. J. (1968). Stereograms in numerical taxonomy. *Systematic Zool.*, **17**, 246–255.

Rohlf, F. J. (1969). GRAFPAC, graphic output subroutines for the GE 635 computer. *Kansas Geol. Surv. Computer Contrib.*, No. 36, 50 pp.

Rohlf, F. J. (1970). Adaptive hierarchical clustering schemes. *Systematic Zool.*, **19**, 58–82.

Rohlf, F. J. (1972). An empirical comparison of three ordination techniques in numerical taxonomy. *Systematic Zool.*, **21**, 271–280.

Rohlf, F. J. (1973a). A new approach to the computation of Jardine and Sibson's B_k clusters. (MS in preparation).

Rohlf, F. J. (1973b). Graphs implied by the Jardine-Sibson overlapping clustering methods, B_k. (in preparation).

Rohlf, F. J. (1973c). Hierarchical clustering using the minimum spanning tree. *Computer J.*, **16** (in press).

Rohlf, F. J., and D. R. Fisher (1968). Tests for hierarchical structure in random data sets. *Systematic Zool.*, **17**, 407–412.

Rohlf, F. J., J. Kishpaugh, and D. Kirk (1971). *NT-SYS. Numerical Taxonomy System of Multivariate Statistical Programs.* Tech. Rep. State University of New York at Stony Brook, New York.

Rohlf, F. J., and R. R. Sokal (1962). The description of taxonomic relationships by factor analysis. *Systematic Zool.*, **11**, 1–16.

Rohlf, F. J., and R. R. Sokal (1965). Coefficients of correlation and distance in numerical taxonomy. *Univ. Kansas Sci. Bull.*, **45**, 3–27.

Rohlf, F. J., and R. R. Sokal (1967). Taxonomic structure from randomly and systematically scanned biological images. *Systematic Zool.*, **16**, 246–260.

Rohlf, F. J., and R. R. Sokal (1969). *Statistical Tables.* W. H. Freeman and Company, San Francisco. 253 pp.

Rollins, R. C. (1965a). On the bases of biological classification. *Taxon*, **14**, 1–6.

Rollins, R. C. (1965b). The role of the university herbarium in research and teaching. *Taxon*, **14**, 115–120.

Rose, M. J. (1964). Classification of a set of elements. *Computer J.*, **7**, 208–211.

Rosen, C. A. (1967). Pattern classification by adaptive machines. *Science*, **156**, 38–44.

Ross, A. S. C. (1950). Philological probability problems. *J. Roy. Statist. Soc., Ser. B*, **12**, 19–41.

Ross, G. J. S. (1969a). Minimum spanning tree. *Appl. Statist.*, **18**, 103–104.

Ross, G. J. S. (1969b). Printing the minimum spanning tree. *Appl. Statist.*, **18**, 105–106.

Ross, G. J. S. (1969c). Single linkage cluster analysis. *Appl. Statist.*, **18**, 106–110.

Ross, G. J. S. (1969d). Classification techniques for large sets of data. In A. J. Cole (ed.), *Numerical Taxonomy*. Proceedings of the Colloquium in Numerical Taxonomy Held in the University of St. Andrews, September 1968, pp. 224–233. Academic Press, London. 324 pp.

Rostański, K. (1968). Próba zastosowania metody dendrytowej w systematyce gatunków rodzaju *Oenothera* L. *Acta Soc. Bot. Polon.*, **37**, 235–244.

Rostański, K. (1969). Die Anwendung der Dendriten-Methode (Wroclawer Taxonomie) in der Systematik der Pflanzen am Beispiel der Gattung *Oenothera* L. *Feddes Repert.*, **80**, 373–381.

Rosypal, S., A. Rosypalová, and J. Hořejš (1966). The classification of micrococci and staphylococci based on their DNA base composition and Adansonian analysis. *J. Gen. Microbiol.*, **44**, 281–292.

Rothfels, K. H. (1956). Blackflies, siblings, sex, and species grouping. *J. Hered.*, **47**, 113–122.

Rouatt, J. W., G. W. Skyring, V. Purkayastha, and C. Quadling (1970). Soil bacteria: numerical analysis of electrophoretic protein patterns developed in acrylamide gels. *Can. J. Microbiol.*, **16**, 202–205.

Rovira, A. D., and P. G. Brisbane (1967). Numerical taxonomy and soil bacteria. In T. R. G. Gray, and D. Parkinson (eds.), *The Ecology of Soil Bacteria: An International Symposium*, pp. 337–350. Liverpool University Press, Liverpool. 681 pp.

Rowell, A. J. (1967). A numerical taxonomic study of the chonetacean brachiopods. In C. Teichert and E. L. Yochelson (eds.), *Essays in Paleontology and Stratigraphy: R. C. Moore Commemorative Volume*, pp. 113–140. University of Kansas Press, Lawrence, Kan. 626 pp.

Rowell, A. J. (1970). The contribution of numerical taxonomy to the genus concept. In E. L. Yochelson (ed.), *Proceedings of the North American Paleontological Convention, Chicago 1969*, vol. 1, part C, pp. 264–293. Allen Press, Lawrence, Kan. 2 vols.

Rowley, G. D. (1967). A numerical survey of the genera of Aloineae. *Nat. Cactus Succulent J.*, **22**, 71–75.

Rowley, G. D. (1969). Better botany by computer. *Taxon*, **18**, 625–628.

Rubin, J. (1966). An approach to organizing data into homogeneous groups. *Systematic Zool.*, **15**, 169–182.

Rucker, J. B. (1967). Paleoecological analysis of cheilostome Bryozoa from Venezuela-British Guiana shelf sediments. *Bull. Marine Sci.*, **17**, 787–839.

Russell, J. S., and A. W. Moore (1967). Use of a numerical method in determining affinities between some deep sandy soils. *Geoderma*, **1**, 47–68.

Russell, J. S., and A. W. Moore (1968). Comparison of different depth weightings in the numerical analysis of anisotropic soil profile data. *Proc. IXth Int. Congr. Soil. Sci.*, **4**, 205–213.

Ruud, J. T. (1954). Vertebrates without erythrocytes and blood pigment. *Nature*, **173**, 848–850.

Rypka, E. W., and R. Babb (1970). Automatic construction and use of an identification scheme. *Med. Res. Engineering*, **9**(2), 9–19.

Rypka, E. W., W. E. Clapper, I. G. Bowen, and R. Babb (1967). A model for the identification of bacteria. *J. Gen. Microbiol.*, **46**, 407–424.

Sackin, M. J. (1967). Comparisons of protein sequences. M.Sc. thesis, University of Leicester. 145 pp.

Sackin, M. J. (1969). Applications of cross-association to an evolutionary study of cytochrome *c*. In A. J. Cole (ed.), *Numerical Taxonomy*. Proceedings of the Colloquium in Numerical Taxonomy Held in the Univesity of St. Andrews, September 1968, pp. 241–256. Academic Press, London. 324 pp.

Sackin, M. J. (1971). Crossassociation: a method of comparing protein sequences. *Biochem. Genet.*, **5**, 287–313.

Sackin, M. J., and D. F. Merriam (1969). Autoassociation, a new geological tool. *Int. J. Math. Geol.*, **1**, 7–16.

Sackin, M. J., P. H. A. Sneath, and D. F. Merriam (1965). ALGOL program for cross-association of nonmetric sequences using a medium-size computer. *Kansas Geol. Surv. Spec. Distrib. Pub.* No. 23. 36 pp.

Saila, S. B., and J. M. Flowers (1969). Geographic morphometric variation in the American lobster. *Systematic Zool.*, **18**, 330–338.

Saito, H., H. Tasaka, and N. Takei (1968). Studies on atypical acid-fast bacilli of group IV mycobacterium: with special reference to the classification of nonphotochromogenic, rapidly growing, acid-fast organisms from natural sources. *Jap. J. Bacteriol.*, **23**, 758–766.

Sakai, S. (1970). *Dermaptorum Catalogus Praeliminaris*. I–II. Daito Bunka University, Tokyo. 91, 177 pp.

Sakai, S. (1971). *Dermaptorum Catalogus Praeliminaris*. III–IV. Ikegami, Tokyo. 68, 14, 162, 265, 210 pp.

Sakazaki, R., C. Z. Gomez, and M. Sebald (1967). Taxonomical studies of the so-called NAG vibrios. *Jap. J. Med. Sci. Biol.*, **20**, 265–280.

Sammon, J. W., Jr. (1969). A nonlinear mapping for data structure analysis. *I.E.E.E. Trans. Computers*, **C-18**, 401–409.

Sandifer, M. G., L. M. Green, and E. Carr-Harris (1966). The construction and comparison of psychiatric diagnosis stereotypes. *Behav. Sci.*, **11**, 471–477.

Sands, D. C., M. N. Schroth, and D. C. Hildebrand (1970). Taxonomy of phytopathogenic pseudomonads. *J. Bacteriol.*, **101**, 9–23.

Sanghvi, L. D. (1953). Comparison of genetical and morphological methods for a study of biological differences. *Amer. J. Phys. Anthropol.*, **11**, 385–404.

Sarkar, P. K., O. W. Bidwell, and L. F. Marcus (1966). Selection of characteristics for numerical classification of soils. *Soil Sci. Soc. Amer. Proc.*, **30**, 269–272.

Sarich, V. M. (1969a). Pinniped origins and the rate of evolution of carnivore albumins. *Systematic Zool.*, **18**, 286–295.

Sarich, V. M. (1969b). Pinniped phylogeny. *Systematic Zool.*, **18**, 416–422.

Sattler, R. (1966). Towards a more adequate approach to comparative morphology. *Phytomorphology*, **16**, 417–429.

Savory, T. (1962). *Naming the Living World: An Introduction to the Principles of Biological Nomenclature*. English Universities Press, London. 128 pp.

Scadding, J. G. (1963). Meaning of diagnostic terms in broncho-pulmonary disease. *Brit. Med. J.*, **1963**(ii), 1425–1430.

Schilder, F. A., and M. Schilder (1951). *Anleitung zu biostatistischen Untersuchungen*. Max Niemeyer, Halle (Saale), Germany. 111 pp.

Schlee, D. (1969). Hennig's principle of phylogenetic systematics, an "intuitive" statistico-phenetic taxonomy? *Systematic Zool.*, **18**, 127–134.

Schmalhausen, I. I. (1949). *Factors of Evolution. The Theory of Stabilizing Selection.* Blakiston, Philadelphia. 327 pp.

Schmid, W. D. (1965). Distribution of aquatic vegetation as measured by line intercept with SCUBA. *Ecology*, **46**, 816–823.

Schnell, G. D. (1969). A phenetic study of the suborder Lari (Aves) using various techniques of numerical taxonomy. Ph.D. thesis, University of Kansas. 170 pp.

Schnell, G. D. (1970a). A phenetic study of the suborder Lari (Aves). I. Methods and results of principal components analyses. *Systematic Zool.*, **19**, 35–57.

Schnell, G. D. (1970b). A phenetic study of the suborder Lari (Aves). II. Phenograms, discussion, and conclusions. *Systematic Zool.*, **19**, 264–302.

Schopf, J. M. (1960). Emphasis on holotype (?) *Science*, **131**, 1043.

Schwartz, J. T. (1961). *Introduction to Matrices and Vectors.* McGraw-Hill, New York. 163 pp.

Scudder, G. G. E. (1963). Adult abdominal characters in the Lygaeoid-Coreoid complex of the Heteroptera, and the classification of the group. *Can. J. Zool.*, **41**, 1–14.

Seal, H. L. (1964). *Multivariate Statistical Analysis for Biologists.* Methuen, London. 209 pp.

Searle, S. R. (1966). *Matrix Algebra for the Biological Sciences (Including Applications in Statistics).* Wiley, New York. 296 pp.

Sebestyen, G. S. (1962). *Decision-Making Processes in Pattern Recognition.* Macmillan, New York. 162 pp.

Seki, T. (1968). A revision of the family Sematophyllaceae of Japan with special reference to a statistical demarcation of the family. *J. Sci. Hiroshima Univ., Ser. B, Div. 2 (Bot.)*, **12**, 1–80.

Selander, R. B., and J. M. Mathieu (1969). Ecology, behavior, and adult anatomy of the Albida Group of the genus *Epicauta* (Coleoptera, Meloidae). *Illinois Biol. Monogr.*, No. 41. 168 pp.

Selander, R. K. (1970). Biochemical polymorphism in populations of the house mouse and old-field mouse. *Symp. Zool. Soc. Lond.*, **26**, 73–91.

Selander, R. K., W. G. Hunt, and S. Y. Yang (1969). Protein polymorphism and genic hetero-zygosity in two European subspecies of the house mouse. *Evolution*, **23**, 379–390.

Selander, R. K., S. Y. Yang, R. C. Lewontin, and W. E. Johnson (1970). Genetic variation in the horseshoe crab (*Limulus polyphemus*), a phylogenetic "relic." *Evolution*, **24**, 402–414.

Selwood, N. H. (1969). Data analysis for leucocyte groups in man. *Transplantation*, **7**, 315–331.

Seyfried, P. L. (1968). An approach to the classification of lactobacilli using computer-aided numerical analysis. *Can. J. Microbiol.*, **14**, 313–318.

Shahi, B. B., H. Morishima, and H. I. Oka, (1969). A survey of variations in peroxidase, acid phosphatase and esterase isozymes of wild and cultivated *Oryza* species. *Jap. J. Genet.*, **44**, 303–319.

Shaw, A. B. (1969). Adam and Eve, paleontology and the non-objective arts. *J. Paleontol.*, **43**, 1085–1098.

Sheals, J. G. (1964). The application of computer techniques to Acarine taxonomy: a preliminary examination with species of the *Hypoaspis-Androlaelaps* complex (Acarina). *Proc. Linn. Soc. Lond.*, **176**, 11–21.

Sheals, J. G. (1969). Computers in acarine taxonomy. *Acarologia*, **11**, 376–394.

Shepard, J. H. (1971). A phenetic analysis of the Luciliini (Diptera, Calliphoridae). *Systematic Zoology*, **20**, 223–232.

Shepard, R. N. (1962). The analysis of proximities: multidimensional scaling with an unknown distance function. I and II. *Psychometrika*, **27**, 125–140, 219–246.

Shepard, R. N. (1966). Metric structures in ordinal data. *J. Math. Psychol.*, **3**, 287–315.

Shepard, R. N., and J. D. Carroll (1966). Parametric representation of nonlinear data structures. In P. R. Krishnaiah (ed.), *Multivariate Analysis.* Proceedings of an International Symposium Held at Dayton, Ohio, June 14–19, 1965, pp. 561–592. Academic Press, New York. 592 pp.

Shepherd, M. J., and A. J. Willmott (1968). Cluster analysis on the Atlas computer. *Computer J.*, **11**, 57–62.

Shewan, J. M., G. Hobbs, and W. Hodgkiss (1960). A determinative scheme for the identification of certain genera of Gram-negative bacteria, with special reference to the Pseudomona-daceae. *J. Appl. Bacteriol.*, **23**, 379–390.

Shipton, W. A., and G. Fleischmann (1969). Taxonomic significance of protein patterns of rust species and formae speciales obtained by disc electrophoresis. *Can. J. Bot.*, **47**, 1351–1358.

Shmidt, V. M. (1962). On the method of taxonomical analysis elaborated by E. S. Smirnov and on some possibilities of its application in botany. *Bot. Zh.*, **47**, 1648–1654. In Russian.

Shmidt, V. M. (1970). On two methods of taxonomic analysis. *Bot. Zh.*, **55**, 386–396. In Russian.

Sibley, C. G. (Chairman) (1969). *Systematic Biology*. Proceedings of an International Conference Conducted at the University of Michigan, Ann Arbor, Michigan, June 14–16, 1967. Nat. Acad. Sci., Wash. Pub. 1962. 632 pp.

Silva, G. A. Nigel da, and J. G. Holt (1965). Numerical taxonomy of certain coryneform bacteria. *J. Bacteriol.*, **90**, 921–927.

Silvestri, L. G., and L. R. Hill (1964). Some problems of the taxometric approach. In V. H. Heywood and J. McNeill (eds.), *Phenetic and Phylogenetic Classification*, pp. 87–103. Syst. Ass. Pub. 6. 164 pp.

Silvestri, L. G., and L. R. Hill (1965). Agreement between deoxyribonucleic acid base composition and taxometric classification of Gram-positive cocci. *J. Bacteriol.*, **90**, 136–140.

Silvestri, L. G., M. Turri, L. R. Hill, and E. Gilardi (1962). A quantitative approach to the systematics of actinomycetes based on overall similarity. In G. C. Ainsworth and P. H. A. Sneath (eds.), *Microbial Classification*. 12th Symposium of the Society for General Microbiology, pp. 333–360. Cambridge University Press, Cambridge. 483 pp.

Simon, J. P., and D. W. Goodall (1968). Relationship in annual species of *Medicago*. VI. Two-dimensional chromatography of the phenolics and analysis of the results by probabilistic similarity methods. *Aust. J. Bot.*, **16**, 89–100.

Simon, Z. (1968). Intermolecular forces pertinent to amino acid-amino acid complementary relations. *Rev. Roumaine Biochim.*, **5**, 319–324.

Simpson, G. G. (1941). Large Pleistocene felines of North America. *Amer. Mus. Novitates*, **1136**, 1–27.

Simpson, G. G. (1944). *Tempo and Mode in Evolution*. Columbia University Press, New York. 237 pp.

Simpson, G. G. (1945). The principles of classification and a classification of mammals. *Bull. Amer. Mus. Natur. Hist.*, **85**, 1–350.

Simpson, G. G. (1953). *The Major Features of Evolution*. Columbia University Press, New York. 434 pp.

Simpson, G. G. (1960). Diagnosis of the classes Reptilia and Mammalia. *Evolution*, **14**, 388–392.

Simpson, G. G. (1961). *Principles of Animal Taxonomy*. Columbia University Press, New York. 247 pp.

Simpson, G. G., A. Roe, and R. C. Lewontin (1960). *Quantitative Zoology*. Harcourt, Brace, New York. 440 pp.

Sims, R. W. (1966). The classification of the megascolecoid earthworms: an investigation of oligochaete systematics by computer techniques. *Proc. Linn. Soc. Lond.*, **177**, 125–141.

Sims, R. W. (1969a). Outline of an application of computer techniques to the problem of the classification of the megascolecoid earthworms. *Pedobiologia*, **9**, 35–41.

Sims, R. W. (1969b). A numerical classification of megascolecoid earthworms. In J. G. Sheals (ed.), *The Soil Ecosystem*, pp. 143–153. Syst. Ass. Pub. 8. 247 pp.

Singer, C. (1959). *A History of Biology*, 3rd ed. Abelard-Schuman, London. 579 pp.

Skaggs, R. H. (1969). Analysis and regionalization of the diurnal distribution of tornadoes in the United States. *Monthly Weather Rev.*, **97**, 103–115.

Skerman, V. B. D. (1967). *A Guide to the Identification of the Genera of Bacteria*. 2nd ed. Williams & Wilkins, Baltimore, Md. 303 pp.

Sklar, A. (1964). On category overlapping in taxonomy. In J. R. Gregg and F. T. C. Harris (eds.), *Form and Strategy in Science*. Studies Dedicated to Joseph Henry Woodger on the Occasion of His Seventieth Birthday, pp. 395–401. D. Reidel Publ. Co., Dordrecht, The Netherlands. 476 pp.

Skyring, G. W., and C. Quadling (1969a). Taxonomy of arthrobacter-coryneform soil isolates in relation to named cultures. *Bacteriol. Proc.*, **1969**, 19.

Skyring, G. W., and C. Quadling (1969b). Soil bacteria: principal component analysis of descriptions of named cultures. *Can. J. Microbiol.*, **15**, 141–158.

Skyring, G. W., and C. Quadling (1969c). Soil bacteria: comparisons of rhizosphere and non-rhizosphere populations. *Can. J. Microbiol.*, **15**, 473–488.

Skyring, G. W., and C. Quadling (1970). Soil bacteria: a principal component analysis and guanine-cytosine contents of some arthrobacter-coryneform soil isolates and of some named cultures. *Can. J. Microbiol.*, **16**, 95–106.

Smirnov, E. (1925). The theory of type and the natural system. *Z. Indukt. Abstamm. Vererbungsl.*, **37**, 28–66.

Smirnov, E. (1927). Mathematische Studien über individuelle und Kongregationenvariabilität. *Verhandl. V. Int. Kongr. Vererbungswiss.*, **2**, 1373–1392.

Smirnov, E. S. (1968). On exact methods in systematics. *Systematic Zool.*, **17**, 1–13.

Smirnov, E. S. (1969). *Taxonomic Analysis*. State University of Moscow, Moscow. 187 pp. In Russian.

Smirnov, E. S., and L. I. Fedoseeva (1967). Taxonomical structure of the genus *Meromyza* Meig. *Zh. Obsch. Biol.*, **28**, 604–611. In Russian.

Smith, D. W. (1969). A taximetric study of *Vaccinium* in northeastern Ontario. *Can. J. Bot.*, **47**, 1747–1759.

Smith, I. W. (1963). The classification of 'Bacterium salmonicida.' *J. Gen. Microbiol.*, **33**, 263–274.

Smith, J. E., and E. Thal (1965). A taxonomic study of the genus *Pasteurella* using a numerical technique. *Acta Pathol. Microbiol. Scand.*, **64**, 213–223.

Sneath, P. H. A. (1957a). Some thoughts on bacterial classification. *J. Gen. Microbiol.*, **17**, 184–200.

Sneath, P. H. A. (1957b). The application of computers to taxonomy. *J. Gen. Microbiol.*, **17**, 201–226.

Sneath, P. H. A. (1958). Some aspects of Adansonian classification and of the taxonomic theory of correlated features. *Ann. Microbiol. Enzimol.*, **8**, 261–268.

Sneath, P. H. A. (1961). Recent developments in theoretical and quantitative taxonomy. *Systematic Zool.*, **10**, 118–139.

Sneath, P. H. A. (1962). The construction of taxonomic groups. In G. C. Ainsworth and P. H. A. Sneath (eds.), *Microbiol Classification*. 12th Symposium of the Society for General Microbiology, pp. 289–332. Cambridge University Press, Cambridge. 483 pp.

Sneath, P. H. A. (1964a). New approaches to bacterial taxonomy: use of computers. *Ann. Rev. Microbiol.*, **18**, 335–346.

Sneath, P. H. A. (1964b). Computers in bacterial classification. *Advance. Sci.*, **20**, 572–582.

Sneath, P. H. A. (1964c). Mathematics and classification from Adanson to the present. In G. H. M. Lawrence (ed.), *Adanson. The Bicentennial of Michel Adanson's "Familles des plantes,"* part 2, pp. 471–498. Hunt Botanical Library, Carnegie Institute of Technology, Pittsburgh, Penna. 635 pp.

Sneath, P. H. A. (1964d). Comparative biochemical genetics in bacterial taxonomy. In C. A. Leone (ed.), *Taxonomic Biochemistry and Serology*, pp. 565–583. Ronald Press, New York. 728 pp.

Sneath, P. H. A. (1965). The application of numerical taxonomy to medical problems. In *Mathematics and Computer Science in Biology and Medicine*, pp. 81–91. Her Majesty's Stationery Office, London. 317 pp.

Sneath, P. H. A. (1966a). Computers in diagnosis. *J. Roy. Coll. Surg. Edinburgh*, **11**, 130–132.

Sneath, P. H. A. (1966b). A comparison of different clustering methods as applied to randomly-spaced points. *Classification Soc. Bull.*, **1**(2), 2–18.

Sneath, P. H. A. (1966c). Estimating concordance between geographical trends. *Systematic Zool.*, **15**, 250–252.

Sneath, P. H. A. (1966d). Relations between chemical structure and biological activity in peptides. *J. Theoret. Biol.*, **12**, 157–195.

Sneath, P. H. A. (1966e). A method for curve seeking from scattered points. *Computer J.*, **8**, 383–391.

Sneath, P. H. A. (1967a). Trend-surface analysis of transformation grids. *J. Zool., Lond.*, **151**, 65–122.

Sneath, P. H. A. (1967b). Some statistical problems in numerical taxonomy. *Statistician*, **17**, 1–12.

Sneath, P. H. A. (1967c). Conifer distributions and continental drift. *Nature*, **215**, 467–470.

Sneath, P. H. A. (1967d). Numerical taxonomy: steps in preparing taxonomic data for the computer. *Classification Soc. Bull.*, **1**(3), 14–18.

Sneath, P. H. A. (1967e). Quality and quantity of available geologic information for studies in time. *Kansas Geol. Surv. Computer Contrib.*, No. 18, 57–61.

Sneath, P. H. A. (1968a). Numerical taxonomic study of the graft chimaera + *Laburnocytisus adamii* (*Cytisus purpureus* + *Laburnum anagyroides*). *Proc. Linn. Soc. Lond.*, **179**, 83–96.

Sneath, P. H. A. (1968b). Goodness of intuitive arrangements into time trends based on complex pattern. *Systematic Zool.*, **17**, 256–260.

Sneath, P. H. A. (1968c). The future outline of bacterial classification. *Classification Soc. Bull.*, **1**(4), 28–45.

Sneath, P. H. A. (1968d). Vigour and pattern in taxonomy. *J. Gen. Microbiol.*, **54**, 1–11.

Sneath, P. H. A. (1969a). Computers in bacteriology. *J. Clin. Pathol.* **22**, Suppl. (College of Pathologists), No. 3, 87–92.

Sneath, P. H. A. (1969b). Evaluation of clustering methods. In A. J. Cole (ed.), *Numerical Taxonomy*. Proceedings of the Colloquium in Numerical Taxonomy Held in the University of St. Andrews, September 1968, pp. 257–271. Academic Press, London. 324 pp.

Sneath, P. H. A. (1969c). Problems of homology in geology and related fields. *Classification Soc. Bull.*, **2**(1), 5–11.

Sneath, P. H. A. (1969d). Recent trends in numerical taxonomy. *Taxon*, **18**, 14–20.

Sneath, P. H. A. (1971a). Theoretical aspects of microbiological taxonomy. In A. Pérez-Miravete and D. Paláez (eds.), *Recent Advances in Microbiology*, pp. 581–586. Asociacion Mexicana de Microbiologia, Mexico City. 623 pp.

Sneath, P. H. A. (1971b). Numerical taxonomy: criticisms and critiques. *Biol. J. Linn. Soc.*, **3**, 147–157.

Sneath, P. H. A. (1972). Computer taxonomy. In J. R. Norris and D. W. Ribbons (eds.), *Methods in Microbiology*, vol. 7A, pp. 29–98. Academic Press, London.

Sneath, P. H. A., and F. E. Buckland (1959). The serology and pathogenicity of the genus *Chromobacterium*. *J. Gen. Microbiol.*, **20**, 414–425.

Sneath, P. H. A., and S. T. Cowan (1958). An electro-taxonomic survey of bacteria. *J. Gen. Microbiol.*, **19**, 551–565.

Sneath, P. H. A., and R. Johnson (1972). The influence on numerical taxonomic similarities of errors in microbiological tests. *J. Gen. Microbiol.*, **72**, 377–392.

Sneath, P. H. A., and R. R. Sokal (1962). Numerical taxonomy. *Nature*, **193**, 855–860.

Sokal, R. R. (1952). Variation in a local population of *Pemphigus*. *Evolution*, **6**, 296–315.

Sokal, R. R. (1958a). Quantification of systematic relationships and of phylogenetic trends. *Proc. Xth Int. Congr. Entomol.*, **1**, 409–415.

Sokal, R. R. (1958b). Thurstone's analytical method for simple structure and a mass modification thereof. *Psychometrika*, **23**, 237–257.

Sokal. R. R. (1959). Comments on quantitative systematics. *Evolution*, **13**, 420–423.

Sokal, R. R. (1961). Distance as a measure of taxonomic similarity. *Systematic Zool.*, **10**, 70–79.

Sokal, R. R. (1962a). Variation and covariation of characters of alate *Pemphigus populi-transversus* in eastern North America. *Evolution*, **16**, 227–245.

Sokal, R. R. (1962b). Typology and empiricism in taxonomy. *J. Theoret. Biol.*, **3**, 230–267.

Sokal, R. R. (1964a). The future systematics. In C. A. Leone (ed.), *Taxonomic Biochemistry and Serology*, pp. 33–48. Ronald Press, New York. 728 pp.

Sokal, R. R. (1964b). Numerical taxonomy and disease classification. In J. A. Jacquez (ed.), *The Diagnostic Process*. The Proceedings of a Conference Sponsored by the Biomedical Data Processing Program of the University of Michigan, 1963, pp. 51–68. University of Michigan, Ann Arbor. 391 pp.

Sokal, R. R. (1965). Statistical methods in systematics. *Biol. Rev.*, **40**, 337–391.

Sokal, R. R. (1966). Numerical taxonomy. *Sci. Amer.*, **215**(6): 106–116.

Sokal, R. R. (1968). Numerical taxonomy: methodology and recent development. *Zh. Obsch. Biol.*, **29**, 297–315. In Russian.

Sokal, R. R. (1969). Animal taxonomy: theory and practice. *Quart. Rev. Biol.*, **44**, 209–211.

Sokal. R. R. (1970). Another new biology. *BioScience*, **20**, 152–159.

Sokal, R. R., and J. H. Camin (1965). The two taxonomies: areas of agreement and conflict. *Systematic Zool.*, **14**, 176–195.

Sokal, R. R., J. H. Camin, F. J. Rohlf, and P. H. A. Sneath (1965). Numerical taxonomy: some points of view. *Systematic Zool.*, **14**, 237–243.

Sokal, R. R., and T. J. Crovello (1970). The biological species concept: a critical evaluation. *Amer. Natur.*, **104**, 127–153.

Sokal, R. R., N. N. Heryford, and J. R. L. Kishpaugh (1971). Changes in microgeographic variation patterns of *Pemphigus populitransversus* over a six-year span. *Evolution*, **25**, 584–590.

Sokal, R. R., and C. D. Michener (1958). A statistical method for evaluating systematic relationships. *Univ. Kansas Sci. Bull.*, **38**, 1409–1438.

Sokal, R. R., and C. D. Michener (1967). The effects of different numerical techniques on the phenetic classification of bees of the *Hoplitis* complex (Megachilidae). *Proc. Linn. Soc. Lond.*, **178**, 59–74.

Sokal, R. R., and R. C. Rinkel (1963). Geographic variation of alate *Pemphigus populi-transversus* in eastern North America. *Univ. Kansas Sci. Bull.*, **44**, 467–507.

Sokal, R. R., and F. J. Rohlf (1962). The comparison of dendrograms by objective methods. *Taxon*, **11**, 33–40.

Sokal, R. R., and F. J. Rohlf (1966). Random scanning of taxonomic characters. *Nature*, **210**, 461–462.

Sokal, R. R., and F. J. Rohlf (1969). *Biometry, the Principles and Practice of Statistics in Biological Research*. W. H. Freeman and Company, San Francisco. 776 pp.

Sokal, R. R., and F. J. Rohlf (1970). The intelligent ignoramus, an experiment in numerical taxonomy. *Taxon*, **19**, 305–319.

Sokal, R. R., and P. H. A. Sneath (1963). *Principles of Numerical Taxonomy*. W. H. Freeman and Company, San Francisco. 359 pp. French translation published by Laboratoire Central, Compagnie Française des Pétroles, Bordeaux. 450 pp.

Sokal, R. R., and P. H. A. Sneath (1966). Efficiency in taxonomy. *Taxon*, **15**, 1–21.

Sokal, R. R., and P. A. Thomas (1965). Geographic variation of *Pemphigus populi-transversus* in eastern North America: stem mothers and new data on alates. *Univ. Kansas Sci. Bull.*, **46**, 201–252.

Soper, J. H. (1964). Mapping the distribution of plants by machine. *Can. J. Bot.*, **42**, 1087–1100.

Soper, J. H., and F. H. Perring (1967). Data processing in the herbarium and museum. *Taxon*, **16**, 13–19.

Sørensen, T. (1948). A method of establishing groups of equal amplitude in plant sociology based on similarity of species content and its application to analyses of the vegetation on Danish commons. *Biol. Skr.*, **5**(4), 1–34.

Soria, V. J., and C. B. Heiser, Jr. (1961). A statistical study of relationships of certain species of the *Solanum nigrum* complex. *Econ. Botany*, **15**, 245–255.

Soulé, M. (1966). Trends in the insular radiation of a lizard. *Amer. Natur.*, **100**, 47–64.

Soulé, M. (1967a). Phenetics of natural populations. I. Phenetic relationships of insular populations of the side-blotched lizard. *Evolution*, **21**, 584–591.

Soulé, M. (1967b). Phenetics of natural populations. II. Asymmetry and evolution in a lizard. *Amer. Natur*, **101**, 141–160.

Soulé, M. (1972). Phenetics of natural populations. III. The sources of morphological variation in insular populations of a lizard. *Amer. Natur.*, **106**, 429–446.

Sparck Jones, K., and D. M. Jackson (1970). The use of automatically-obtained keyword classifications for information retrieval. *Information Storage and Retrieval*, **5**, 175–201.

Spence, N. A. (1968). A multifactor uniform regionalization of British counties on the basis of employment data for 1961. *Regional Stud.*, **2**, 87–104.

Spence, N. A., and P. J. Taylor (1970) Quantitative methods in regional taxonomy. *Progr. Geogr.*, **2**, 1–64.

Spiegel, M. (1960). Protein changes in development. *Biol. Bull.*, **118**, 451–462.

Splittstoesser, D. F., M. Mautz, and R. R. Colwell (1968). Numerical taxonomy of catalase-negative cocci isolated from frozen vegetables. *Appl. Microbiol.*, **16**, 1024–1028.

Splittstoesser, D. F., M. Wexler, J. White, and R. R. Colwell (1967). Numerical taxonomy of Gram-positive and catalase-positive rods isolated from frozen vegetables. *Appl. Microbiol.*, **15**, 158–162.

Stafleu, F. A. (1963). Adanson and the "Familles des plantes." In G. H. M. Lawrence (ed.), *Adanson, the Bicentennial of Michel Adanson's "Familles des plantes,"* part 1, pp. 123–264. Hunt Botanical Library, Carnegie Institute of Technology, Pittsburgh, Penna. 635 pp.

Stalker, H. D. (1966). The phylogenetic relationships of the species in the *Drosophila melanica* group. *Genetics*, **53**, 327–342.

Stallings, D. B., and J. R. Turner (1957). A review of the Megathymidae of Mexico, with a synopsis of the classification of the family. *Lepidopterists' News*, **11**, 113–137.

Stant, M. Y. (1964). Anatomy of the Alismataceae. *J. Linn. Soc. Bot.*, **59**, 1–42.

Stant, M. Y. (1967). Anatomy of the Butomaceae. *J. Linn. Soc. Bot.*, **60**, 31–60.

Stearn, W. T. (1956). Keys, botanical, and how to use them. In P. M. Synge (ed.), *Supplement to the Dictionary of Gardening, a Practical and Scientific Encyclopaedia of Horticulture*, pp. 251–253. Clarendon Press, Oxford. 334 pp.

Stearn, W. T. (1961). Botanical gardens and botanical literature in the eighteenth century. In *Catalogue of Botanical Books in the Collection of R. M. M. Hunt*, vol. 2, pp. xli–cxl. Hunt Foundation, Pittsburgh, Penna.

Stearn, W. T. (1968). Observations on a computer-aided survey of the Jamaican species of *Columnea* and *Alloplectus*. In V. H. Heywood (ed.), *Modern Methods in Plant Taxonomy*, pp. 219–224. Academic Press, London. 312 pp.

Stearn, W. T. (1969). The Jamaican species of *Columnea* and *Alloplectus* (Gesneriaceae). *Bull. Brit. Mus. Natur. Hist. (Bot.)*, **4**, 179–236.

Stearns, F. (1968). A method for estimating the quantitative reliability of isoline maps. *Ann. Ass. Amer. Geogr.*, **58**, 590–600.

Stebbins, G. L., Jr (1950). *Variation and Evolution in Plants*. Columbia University Press, New York. 643 pp.

Steel, K. J. (1965). Microbiol identification. *J. Gen. Microbiol.*, **40**, 143–148.

Steere, W. C. (Chairman) (1971). *The Systematic Biology Collections of the United States: An Essential Resource. I. The Great Collections*. Report to the National Science Foundation. New York Botanical Gardens, New York. 33 pp.

Stephenson, W. (1936). The inverted factor technique. *Brit. J. Psychol.*, **26**, 344–361.

Stephenson, W., W. T. Williams, and G. N. Lance (1968). Numerical approaches to the relationships of certain American swimming crabs (Crustacea: Portunidea). *Proc. U.S. Nat. Mus.*, **124**(3645), 1–25.

Stevens, M. (1969). Development and use of multi-inoculation test methods for a taxonomic study. *J. Med. Lab. Technol.*, **26**, 253–263.

Stevens, M. E., V. E. Giuliano, and L. B. Heilprin (eds.) (1965). *Statistical Association Methods for Mechanized Documentation*. U.S. Govt. Printing Office, Washington, D.C. 261 pp.

Stevens, P. F. (1970). *Calluna, Cassiope* and *Harrimanella*: a taxonomic and evolutionary problem. *New Phytol.*, **69**, 1131–1148.

Steward, C. C. (1968). Numerical classification of the Canadian species of the genus *Aedes* (Diptera: Culicidae). *Systematic Zool.*, **17**, 426–437.

Steyskal, G. C. (1968). The number and kind of characters needed for significant numerical taxonomy. *Systematic Zool.*, **17**, 474–477.

Stone, D. E., G. A. Adrouny, and R. H. Flake (1969). New World Juglandaceae. II. Hickory nut oils, phenetic similarities, and evolutionary implications in the genus *Carya*. *Amer. J. Bot.*, **56**, 928–935.

Strauss, J. S., J. J. Bartko, and W. T. Carpenter, Jr. (1972). The use of clustering techniques for the classification of psychiatric patients. *Brit. J. Psychiatry* (in press).

Stringer, P. (1967). Cluster analysis of non-verbal judgements of facial expressions. *Brit. J. Math. Statist. Psychol.*, **20**, 71–79.

Stroud, C. P. (1953). An application of factor analysis to the systematics of *Kalotermes*. *Systematic Zool.*, **2**, 76–92.

Sturtevant, A. H. (1939). On the subdivision of the genus *Drosophila*. *Proc. Nat. Acad. Sci. U.S.A.*, **25**, 137–141.

Sturtevant, A. H. (1942). The classification of the genus *Drosophila*, with descriptions of nine new species. *Univ. Texas Pub.*, No. 4213. 51 pp.

Stutz, E., and H. Noll (1967). Characterization of cytoplasmic and chloroplast polysomes in plants: evidence for three classes of ribosomal RNA in nature. *Proc. Nat. Acad. Sci. U.S.A.*, **57**, 774–781.

Styron, C. E. (1969). Taxonomy of two populations of an aquatic isopod, *Lirceus fontinalis* Raf. *Amer. Midland Natur.*, **82**, 402–416.

Subak-Sharpe, H., R. R. Bürk, L. V. Crawford, J. M. Morrison, J. Hay, and H. M. Keir (1966). An approach to evolutionary relationships of mammalian DNA viruses through analysis of the pattern of nearest neighbor base sequences. *Cold Spring Harbor Symp. Quant. Biol.*, **31**, 737–748.

Sundman, V. (1970). Four bacterial soil populations characterized and compared by a factor analytical method. *Can. J. Microbiol.*, **16**, 455–464.

Sundman, V., and G. Carlberg (1967). Application of factor analysis in microbiology. 4. The value of geometric parameters in the numerical description of bacterial soil populations. *Ann. Acad. Sci. Fenn.*, *Ser. A. IV. Biol.*, No. 115, 12 pp.

Sundman, V., and H. G. Gyllenberg (1967). Application of factor analysis in microbiology. 1. General aspects on the use of factor analysis in microbiology. *Ann. Acad. Sci. Fenn.*, *Ser. A. IV. Biol.*, No. 112, 32 pp.

Svartvik, J. (1966). *On Voice in the English Verb*. Mouton, The Hague. 200 pp.

Swinnerton, H. H. (1932). Unit characters in fossils. *Biol Rev.*, **7**, 321–335.

Switzer, P. (1970). Numerical classification. In D. F. Merriam (ed.), *Geostatistics, a Colloquium*, pp. 31–43. Plenum Press, New York. 177 pp.

Sylvester-Bradley, P. C. (ed.) (1956). *The Species Concept in Palaeontology*. Syst. Ass. Pub. 2. 145 pp.

Takade, S. G. (1971). The analysis of a dimorphism in a local population of *Pemphigus populi-transversus* Riley. M.A. thesis, University of Kansas. 44 pp.

Takakura, S. (1962). Some statistical methods of classification by the theory of quantification. *Mem. Inst. Statist. Math.*, **9**, 81–105. In Japanese.

Takeya, K., Y. Nakayama, and H. Nakayama. (1967). Relationship between *Mycobacterium fortuitum* and *Mycobacterium runyonii*. *Amer. Rev. Respiratory Dis.*, **96**, 532–535.

Talbot, J. M., and P. H. A. Sneath (1960). A taxonomic study of *Pasteurella septica*, especially of strains isolated from human sources. *J. Gen. Microbiol.*, **22**, 303–311.

Talkington, L. (1967). A method of scaling for a mixed set of discrete and continuous variables. *Systematic Zool.*, **16**, 149–152.

Tanner, J. M., F. E. Johnston, R. W. Whitehouse, and P. H. A. Sneath (1967). A mathematical analysis of shape changes in growing children. *Amer. J. Phys. Anthropol.*, **27**, 246.

Tarrant, J. R. (ed.) (1972) *Computers in the Environmental Sciences*. Geo Abstracts, University of East Anglia, Norwich, England. 147 pp.

Taylor, R. J. (1971). Intraindividual phenolic variation in the genus *Tiarella* (Saxifragaceae): its genetic regulation and application to systematics. *Taxon*, **20**, 467–472.

Taylor, R. J., and D. Campbell (1969). Biochemical systematics and phylogenetic interpretations in the genus *Aquilegia*. *Evolution*, **23**, 153–162.

Taylor, R. L. (1966). Taximetrics as applied to the genus *Lithophragma* (Saxifragaceae). *Amer. J. Bot.*, **53**, 372–377.

Teissier, G. (1960). Relative growth. In T. H. Waterman (ed.), *The Physiology of the Crustacea*, vol. 1, pp. 537–560. Academic Press, New York. 2 vols.

Terentjev, P. V. (1931). Biometrische Untersuchungen über die morphologischen Merkmale von *Rana ridibunda* Pall. (Amphibia, Salientia). *Biometrika*, **23**, 23–51.

Terentjev, P. V. (1959). The method of correlation pleiades. *Vestn. Leningrad Univ. Ser. Biol.*, **9**, 127–141. In Russian.

Thielges, B. A. (1969). A chromatographic investigation of interspecific relationships in *Pinus* (subsection *Sylvestres*). *Amer. J. Bot.*, **56**, 406–409.

Thomas, P. A. (1968a). Geographic variation of the rabbit tick, *Haemaphysalis leporispalustris* in North America. *Univ. Kansas Sci. Bull.*, **47**, 787–828.

Thomas, P. A. (1968b). Variation and covariation in characters of the rabbit tick, *Haemaphysalis leporispalustris*. *Univ. Kansas Sci. Bull.*, **47**, 829–862.

Thompson, D. W. (1917). *Growth and Form*. Cambridge University Press, Cambridge. 793 pp.

Thompson, H. K., Jr., and M. A. Woodbury (1970). Clinical data representation in multidimensional space. *Computers and Biomedical Res.*, **3**, 58–73.

Thompson, W. R. (1952). The philosophical foundations of systematics. *Can. Entomol.*, **84**, 1–16.

Thornley, M. J. (1960). Computation of similarities between strains of *Pseudomonas* and *Achromobacter* isolated from chicken meat. *J. Appl. Bacteriol.*, **23**, 395–397.

Thornley, M. J. (1967). A taxonomic study of *Acinetobacter* and related genera. *J. Gen. Microbiol.*, **49**, 211–257.

Thornley, M. J. (1968). Properties of *Acinetobacter* and related genera. In B. M. Gibbs and D. A. Shapton (eds.), *Identification Methods for Microbiologists*, part B, pp. 29–50. Academic Press, London. 212 pp.

Thornton, I. W. B., and S. K. Wong (1967). A numerical taxonomic analysis of the Peripsocidae of the Oriental Region and the Pacific Basin. *Systematic Zool.*, **16**, 217–240.

Throckmorton, L. H. (1965). Similarity *versus* relationship in *Drosophila*. *Systematic Zool.*, **14**, 221–236.

Throckmorton, L. H. (1968). Concordance and discordance of taxonomic characters in *Drosophila* classification. *Systematic Zool.*, **17**, 355–387.

Tobler, W. R., H. W. Mielke, and T. R. Detwyler (1970). Geobotanical distance between New Zealand and neighboring islands. *BioScience*, **20**, 537–542.

Tremaine, J. H. (1970). Physical, chemical, and serological studies on carnation mottle virus. *Virology*, **42**, 611–620.

Tremaine, J. H., and E. Argyle (1970). Cluster analysis of viral proteins. *Phytopathology*, **60**, 654–659.

True, D. L., and R. G. Matson (1970). Cluster analysis and multidimensional scaling of archeological sites in Northern Chile. *Science*, **169**, 1201–1203.

Trueman, A. E. (1930). Results of some recent statistical investigations of invertebrate fossils. *Biol. Rev.*, **5**, 296–308.

Tryon, R. C., and D. E. Bailey (1970). *Cluster Analysis*. McGraw-Hill, New York. 347 pp.

Tschulok, S (1922). *Deszendenzlehre*. Gustav Fischer, Jena. 324 pp.

Tsukamura, M. (1966). Adansonian classification of mycobacteria. *J. Gen. Microbiol.*, **45**, 253–273.

Tsukamura, M. (1967a). Identification of mycobacteria. *Tubercle*, **48**, 311–338.

Tsukamura, M. (1967b). *Mycobacterium chitae*: a new species. *Jap. J. Microbiol.*, **11**, 43–47.

Tsukamura, M. (1967c). Two types of slowly growing, nonphotochromogenic mycobacteria obtained from soil by the mouse passage method: *Mycobacterium terrae* and *Mycobacterium novum*. *Jap. J. Microbiol.*, **11**, 163–172.

Tsukamura, M. (1967d). A statistical approach to the definition of bacterial species. *Jap. J. Microbiol.*, **11**, 213–220.

Tsukamura, M. (1968). Classification of scotochromogenic mycobacteria. *Jap. J. Microbiol.*, **12**, 63–75.

Tsukamura, M. (1969). Numerical taxonomy of the genus *Nocardia*. *J. Gen. Microbiol.*, **56**, 265–287.

Tsukamura, M., and S. Mizuno (1968). "Hypothetical Mean Organisms" of mycobacteria. A study of classification of mycobacteria. *Jap. J. Microbiol.*, **12**, 371–384.

Tsukamura, M., and S. Mizuno (1969). Taxonomy of subgroup "V" of the group III nonphotochromogenic mycobacteria. *Kekkaku*, **44**, 13–17. In Japanese.

Tsukamura, M., S. Mizuno, and S. Tsukamura (1967). Numerical classification of atypical mycobacteria. *Jap. J. Microbiol.*, **11**, 233–241.

Tsukamura, M., and S. Tsukamura (1966). Classification and identification of slowly growing mycobacteria. I. Adansonian classification of slowly growing mycobacteria by forty-three characters. *Jap. J. Bacteriol.*, **21**, 217–221. (In Japanese.)

Tsukamura, M., S. Tsukamura, and S. Mizuno (1967). Numerical taxonomy of *Mycobacterium fortuitum*. *Jap. J. Microbiol.*, **11**, 243–252.

Tugby, D. J. (1958). A typological analysis of axes and choppers from Southeast Australia. *Amer. Antiquity*, **24**, 24–33.

Tugby, D. J. (1965). Archaeological objectives and statistical methods: a frontier in archaeology. *Amer. Antiquity*, **31**, 1–16.

Turner, B. L. (1969). Chemosystematics: recent developments. *Taxon*, **18**, 134–151.

Turner, B. L. (1971). Training of systematists for the seventies. *Taxon*, **20**, 123–130.

Turner, M. E. (1969). Credibility and cluster. *Ann. New York Acad. Sci.*, **161**, 680–688.

Uhr, L. (ed.) (1966). *Pattern Recognition*. Wiley, New York. 393 pp.

Ukoli, F. M. A. (1967). On *Apharyngostrigea (Apharyngostrigea) simplex* (Johnston, 1904) new comb. and *A. (Apharyngostrigea) serpentia* n. sp. (Strigeidae: Trematoda) with an evaluation of the taxonomy of the genus *Apharyngostrigea* Ciurea, 1927 by the method of numerical taxonomy. *J. Helminthol.*, **41**, 235–156.

Underwood, R. (1969). The classification of constrained data. *Systematic Zool.*, **18**, 312–317.

Unger, S. H. (1959). Pattern detection and recognition. *Proc. Inst. Radio Eng.*, **47**, 1737–1752.

Valentine, J. W. (1966). Numerical analysis of marine molluscan ranges on the extratropical Northeastern Pacific Shelf. *Limnol. Oceanogr.*, **11**, 198–211.

Valentine, J. W., and R. G. Peddicord (1967). Evaluation of fossil assemblages by cluster analysis. *J. Paleontol.*, **41**, 502–507.

Van de Geer, J. P. (1971). *Introduction to Multivariate Analysis for the Social Sciences*. W. H. Freeman and Company, San Francisco. 293 pp.

Vandenberg, S. G., and H. H. Strandskov (1964). A comparison of identical and fraternal twins on some anthropometric measures. *Human Biol.*, **36**, 45–52.

Van Valen, L. (1964). An analysis of some taxonomic concepts. In J. R. Gregg and F. T. C. Harris (eds.), *Form and Strategy in Science*. Studies Dedicated to Joseph Henry Woodger on the Occasion of His Seventieth Birthday, pp. 402–415. D. Reidel Publ. Co., Dordrecht, the Netherlands. 476 pp.

Van Valen, L. (1969). Variation genetics of extinct animals. *Amer. Natur.*, **103**, 193–224.

Varma, A., A. J. Gibbs, and R. D. Woods (1970). A comparative study of red clover vein mosaic virus and some other plant viruses. *J. Gen. Virol.*, **8**, 21–32.

Varty, A., and D. White (1964). Application of multivariate analysis to montmorillonite. *Clay Minerals Bull.*, **5**, 465–473.

Vaughan, J. G., and K. E. Denford (1968). An acrylamide gel electrophoretic study of the seed proteins of *Brassica* and *Sinapis* species, with special reference to their taxonomic value. *J. Exp. Bot.*, **19**, 724–732.

Vaughan, J. G., K. E. Denford, and E. I. Gordon (1970). A study of the seed proteins of synthesized *Brassica napus* with respect to its parents. *J. Exp. Bot.*, **21**, 892–898.

Veevers, J. J. (1968). Identification of reef facies by computer classification. *J. Geol. Soc. Aust.*, **15**, 209–215.

Véron, M. (1966a). Taxonomie numérique des vibrions et de certaines bactéries comparables. I. Méthode et matériel taxométriques utilisés. *Ann. Inst. Pasteur*, **111**, 314–333.

Véron, M. (1966b). Taxonomie numérique des vibrions et de certaines bactéries comparables. II. Corrélation entre les similitudes phénétiques et la composition en bases de l'ADN. *Ann. Inst. Pasteur*, **111**, 671–709.

Véron, M. (1969). Taxonomie numérique et classification des bactéries. *Bull. Inst. Pasteur*, **67**, 2739–2766.

Vicq-d'Azyr, F. (1792). Quadrupèdes. Discours préliminaire. In *Encyclopédie méthodique*, vol. 2, pp. i–cliv. Panckoucke, Paris. clxiv, 632 pp.

Vinod, H. D. (1969). Integer programming and the theory of grouping. *J. Amer. Statist. Ass.*, **64**, 506–519.

Voss, E. G. (1952). The history of keys and phylogenetic trees in systematic biology. *J. Sci. Lab. Denison Univ.*, **43**, 1–25.

Vyas, G. N., H. M. Bhatia, D. D. Banker, and N. M. Purandare (1958). Study of blood groups and other genetical characters in six Gujarati endogamous groups in Western India. *Ann. Human Genet.*, **22**, 185–199.

Wagner, W. H., Jr. (1963). Biosystematics and taxonomic categories in lower vascular plants. *Regnum Vegetabile*, **27**, 63–71.

Wagner, W. H., Jr. (1969). The construction of a classification. In C. G. Sibley (Chairman), *Systematic Biology*. Proceedings of an International Conference Conducted at the University of Michigan, Ann Arbor, Michigan, June 14–16, 1967, pp. 67–90. Nat. Acad. Sci., Wash. Pub. 1692. 632 pp.

Wahlstedt, W. J., and J. C. Davis (1968). FORTRAN IV program for computation and display of principal components. *Kansas Geol. Surv. Computer Contrib.*, No. 21, 27 pp.

Wainstein, B. A. (1968). On numerical taxonomy. *Zh. Obsch. Biol.*, **29**, 153–167. In Russian.

Walker, D., P. Milne, J. Guppy, and J. Williams (1968). The computer assisted storage and retrieval of pollen morphological data. *Pollen et Spores*, **10**, 251–262.

Walker, P. M. B. (1969). The specificity of molecular hybridization in relation to studies on higher organisms. *Progr. Nucleic Acid Res. Mol. Biol.*, **9**, 301–326.

Wallace, C. S., and D. M. Boulton (1968). An information measure for classification. *Computer J.*, **11**, 185–194.

Wallace, J. T., and R. S. Bader (1967). Factor analysis in morphometric traits of the house mouse. *Systematic Zool.*, **16**, 144–148.

Walraven, W. C. (1970). A statistical analysis of sixteen taxa of *Rhynchosia* (Leguminosae) in the United States. *Brittonia*, **22**, 85–92.

Walters, S. M. (1965). 'Improvement' versus stability in botanical classification. *Taxon*, **14**, 6–10.

Wang, H.-H. (1968). Further studies on numerical taxonomy of glutamate-accumulating bacteria. *Chinese J. Microbiol.*, **1**, 64–83.

Wanke, A. (1953). Metoda badań częstości występowania zespołow cech czyli metoda stochastycznej korelacji wielorakiej. *Przegląd antropologiczny*, **19**, 106–147.

Ward, J. H., Jr. (1963). Hierarchical grouping to optimize an objective function. *J. Amer. Statist. Ass.*, **58**, 236–244.

Warner, H. R., A. F. Toronto, L. G. Veasey, and R. Stephenson (1961). A mathematical approach to medical diagnosis: application to congenital heart disease. *J. Amer. Med. Ass.*, **177**, 177–183.

Watson, L., W. T. Williams, and G. N. Lance (1966). Angiosperm taxonomy: a comparative study of some novel numerical techniques. *J. Linn. Soc. Bot.*, **59**, 491–501.

Watson, L., W. T. Williams, and G. N. Lance (1967). A mixed-data numerical approach to angiosperm taxonomy: the classification of Ericales. *Proc. Linn. Soc. Lond.*, **178**, 25–35.

Wayne, L. G. (1966). Classification and identification of mycobacteria. III. Species within Group III. *Amer. Rev. Respiratory Dis.*, **93**, 919–928.

Wayne, L. G. (1967). Selection of characters for an Adansonian analysis of mycobacterial taxonomy. *J. Bacteriol.*, **93**, 1382–1391.

Wayne, L. G., J. R. Doubek, and G. A. Diaz (1967). Classification and identification of mycobacteria. IV. Some important scotochromogens. *Amer. Rev. Respiratory Dis.*, **96**, 88–95.

Webb, L. J., J. G. Tracey, W. T. Williams, and G. N. Lance (1967a). Studies in the numerical analysis of complex rain-forest communities. I. A comparison of methods applicable to site/species data. *J. Ecol.*, **55**, 171–191.

Webb, L. J., J. G. Tracey, W. T. Williams, and G. N. Lance (1967b). Studies in the numerical analysis of complex rain-forest communities. II. The problem of species-sampling. *J. Ecol.*, **55**, 525–538.

Webb, L. J., J. G. Tracey, W. T. Williams, and G. N. Lance (1970). Studies in the numerical analysis of complex rain-forest communities. V. A. comparison of the properties of floristic and physiognomic-structural data. *J. Ecol.*, **58**, 203–232.

Webster, R. (1968). Fundamental objections to the 7th approximation. *J. Soil Sci.*, **19**, 354–366.

Webster, R., and I. F. T. Wong (1969). A numerical procedure for testing soil boundaries interpreted from air photographs. *Photogrammetria*, **24**, 59–72.

Weimarck, G. (1970). Spontaneous and induced variation in some chemical leaf constituents in *Hierochloë* (Gramineae). *Bot. Notiser*, **123**, 231–268.

West, N. E. (1966). Matrix cluster analysis of montane forest vegetation of the Oregon Cascades. *Ecology*, **47**, 975–980.

Westoll, T. S. (1949). On the evolution of the Dipnoi. In G. L. Jepsen, E. Mayr, and G. G. Simpson (eds.), *Genetics, Paleontology, and Evolution*, pp. 121–184. Princeton University Press, Princeton, N. J. 474 pp.

Whewell, W. (1840). *The Philosophy of the Inductive Sciences, Founded upon Their History*, 2 vols. Parker, London, and Deighton, Cambridge. Cited material in vol. 1, pp. 449–523

Whitehead, F. H., and R. P. Sinha (1967). Taxonomy and taximetrics of *Stellaria media* (L.) Vill., *S. neglecta* Weihe and *S. pallida* (Dumort.) Piré. *New Phytol.*, **66**, 769–784.

Whitehouse, R. N. H. (1969). An application of canonical analysis to plant breeding. *Genetica Agraria*, **23** (Suppl.), 61–96.

Whitehouse, R. N. H. (1971). Canonical analysis as an aid to plant breeding. In R. A. Nilan (ed.), *Barley Genetics II*. Proceedings of 2nd International Barley Genetics Symposium, pp. 269–282. Washington State University Press, Pullman, Wash. 621 pp.

Whitney, P. J., J. G. Vaughan, and J. B. Heale (1968). A disc electrophoretic study of the proteins of *Verticillium albo-atrum*, *Verticillium dahliae*, and *Fusarium oxysporum* with reference to their taxonomy. *J. Exp. Bot.*, **19**, 415–426.

Whitney, V. K. M. (1972). Algorithm 422. Minimal spanning tree. *Commun. Ass. Comput. Mach.*, **15**, 273–274.

Whittaker, R. H. (ed.) (1972). *Ordination and Classification of Communities. Handbook of Vegetation Science*. Part V. Junk, The Hague. (in press).

Wied, G. L., G. F. Bahr, and P. H. Bartels (1970). Automatic analysis of cell images by TICAS. In G. L. Wied and G. F. Bahr (eds.), *Automated Cell Identification and Cell Sorting*, pp. 195–360. Academic Press, New York. 403 pp.

Wierciński, A. (1962). The racial analysis of human populations in relation to their ethnogenesis. *Current Anthropol.*, **3**,2, 9–46.

Wilhelmi, R. W. (1940). Serological reactions and species specificity of some helminths. *Biol. Bull.*, **79**, 64–90.

Wilkins, D. A., and M. C. Lewis (1969). An application of ordination to genecology. *New Phytol.*, **68**, 861–871.

Wilkinson, C. (1967). A taxonomic revision of the genus *Teldenia* Moore (Lepidoptera: Drepanidae, Drepaninae). *Trans. Roy. Entomol. Soc. Lond.*, **119**, 303–362.

Wilkinson, C. (1968). A taxonomic revision of the genus *Ditrigona* (Lepidoptera: Drepanidae: Drepaninae). *Trans. Zool. Soc. Lond.*, **31**, 403–517.

Wilkinson, C. (1970a). Numerical taxonomic methods applied to some Indo-Australian Drepanidae: Lepidoptera. *J. Natur. Hist.*, **4**, 269–288.

Wilkinson, C. (1970b). Adding a point to a principal coordinates analysis. *Systematic Zool.*, **19**, 258–263.

Williams, C. B. (1964). *Patterns in the Balance of Nature and Related Problems in Quantitative Ecology*. Academic Press, New York. 324 pp.

Williams, S. T., F. L. Davies, and D. M. Hall (1969). A practical approach to the taxonomy of actinomycetes isolated from soil. In J. G. Sheals (ed.), *The Soil Ecosystem*, pp. 107–115. Syst. Ass. Pub. 8. 247 pp.

Williams, W. T. (1964). The analysis of large scale botanical survey data. *Abstr. Pap. X Int. Bot. Congr.*, Edinburgh, p. 131.

Williams, W. T. (1967a). The computer botanist. *Aust. J. Sci.*, **29**, 266–271.

Williams, W. T. (1967b). Numbers, taxonomy and judgement. *Bot. Rev.*, **33**, 379–386.

Williams, W. T. (1969). The problem of attribute-weighting in numerical classification. *Taxon*, **18**, 369–374.

Williams, W. T., and H. T. Clifford (1971). On the comparison of two classifications of the same set of elements. *Taxon*, **20**, 519–522.

Williams, W. T., and M. B. Dale (1962). Partition correlation matrices for heterogeneous quantitative data. *Nature*, **196**, 602.

Williams, W. T., and M. B. Dale (1965). Fundamental problems in numerical taxonomy. *Advance. Bot. Res.*, **2**, 35–68.

Williams, W. T., M. B. Dale, and P. Macnaughton-Smith (1963). An objective method of weighting in similarity analysis. *Nature*, **201**, 426.

Williams, W. T., and J. M. Lambert (1959). Multivariate methods in plant ecology. I. Association-analysis in plant communities. *J. Ecol.*, **47**, 83–101.

Williams, W. T., and J. M. Lambert (1960). Multivariate methods in plant ecology. II. The use of an electronic digital computer for association-analysis. *J. Ecol.*, **48**, 689–710.

Williams, W. T., and J. M. Lambert (1961a). Multivariate methods in plant ecology. III. Inverse association-analysis. *J. Ecol.*, **49**, 717–729.

Williams, W. T., and J. M. Lambert (1961b). Nodal analysis of associated populations. *Nature*, **191**, 202.

Williams, W. T., J. M. Lambert, and G. N. Lance (1966). Multivariate methods in plant ecology. V. Similarity analyses and information-analysis. *J. Ecol.*, **54**, 427–445.

Williams, W. T., and G. N. Lance (1965). Logic of computer-based intrinsic classifications. *Nature*, **207**, 159–161.

Williams, W. T., and G. N. Lance (1968). The choice of strategy in the analysis of complex data. *Statistician*, **18**, 31–43.

Williams, W. T., and G. N. Lance (1969). Application of computer classification techniques to problems in land survey. *Bull. Int. Statist. Inst.*, **42**, 345–354.

Williams, W. T., G. N. Lance, L. J. Webb, J. G. Tracey, and M. B. Dale (1969a). Studies in the numerical analysis of complex rain-forest communities. III. The analysis of successional data. *J. Ecol.*, **57**, 515–535.

Williams, W. T., G. N. Lance, L. J. Webb, J. G. Tracey, and J. H. Connell (1969b). Studies in the numerical analysis of complex rain-forest communities. IV. A method for the elucidation of small-scale forest pattern. *J. Ecol.*, **57**, 635–654.

Williams, W. T., and L. J. Webb (1968). The computer and the tropical rainforest. *Aust. Natur. Hist.*, **16**, 92–96.

Williamson, M. H. (1961). An ecological survey of a Scottish herring fishery. IV. Changes in the plankton during the period 1949 to 1959. *Bull. Marine Ecol.*, **5**(48), 207–229.

Willis, H. L. (1971). Numerical cladistics: the *Ellipsoptera* group of the genus *Cicindela*. *Cicindela*, **3**, 13–20.

Willis, J. C. (1922). *Age and Area. A Study in Geographical Distribution and Origin of Species.* Cambridge University Press, Cambridge. 259 pp.

Willis, J. C., and G. U. Yule (1922). Some statistics of evolution and geographical distribution in plants and animals, and their significance. *Nature*, **109**, 177–179.

Willmott, A. J., and P. N. Grimshaw (1969). Cluster analysis in social geography. In A. J. Cole (ed.), *Numerical Taxonomy.* Proceedings of the Colloquium in Numerical Taxonomy Held in the University of St. Andrews, September 1968, pp. 272–281. Academic Press, London. 324 pp.

Wilson, E. O. (1965). A consistency test for phylogenies based on contemporaneous species. *Systematic Zool.*, **14**, 214–220.

Wilson, E. O., and W. L. Brown, Jr. (1953). The subspecies concept and its taxonomic application. *Systematic Zool.*, **2**, 97–111.

Wirth, M., G. F. Estabrook, and D. J. Rogers (1966). A graph theory model for systematic biology, with an example for the Oncidiinae (Orchidaceae). *Systematic Zool.*, **15**, 59–69.

Wishart, D. (1969a). An algorithm for hierarchical classifications. *Biometrics*, **22**, 165–170.

Wishart, D. (1969b). Mode analysis, a generalization of nearest neighbour which reduces chaining effects. In A. J. Cole (ed.), *Numerical Taxonomy.* Proceedings of the Colloquium in Numerical Taxonomy Held in the University of St. Andrews, September 1968, pp. 282–311. Academic Press, London. 324 pp.

Wishart, D. (1969c). Numerical classification method for deriving natural classes. *Nature*, **221**, 97–98.

Wishart, D. (1969d). The use of cluster analysis in the classification of diseases. *Scottish Med. J.*, **14**, 96.

Wishart, D. (1969e). FORTRAN II programs for 8 methods of cluster analysis (CLUSTAN I). *Kansas Geol. Surv. Computer Contrib.*, No. 38, 112 pp.

Wishart, D. (1970). Some problems in the theory and application of the methods of numerical taxonomy. Ph.D. thesis, University of St. Andrews, Scotland. 436 pp.

Wishart, D., and S. V. Leach (1970). A multivariate analysis of Platonic prose rhythm. *Computer Stud. Humanities Verbal Behav.* **3**, 90–99.

Withers, R. F. J. (1964). Morphological correspondence and the concept of homology. In J. R. Gregg and F. T. C. Harris (eds.), *Form and Strategy in Science*. Studies Dedicated to Joseph Henry Woodger on the Occasion of His Seventieth Birthday, pp. 378–394. D. Reidel Publ. Co., Dordrecht, The Netherlands. 476 pp.

Wolfe, J. H. (1970). Pattern clustering by multivariate mixture analysis. *Multivariate Behav. Res.*, **5**, 329–350.

Woodger, J. H. (1937). *The Axiomatic Method in Biology*. Cambridge University Press, Cambridge. 174 pp.

Woodger, J. H. (1945). On biological transformations. In W. E. Le Gros Clark and P. B. Medawar (eds.), *Essays on Growth and Form Presented to D'Arcy Wentworth Thompson*, pp. 94–120. Clarendon Press, Oxford. 408 pp.

Woodger, J. H. (1951). Science without properties. *Brit. J. Phil. Sci.*, **2**, 193–216.

Woodger, J. H. (1952). From biology to mathematics. *Brit. J. Phil. Sci.*, **3**, 1–21.

Woźnicka, W. (1967). Trials of classification of the "yellow series." II. Taxonometric studies. *Exp. Med. Microbiol.* **19**, 23–29.

Wrenn, W. J. (1972). A phenetic study of the larvae and nymphs of the chigger genus *Euschoengastia* (Acarina: Trombiculidae) using numerical taxonomy. Ph.D. Thesis, Univ. of Kansas, Lawrence, 339 pp.

Wright, S. (1941). The "age and area" concept extended. *Ecology*, **22**, 345–347.

Wuenscher, J. E. (1969). Niche specification and competition modeling. *J. Theoret. Biol.*, **25**, 436–443.

Young, D. J., and L. Watson (1969). Softwood structure and the classification of conifers. *New Phytol.*, **68**, 427–432.

Young, D. J., and L. Watson (1970). The classification of dicotyledons: a study of the upper levels of the hierarchy. *Aust. J. Bot.*, **18**, 387–433.

Yourassowsky, E., W. Hansen, M. Labbe, and J. Van Molle (1965). Problèmes taxinomiques. Orientation mechanographique du diagnostic des éspèces microbiennes. *Acta Clin. Belg.*, **20**, 279–285.

Yule, G. U. (1924). A mathematical theory of evolution, based on the conclusions of Dr. J. C. Willis, F.R.S. *Phil. Trans. Roy. Soc. Lond. Ser. B*, **213**, 21–87.

Zangerl, R. (1948). The methods of comparative anatomy and its contribution to the study of evolution. *Evolution*, **2**, 351–374.

Zarapkin, S. R. (1934). Zur Phänoanalyse von geographischen Rassen und Arten. *Arch. Naturgesch. N. F.*, **3**, 161–186.

Zarapkin, S. R. (1939). Das Divergenzprinzip in der Bestimmung kleiner systematischer Kategorien. *Verhandl. VII. Int. Kongr. Entomol.*, **1**, 494–518.

Zarapkin, S. R. (1943). Die Hand des Menschen und der Menschenaffen. *Z. Menschl. Vererb. Konstitutionsl.*, **27**, 390–414.

Zhantiev, R. D. (1967). A contribution to the taxonomic analysis of the genus *Dermestes* (Coleoptera, Dermestidae). *Zool. Zh.*, **46**, 1350–1356. In Russian.

Zinsser, H. H. (1964). Patient derived autodefinition of a disease: pyelonephritis. In J. A. Jacquez (ed.), *The Diagnostic Process*. Proceedings of a Conference Sponsored by the Biomedical Training Program of the University of Michigan, 1963, pp. 103–128. University of Michigan, Ann Arbor. 391 pp.

Ziswiler, V. (1967). Numerische Taxonomie und ornithologische Systematik. *J. Ornithol.*, **108**, 474–479.

Zuckerkandl, E., and L. Pauling (1965a). Molecules as documents of evolutionary history. *J. Theoret. Biol.*, **8**, 357–366.

Zuckerkandl, E., and L. Pauling (1965b). Evolutionary divergence and convergence in proteins. In V. Bryson and H. J. Vogel (eds.), *Evolving Genes and Proteins*, pp. 97–166. Academic Press, New York. 629 pp.

Author Index

Hall, E. C., 500
Hall, O., 371, 515
Haltenorth, T., 13, 123, 509
Hamann, U., 81, 133, 470, 509
Hanan, M., 327, 509
Handler, P., 451, 509
Hansell, R., 320
Hansen, A. J., 477, 509
Hansen, B., 390, 509
Hansen, E. W., 547
Hanson, W. D., 372, 469, 494
Harary, F., 253, 509
Harbaugh, J. W., 377, 446, 509
Harborne, J. B., 468, 526
Harding, J. O., 199, 509
Harman, H. H., 246, 247, 248, 487, 509
Harrington, B. J., 475, 509
Harrison, B. D., 504
Harrison, G. A., 3, 13, 14, 28, 29, 37, 55, 58, 71, 104, 110, 123, 125, 136, 153, 181, 458, 494
Harrison, P. J., 211, 449, 509
Hartigan, J. A., 66, 200, 210, 281, 282, 287, 501, 509
Hartman, P. A., 240, 385, 391, 472, 475, 483, 511, 522
Hartzig, R. J., 126
Haswell, W. A., 162
Hauser, M. M., 475, 509
Hawkes, J. G., 93, 509
Hawkins, D., 18, 444, 509
Hawksworth, F. G., 144, 258, 467, 509
Hay, J., 541
Hayashi, C., 250, 509
Hayashi, K., 196, 472, 475, 477, 510
Hayes, P. R., 261, 477, 503
Hayhoe, F. G. J., 441, 510
Hazel, J. E., 117, 129, 361, 495, 510
Heale, J. B., 299, 472, 545
Healy, M. J. R., 161
Heatwole, H., 438, 510
Heberlein, G. T., 15, 302, 477, 510
Hecht, A. D., 447, 510
Hedge, I. C., 424, 425, 430, 493
Hedges, S. R., 459, 510
Hedgpeth, J. W., 461, 504
Hedrick, P. W., 187, 510
Hegenauer, R., 91, 510
Heilprin, L. B., 448, 540
Heincke, F., 13, 126, 305, 510
Heiser, C. B., Jr., 137, 371, 372, 432, 467, 468, 510, 539
Heisey, W. M., 374, 495
Hendrickson, J. A., Jr., 101, 137, 246, 269, 273, 274, 307, 319, 335, 341, 351, 459, 461, 465, 510
Hennig, W., 13, 16, 29, 30, 38–44, 46, 49, 50, 53, 54, 59, 61, 63, 69, 75, 76, 95, 97, 296, 298, 319, 322, 347, 510, 511
Henningsmoen, G., 358, 511
Herrin, C. S., 461, 511
Heryford, N. N., 378, 379, 539,
Heslop-Harrison, J., 60, 373, 505, 511
Hesser, J. M., 475, 511
Hester, D. J., 475, 511
Heywood, V. H., 17, 27, 30, 61, 63, 75, 91, 145, 244, 291, 307, 352, 364, 373, 374, 383, 388, 415, 416, 430, 452, 457, 467, 499, 511, 523
Hickman, J. C., 369, 379, 467, 511

Hiernaux, J., 186, 511
Hildebrand, D. C., 479, 534
Hill, L. R., 213, 253, 287, 301, 394, 407, 416, 429, 473, 475, 476, 504, 511, 536
Hill, R. H., 458, 493
Hill, R. L., 301, 493
Hillis, W. E., 379, 466, 489
Himmelmann, G., 513
Hinz, U., 513
Hironaka, M., 440, 511
Hobbs, G., 479, 535
Hodgkiss, W., 477, 479, 511, 535
Hodson, F. R., 126, 251, 288, 434, 445, 446, 450, 500, 511
Hoff, K. M., 512
Hole, F., 250, 511
Hole, F. D., 439, 440, 491, 511
Holloway, J. D., 379, 433, 438, 511, 512
Holm, R. W., 17, 101, 173, 282, 283, 367, 376, 418, 431, 432, 453, 467, 501, 515, 531
Holt, J. G., 94, 476, 478, 522, 536
Holttum, R. E., 35, 512
Honigfeld, G., 406, 502
Hopkins, J. W., 158, 385, 473, 512, 527, 530
Hořejš, J., 476, 533
Horne, S. L., 343, 459, 512
Hotelling, H. O., 287
Householder, A. S., 246
Howarth, R. J., 438, 512
Howd, F. H., 447, 512
Howden, H. F., 411, 512
Howells, W. W., 444, 512
Hoyer, B. H., 296, 301, 314, 512
Hubac, J. M., 14, 269, 369, 371, 372, 457, 467, 469, 491, 512
Hubálek, Z., 475, 512
Hubby, J. L., 299, 461, 512
Huber, I., 431, 461, 484, 512
Huckabay, J. P., 470, 500
Hudson, G. E., 14, 101, 155, 282, 301, 302, 430, 431, 459, 512
Hudson, L., 444, 512
Hughes, N. F., 361, 512
Hughes, R. E., 439, 512
Hughes, W. L., 443, 512
Huizinga, J., 128, 370, 445, 512
Hull, D. L., 18, 22, 65, 80, 412, 413, 419, 422, 493, 513
Humphries, A. W., 439, 526
Hunt, W. G., 460, 535
Hureau, J. C., 139, 458, 459, 499, 513
Hurlbutt, H. W., 434, 437, 461, 513
Hutchinson, G. E., 375, 433, 513
Hutchinson, M., 285, 306, 396, 478, 513
Huxley, J. S., 16, 157, 311, 418, 513
Hybner, C. J., 531
Hyvärinen, L., 145, 175, 211, 513

Ibara, R. M., 501
IBM, 468, 513
Ibrahim, F. M., 95, 151, 179, 433, 471, 513
Ihm, P., 511, 513
Iizuka, H., 478, 518
Ikeda, A. G., 391, 492
Imaizumi, Y., 431, 459, 513
Imbrie, J., 446, 447, 514

Inger, R. F., 63, 418, 514
Ingle, J., 55, 522
Inglis, W. G., 79, 82, 423, 514
Ingram, V. M., 86, 301, 514
Irwin, H. S., 467, 514
Ising, G., 131, 371, 372, 470, 514
Ivimey-Cook, R. B., 145, 253, 307, 430, 436, 467, 514
Izzo, N. F., 443, 514

Jaccard, P., 131, 514
Jackson, D. M., 168, 430, 448, 514, 528, 539
Jacob, F., 54, 514
Jacobi, C. G. J., 246
Jacobs, M., 23, 514
Jacquez, J. A., 442, 514
Jago, N. D., 462, 514
Jahn, T. L., 409, 410, 413, 514
James, A. T., 445, 501
James, M. T., 14, 514
James, W. H., 443, 514
Jamieson, B. G. M., 465, 514
Jancey, R. C., 213, 467, 514
Janetschek, H., 419, 515
Jardine, C. J., 15, 65, 77, 78, 83, 199, 219, 318, 343, 424, 458, 459, 515
Jardine, N., 4, 15, 17, 20, 48, 61, 63, 65, 66, 71, 76–78, 81–84, 108, 145, 199, 201, 206–208, 219, 256, 269, 280, 281, 305, 318, 343, 357, 379, 421, 423, 424, 433, 438, 445, 458, 459, 467, 487, 512, 515
Jarvis, B. D. W., 302, 475, 515
Jaworska, H., 299, 467, 515
Jeffers, J. N. R., 13, 471, 515
Jellett, L. B., 442, 494
Jensen, R. E., 486, 515
Jevons, W. S., 21, 515
Jewsbury, J. M., 128, 433, 465, 515
Jičín, R., 180, 385, 515
Jizba, Z. V., 111, 446, 515
Johnson, B. L., 299, 371, 467, 515
Johnson, L., Jr., 250, 251, 498, 515
Johnson, L. A. S., 120, 419, 420, 423, 424, 427, 428, 451, 515
Johnson, M. P., 101, 173, 282, 283, 369, 379, 431, 432, 467, 511, 515
Johnson, R., 167, 168, 538
Johnson, R. G., 433, 438, 515, 516
Johnson, R. M., 472, 516
Johnson, S. C., 198, 206, 218, 222, 516
Johnson, W. E., 535
Johnston, D. E., 184, 487, 516
Johnston, F. E., 541
Johnston, R. F., 258, 369, 378, 379, 459, 506, 516
Johnston, R. J., 448, 516
Johnstone, K. I., 285, 306, 396, 478, 513
Jolicoeur, P., 158, 516
Joly, P., 145, 471, 491, 516
Jones, D., 301, 302, 352, 371, 373, 431, 516
Jones, J. H., 441, 516
Jones, K. J., 248
Jones, L. A., 473, 516
Jones, W. D., Jr., 519
Jorgensen, J. G., 445, 516
Joyce, T., 449, 516
Joysey, K. A., 35, 357, 516

Subject Index

Boldface numbers indicate definitions, descriptions, or important sections; *italic* numbers indicate Figures.